STUDY GUIDE WITH SELECTED SOLUTIONS
TO ACCOMPANY

PHYSICS
FOR SCIENTISTS & ENGINEERS

WITH MODERN PHYSICS

THIRD EDITION

RAYMOND A. SERWAY
JOHN R. GORDON
James Madison University

SAUNDERS
HBJ

Saunders College Publishing
A Harcourt Brace Jovanovich College Publisher
Fort Worth Philadelphia San Diego New York Orlando Austin San Antonio
Toronto Montreal London Sydney Tokyo

Printed in the United States of America.

Serway: Study Guide with Selected Solutions to accompany
    PHYSICS FOR SCIENTISTS AND ENGINEERS WITH MODERN PHYSICS

ISBN 0-03-074493-8

 23 014 9876543

Preface

This study guide has been written to accompany the third edition of the textbooks *Physics for Scientists and Engineers* and *Physics for Scientists and Engineers with Modern Physics* by Raymond A. Serway. The purpose of this study guide is to provide students with a convenient review of the basic concepts and applications presented in the textbook, together with further drill on problem solving methodology using a number of worked examples. The study guide is not an attempt to rewrite the textbook in a condensed fashion. Rather, emphasis is placed upon discussing troublesome points, and providing guidance in reviewing concepts and in problem-solving strategies.

Each chapter of the study guide is divided into several parts, and every textbook chapter has a matching chapter in the study guide. The study guide contains various features which have been included to make the guide a useful supplement to the textbook. Most chapters contain the following sections:

A. **Objectives:** This is a list of topics and techniques you should understand after reading the chapter and working the assigned problems.

B. **Skills:** The purpose of this section is to provide the student with any or all of the following: (a) a review of mathematical techniques used in that chapter, (b) a discussion of problem-solving methodology, and (c) a review of sign conventions or other material unique to that chapter.

C. **Notes From Selected Chapter Sections:** These notes briefly describe the major ideas discussed in each section of that chapter.

D. **Equations and Concepts:** This represents a review of the chapter, with emphasis on highlighting important concepts and discussing important equations and formalisms.

E. **Examples:** Here we present one or two worked examples which can serve as models for working end-of-chapter problems.

F. **Answers to Selected Questions:** These are answers to selected questions found at the end of each chapter of the text.

G. **Solutions to Selected Problems:** This new feature represents solutions to selected odd numbered problems found at the end of each chapter of the text. Solutions are given to only 7 to 10 problems per chapter. Problem statements are given before each problem, and appropriate figures are included, where necessary. We strongly urge you to refer to these solutions as a check of your own work, and only after you have made a serious attempt to work the problem on your own.

We sincerely hope that this study guide will be useful to you in reviewing the material presented in the text, and in improving your ability to solve problems and score well on exams. We welcome any comments or suggestions from users of this study guide which might help improve its presentation in future printings or editions, and we wish you the best of luck.

Raymond A. Serway
John R. Gordon
Department of Physics
James Madison University
Harrisonburg, VA 22807

Acknowledgments

We gratefully acknowledge the dedicated work of Mrs. Linda Miller in typing this study guide and for her careful proofreading of the entire manuscript. We are grateful to Professors Lou Cadwell and Steve Van Wyk for pointing out a number of errors in the Study Guide. We also thank the professional staff at Saunders, especially Doris Bruey for inserting all the figures and photographs, and Ellen Newman for managing all phases of the project. Finally, we express our appreciation to our families for their inspiration, patience, and encouragement.

Suggestions for Study

Very often students ask us "How should I study this subject and prepare for examinations?" Although there is no simple answer to this question, we would like to offer some suggestions which may be useful to you.

1. It is essential that you understand the basic concepts and principles before attempting to answer questions or solve assigned problems. This is best accomplished through a careful reading of the textbook before attending your lectures on that material, jotting down certain points which you do not understand, taking careful notes in class, and asking questions. You should reduce memorization of material to a minimum. Memorizing sections of a text, equations, and derivations does not necessarily mean that you understand the material. Perhaps the best test of your understanding of the subject will be your ability to answer questions and solve problems in the text or those given on exams.

2. Solve as many problems at the end of each chapter of the text as possible. You will be able to check the accuracy of your calculations to the odd-numbered problems whose answers are included at the end of your text. Furthermore, you can check solutions to selected odd-numbered problems with those given in this study guide.

3. The end of each chapter contains a set of questions which are included to test your understanding of the concepts presented. The answers to some of these questions are given in the study guide. Once you have attempted to answer the question on your own, check it against that given in the study guide. Many of the questions serve as reviews and can be answered by referring to the text itself. You should discuss the question with your instructor or a classmate for further guidance.

4. The method of solving problems should be carefully planned. First, read the problem several times until you are confident you understand what is being asked. Look for key words which will help simplify the problem and perhaps allow you to make certain assumptions. You should pay special attention to the information provided in the problem. It is a good idea to write down the given information before proceeding with a solution. After you have decided on a method you think is appropriate for the problem, proceed with your solution.

5. If you are having difficulty in working problems, we suggest that you read the text and your lecture notes in order to develop a good understanding of the physical concepts and models. It may take several readings before you are ready to solve certain problems. We also suggest that you carefully read the problem-solving strategy sections that are included in selected chapters of the textbook.

6. After reading a chapter, you should be able to define any new quantities that were introduced and discuss the basic principles that were used to derive certain fundamental formulas. A review is provided in each chapter of the study guide for this purpose, and the marginal notes in the textbook (or the index) will help you locate these topics. You should be able to correctly associate with each physical quantity the symbol used to represent that quantity (including vector notation if appropriate) and the SI unit associated with that quantity. Furthermore, you should be able to express each important equation in a concise and accurate prose statement.

Table of Contents

*Chapters 41 through 47 correspond to those chapters in the extended version of the text, PHYSICS FOR SCIENTISTS AND ENGINEERS WITH MODERN PHYSICS.

Introduction: Physics and Measurement

OBJECTIVES

1. Discuss the units of length, mass and time and the standards for these quantities in SI units.

2. Describe (a) the *density* (mass per unit volume) of a substance, (b) the *atomic weight* of a substance and (c) the method for calculating atomic mass.

3. Perform a *dimensional analysis* of an equation containing physical quantities whose individual units are known.

4. *Convert units* from one system to another.

5. Carry out *order-of-magnitude calculations* or guesstimates.

6. Understand *significant figures* and how to handle them when carrying out simple arithmetic manipulations.

7. Become familiar with the meaning of various mathematical symbols and Greek letters.

SKILLS

You should be familiar with . . .

1. Using powers of ten in expressing such numbers as $0.00058 = 5.8 \times 10^{-4}$. Appendix B.1 gives a brief review of such notation, and the algebraic operations of numbers using powers of ten.

2. Basic algebraic operations such as factoring, handling fractions, solving quadratic equations, and solving linear equations. For your convenience, Appendix B.2 reviews some of these techniques.

EQUATIONS AND CONCEPTS

$$\text{Density} = \frac{\text{Mass}}{\text{Volume}}$$

$$\rho \equiv \frac{m}{V}$$

$$\text{Atomic Mass} = \frac{\text{Atomic Weight}}{\text{Avogadro's number}}$$

$$m = \frac{\text{atomic weight}}{N_A}$$

EXAMPLE PROBLEM SOLUTIONS

Example 1.1 Perform a dimensional analysis of the expression

$$x = v_0 t + \frac{1}{2} at^3$$

where x is a displacement and has units of length, v_0 is velocity, a is acceleration, and t is time.

Solution

Since x has units of length, both terms on the right side must also have units of length. The dimensions of $v_0 t$ are

$$[v_0 t] = \left(\frac{L}{T}\right)(T) = L$$

which is consistent with the dimensions of x. The dimensions of the term $\frac{1}{2} at^3$ are

$$\left[\frac{1}{2} at^3\right] = \left(\frac{L}{T^2}\right)(T^3) = LT$$

which is *not* consistent with the dimensions of x. Therefore, the expression is dimensionally *incorrect*. The correct expression is given by $x = v_0 t + \frac{1}{2} at^2$, where the last term has dimensions of L. A detailed list of physical quantities and their dimensions is given in Appendix A of the text.

Example 1.2 Show that the expression $v^2 = v_0^2 + 2ax$ is dimensionally correct, where v_0 and v represent velocities, a is acceleration and x is the displacement.

Solution

The quantities v^2 and v_0^2 both have dimensions of $\frac{L^2}{T^2}$. Therefore, the term 2ax must have the same dimensions. Since $[a] = \frac{L}{T^2}$ and $[x] = L$, we see that $[2ax] = \left(\frac{L}{T^2}\right)(L) = \frac{L^2}{T^2}$, so the expression is dimensionally correct.

Example 1.3 *Converting Units:* A car moves at a speed of 60 mi/h. Find its speed in m/s.

Solution

Since 1 mi = 1610 m and 1 h = 3600 s, we see that

$$60 \frac{mi}{h} = 60 \frac{mi}{h} \times 1610 \frac{m}{mi} \times \frac{1}{3600} \frac{h}{s} = 26.8 \text{ m/s}$$

ANSWERS TO SELECTED QUESTIONS_____

2. The height of a horse is sometimes given in units of "hands." Why do you suppose this is a poor standard of length?

Answer: The size of a hand differs from person to person.

3. Express the following quantities using the prefixes given in Table 1.4: (a) 3×10^{-4} m, (b) 5×10^{-5} s, (c) 72×10^2 g.

Answer: (a) 0.3 mm   (b) 50 µs   (c) 7.2 kg

5. Suppose that two quantities A and B have different dimensions. Determine which of the following arithmetic operations *could* be physically meaningful: (a) A + B, (b) A/B, (c) B – A, (d) AB.

Answer: The sum or difference [(a) and (b)] are not meaningful if A and B have different dimensions. However, the ratio or product of A and B could be meaningful.

6. What accuracy is implied in an order-of-magnitude calculation?

Answer: An order-of-magnitude calculation implies that the answer should be reliable to within a factor of 10 of the "correct" answer.

10. Is it possible to use length, density, and time as three fundamental units rather than length, mass, and time? If so, what could be used as a standard of density?

Answer: Yes. A known volume of pure water at some standard pressure and temperature could be used as the standard of density.

SOLUTIONS TO SELECTED END-OF-CHAPTER PROBLEMS_____

5. Calculate the mass of an atom of (a) helium, (b) iron, and (c) lead. Give your answers in atomic mass units and in grams. The atomic weights are 4, 56, and 207, respectively, for the atoms given.

Solution

To solve this problem, use $m = \dfrac{\text{atomic weight}}{N_A}$ and the conversion $1 \text{ u} = 1.66 \times 10^{-24}$ g.

(a) For He, $m = \dfrac{4 \text{ g/mol}}{6.02 \times 10^{23} \text{ molecules/mol}} = \boxed{6.64 \times 10^{-24} \text{ g} = 4 \text{ u}}$

(b) For Fe, $m = \dfrac{56 \text{ g/mol}}{6.02 \times 10^{23} \text{ molecules/mol}} = \boxed{9.30 \times 10^{-23} \text{ g} = 56 \text{ u}}$

(c) For Pb, $m = \dfrac{207 \text{ g/mol}}{6.02 \times 10^{23} \text{ molecules/mol}} = \boxed{3.44 \times 10^{-22} \text{ g} = 207 \text{ u}}$

6. A small particle of iron in the shape of a cube is observed under a microscope. The edge of the cube is 5×10^{-6} cm long. Find (a) the mass of the cube and (b) the number of iron atoms in the particle. The atomic weight of iron is 56, and it density is 7.86 g/cm^3.

Solution   Since $\rho = M/V$, and $V = L^3$, we have $M = \rho L^3$

(a) $M = \rho L^3 = (7.86 \text{ g/cm}^3)(5 \times 10^{-6} \text{ cm})^3 = \boxed{9.83 \times 10^{-16} \text{ g}}$

(b) $N = M \left(\dfrac{N_A}{\text{Atomic weight}} \right) = (9.83 \times 10^{-16} \text{ g}) \left(\dfrac{6.02 \times 10^{23} \text{ atoms/mol}}{56 \text{ g/mol}} \right) = \boxed{1.06 \times 10^7 \text{ atoms}}$

10. The displacement of a particle when moving under uniform acceleration is some function of the time and the acceleration. Suppose we write this displacement $s = ka^m t^n$, where k is a dimensionless constant. Show by dimensional analysis that this expression is satisfied if $m = 1$ and $n = 2$. Can this analysis give the value of k?

Solution $[s] = L$ $[a] = LT^{-2}$ $[t] = T$

$s = ka^m t^n$, where k has no dimensions

$[s] = [a]^m [t]^n = (LT^{-2})^m T^n = L^m T^{-2\,m + n}$

$L = L^m T^{-2\,m + n}$

Comparing the left and right sides, we have $\boxed{m = 1}$ and $-2\,m + n = 0$

Therefore, $\boxed{n = 2m = 2}$ or $s = kat^2$

The value of the dimensionless constant k *cannot* be determined by this analysis.

15. Suppose that the displacement of a particle is related to time according to the expression $s = ct^3$. What are the dimensions of the constant c?

Solution

$[s] = L$ $[t] = T$

Since $c = \dfrac{s}{t^3}$, $[c] = \dfrac{[s]}{[t^3]} = \dfrac{L}{T^3} = \boxed{LT^{-3}}$

23. A solid piece of lead has a mass of 23.94 g and a volume of 2.10 cm³. From these data, calculate the density of lead in SI units (kg/m³).

Solution

Use the conversions $1\ g = 10^{-3}\ kg$ and $1\ cm^3 = 10^{-6}\ m^3$ to find $M = 23.94\ g = 23.94 \times 10^{-3}\ kg$ and $V = 2.10\ cm^3 = 2.10 \times 10^{-6}\ m^3$. Therefore,

$$\rho = \frac{M}{V} = \frac{23.94 \times 10^{-3}\ kg}{2.10 \times 10^{-6}\ m^3} = \boxed{1.14 \times 10^4\ \frac{kg}{m^3}}$$

25. Estimate the age of the earth in years using the data in Table 1.3 and the appropriate conversion factors.

Solution

From Table 1.3, $T = 1.3 \times 10^{17}\ s$. Since $1\ y = 3.16 \times 10^7\ s$,

$$T = \frac{1.3 \times 10^{17}\ s}{3.16 \times 10^7\ s/y} = \boxed{4.1 \times 10^9\ y}$$

27. Using the fact that the speed of light in free space is about 3.00×10^8 m/s, determine how many miles a pulse from a laser beam will travel in one hour.

Solution

$t = 1\ h$ $v = 3 \times 10^8$ m/s

Use the conversions $1\ mi = 1609\ m$; $1\ h = 60\ min$; $1\ min = 60\ s$

$$s = vt = \left(3 \times 10^8\ \frac{m}{s}\right)(1\ h) = \left(3 \times 10^8\ \frac{m}{s}\right)\left(60\ \frac{s}{min}\right)\left(60\ \frac{min}{h}\right)(1\ h)\left(\frac{1\ mi}{1609\ m}\right) = \boxed{6.71 \times 10^8\ mi}$$

41. Estimate the number of Ping-Pong balls that can be packed into an average-size room (without crushing them).

Solution

A Ping-Pong ball has a diameter of about 3 cm and can be thought of as an object which occupies a cube of volume $3 \times 3 \times 3$ cm^3 = 27 cm^3. That is, V_{Ball} = 27 cm^3. A typical room has dimensions 12 ft. $\times$ 15 ft $\times$ 8 ft. Using the conversion, 1 ft $\cong$ 30 cm, we find

$$V_{Room} = 12 \times 15 \times 8 \text{ ft}^3 = 1440 \text{ ft}^3 = (1440 \text{ ft}^3)\left(\frac{30 \text{ cm}}{\text{ft}}\right)^3 \cong 3.9 \times 10^7 \text{ cm}^3$$

The number of Ping-Pong balls which can fill the room is

$$N = \frac{V_{Room}}{V_{Ball}} \cong \frac{4 \times 10^7 \text{ cm}^3}{27 \text{ cm}^3} = \boxed{1.5 \times 10^6 \text{ balls}}$$

Therefore, a typical room can hold about 10^6 Ping-Pong balls.

⸻

43. Soft drinks are commonly sold in aluminum containers. Estimate the number of such containers thrown away each year by U.S. consumers. Approximately how many tons of aluminum does this represent?

Solution

Assuming an average family of four people consumes two 6-packs per week, and a total population of 200 million people, we estimate 30 billion cans per year. Taking the mass of one can as 5 g, we estimate a total mass of 1.5×10^8 kg, corresponding to about 10^5 tons.

⸻

49. Estimate the number of piano tuners living in New York City. This question was raised by the physicist Enrico Fermi, who was well known for making order-of-magnitude calculations.

Solution

In this problem, assume a total population of 10^7 people. Also, let us estimate that one family in ten owns a piano, with four people per family. Thus, for each 100 people we estimate there are 3 pianos. In addition, in one year a single piano tuner can service about 1000 pianos (about 4 per day for 250 weekdays, assuming each piano is tuned once per year). Therefore,

$$\text{The number of tuners} = \left(\frac{1 \text{ tuner}}{1000 \text{ pianos}}\right)\left(\frac{3 \text{ pianos}}{100 \text{ people}}\right)(10^7 \text{ people}) \cong \boxed{300}$$

⸻

51. Calculate (a) the circumference of a circle of radius 3.5 cm and (b) the area of a circle of radius 4.65 cm.

Solution

(a) $C = 2\pi r = 2\pi(3.5 \text{ cm}) = 21.99115$ cm. Since the radius r has only 2 significant figures, we must round off to 2 significant figures. That is,

$$C = \boxed{22 \text{ cm}}$$

(b) $A = \pi r^2 = (3.14159)(4.65 \text{ cm})^2 = 67.92903$ cm². Since r has only 3 significant figures, we must round off to 3 significant figures. Thus,

$$A = \boxed{67.9 \text{ cm}^2}$$

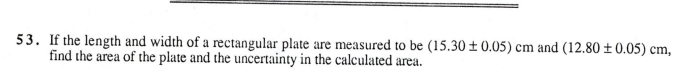

53. If the length and width of a rectangular plate are measured to be (15.30 ± 0.05) cm and (12.80 ± 0.05) cm, find the area of the plate and the uncertainty in the calculated area.

Solution

Referring to the sketch we have,

$A = Lw = (15.30 \pm 0.05)(12.80 \pm 0.05)$ cm²

$A = [(15.30)(12.80) \pm (15.30)(0.05) \pm (0.05)(12.80)]$

$\quad = (195.8 \pm 1.4)$ cm²

In the above we have our area ± an error term. However, we neglected the term (0.05)(0.05) or (error)². From the sketch, it is clearly insignificant.

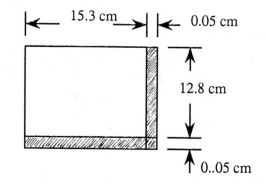

Vectors

OBJECTIVES

1. Describe the coordinates of a point in space using both cartesian coordinates and polar coordinates. (Section 2.1)

2. Distinguish between vector quantities and scalar quantities. (Section 2.2)

3. Understand and describe the basic properties of vectors such as the rules of vector addition and graphical solutions for addition of two or more vectors. (Section 2.3)

4. Resolve a vector into its rectangular components. Determine the magnitude and direction of a vector from its rectangular components. (Section 2.4)

5. Understand the use of unit vectors and describe any vector in terms of its components. (Section 2.4)

6. Become familiar with the concept of force, its vector nature, and the technique of resolving a force into its rectangular components. (Section 2.5)

SKILLS

You should be familiar with . . .

1. The fundamentals of plane and solid geometry--including the ability to graph functions, calculate the areas and volumes of standard geometric figures and recognize the equations and graphs of a straight line, a circle, an ellipse, a parabola and a hyperbola.

2. The basic ideas of trigonometry--definitions and properties of the sine, cosine, and tangent functions; the Pythagorean Theorem, the law of cosines, the law of sines, and some of the basic trigonometric identities.

For your convenience, reviews of geometry and trigonometry are given in Appendix B of the text.

NOTES FROM SELECTED CHAPTER SECTIONS

2.1 COORDINATE SYSTEMS AND FRAMES OF REFERENCE

A *coordinate system* used to specify locations in space consists of:

1. A fixed reference point called the *origin*.

2. A set of specified *axes* or directions.

3. Instructions (properties of the particular coordinate system) which describe how to label a point in space relative to the origin and the axes.

Convenient coordinate systems include *cartesian* and *plane polar* coordinate systems.

2.2 VECTORS AND SCALARS

A *vector* is a physical quantity that must be specified by both a magnitude and direction.

A *scalar* quantity has only magnitude.

2.4 COMPONENTS OF A VECTOR AND UNIT VECTORS

Any vector can be completely described by its *components*.

A *unit vector* is a dimensionless vector one unit in length used to *specify a given direction*.

When two or more vectors are to be added, the following step-by-step procedure is recommended:

1. Select a coordinate system.
2. Draw a sketch of the vectors to be added (or subtracted), with a label on each vector.
3. Find the x and y components of all vectors.
4. Find the resultant components (the algebraic sum of the components) in both the x and y directions.
5. Use the Pythagorean theorem to find the magnitude of the resultant vector.
6. Use a suitable trigonometric function to find the angle the resultant vector makes with the x-axis.

To add vector **A** to vector **B** graphically, first construct **A**, and then draw **B** such that the tail of **B** starts at the head of **A**. The sum **A + B** is the vector that completes the triangle as shown in Fig. 2.1a. The procedure for adding more than two vectors (the polygon rule) is illustrated in Fig. 2.1b.

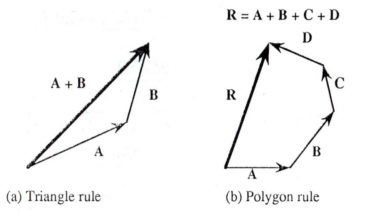

(a) Triangle rule   (b) Polygon rule

Figure 2.1   Adding vectors by (a) the triangle rule and (b) the polygon rule.

In order to subtract two vectors *graphically*, recognize that **A − B** is equivalent to the operation **A + (-B)**. Since the vector **−B** is a vector whose magnitude is B and is opposite to **B**, the construction shown in Fig. 2.2 is obtained:

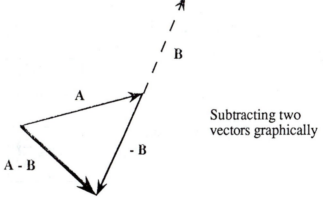

Subtracting two vectors graphically

Figure 2.2

EQUATIONS AND CONCEPTS

Relationship between cartesian and polar coordinates.

$$x = r \cos \theta \tag{2.1}$$

$$y = r \sin \theta \tag{2.2}$$

$$\tan \theta = \frac{y}{x} \tag{2.3}$$

$$r^2 = \sqrt{x^2 + y^2} \tag{2.4}$$

Commutative Law of Addition.

$$\mathbf{A} + \mathbf{B} = \mathbf{B} + \mathbf{A} \tag{2.6}$$

Associative Law of Addition.

$$\mathbf{A} + (\mathbf{B} + \mathbf{C}) = (\mathbf{A} + \mathbf{B}) + \mathbf{C} \tag{2.7}$$

Vector Subtraction.

$$\mathbf{A} - \mathbf{B} = \mathbf{A} + (-\mathbf{B}) \tag{2.8}$$

Magnitude (length) of a vector.

$$|\mathbf{A}| = A$$

x and y components of a vector (see Fig. 2.3).

$$A_x = A \cos \theta$$
$$A_y = A \sin \theta$$

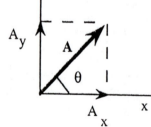

Figure 2.3   Components of a vector

Magnitude of a vector.

$$A = \sqrt{A_x^2 + A_y^2} \tag{2.10}$$

Direction of a vector relative to the positive x-axis.

$$\tan \theta = \frac{A_y}{A_x} \tag{2.11}$$

The unit vectors **i**, **j**, and **k** form a set of *mutually perpendicular* vectors as illustrated in Fig. 2.4.

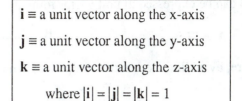

i ≡ a unit vector along the x-axis
j ≡ a unit vector along the y-axis
k ≡ a unit vector along the z-axis
where $|\mathbf{i}| = |\mathbf{j}| = |\mathbf{k}| = 1$

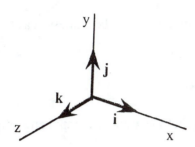

Figure 2.4 Unit vectors

A vector **A** lying in the xy plane, having rectangular components A_x and A_y, can be expressed in unit vector notation.

$$\mathbf{A} = A_x\mathbf{i} + A_y\mathbf{j} \tag{2.12}$$

EXAMPLE PROBLEM SOLUTION

Example 2.1   Vector **A** has a magnitude of 3 m and is directed along the x-axis as shown in Fig. 2.5. Vector **B** has a magnitude of 5 m and makes an angle of 65° with the x-axis.

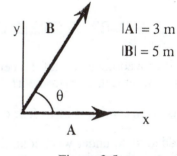

|A| = 3 m
|B| = 5 m

Figure 2.5

(a) Graphical solutions for **A** + **B** and **A** − **B** are sketched below.

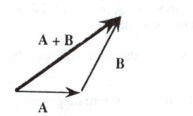

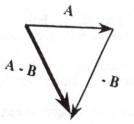

(b) The rectangular components of **A** and **B** are as follows:

$A_x = 3$ m $B_x = (5 \text{ m}) \cos 65° = 2.11$ m

$A_y = 0$ $B_y = (5 \text{ m}) \sin 65° = 4.53$ m

(c) **A** and **B** can be expressed in unit vector form as

$$\mathbf{A} = 3\mathbf{i} \text{ m} \qquad\qquad \mathbf{B} = (2.11\mathbf{i} + 4.53\mathbf{j}) \text{ m}$$

Therefore, by adding the x and y components, the sum and differences can be expressed as follows:

$$\mathbf{A} + \mathbf{B} = (5.11\mathbf{i} + 4.53\mathbf{j}) \text{ m}$$

$$\mathbf{A} - \mathbf{B} = (0.89\mathbf{i} - 4.53\mathbf{j}) \text{ m}$$

(d) To find the angle θ that **A** + **B** makes with the x-axis, note that the x and y components of **A** + **B** are 5.11 and 4.53, respectively.

Therefore, from trigonometry,

$$\tan \theta = \frac{4.53}{5.11} = 0.887$$

$$\theta = 41.6°$$

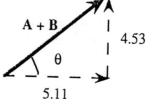

ANSWERS TO SELECTED QUESTIONS

1. A book is moved once around the perimeter of a table of dimensions 1 m × 2 m. If the book ends up at its initial position, what is its displacement? What is the distance traveled?

Answer: Its displacement is zero. The distance traveled equals 6 m.

2. If **B** is added to **A**, under what condition does the resultant vector have a magnitude equal to A + B? Under what conditions is the resultant vector equal to zero?

Answer: The resultant has a magnitude A + B when **A** is oriented in the *same* direction as **B**. The resultant vector is zero when **A** is oriented in the direction opposite to **B**, and when A = B.

5. A vector **A** lies in the xy plane. For what orientations of **A** will both of its rectangular components be negative? For what orientations will its components have opposite signs?

Answer: Both of its components are negative when **A** lies in the third quadrant. Its components are opposite in sign when **A** lies in either the second or fourth quadrant.

6. Can a vector have a component equal to zero and still have a nonzero magnitude? Explain.

Answer: Yes. For example, a vector oriented along the x axis has no y or z component, yet its magnitude is nonzero.

7. If one of the components of a vector is not zero, can its magnitude be zero? Explain.

Answer: No. The magnitude of a vector $\mathbf{A}$ is equal to $\sqrt{A_x^2 + A_y^2 + A_z^2}$. Therefore, if any component is nonzero, $\mathbf{A}$ cannot be zero.

8. If the component of vector $\mathbf{A}$ along the direction of vector $\mathbf{B}$ is zero, what can you conclude about the two vectors?

Answer: The two vectors must be *perpendicular* to each other.

10. Can the magnitude of a vector have a negative value? Explain.

Answer: No. The magnitude of a vector is equal to the absolute value (length) of a vector, which is always positive. However, a component of a vector can be negative.

11. If $\mathbf{A} + \mathbf{B} = 0$, what can you say about the components of the two vectors?

Answer: $\mathbf{A} = -\mathbf{B}$, therefore the components must have *opposite* signs and equal magnitudes.

12. Which of the following are vectors and which are not: force, temperature, the volume of water in a can, ratings of a TV show, the height of a building, the velocity of a sports car, the age of the universe?

Answer: The vectors are force and velocity. The others are not.

13. Under what circumstances would a vector have components that are equal in magnitude?

Answer: A vector at an angle of 45 degrees with both the x and y axes would have equal components along these directions.

14. Is it possible to add a vector quantity to a scalar quantity? Explain.

Answer: Vector and scalar quantities are distinctly different and cannot be added or subtracted from each other.

15. Two vectors have unequal magnitudes. Can their sum be zero? Explain.

Answer: No. Two vectors can cancel only if they are equal in magnitude and opposite in direction.

SOLUTIONS TO SELECTED END-OF-CHAPTER PROBLEMS

3. The polar coordinates of a point are r = 5.50 m and θ = 240°. What are the cartesian coordinates of this point?

Solution

$x = r \cos \theta = (5.50 \text{ m}) \cos 240° = (5.50 \text{ m})(-0.500) = \boxed{-2.75 \text{ m}}$

$y = r \sin \theta = (5.50 \text{ m}) \sin 240° = (5.50 \text{ m})(-0.866) = \boxed{-4.76 \text{ m}}$

5. A certain corner of a room is selected as the origin of a rectangular coordinate system. If a fly is crawling on an adjacent wall at a point having coordinates (2.0, 1.0), where the units are in meters, what is the distance of the fly from the corner of the room?

Solution

The fly crawls from the origin (0, 0) to the point whose coordinates are (2.0 m, 1.0 m). Hence, the distance from the fly to the origin is

$$d = \sqrt{(\Delta x)^2 + (\Delta y)^2} = \sqrt{(2.0 - 0)^2 + (1.0 - 0)^2} = \boxed{\sqrt{5} \text{ m}}$$

---

11. Vector **A** is 3 units in length and points along the positive x axis. Vector **B** is 4 units in length and points along the negative y axis. Use graphical methods to find the magnitude and direction of the vectors (a) **A** + **B**, (b) **A** − **B**.

Solution

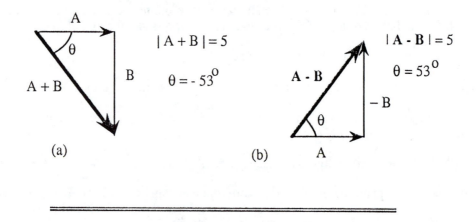

$|A + B| = 5$

$\theta = -53^{\circ}$

(a)

$|A - B| = 5$

$\theta = 53^{\circ}$

(b)

---

13. A person walks along a circular path of radius 5 m, around one half of the circle. (a) Find the magnitude of the displacement vector. (b) How far did the person walk? (c) What is the magnitude of the displacement if the circle is completed?

Solution

See the sketch on the right.

(a) $|\mathbf{d}| = |-10\mathbf{i}| = \boxed{10 \text{ m}}$ since the displacement is a straight line from point A to point B.

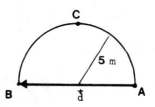

(b) The actual distance walked is not equal to the straight-line displacement. The distance follows the curved path of the semicircle

(ACB). $s = \left(\frac{1}{2}\right)(2\pi r) = 5\pi = \boxed{15.7 \text{ m}}$

(c) If the circle is complete, d begins and ends at point A. Hence, $|\mathbf{d}| = 0$.

16

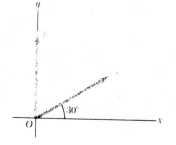

15. Each of the displacement vectors **A** and **B** shown in Figure 2.22 has a magnitude of 3 m. Find graphically (a) **A** + **B**, (b) **A** − **B**, (c) **B** − **A**, (d) **A** − 2**B**.

Solution

To find these vector expressions graphically, we draw each set of vectors as indicated by the drawings below. Measurements of the results are taken using a ruler and protractor.

Figure 2.22

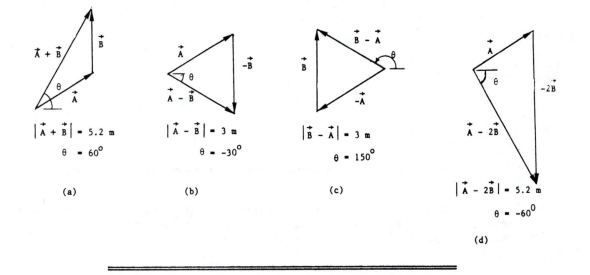

29. Two vectors are given by **A** = 3i − 2j and **B** = −i − 4j. Calculate (a) **A** + **B**, (b) **A** − **B**, (c) |**A** + **B**|, (d) |**A** − **B**|, (e) the direction of **A** + **B** and **A** − **B**.

Solution

(a) $\mathbf{A} + \mathbf{B} = (3i - 2j) + (-i - 4j) = \boxed{2i - 6j}$

(b) $\mathbf{A} - \mathbf{B} = (3i - 2j) - (-i - 4j) = \boxed{4i + 2j}$

(c) $|\mathbf{A} + \mathbf{B}| = \sqrt{2^2 + (-6)^2} = \boxed{6.32}$

(d) $|\mathbf{A} - \mathbf{B}| = \sqrt{4^2 + 2^2} = \boxed{4.47}$

(e) For $\mathbf{A} + \mathbf{B}$, $\theta = \tan^{-1}\left(\frac{-6}{2}\right) = -71.6° = \boxed{288°}$

For $\mathbf{A} - \mathbf{B}$, $\theta = \tan^{-1}\left(\frac{2}{4}\right) = \boxed{26.6°}$

33. A particle undergoes the following consecutive displacements: 3.5 m south, 8.2 m northeast, and 15.0 m west. What is the resultant displacement?

Solution

$\mathbf{d}_1 = (-3.50 \text{ m})\mathbf{j}$

$\mathbf{d}_2 = (8.20 \text{ m}) \cos (45°)\mathbf{i} + (8.20 \text{ m}) \sin (45°)\mathbf{j} = (5.80 \text{ m})\mathbf{i} + (5.80 \text{ m})\mathbf{j}$

$\mathbf{d}_3 = (-15.0 \text{ m})\mathbf{i}$

$\mathbf{R} = \mathbf{d}_1 + \mathbf{d}_2 + \mathbf{d}_3 = (-15.0 + 5.8)\mathbf{i} + (5.8 - 3.5)\mathbf{j} = (-9.20 \text{ m})\mathbf{i} + (2.30)\mathbf{j}$

$|\mathbf{R}| = \sqrt{(9.20 \text{ m})^2 + (2.30 \text{ m})^2} = \boxed{9.48 \text{ m}}$

$\phi = \left|\tan^{-1}\left(\frac{R_y}{R_x}\right)\right| = \left|\tan^{-1}\left(\frac{2.30 \text{ m}}{-9.20 \text{ m}}\right)\right| = 14°$ from -x axis

so $\theta = \boxed{166°}$ from +x axis

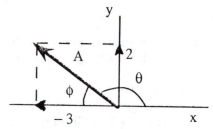

Problem 33

43. A vector **A** has a negative x component 3 units in length and a positive y component 2 units in length. (a) Determine an expression for **A** in unit-vector notation. (b) Determine the magnitude and direction of **A**. (c) What vector **B** when added to **A** gives a resultant vector with no x component and a negative y component 4 units in length?

Solution

$A_x = -3$ units, $A_y = 2$ units

(a) $\mathbf{A} = A_x\mathbf{i} + A_y\mathbf{j} = -3\mathbf{i} + 2\mathbf{j}$ units

(b) $|\mathbf{A}| = \sqrt{A_x^2 + A_y^2} = \sqrt{(-3)^2 + (2)^2} = 3.61$ units

$\tan \phi = \left|\frac{A_y}{A_x}\right| = \left|\frac{2}{-3}\right| = 0.667 \Rightarrow \phi = 33.7°$ (relative to the -x axis)

A is in the *second quadrant*; $\theta = 180° - \phi = \boxed{146°}$

(c) $R_x = 0$, $R_y = -4$;   since $\mathbf{R} = \mathbf{A} + \mathbf{B}$, $\mathbf{B} = \mathbf{R} - \mathbf{A}$

$B_x = R_x - A_x = 0 - (-3) = 3$

$B_y = R_y - A_y = -4 - 2 = -6$

Therefore, $\mathbf{B} = B_x\mathbf{i} + B_y\mathbf{j} = \boxed{(3\mathbf{i} - 6\mathbf{j}) \text{ units}}$

Problem 43

45. Find the magnitude and direction of a displacement vector having x and y components of – 5 m and 3 m, respectively.

Solution

$D_x = -5$ m $D_y = 3$ m

$D = \sqrt{D_x^2 + D_y^2}$

$D = \sqrt{(-5 \text{ m})^2 + (3 \text{ m})^2} = \boxed{5.83 \text{ m}}$

$\phi = \tan^{-1}\left|\dfrac{D_y}{D_x}\right| = \tan^{-1}\left(\dfrac{3 \text{ m}}{5 \text{ m}}\right) = 31°$

$\theta = 180° - \phi = \boxed{149°}$ from +x axis

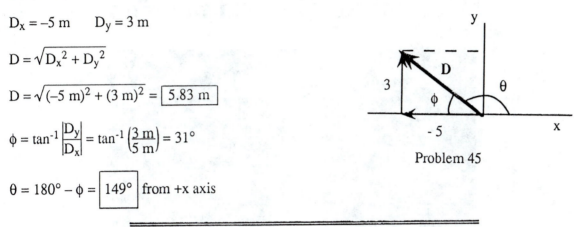

Problem 45

51. A particle moves from a point in the xy plane having cartesian coordinates (–3, –5) m to a point with coordinates (–1, 8) m. (a) Write vector expressions for the position vectors in unit-vector form for these two points. (b) What is the displacement vector? (See Problem 54 for definition.)

Solution

(a) $\mathbf{R}_1 = x_1\mathbf{i} + y_1\mathbf{j} = \boxed{(-3\mathbf{i} - 5\mathbf{j}) \text{ m}}$

   $\mathbf{R}_2 = x_2\mathbf{i} + y_2\mathbf{j} = \boxed{(-\mathbf{i} + 8\mathbf{j}) \text{ m}}$

(b) Displacement $= \Delta\mathbf{R} = \mathbf{R}_2 - \mathbf{R}_1$

   $\Delta\mathbf{R} = (x_2 - x_1)\mathbf{i} + (y_2 - y_1)\mathbf{j} = -\mathbf{i} - (-3\mathbf{i}) + 8\mathbf{j} - (-5\mathbf{j}) = \boxed{(2\mathbf{i} + 13\mathbf{j}) \text{ m}}$

Motion in One Dimension

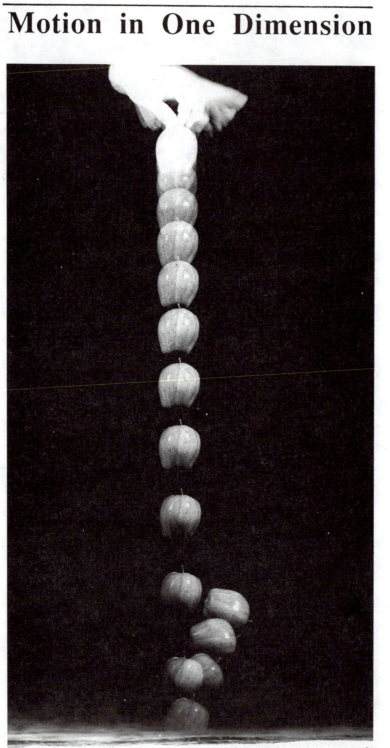

OBJECTIVES

1. Define the displacement and average velocity of a particle in motion.

2. Define the instantaneous velocity and understand how this quantity differs from average velocity.

3. Define average acceleration and instantaneous acceleration.

4. Construct position versus time and velocity versus time graphs for a particle in motion along a straight line. From these graphs, you should be able to determine both average and instantaneous values of velocity and acceleration.

5. Obtain the instantaneous velocity and instantaneous acceleration if the position of a particle is given as a function of time. To do this, you should know how to take a derivative of a function such as $x = At^2 + Bt$. (See Appendix B.5)

6. Recognize that the equations of kinematics apply when motion occurs under *constant* acceleration--and be able to derive the equations of kinematics from the definitions of acceleration, velocity, and displacement.

7. Describe what is meant by a body in *free fall* (one moving under the influence of gravity--where air resistance is neglected). Recognize that the equations of kinematics apply directly to a freely falling object--where the acceleration is given by $a = -g$ (where $g = 9.8$ m/s^2).

8. Apply the equations of kinematics to any situation where the motion occurs under constant acceleration.

SKILLS

You should be able to make plots of position versus time (given a function such as $x = 5 + 3t - 2t^2$) and velocity versus time. From these graphs, you should know how to find the average and instantaneous values of the velocity and acceleration.

You should be familiar with the operation of differentiation--as a limiting process $\left(e.g., \ v = \lim_{\Delta t \to 0} \dfrac{\Delta x}{\Delta t} \right)$ and the graphical interpretation of a derivative as the slope of a graph (e.g., v is the slope of the x versus t graph at a particular value of t). Appendix B.5 gives a brief review of differential calculus.

NOTES FROM SELECTED CHAPTER SECTIONS

3.2   INSTANTANEOUS VELOCITY

The slope of the tangent to the x versus t curve at any instant equals the instantaneous velocity. The area under the v versus t curve equals the displacement during that interval.

3.3   ACCELERATION

Acceleration equals the slope of the tangent to the v vs. t curve.

3.5   FREELY FALLING BODIES

An object thrown upward (or downward) will experience the same acceleration as an object released from rest. In *free fall*, an object will have an acceleration *downward* equal to the acceleration due to gravity.

EQUATIONS AND CONCEPTS

The *displacement* Δx of a particle moving from position x_i to position x_f equals the final coordinate minus the initial coordinate.

$$\Delta x = x_f - x_i$$

The *average velocity* $\bar{v}$ is defined as the ratio of the displacement to the time interval Δt.

$$\bar{v} \equiv \frac{\Delta x}{\Delta t} \qquad (3.1)$$

The *instantaneous velocity* v is defined as the limit of the ratio $\Delta x/\Delta t$ as Δt approaches zero.

$$v \equiv \lim_{\Delta t \to 0} \frac{\Delta x}{\Delta t} = \frac{dx}{dt} \qquad (3.3)$$

The *average acceleration* $\bar{a}$ is defined as the ratio of the change in velocity Δv to the time interval Δt.

$$\bar{a} \equiv \frac{\Delta v}{\Delta t} \qquad (3.4)$$

The *instantaneous acceleration* a is defined as the limit of the ratio $\Delta v/\Delta t$ as Δt approaches zero.

$$a \equiv \lim_{\Delta t \to 0} \frac{\Delta v}{\Delta t} = \frac{dv}{dt} \qquad (3.5)$$

The *equations of kinematics* for a particle moving along the x-axis with *constant acceleration*.

$$v = v_o + at \qquad (3.7)$$

$$x - x_o = \frac{1}{2}(v_o + v)t \qquad (3.9)$$

$$x - x_o = v_o t + \frac{1}{2}at^2 \qquad (3.10)$$

$$v^2 = v_o^2 + 2a(x - x_o) \qquad (3.11)$$

Kinematic equations for a body in *free fall*.

$$v = v_o - gt \qquad (3.12)$$

$$y - y_o = \frac{1}{2}(v_o + v)t \qquad (3.13)$$

$$y - y_o = v_o t - \frac{1}{2}gt^2 \qquad (3.14)$$

$$v^2 = v_o^2 - 2g(y - y_o) \qquad (3.15)$$

EXAMPLE PROBLEM SOLUTION_____

Example 3.1 The coordinate of a particle moving along the x-axis depends on time according to the expression

$$x = 5t^2 - 2t^3$$

where x is in meters and t is in seconds.

(a) Find the velocity and acceleration of the particle as a function of time.

Solution

The velocity and acceleration can be obtained by using Eqs. 3.3 and 3.5. Taking first and second derivatives of x given above, we have

$$v = \frac{dx}{dt} = \frac{d}{dt}\left[5t^2 - 2t^3\right] = 10t - 6t^2$$

$$a = \frac{dv}{dt} = \frac{d}{dt}\left[10t - 6t^2\right] = 10 - 12t$$

(b) Find the time it takes the particle to reach its maximum positive x-coordinate.

Solution

When the particle has reached its maximum x-coordinate, $v = 0$ [that is, it stops and heads back towards the origin]. Therefore, from part (a),

$$10t - 6t^2 = t(10 - 6t) = 0$$

This has two solutions, $t = 0$ corresponding to the beginning of motion where $x = 0$ and $v = 0$; and $t = \frac{5}{3}$ s, which is the time in question, or the turning point, where $x = x_{max}$ and $v = 0$.

(c) Find the displacement during the first 2 seconds.

Solution

Using the given expression for x, we find that the coordinate of the particle at $t = 2$ s is

$$x = 5(2)^2 - 2(2)^3 = 4 \text{ m}$$

Since the particle starts from the origin, $x_0 = 0$ at $t_0 = 0$, the displacement of the particle is also 4 m. Note, however, that the total distance traveled in 2 s is *greater* than 4 m. In fact, the total distance traveled in 2 s equals $x_{max} + (x_{max} - 4)$, where x_{max} is the maximum positive x coordinate which occurs at $t = \frac{5}{3}$ s [see part (b)]. You should show that this *total distance traveled* in 2 s is 5.26 m.

(d) Find the velocity and acceleration of the particle after 2 seconds.

Solution

Using the results of part (a), we have

at $t = 2$ s, $\qquad\qquad v = 10(2) - 6(2)^2 = -4$ m/s

$$a = 10 - 12(2) = -14 \text{ m/s}^2$$

At some later time, both the velocity and acceleration will be larger in magnitude.

ANSWERS TO SELECTED QUESTIONS

1. Average velocity and instantaneous velocity are generally different quantities. Can they ever be equal for a specific type of motion? Explain.

Answer: Yes. If a body moves with *constant* velocity, $\overline{v} = v$.

2. If the average velocity is nonzero for some time interval, does this mean that the instantaneous velocity is never zero during this interval? Explain.

Answer: No. The average velocity v may be nonzero, but the particle may come to rest at some instant during this interval. This happens, for example, when the particle reaches a turning point in its motion.

6. If the velocity of a particle is zero, can its acceleration ever be nonzero? Explain.

Answer: Yes. If a particle moves with *constant* velocity, then the acceleration is not zero.

8. A ball is thrown vertically upward. What are its velocity and acceleration when it reaches its maximum altitude? What is its acceleration just before it strikes the ground?

Answer: At the peak of its motion, $v = 0$ and $a = -g$. Just before it strikes the ground, $a = -g$. *Its acceleration is $-g$ during its entire flight!*

9. A stone is thrown upward from the top of a building. Does the stone's displacement depend on the location of the origin of the coordinate system? Does the stone's velocity depend on the origin? (Assume that the coordinate system is stationary with respect to the building.) Explain.

Answer: No. The displacement (final minus initial coordinate) is *independent of the origin*. Likewise, the stone's velocity is independent of the origin--assuming the coordinate system is stationary.

10. A child throws a marble into the air with an initial velocity v_0. Another child drops a ball at the same instant. Compare the accelerations of the two objects while they are in flight.

Answer: Both have an acceleration of $-g$, since *both are freely falling*.

11. A student at the top of a building of height h throws one ball upward with an initial speed v_0 and then throws a second ball downward with the same initial speed. How do the final velocities of the balls compare when they reach the ground?

Answer: They are the same.

14. If the average velocity of an object is zero in some time interval, what can you say about the displacement of the object for that interval?

Answer: The displacement is *zero*, since the displacement is proportional to average velocity.

SOLUTIONS TO SELECTED END-OF-CHAPTER PROBLEMS

5. An athlete swims the length of a 50-m pool in 20 s and makes the return trip to the starting position in 22 s. Determine his average velocity in (a) the first half of the swim, (b) the second half of the swim, and (c) the round trip.

Solution

(a) $\bar{v}_1 = \dfrac{d}{t_1} = \dfrac{50\ m}{20\ s} = \boxed{2.50\ m/s}$ (b) $\bar{v}_2 = \dfrac{d}{t_2} = \dfrac{-50\ m}{22\ s} = \boxed{-2.27\ m/s}$

(c) Since the displacement is *zero* for the round trip, $\bar{v} = 0$.

7. A car makes a 200-km trip at an average speed of 40 km/h. A second car starting 1 h later arrives at their mutual destination at the same time. What was the average speed of the second car?

Solution

$t_1 = \dfrac{d}{v_1} = \dfrac{200\ km}{40\ km/h} = 5\ h$ for car 1 $t_2 = t_1 - 1 = 4\ h$ for car 2

$\bar{v}_2 = \dfrac{d}{t_2} = \dfrac{200\ km}{4\ h} = \boxed{50.0\ km/h}$

9. The position-time graph for a particle moving along the x axis is as shown in Figure 3.17. (a) Find the average velocity in the time interval t = 1.5 s to t = 4 s. (b) Determine the instantaneous velocity at t = 2 s by measuring the slope of the tangent line shown in the graph. (c) At what value of t is the velocity zero?

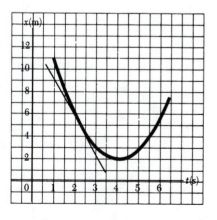

Figure 3.17

Solution

(a)   At point A, where $t_i = 1.5$ s, $x_i = 8$ m

   At point B, where $t_f = 4$ s, $x_f = 2$ m

   Therefore, $\bar{v} = \dfrac{x_f - x_i}{t_f - t_i} = \dfrac{2\ m - 8\ m}{4\ s - 1.5\ s} = \boxed{-2.4\ m/s}$

(b)   The slope of the tangent line is found from points C and D ($x_c = 9$ m, $t_c = 1$ s);

   ($x_D = 1$ m, $t_D = 3.5$ s), thus $v \cong \boxed{-3.2\ m/s}$

(c)   The velocity is zero when x is a minimum. That is, at $t \cong \boxed{4\ s}$

---

17. A particle moves along the x axis according to the equation $x = 2t + 3t^2$, where x is in m and t is in s. Calculate the instantaneous velocity and instantaneous acceleration at $t = 3$ s.

Solution

$x = 2t + 3t^2 \qquad v(t) = \dfrac{dx}{dt} = (2 + 6t)\ m/s \qquad a = \dfrac{dv}{dt} = 6\ m/s^2$

Therefore, at $t = 3$ s, $v(3) = 2 + 6(3) = \boxed{20.0\ m/s}$ and $\boxed{a = 6.00\ m/s^2}$

---

25. A body moving with uniform acceleration has a velocity of 12 cm/s when its x coordinate is 3 cm. If its x coordinate 2 s later is –5 cm, what is the magnitude of its acceleration?

Solution

At $t = 0$, you are given that $x_o = 3$ cm and $v_o = 12$ cm/s. Also, at $t = 2$ s, $x = -5$ cm. To find a, we can use the kinematic relation

$$x - x_o = v_o t + \frac{1}{2} a t^2$$

$$-5 - 3 = 12(2) + \frac{1}{2} a(2)^2$$

$$-8 = 24 + 2a$$

$$a = -\frac{32}{2} = \boxed{-16.0\ cm/s^2}$$

39. A car moving at a constant speed of 30 m/s suddenly stalls at the bottom of a hill. The car undergoes a constant acceleration of –2 m/s² (opposite its motion) while ascending the hill. (a) Write equations for the position and the velocity as functions of time, taking x = 0 at the bottom of the hill where v_o = 30 m/s. (b) Determine the maximum distance traveled by the car up the hill after stalling.

Solution

At t = 0, you are given v_o = 30 m/s, x_o = 0, and a = –2 m/s² = constant

(a) $x = x_o + v_o t + \frac{1}{2} at^2 =$ $\boxed{(30t - t^2) \text{ m}}$ (1)

    $v = v_o + at =$ $\boxed{(30 - 2t) \text{ m/s}}$ (2)

(b) x reaches a *maximum* when v = 0 (turning point), or when 30 – 2t = 0. This gives t = 15 s. The value of x at t = 15 s [from (1)] is

$$x_{max} = 30(15) - (15)^2 = \boxed{225 \text{ m}}$$

---

47. A student throws a set of keys vertically upward to her sorority sister in a window 4.0 m above. The keys are caught 1.5 s later by the sister's outstretched hand. (a) With what initial velocity were the keys thrown? (b) What was the velocity of the keys just before they were caught?

Solution

(a) Taking y_o = 0 (at the position of the thrower) and given that y = 4 m at t = 1.5 s, we find (with a = –9.80 m/s²)

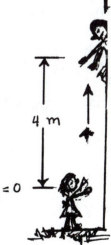

$$y = v_o t + \frac{1}{2} at^2$$

$$4 = 1.5\, v_o - 4.90(1.5)^2$$

$$v_o = \boxed{10.0 \text{ m/s}}$$

(b) The velocity at any time t > 0 is given by v = v_o + at.

Therefore, at t = 1.5 s,

$$v = 10.0 \text{ m/s} - (9.80 \text{ m/s}^2)(1.5 \text{ s}) = \boxed{-4.68 \text{ m/s}}$$

Note that the keys are moving *downward* just before they are caught.

---

53. A stone is thrown upwards from the edge of a cliff 18 m high. It just misses the cliff on the way down and hits the ground below with a speed of 18.8 m/s. (a) With what velocity was it released? (b) What is its maximum distance from the ground during its flight?

Solution

Take $y_0 = 0$ at the top of the cliff. We are given $v = -18.8$ m/s at $y = -18.0$ m. Also, $a = -9.80$ m/s^2.

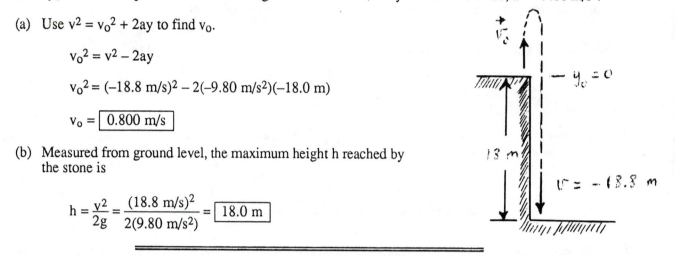

(a) Use $v^2 = v_0^2 + 2ay$ to find v_0.

$$v_0^2 = v^2 - 2ay$$

$$v_0^2 = (-18.8 \text{ m/s})^2 - 2(-9.80 \text{ m/s}^2)(-18.0 \text{ m})$$

$$v_0 = \boxed{0.800 \text{ m/s}}$$

(b) Measured from ground level, the maximum height h reached by the stone is

$$h = \frac{v^2}{2g} = \frac{(18.8 \text{ m/s})^2}{2(9.80 \text{ m/s}^2)} = \boxed{18.0 \text{ m}}$$

59. A particle is moving along the x axis. Its velocity as a function of time is given by $v = 5 + 10t$, where v is in m/s. The position of the particle at $t = 0$ is 20 m. Find (a) the acceleration as a function of time, (b) the position as a function of time, and (c) the velocity of the particle at $t = 0$.

Solution

At $t = 0$, $x_0 = 20$ m $\qquad v = (5 + 10t)$ m/s

(a) $a = \dfrac{dv}{dt} = \dfrac{d}{dt}(5 + 10t) = \boxed{10.0 \text{ m/s}^2}$

(b) Using $v = \dfrac{dx}{dt}$, we find $dx = vdt$, and integrating gives

$$\int_{x_0=20}^{x} dx = \int_{t=0}^{t} vdt = \int_{0}^{t} (5 + 10t)dt$$

$$x - 20 = 5t + 5t^2$$

or

$$x = \boxed{(20 + 5t + 5t^2) \text{ m}}$$

(c) Since $v = (5 + 10t)$ m/s, at $t = 0$, $v_0 = \boxed{5.00 \text{ m/s}}$

69. A young woman named Kathy Kool buys a superdeluxe sports car that can accelerate at the rate of 16 ft/s². She decides to test the car by dragging with another speedster, Stan Speedy. Both start from rest, but experienced Stan leaves 1 s before Kathy. If Stan moves with a constant acceleration of 12 ft/s² and Kathy maintains an acceleration of 16 ft/s², find (a) the time it takes Kathy to overtake Stan, (b) the distance she travels before she catches him, and (c) the velocities of both cars at the instant she overtakes him.

Solution

(a) Let x_k = Kathy's position and x_s = Stan's position. We require that $x_s = x_k$ when Kathy overtakes Stan.

$$x_k = \frac{1}{2} a_k t^2 = \frac{1}{2} (16)t^2 = 8t^2 \tag{1}$$

Since Stan leaves 1 s *before* Kathy, and taking $t = 0$ to be the time Kathy leaves the origin, then Stan's position at $t = 0$ is

$$x_{so} = \frac{1}{2} a_s(1)^2 = \frac{1}{2} (12)(1)^2 = 6 \text{ ft.}$$

Also, at $t = 1$ s, Stan has a velocity

$$v_{so} = a_s t = 12(1) = 12 \text{ ft/s}$$

Therefore,

$$x_s = x_{so} + v_{so}t + \frac{1}{2} a_s t^2 = 6 + 12t + 6t^2 \tag{2}$$

Setting $x_s = x_k$, using (1) and (2) we find

$$6 + 12t + 6t^2 = 8t^2$$

$$2t^2 - 12t - 6 = 0$$

$$t = \boxed{6.46 \text{ s}}$$

(b) $x_k = \frac{1}{2} a_k t^2 = \frac{1}{2} (16 \text{ ft/s}^2)(6.46 \text{ s})^2 = \boxed{334 \text{ ft}}$

(c) $v_k = a_k t = (16 \text{ ft/s}^2)(6.46 \text{ s}) = \boxed{103 \text{ ft/s}}$

$v_s = a_s(t + 1) = (12 \text{ ft/s}^2)(7.46 \text{ s}) = \boxed{89.6 \text{ ft/s}}$

4

Motion in Two Dimensions

OBJECTIVES

1. Describe the displacement, velocity, and acceleration of a particle moving in the xy plane.

2. Derive expressions for the velocity and displacement as functions of time for a particle moving in a plane with *constant* acceleration.

3. Recognize that two-dimensional motion in the xy plane with constant acceleration is equivalent to two independent motions along the x and y directions with constant acceleration components a_x and a_y.

4. Discuss the assumptions used in describing projectile motion; that is, two-dimensional motion in the presence of gravity.

5. Develop expressions for the velocity components and coordinates of a projectile at any time t, in terms of its initial velocity components v_{xo} and v_{yo}.

6. Recognize the fact that if the initial speed v_o and initial angle θ_o of a projectile are known at a given point at $t = 0$, the velocity components and coordinates can be found at any later time t. Furthermore, one can also calculate the horizontal range R and maximum height h if v_o and θ_o are known.

7. Understand the nature of the acceleration of a particle moving in a circle with constant speed. In this situation, note that although $|v|$ = constant, the *direction* of v varies in time, which is the origin of the radial, or centripetal acceleration.

8. Describe the components of acceleration for a particle moving on a curved path, where both the magnitude and direction of v are changing with time. In this case, the particle has a tangential component of acceleration and a radial component of acceleration.

9. Realize that the outcome of a measurement of the motion of a particle (its position, velocity and acceleration) depends on the frame of reference of the observer.

SKILLS

1. You should be familiar with the mathematical expression for a *parabola*. In particular, the equation which describes the trajectory of a projectile moving under the influence of gravity is given by

$$y = ax - bx^2$$

where $a = \tan \theta_o$ and $b = \dfrac{g}{2v_o^2 \cos^2 \theta_o}$

Note that this expression for y assumes that the particle leaves the origin at $t = 0$, with a velocity v_o. A sketch of y versus x for this situation is shown in Fig. 4.1.

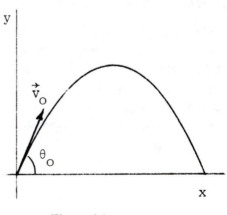

Figure 4.1

31

2. If you are given v_0 and θ_0, you should be able to make a point-by-point plot of the trajectory using the expressions for $x(t)$ and $y(t)$. Furthermore, you should know how to calculate the velocity component v_y at any time t. (Note that the component $v_x = v_{xo} = v_0 \cos \theta_0 = $ constant, since $a_x = 0$.)

3. Assuming that you have values for x and y at any time $t > 0$, you should be able to write an expression for the position vector $\mathbf{r}$ at that time using the relation $\mathbf{r} = x\mathbf{i} + y\mathbf{j}$. From this you can find the *displacement* r, where $r = \sqrt{x^2 + y^2}$. Likewise, if v_x and v_y are known at any time $t > 0$, you can express the velocity vector $\mathbf{v}$ in the form $\mathbf{v} = v_x\mathbf{i} + v_y\mathbf{j}$. From this, you can find the *speed* at any time, since $v = \sqrt{v_x^2 + v_y^2}$.

NOTES FROM SELECTED CHAPTER SECTIONS

4.1 DISPLACEMENT, VELOCITY, AND ACCELERATION VECTORS

Displacement vector is the final position vector minus the initial position vector. Note that the magnitude of the displacement vector is in general less than the distance measured along the actual path of travel.

It is important to recognize that a particle experiences an *acceleration* when:

(1) the magnitude of the velocity (speed) changes while the direction remains constant (i.e., a sphere rolling down an inclined plane;
(2) the magnitude of the velocity remains constant while the direction of the velocity changes (i.e., a particle moving at constant speed around a circle of constant radius);
(3) both the magnitude and direction of the velocity changes (i.e., a mass vibrating up and down on the end of a spring).

4.3 PROJECTILE MOTION

Projectile motion is a common example of motion in two dimensions under constant acceleration. When dealing with a projectile motion in the presence of earth's gravity, we simplify matters by making the following assumptions:

(1) The acceleration due to gravity, g, is constant over the range of motion and directed downward.
(2) Air resistance is negligible.
(3) The earth's rotation does not affect the projectile's motion.

The trajectory of a projectile is a *parabola* as shown in Fig.4.2 on the following page. We choose the motion to be in the xy plane, and take the *initial* velocity of the projectile to have a magnitude of v_0, directed at an angle θ_0 with the horizontal. The parabolic path of travel is completely determined when the magnitude, v_0, and the direction, θ_0, of the initial velocity vector are given. The actual motion of the projectile is a superposition of two motions: (1) motion of a freely-falling body in the vertical direction with constant acceleration, $-g$, and (2) motion in the horizontal direction with constant velocity, $v_0 \cos \theta_0$.

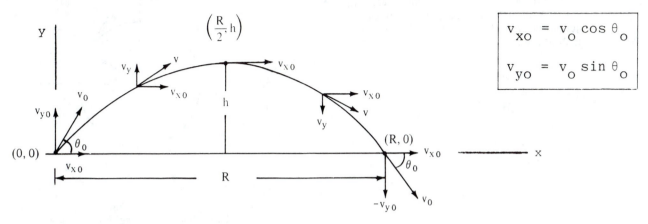

$$v_{xo} = v_o \cos \theta_o$$

$$v_{yo} = v_o \sin \theta_o$$

Figure 4.2 Trajectory of a Projectile

The *maximum height* h and *horizontal range* R
are *special* coordinates defined in Fig. 4.3.
These can be obtained from Eqs. 4.10 - 4.13.

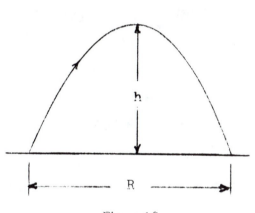

Figure 4.3

4.4 UNIFORM CIRCULAR MOTION

The *centripetal* acceleration is the acceleration experienced by a mass which moves uniformly in a circular path of constant radius. The *direction* of the centripetal acceleration is always toward the center of the circular path.

Uniform circular motion is the motion of an object moving in a circular path with *constant linear* speed. The velocity vector is always tangent to the path of the moving body and therefore *perpendicular* to the radius.

4.5 TANGENTIAL AND RADIAL ACCELERATION IN CURVILINEAR MOTION

Tangential acceleration of a particle moving in a circular path is due to a change in the speed (magnitude of the velocity vector) of the particle. The direction of the tangential acceleration at any instant is tangent to the circular path (perpendicular to the radius).

If a particle moves in a circle such that the speed v is *not* constant, the components of acceleration and the total acceleration at some instant are as shown in Fig. 4.4.

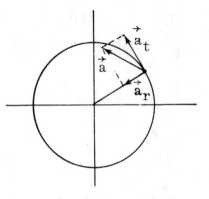

Figure 4.4
Components of acceleration when
$|\vec{v}| \neq$ constant

EQUATIONS AND CONCEPTS

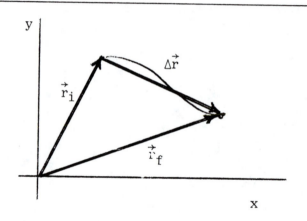

Figure 4.5

A particle whose position vector changes from r_i to r_f undergoes a *displacement* $\Delta r \equiv r_f - r_i$.

$$\Delta r \equiv r_f - r_i \qquad (4.1)$$

The *average velocity* of a particle which undergoes a displacement Δr in a time interval Δt equals the ratio $\Delta r/\Delta t$.

$$\overline{v} \equiv \frac{\Delta r}{\Delta t} \qquad (4.2)$$

The *instantaneous velocity* of a particle equals the limit of the average velocity as $\Delta t \to 0$.

$$v \equiv \lim_{\Delta t \to 0} \frac{\Delta r}{\Delta t} = \frac{dr}{dt} \qquad (4.3)$$

The *average acceleration* of a particle which undergoes a change in velocity Δv in a time interval Δt equals the ratio $\Delta v / \Delta t$.

$$\bar{a} \equiv \frac{\Delta v}{\Delta t} \qquad (4.4)$$

The *instantaneous acceleration* is defined as the limit of the average velocity as $\Delta t \rightarrow 0$.

$$a \equiv \lim_{\Delta t \rightarrow 0} \frac{\Delta v}{\Delta t} = \frac{dv}{dt} \qquad (4.5)$$

The *velocity* of a particle as a function of time moving with *constant acceleration* : (at $t = 0$, the velocity is v_o).

$$v = v_o + at \qquad (4.8)$$

The *position vector* as a function of time for a particle moving with *constant acceleration* : (at $t = 0$, the position vector is r_o and the velocity is v_o).

$$r = r_o + v_o t + \frac{1}{2} at^2 \qquad (4.9)$$

x and y components of velocity versus time for a projectile.

$$v_x = v_o \cos \theta_o = \text{constant} \qquad (4.10)$$
$$v_y = v_o \sin \theta_o - gt \qquad (4.11)$$

x and y coordinates versus time for a projectile.

$$x = (v_o \cos \theta_o) t \qquad (4.12)$$
$$y = (v_o \sin \theta_o) t - \frac{1}{2} gt^2 \qquad (4.13)$$

The *maximum height* of a projectile in terms of v_o and θ_o.

$$h = \frac{v_o^2 \sin^2 \theta_o}{2g} \qquad (4.17)$$

The *horizontal range* of a projectile in terms of v_o and θ_o.

$$R = \frac{v_o^2 \sin 2\theta_o}{g} \qquad (4.18)$$

A particle moving in a circle of radius r with speed v undergoes a *centripetal acceleration* a_r equal in magnitude to v^2/r.

$$a_r = \frac{v^2}{r} \qquad (4.19)$$

If the speed of a particle moving on a curved path changes with time, the particle has a *tangential* component of acceleration equal in magnitude to dv/dt.

$$a_t = \frac{dv}{dt}$$

(4.21)

In general, a particle moving on a curved path can have both a centripetal component and tangential component of acceleration, where a_r is directed towards the center of curvature and a_t is tangent to the path.

$$a = a_r + a_t$$

(4.20)

EXAMPLE PROBLEM SOLUTIONS

Example 4.1 A right-fielder throws a ball directly towards home plate with an initial speed of 50 m/s at an angle of 10° with the horizontal. Home plate is at a distance of 100 m from the fielder.

(a) Where does the ball land relative to home plate?

Solution

The initial components of the velocity are given by

$$v_{xo} = v_o \cos \theta_o = (50 \text{ m/s}) \cos (10°) = 49 \text{ m/s}$$

$$v_{yo} = v_o \sin \theta_o = (50 \text{ m/s}) \sin (10°) = 8.7 \text{ m/s}$$

We can calculate the horizontal range of the ball using Eq. 4.18:

$$R = \frac{v_o^2 \sin 2\theta_o}{g} = \frac{(50 \text{ m/s})^2 \sin (20°)}{9.8 \text{ m/s}^2} = 87 \text{ m}$$

Therefore, the ball lands 13 m short of home plate. (Assuming that the ball is thrown from ground level.)

(b) How long is the ball in the air before striking the ground?

Solution

When it strikes the ground, x = R, therefore using Eq. 4.12 we can find t:

$$t = \frac{R}{v_o \cos \theta_o} = \frac{R}{v_{xo}} = \frac{87 \text{ m}}{49 \text{ m/s}} = 1.8 \text{ s}$$

(c) What is the maximum height reached by the ball?

Solution

Using Eq. 4.17, we get

$$h = \frac{v_o^2 \sin^2 \theta_o}{2g} = \frac{(50 \text{ m/s})^2 \sin^2 (10°)}{2 \times 9.8 \text{ m/s}^2} = 3.8 \text{ m}$$

Example 4.2 A place-kicker in a football game is located 40 m from the goal post. The horizontal bar to be cleared is located 3 m above the ground (Fig. 4.6). If the ball is kicked with an initial speed of 30 m/s at an angle of 20° with the horizontal, does it clear the bar?

Solution

First, let us calculate the initial velocity components.

$$v_{xo} = v_o \cos \theta_o = (30 \text{ m/s}) \cos (20°) = 28.2 \text{ m/s}$$

$$v_{yo} = v_o \sin \theta_o = (30 \text{ m/s}) \sin (20°) = 10.3 \text{ m/s}$$

We can determine the time it takes to reach the bar using Eq. 4.12, with x = 40 m:

$$x = (v_o \cos \theta_o)t$$

or

$$t = \frac{x}{v_{xo}} = \frac{40 \text{ m}}{28.2 \text{ m/s}} = 1.42 \text{ s}$$

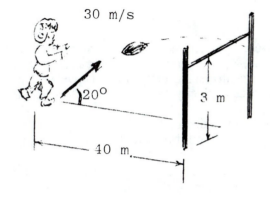

Figure 4.6

Substituting this value of t into Eq. 4.13, we can find the height y when x = 40 m:

$$y = v_{yo}t - \frac{1}{2} gt^2 = (10.3 \text{ m/s})(1.42 \text{ s}) - (4.9 \text{ m/s}^2)(1.42)^2 = 4.7 \text{ m}$$

Since the bar is 3 m high, the ball clears the bar by 1.7 m.

ANSWERS TO SELECTED QUESTIONS

1. If the average velocity of a particle is zero in some time interval, what can you say about the displacement of the particle for that interval?

Answer: The displacement is *zero*, since the displacement is proportional to $\bar{v}$.

2. If you know the position vectors of a particle at two points along its path and also know the time it took to get from one point to the other, can you determine the particle's instantaneous velocity? Its average velocity?

Answer: Its instantaneous velocity cannot be determined from this information. However, the average velocity can be determined from its definition and the given information.

3. Describe a situation in which the velocity of a particle is perpendicular to the position vector.

Answer: A particle moving in a circular path, where the origin of **r** is at the center of the circle.

4. Can a particle accelerate if its speed is constant?
Can it accelerate if its velocity is constant?

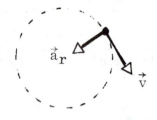

Answer: Yes. Its speed may be constant, but the *direction* of **v** may change -- causing an acceleration. For example, a particle moving in a circle with constant speed has an acceleration directed toward the center of the circle (a centripetal acceleration). However, a particle has *zero* acceleration when its *velocity* is constant. Note that constant velocity means that both the *direction* and *magnitude* of **v** remain constant.

5. Explain whether or not the following particles have an acceleration: (a) a particle moving in a straight line with constant speed and (b) a particle moving around a curve with constant speed.

Answer: (a) The acceleration is zero, since |**v**| and its direction remain constant. (b) The particle has an acceleration since the direction of **v** changes.

8. A student argues that as a satellite orbits the earth in a circular path, it moves with constant velocity and therefore has no acceleration. What is wrong with the student's argument?

Answer: The velocity is not constant since its direction *changes* with time. The satellite has a centripetal acceleration supplied by gravity.

9. What is the fundamental difference between the unit vectors $\hat{r}$ and $\hat{\theta}$ defined in Fig. 4.15 and the unit vectors **i** and **j**?

Answer: The unit vectors $\hat{r}$ and $\hat{\theta}$ move with the particle being described, so their orientations change with time relative to a fixed coordinate system. The unit vectors **i** and **j** are fixed in some frame of reference.

SOLUTIONS TO SELECTED END-OF-CHAPTER PROBLEMS

7. A fish swimming horizontally has velocity $v_0 = (4i + j)$ m/s at a point in the ocean whose distance from a certain rock is $r_0 = (10i - 4j)$ m. After swimming with constant acceleration for 20.0 s, its velocity is $v = (20i - 5j)$ m/s. (a) What are the components of the acceleration? (b) What is the direction of the acceleration with respect to unit vector **i**? (c) Where is the fish at t = 25 s and in what direction is it moving?

Solution

At t = 0, $v_0 = (4i + j)$ m/s and $r_0 = (10i - 4j)$ m.   At t = 20 s, $v = (20i - 5j)$ m/s

(a) $a_x = \dfrac{\Delta v_x}{\Delta t} = \dfrac{20 \text{ m/s} - 4 \text{ m/s}}{20 \text{ s}} = \boxed{0.800 \text{ m/s}^2}$

$a_y = \dfrac{\Delta v_y}{\Delta t} = \dfrac{-5 \text{ m/s} - 1 \text{ m/s}}{20 \text{ s}} = \boxed{-0.300 \text{ m/s}^2}$

(b) $\theta = \tan^{-1}\left(\dfrac{a_y}{a_x}\right) = \tan^{-1}\left(\dfrac{-0.300 \text{ m/s}^2}{0.800 \text{ m/s}^2}\right) = -20.6°$ or $\boxed{339°}$ from the +x axis

(c) At $t = 25$ s, its coordinates are

$$x = x_o + v_{xo}t + \frac{1}{2}a_xt^2 = 10 \text{ m} + (4 \text{ m/s})(25 \text{ s}) + \frac{1}{2}(0.800 \text{ m/s}^2)(25 \text{ s})^2 = \boxed{360 \text{ m}}$$

$$y = y_o + v_{yo}t + \frac{1}{2}a_yt^2 = -4 \text{ m} + (1 \text{ m/s})(25 \text{ s}) + \frac{1}{2}(-0.300 \text{ m/s}^2)(25 \text{ s})^2 = \boxed{-72.8 \text{ m}}$$

$$\theta = \tan^{-1}\left(\frac{v_y}{v_x}\right) = \tan^{-1}\left(\frac{-6.5 \text{ m/s}}{24 \text{ m/s}}\right) = \boxed{-15°} \quad \text{(where } v_x \text{ and } v_y \text{ were evaluated at } t = 25 \text{ s)}$$

---

9. A particle initially located at the origin has an acceleration of $\mathbf{a} = 3\mathbf{j}$ m/s^2 and an initial velocity of $\mathbf{v_0} = 5\mathbf{i}$ m/s. Find (a) the vector position and velocity at any time t and (b) the coordinates and speed of the particle at $t = 2$ s.

Solution

Given $\quad \mathbf{a} = 3\mathbf{j}$ m/s^2 $\quad \mathbf{v_0} = 5\mathbf{i}$ m/s $\quad \mathbf{r_0} = 0\mathbf{i} + 0\mathbf{j}$

(a) $\mathbf{r} = \mathbf{r_0} + \mathbf{v_0}t + \frac{1}{2}\mathbf{a}t^2 = \left(5t\mathbf{i} + 1.5t^2\mathbf{j}\right)$ m

$\quad\quad \mathbf{v} = \mathbf{v_0} + \mathbf{a}t = \boxed{(5\mathbf{i} + 3t\mathbf{j}) \text{ m/s}}$

(b) At $t = 2$ s, we find

$$\mathbf{r} = 5(2)\mathbf{i} + 1.5(2)^2\mathbf{j} = (10\mathbf{i} + 6\mathbf{j}) \text{ m}$$

That is, $\quad\quad (x,y) = \boxed{(10 \text{ m}, 6 \text{ m})}$

$$\mathbf{v} = 5\mathbf{i} + 3(2)\mathbf{j} = (5\mathbf{i} + 6\mathbf{j}) \text{ m/s}$$

so $\quad\quad\quad v = |\mathbf{v}| = \sqrt{v_x^2 + v_y^2} = \sqrt{5^2 + 6^2} \text{ m/s} = \boxed{7.81 \text{ m/s}}$

---

15. A ball is thrown horizontally from the top of a building 35 m high. The ball strikes the ground at a point 80 m from the base of the building. Find (a) the time the ball is in flight, (b) its initial velocity, and (c) the x and y components of velocity just before the ball strikes the ground.

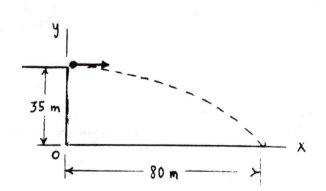

Solution

Using the coordinate system shown, we have $x_0 = 0$ and $y_0 = 35$ m. Furthermore, $v_{xo} = v_o$, $v_{yo} = 0$, $a_x = 0$, and $a_y = -9.80$ m/s^2.

(a) When the ball reaches the ground, x = 80 m and y = 0. To find the time it takes to reach the ground, use

$$y = y_o + v_{yo}t + \frac{1}{2}a_yt^2 = 35 - 4.9t^2 = 0$$

Thus,

$$t^2 = \frac{35}{4.9} \quad \text{or} \quad t = \boxed{2.67 \text{ s}}$$

(b) Using $x = x_o + v_{xo}t = v_ot$ with $t = 2.67$ s gives

$$80 \text{ m} = v_o(2.67 \text{ s})$$

$$v_o = \frac{80 \text{ m}}{2.67 \text{ s}} = \boxed{29.9 \text{ m/s}}$$

(c) $v_x = v_{xo} = \boxed{29.9 \text{ m/s}}$

$$v_y = v_{yo} - gt = 0 - 9.80t = (-9.80 \text{ m/s}^2)(2.67 \text{ s}) = \boxed{-26.2 \text{ m/s}}$$

---

23. A projectile is fired in such a way that its horizontal range is equal to three times its maximum height. What is the angle of projection?

Solution

In this problem, we want to find θ_o such that $R = 3h$. We can use

$$R = \frac{v_o^2 \sin(2\theta_o)}{g} \quad \text{and} \quad h = \frac{v_o^2 \sin^2\theta_o}{2g}$$

Since we require $R = 3h$,

$$\frac{v_o^2 \sin(2\theta_o)}{g} = \frac{3v_o^2 \sin^2\theta_o}{2g}$$

or,

$$\frac{2}{3} = \frac{\sin^2\theta_o}{\sin(2\theta_o)} = \frac{\sin^2\theta_o}{2\sin\theta_o\cos\theta_o} = \frac{\tan\theta_o}{2}$$

$$\theta_o = \tan^{-1}\left(\frac{4}{3}\right) = \boxed{53.1°}$$

---

33. A particle moves in a circular path 0.4 m in radius with constant speed. If the particle makes five revolutions in each second of its motion, find (a) the speed of the particle and (b) its acceleration.

Solution

(a) Since r = 0.4 m, the particle travels a distance of $2\pi r = 2\pi(0.4 \text{ m}) = 2.51$ m in each revolution. Therefore, since it makes 5 rev each second, it travels a distance of 5(2.51 m) = 12.57 m each second. Hence,

$$v = \frac{12.57 \text{ m}}{1 \text{ s}} = \boxed{12.6 \text{ m/s}}$$

(b) $a = \dfrac{v^2}{r} = \dfrac{(12.6 \text{ m/s})^2}{0.4 \text{ m}} = \boxed{395 \text{ m/s}^2}$

a is directed toward the *center* of the circle.

---

37. A train slows down as it rounds a sharp horizontal turn, slowing from 90 km/h to 50 km/h in the 15 s that it takes to round the bend. The radius of the curve is 150 m. Compute the acceleration at the moment the train speed reaches 50 km/h.

Solution

$$50 \text{ km/h} = \left(50 \frac{\text{km}}{\text{h}}\right)\left(10^3 \frac{\text{m}}{\text{km}}\right)\left(\frac{1 \text{ h}}{3600 \text{ s}}\right) = 13.89 \text{ m/s}$$

$$90 \text{ km/h} = \left(90 \frac{\text{km}}{\text{h}}\right)\left(10^3 \frac{\text{m}}{\text{km}}\right)\left(\frac{1 \text{ h}}{3600 \text{ s}}\right) = 25.0 \text{ m/s}$$

Therefore, when v = 13.89 m/s,

$$a_r = \frac{v^2}{r} = \frac{(13.89 \text{ m/s})^2}{150 \text{ m}} = 1.29 \text{ m/s}^2$$

$$a_t = \frac{\Delta v}{\Delta t} = \frac{13.89 \text{ m/s} - 25.0 \text{ m/s}}{15 \text{ s}} = -0.741 \text{ m/s}^2$$

$$a = \sqrt{a_r^2 + a_t^2} = \sqrt{(1.29 \text{ m/s})^2 + (-0.741 \text{ m/s})^2} = \boxed{1.48 \text{ m/s}^2}$$

---

49. A car travels due east with a speed of 50 km/h. Rain is falling vertically with respect to the earth. The traces of the rain on the side windows of the car make an angle of 60° with the vertical. Find the velocity of the rain with respect to (a) the car and (b) the earth.

Solution

Let $\mathbf{v}_{rg}$ = velocity of the rain relative to ground, $\mathbf{v}_{rc}$ = velocity of rain relative to the car, and $\mathbf{v}_{cg}$ = velocity of the car relative to ground. These vectors are related as shown:

$$\mathbf{v}_{rg} = \mathbf{v}_{rc} + \mathbf{v}_{cg}.$$

Therefore,

(a) $v_{rc} = \dfrac{v_{cg}}{\cos 30°} = \dfrac{50 \text{ km/h}}{\cos 30°} = \boxed{57.7 \text{ km/h}}$

(b) $v_{rg} = v_{cg} \tan 30° = (50 \text{ km/h})(\tan 30°) = \boxed{28.9 \text{ km/h}}$ and is *downward*.

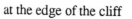

55. A car is parked on a steep incline overlooking the ocean, where the incline makes an angle of 37° with the horizontal. The negligent driver leaves the car in neutral, and the parking brakes are defective. The car rolls from rest down the incline with a constant acceleration of 4 m/s² and travels 50 m to the edge of the cliff. The cliff is 30 m above the ocean. Find (a) the speed of the car when it reaches the cliff and the time it takes to get there, (b) the velocity of the car when it lands in the ocean, (c) the total time the car is in motion, and (d) the position of the car relative to the base of the cliff when the car lands in the ocean.

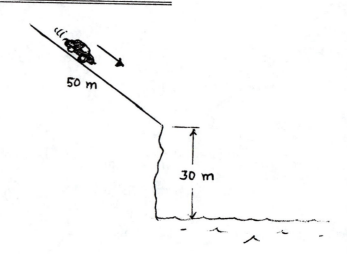

Solution Refer to the sketch.

(a) While on the incline,

$$v^2 - v_0^2 = 2a\Delta x$$

$$v^2 - 0 = 2(4 \text{ m/s}^2)(50 \text{ m})$$

$$v = \boxed{20 \text{ m/s}}$$

$$v - v_0 = at$$

$$20 \text{ m/s} - 0 = (4 \text{ m/s}^2)t$$

$$t = \boxed{5 \text{ s}}$$

(b) Initial free-flight conditions give us

$$v_{xo} = (20 \text{ m/s}) \cos 37 = 16 \text{ m/s}$$
$$v_{yo} = (-20 \text{ m/s}) \sin 37 = -12 \text{ m/s}$$
at the edge of the cliff

$$v_x = v_{xo} \text{ since } a_x = 0$$

$$v_y = \pm \sqrt{2\, a_y \Delta y + v_{yo}^2} = \pm \sqrt{2(-9.8 \text{ m/s}^2)(-30 \text{ m}) + (-12 \text{ m/s})^2} = -27 \text{ m/s}$$

where the negative root is the physically meaningful solution.

$$v = \sqrt{v_x^2 + v_y^2} = \sqrt{16^2 + (-27)^2} = \boxed{31 \text{ m/s}}$$

(c) $t_1 = 5 \text{ s}$

$$t_2 = \frac{v_y - v_{yo}}{a_y} = \frac{(-27 \text{ m/s}) - (-12 \text{ m/s})}{(-9.80 \text{ m/s}^2)} = 1.55 \text{ s}$$

$$t = t_1 + t_2 = \boxed{6.55 \text{ s}}$$

(d) $\Delta x = v_{xo}\, t_1 = (16 \text{ m/s})(1.5 \text{ s}) = \boxed{24 \text{ s}}$

---

66. A home run in a baseball game is hit in such a way that the ball just clears a wall 21 m high, located 130 m from home plate. The ball is hit at an angle of 35° to the horizontal, and air resistance is negligible. Find (a) the initial speed of the ball, (b) the time it takes the ball to reach the wall, and (c) the velocity components and the speed of the ball when it reaches the wall. (Assume the ball is hit at a height of 1 m above the ground.)

Solution

$x_o = 0 \qquad y_o = 1 \text{ m} \qquad v_o = ? \qquad \theta_o = 35°$

When $x = 130 \text{ m}$, $y = 21 \text{ m}$

$v_{xo} = v_o \cos 35°$

$v_{yo} = v_o \sin 35°$

(a), (b) $x = x_o + v_{xo}\, t = v_o t \cos 35°$ (1)

$y = y_o + v_{yo}\, t - \dfrac{1}{2} g t^2 = 1 + (v_o \sin 35°)t - 4.9t^2$ (2)

at $x = 130 \text{ m}$, $y = 21 \text{ m}$ so $v_o t \cos 35° = 130 \text{ m} \rightarrow v_o t = \dfrac{130}{\cos 35°} = 158.7 \text{ m}$

$1 + v_o t \sin 35° - 4.9t^2 = 21$

$158.7 \sin 35° - 4.9t^2 = 20$

$4.9t^2 = 91.0 - 20 = 71 \rightarrow t = \sqrt{\dfrac{71}{4.9}} = \boxed{3.81 \text{ s}}$

$v_o = \dfrac{158.7 \text{ m}}{3.81 \text{ s}} = \boxed{41.7 \text{ m/s}}$

(c) $v_x = v_{xo} = v_o \cos 35° = (41.7 \text{ m/s}) \cos 35° = \boxed{34.1 \text{ m/s}}$

$v_y = v_{yo} - gt = v_o \sin 35° - gt = 23.9 - 9.8t$

When $t = 3.81$ s, $v_y = 23.9 \text{ m/s} - (9.80 \text{ m/s}^2)(3.81 \text{ s}) = \boxed{-13.4 \text{ m/s}}$

$v_x = 34.1 \text{ m/s}$, $v_y = -13.4 \text{ m/s}$

$|v| = \sqrt{v_x^2 + v_y^2} = \sqrt{34.1^2 + (-13.4)^2} = \boxed{36.6 \text{ m/s}}$

79. A skier leaves the ramp of a ski jump with a velocity of 10 m/s, 15° above the horizontal, as in Figure 4.28. The slope is inclined at 50°, and air resistance is negligible. Find (a) the distance that the jumper lands down the slope and (b) the velocity components just before landing. (How do you think the results might be affected if air resistance were included? Note that jumpers lean forward in the shape of an airfoil with their hands at their sides to increase their distance. Why does this work?)

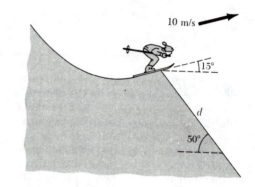

Solution

$x_o = y_o = 0$ $v_o = 10$ m/s at 15°

$v_{xo} = v_o \cos \theta = (10 \text{ m/s}) \cos 15° = 9.66 \text{ m/s}$

$v_{yo} = v_o \sin \theta = (10 \text{ m/s}) \sin 15° = 2.59 \text{ m/s}$

$x = v_{xo}t$ $v_x = v_{xo}$
$y = v_{yo}t - \dfrac{1}{2}gt^2$ $v_y = v_{yo} - gt$

(a) $x = v_{xo}t = 9.66t$ $y = v_{yo}t - \dfrac{1}{2}gt^2 = 2.59t - 4.9t^2$

The skier hits the slope when $\dfrac{y}{x} = \tan(-50°) = -\tan 50°$ so $y = -x \tan 50°$

or $2.59 t - 4.9t^2 = -1.19 (9.66t)$

$4.9t^2 = [2.59 + 1.19 \times 9.66]t = 14.1t$ $\rightarrow$ $t = \dfrac{14.1}{4.9} = 2.88$ s

Then $x = 9.66t = 27.8$ m. But $x = d \cos 50°$ so $d = \dfrac{x}{\cos 50°} = \dfrac{27.8}{\cos 50°} = \boxed{43.2 \text{ m}}$

(b) $v_x = v_{xo} = \boxed{9.66 \text{ m/s}}$ $v_y = v_{yo} - gt = 2.59 - 9.8t$

When $t = 2.88$ s, $v_x = \boxed{9.66 \text{ m/s}}$ $v_y = \boxed{-25.6 \text{ m/s}}$

The Laws of Motion

CHAPTER 5

OBJECTIVES

1. Discuss the concept of force and the effect of an unbalanced force on the motion of a body.

2. Distinguish between contact forces (such as the tension in a rope) and action-at-a-distance forces (such as gravitational and electrostatic forces) -- and be able to identify the four fundamental forces in nature.

3. Write, in your own words, a description of Newton's laws of motion -- and give physical examples of each law.

4. Discuss the concepts of mass and inertia and understand the difference between mass (a scalar) and weight (a vector).

5. Become familiar with the SI units of force (N), mass (kg) and acceleration (m/s^2), and the relation of these units to the English units. For example, 1 N = 0.2248 lb.

6. Realize that the laws of static and kinetic friction are *empirical* in nature (that is, based on observations), and recognize that the *maximum* force of static friction and the force of kinetic friction are both proportional to the normal force on a body.

7. Apply Newton's laws of motion to various mechanical systems using the recommended procedure discussed in Section 5.8. Most important, you should identify all external forces acting on the system, draw the *correct* free-body diagrams which apply to each body of the system, and apply Newton's second law, $\Sigma F = ma$, in *component* form.

SKILLS

Since many of the applications of Newton's laws involve forces with x, y, and z components, you should be familiar with solving several linear equations simultaneously for the unknown quantities. Recall that you must have as many *independent* equations as you have unknowns. For example, you should be able to solve the equations $3F - 2T = 6$ and $F + 3T = 4$ for the unknowns F and T. (Answer: $F = 26/11$; $T = 6/11$)

Much of your study in this and the following chapter will have to do with the application of Newton's second law to systems of one or more masses which are in equilibrium or undergoing uniform acceleration. The following problem-solving strategy is repeated here from your text book (Section 5.8) and will be an important guide to you as you gain confidence in this technique.

The following procedure is recommended when dealing with problems involving the application of Newton's laws:

1. Draw a simple, neat diagram of the system.
2. Isolate the object of interest whose motion is being analyzed. Draw a free-body diagram for this object; that is, a diagram showing *all external forces acting on the object*. For systems containing more than one object, draw *separate* diagrams for each object. *Do not* include forces that the object exerts on its surroundings.
3. Establish convenient coordinate axes for each body and find the components of the forces along these axes.

   Now, apply Newton's second law, $\Sigma F = ma$, in *component* form. Check your dimensions to make sure that all terms have units of force.
4. Solve the component equations for the unknowns. Remember that you must have as many independent equations as you have unknowns in order to obtain a complete solution.
5. It is a good idea to check the predictions of your solutions for extreme values of the variables. You can often detect errors in your results by doing so.

32222

NOTES FROM SELECTED CHAPTER SECTIONS_____

5.2 THE CONCEPT OF FORCE

Equilibrium is the condition under which the *net force* (vector mass of all forces) acting on an object is zero. An object in equilibrium has a zero acceleration (velocity is constant or equals zero).

Fundamental forces in nature are: (1) gravitational (attractive forces between objects due to their masses), (2) electromagnetic forces (between electric charges at rest or in motion), (3) strong nuclear forces (between subatomic particles), and (4) weak nuclear forces (accompanying the process of radioactive decay).

Classical physics is concerned with contact forces (which are the result of physical contact between two or more objects) and action-at-a-distance forces (which act through empty space and do not involve physical contact).

5.3 NEWTON'S FIRST LAW AND INERTIAL FRAMES

Newton's first law is called the *law of inertia* and states that an object at rest will remain at rest and an object in motion will remain in motion with a constant velocity unless acted on by a *net external force*.

5.4 INERTIAL MASS

Mass and *weight* are two different physical quantities. The weight of a body is equal to the *force of gravity* acting on the body and varies with location in the earth's gravitational field. Mass is an *inherent property* of a body and is a measure of the body's inertia (resistance to change in its state of motion). The SI unit of mass is the *hologram* and the unit of weight is the *newton*.

5.5 NEWTON'S SECOND LAW

Newton's second law, the *law of acceleration*, states that the acceleration of an object is directly proportional to the net force acting on it and inversely proportional to its mass.

5.7 NEWTON'S THIRD LAW

Newton's third law, the *law of action-reaction*, states that when two bodies interact, the force which body "A" exerts on body "B" (the *action force*) is equal in magnitude and opposite in direction to the force which body "B" exerts on body "A" (the *reaction force*). A consequence of the third law is that forces occur in *pairs*. Remember that the action force and the reaction force act on *different objects*.

5.8 SOME APPLICATIONS OF NEWTON'S LAWS

Construction of a *free-body diagram* is an important step in the application of Newton's laws of motion to the solution of problems involving bodies in equilibrium or accelerating under the action of external forces. The diagram should include an arrow labeled to identify each of the external forces acting on the body whose motion (or condition of equilibrium) is to be studied. Forces which are the *reactions* to these external forces must *not* be included. When a system consists of more than one body or mass, you must construct a free-body diagram for each mass.

EQUATIONS AND CONCEPTS

If a force acting on a mass m_1 produces an acceleration a_1, and the same force acting on a mass m_2 produces an acceleration a_2, the ratio of the two masses is the reciprocal of the ratio of the accelerations.

$$\frac{m_1}{m_2} \equiv \frac{a_2}{a_1}$$

(5.1)

If a mass m is in a location where the acceleration of gravity is **g**, the *weight* of the body will be determined by the mass and the acceleration due to gravity at the specific location.

$$W = mg$$

(5.7)

Newton's second law of motion states that the acceleration of an object is directly proportional to the resultant force acting on it and inversely proportional to its mass.

$$\Sigma F = ma$$

Newton's third law, sometimes called the law of action-reaction, says that the force exerted by body 1 on body 2 is equal and opposite the force exerted by body 2 on body 1.

$$F_{12} = -F_{21}$$

(5.11)

The force of *static friction* between any two surfaces in contact depends on the normal force and the coefficient of static friction. The equality sign in Eq. 5.9 applies when the object is *on the verge of slipping*.

$$f_s \leq \mu_s N$$

(5.9)

The *force of kinetic friction* between an object and the surface it slides on opposes the motion and has a magnitude which is usually less than the force of static friction.

$$f_k = \mu_k N$$

(5.10)

EXAMPLE PROBLEM SOLUTIONS

Example 5.1 A man pulls a 100 N crate horizontally with a light rope across a rough surface as in Figure 5.1. The man exerts a force of 75 N on the rope, and the coefficient of sliding friction between the crate and surface is 0.5.

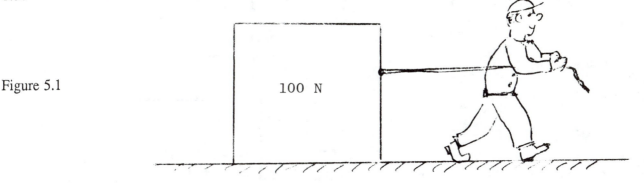

Figure 5.1
100 N

(a) What is the reaction force to the man's force on the rope?

Solution

The reaction force is the force of the rope on the man, or the force of *tension* in the rope. This force is to the left and equals 75 N.

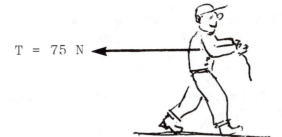

T = 75 N

(b) What force pulls the crate?

Solution

The force of tension in the rope pulls the crate. Ropes or strings *always* pull on objects. (Try pushing a crate with a rope!) This force is to the right and equals 75 N.

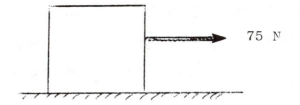

75 N

(c) Draw a free-body diagram for the crate and identify all the forces involved.

Solution

The 75 N force is the force of the rope on the crate; the force **W** is the force of the earth on the crate, or the weight of the crate; the force **N** is the force of the floor on the crate; **f** is the force of friction acting on the crate, which is opposite to the direction of motion. The free-body diagram is shown below. Note that **N** *is not* the reaction force to **W**. The reaction force to **W** is the force of the crate on the earth (upward). The reaction of **N** is the force of the crate on the floor (downward). These reaction forces are not indicated in the free body diagram. Since we are interested in the motion of the crate, we only need to concern ourselves with the action forces, or the external forces acting on the crate.

Free body diagram
for the crate

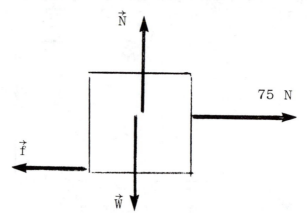

(d) What is the resultant force on the crate?

Solution

Since there is no motion in the vertical direction, we see from the free body diagram that $N = W = 100$ N. The *maximum* frictional force is given by

$$f = \mu_k N = \mu_k W = 0.5 \ (100 \ N) = 50 \ N$$

Since this force is to the left, and the force of the rope on the crate is to the right, the resultant force on the crate is to the right.

$$\Sigma F \text{ on the crate} = 75 \ N - 50 \ N = \boxed{25 \ N}$$

(e) What is the acceleration of the crate?

Solution

From Newton's second law applied to the crate, we have

$$\textit{Resultant} \text{ force on crate} = \text{mass of crate} \times \text{acceleration of crate}$$

$$\Sigma F = ma$$

$$25 \ N = \left(\frac{100}{9.8} \ kg\right) a$$

$$a = \frac{25 \ N}{10.2 \ kg} = \boxed{2.45 \ m/s^2}$$

where the direction of **a** is to the right.

Example 5.2 Two blocks having masses of 2 kg and 3 kg are in contact on a fixed *smooth* inclined plane as in Figure 5.2.

Figure 5.2

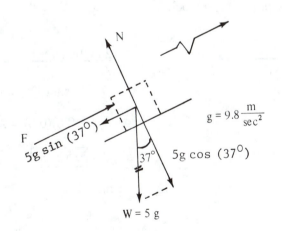

(a) Treating the two blocks as a composite system, calculate the force F that will accelerate the blocks *up* the incline with an acceleration of 2 m/s^2.

Solution

We can replace the two blocks by an equivalent 5 kg block. Letting the x axis be along the incline, the resultant force *on the system* (the two blocks) in the x direction gives

$$\Sigma F_x = F - W \sin (37°) = ma_x$$

$$F - 5g(0.6) = 5(2) \text{ N}$$

$$F = \boxed{39.4 \text{ N}}$$

(b) Draw free-body diagrams for each block, and identify all the forces.

Solution

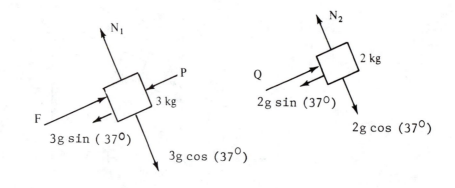

N_1 = force of incline on 3 kg block.
F   = applied force
P   = force of 2 kg block on 3 kg block (contact force)
N_2 = force of incline on 2 kg block
Q   = force of 3 kg block on 2 kg block

The remaining forces are the components of weight along the x and y axes.

(c) Calculate the contact force between the blocks, and show that the acceleration of *each* block is consistent with the resultant force acting on them.

Solution

From Newton's third law, we would expect that P = Q. Let us check this. We can calculate the contact force by applying the second law, $\Sigma F_x = ma_x$, to each block:

3 kg block

$F - P - 3g \sin (37°) = m_3 a_x$

$39.4 - P - 3(9.8)(0.6) = 3(2)$

$P = (39.4 - 6 - 17.6)$ N

$P = \boxed{15.8 \text{ N}}$

2 kg block

$Q - 2g \sin (37°) = m_2 a_x$

$Q - 2(9.8)(0.6) = 2(2)$

$Q = \boxed{15.8 \text{ N}}$

or,   Q = P

Therefore, the two calculations check. That is, the acceleration of *each* block considered separately is consistent with the resultant force acting on them. Note that it is the *combination of F, gravity, and the countering contact force* which accelerate the 3 kg block, while *only the contact force and gravity* accelerate the 2 kg block.

Example 5.3 Two blocks are connected by a light string over a frictionless pulley as in Figure 5.3. The coefficient of sliding friction between m_1 and the surface is μ. Find the acceleration of the two blocks and the tension in the string.

Figure 5.3

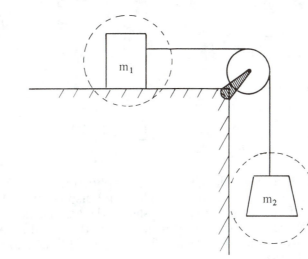

Solution

First, let us isolate each block indicated by dotted lines and determine the external forces acting on each.

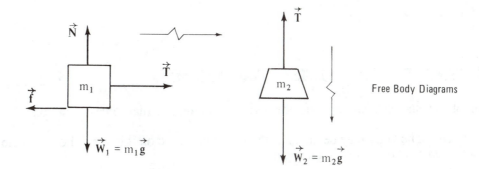

Free Body Diagrams

Consider the motion of m_1. Since the motion is to the right, then $T > f$ (assuming the initial velocity is zero). If T were less than f, the blocks would remain stationary.

$$\text{Equations of Motion for } m_1 \quad \begin{cases} \Sigma F_x \text{ on } m_1 = T - f = m_1 a & (1) \\ \Sigma F_y \text{ on } m_1 = N - m_1 g = 0 & (2) \end{cases}$$

Since $f = \mu N = \mu m_1 g$, we have from (1)

$$T = m_1(a + \mu g) \qquad (3)$$

For m_2, the motion is downward, therefore $m_2 g > T$. Note that the force T is uniform through the rope. That is, the force which acts on m_1 to the right is also the force which keeps m_2 from falling freely. The equation of motion for m_2 is:

$$\Sigma F_y \text{ on } m_2 = T - m_2 g = -m_2 a; \quad T = m_2(g - a) \qquad (4)$$

where we have called the upward direction positive. Subtracting (4) from (3) gives

$$m_1(a + \mu g) - m_2(g - a) = 0$$

$$a = \left(\frac{m_2 - \mu m_1}{m_1 + m_2} \right) g \tag{5}$$

Substituting (5) into (4) gives T.

$$T = m_2 \left(1 - \frac{m_2 - \mu m_1}{m_1 + m_2} \right) g = \frac{m_1 m_2 (1 + \mu) g}{m_1 + m_2} \tag{6}$$

Comments: If the surface were frictionless, we would simply set $\mu = 0$ in (5) and (6). Of course, the condition for $a > 0$ is $m_2 > \mu m_1$, as we can see from (5). We could also obtain (5) by viewing the two blocks as a composite system. The "net force" accelerating the system is the gravitational force on m_2 less the frictional force on m_1. That is, the unbalanced force is $m_2 g - \mu m_1 g$. Setting this equal to the "mass of the system" times the acceleration gives (5) directly. However, to obtain T, we would still have to analyze each block independently. *The first procedure is usually preferred.*

=====

ANSWERS TO SELECTED QUESTIONS

1. If an object is at rest, can we conclude that there are no external forces acting on it?

Answer: No. The body may be at rest if the *resultant* force on it is zero. For example, the force of gravity and the normal force act on a body at rest on a table.

2. If gold were sold by weight, would you rather buy it in Denver or in Death Valley? If sold by mass, in which of the two locations would you prefer to buy it? Why?

Answer: In Denver the gold is farther from the center of the earth, and as a result, the force of gravitational attraction exerted by the earth on the gold is less than it would be at the lower altitudes of Death Valley. Mass, however, remains the same whether in Death Valley or Denver, so it would make no difference where you buy it.

3. A passenger sitting in the rear of a bus claims that he was injured when the driver slammed on the brakes causing a suitcase to come flying toward the passenger from the front of the bus. If you were the judge in this case, what disposition would you make? Why?

Answer: The inertia of the suitcase would tend to keep it moving forward as the bus stops. There would be no tendency for the suitcase to be thrown backward. Throw the case out of court!

4. A space explorer is in a spaceship moving through space far from any planet or star. She notices a large rock, taken as a specimen from an alien planet, floating around the cabin of the spaceship. Should she push it gently toward a storage compartment or kick it toward the compartment? Why?

Answer: Regardless of the location of the rock, it still has mass, and a large force is necessary to move it. Newton's third law says that if he kicks it hard enough to provide the large force, the force back on his toe will be very unpleasant.

5. How much does an astronaut weigh out in space, far from any planets?

Answer: Zero. Since w = mg, and g = 0 in space, then w = 0.

6. Although the frictional force between two surfaces may decrease as the surfaces are smoothed, the force will again increase if the surfaces are made extremely smooth and flat. How do you explain this?

Answer: The attractive force between atoms in the two surfaces increases as the separation between the surfaces decreases. Ultimately, the surfaces can "cold-weld" together. This process can be used to weld smooth metal objects in space.

7. Why is it that the frictional force involved in the rolling of one body over another is less than for sliding motion?

Answer: The number of "rough spots" which are involved in two flat surfaces is much greater than the number for a round surface (somewhat deformed) against a flat surface.

8. A massive metal object on a rough metal surface may undergo contact welding to that surface. Discuss how this affects the frictional forces between the object and the surface.

Answer: In order to move the object, these "welds" must be broken. This serves to increase the frictional force.

10. Identify the action-reaction pairs in the following situations: (a) a man takes a step; (b) a snowball hits a girl in the back; (c) a baseball player catches a ball; (d) a gust of wind strikes a window.

Answer: (a) As a man takes a step, the action is the force his foot exerts on the earth; the reaction is the force of the earth on his foot. (b) The action is the force exerted on the girl's back by the snowball; the reaction is the force exerted on the snowball by the girl's back. (c) The action is the force of the glove on the ball; the reaction is the force of the ball on the glove. (d) The action is the force exerted on the window by the air molecules; the reaction is the force on the air molecules exerted by the window.

11. While a football is in flight, what forces act on it? What are the action-reaction pairs while the football is being kicked, and while it is in flight?

Answer: When a football is in flight, the only force acting on it is its weight, assuming that we neglect air resistance. While it is being kicked, the forces acting on it are its weight and the force exerted on it by the kicker's foot. The reaction to the weight is the gravitational force exerted on the earth by the football. The reaction to the force exerted by the kicker's foot is the force exerted on the foot by the football.

12. A ball is held in a person's hand. (a) Identify all the external forces acting on the ball and the reaction to each. (b) If the ball is dropped, what force is exerted on it while it is falling? Identify the reaction force in this case. (Neglect air resistance.)

Answer: (a) The external forces on a ball held in a person's hand are its weight and the force of the hand upward on the ball. The reaction to the weight is the upward pull of the ball on the earth because of gravitational attraction. The reaction to the force on the ball by the hand is the downward force on the hand exerted by the ball. (b) When the ball is falling, the only force on it is its weight. The reaction force is the upward force on the earth exerted by the ball because of gravitational attraction.

13. Identify all the action-reaction pairs which exist for a horse pulling a cart. Include the earth in your examination.

Answer: Two forces act on the horse: a forward force by the earth against its feet and a backward pull by the cart. The reaction to these are a backward force on the earth by the horse's feet and a forward pull on the cart. The horizontal force on the cart is a forward pull by the horse; the reaction to this is a backward pull on the horse.

14. If a car is traveling westward with a constant speed of 20 m/s, what is the resultant force acting on it?

Answer: If an object moves with constant velocity, the net force on it is zero.

15. A large crate is placed on the bed of a truck without being tied to the truck. (a) As the truck accelerates forward, the crate remains at rest relative to the truck. What force causes the crate to accelerate? (b) If the truck driver slams on the brakes, what could happen to the crate?

Answer: (a) The crate accelerates because of a friction force exerted on it by the floor of the truck. (b) If the driver slams on the brakes, the inertia of the crate would tend to keep it moving forward.

16. A child pulls a wagon with some force, causing it to accelerate. Newton's third law says that the wagon exerts an equal and opposite reaction force on the child. How can the wagon accelerate?

Answer: An object accelerates because of the forces which act on it. If the only horizontal force on it is that of the boy, the wagon will accelerate.

17. A rubber ball is dropped onto the floor. What force causes the ball to bounce back into the air?

Answer: When the ball hits the earth, it is compressed. As the ball returns to its original shape, it exerts a force on the earth, and the reaction to this thrusts it back into the air.

18. What is wrong with the statement, "Since the car is at rest, there are no forces acting on it."? How would you correct this sentence?

Answer: Just because an object is at rest does not mean that no forces act on it. For example, its weight always acts on it. The correct sentence would read, "Since the car is at rest, there is no *net* force acting on it."

19. Suppose you are driving a car along a highway at a high speed. Why should you avoid slamming on your brakes if you want to stop in the shortest distance?

Answer: The brakes may lock and the car will slide farther since the coefficient of sliding friction is less than the coefficient of static friction. If the wheels continue to roll, the force of static friction will decelerate the car.

SOLUTIONS TO SELECTED END-OF-CHAPTER PROBLEMS_____

18. Two forces, F_1 and F_2, act on a 5-kg mass. If $F_1 = 20$ N and $F_2 = 15$ N, find the acceleration in (a) and (b) of Figure 5.21.

Solution

$m = 5$ kg, $\Sigma F = ma$

(a) $\Sigma F = F_1 + F_2 = (20\mathbf{i} + 15\mathbf{j})$ N

$\Sigma F = ma, \quad 20\mathbf{i} + 15\mathbf{j} = 5a$

$\mathbf{a} = (4\mathbf{i} + 3\mathbf{j})$ m/s^2 or $\boxed{a = 5 \text{ m/s}^2}$

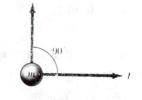

Figure 5.21 (a)

(b) $F_{2x} = 15 \cos 60° = 7.5$ N

$F_{2y} = 15 \sin 60° = 13$ N

$F_2 = (7.5\mathbf{i} + 13\mathbf{j})$ N

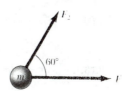

$\Sigma F = F_1 + F_2 = (27.5\mathbf{i} + 13\mathbf{j})$ N $=$ ma $=$ 5a

$\mathbf{a} = (5.5\mathbf{i} + 2.6\mathbf{j})$ m/s^2 or $\boxed{a = 6.08 \text{ m/s}^2}$

Figure 5.21 (b)

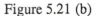

20. A ball is held in a person's hand. (a) Identify all the external forces acting on the ball and the reaction to each of these forces. (b) If the ball is dropped, what force is exerted on it while it is in "flight"? Identify the reaction force in this case. (Neglect air resistance.)

Solution

(a) The external forces acting on the ball are
   (1) F_H, the force which the hand exerts on the ball.
   (2) W, the force of gravity exerted on the ball by the earth.

   The reaction forces are
   (1) To F_H: The force which the ball exerts on the hand.
   (2) To W: The gravitational force which the ball exerts on the earth.

(b) When the ball is in free fall, the only force exerted on it is its *weight,* W, which is exerted by the earth. The reaction force is the gravitational force which *the ball exerts on the earth.*

24. An electron of mass 9.1×10^{-31} kg has an initial speed of 3.0×10^5 m/s. It travels in a straight line, and its speed increases to 7.0×10^5 m/s in a distance of 5.0 cm. Assuming its acceleration is constant, (a) determine the force on the electron and (b) compare this force with the weight of the electron, which we neglected.

Solution

(a) $F = ma$ and $v^2 = v_0^2 + 2ax$ or $a = \dfrac{(v^2 - v_0^2)}{2x}$

Therefore,

$$F = \frac{m(v^2 - v_0^2)}{2x} = \frac{(9.1 \times 10^{-31} \text{ kg})\left[(7 \times 10^5 \text{ m/s})^2 - (3 \times 10^5 \text{ m/s})^2\right]}{(2)(0.05 \text{ m})} = \boxed{3.6 \times 10^{-18} \text{ N}}$$

(b) The weight of the electron is

$$W = mg = (9.1 \times 10^{-31} \text{ kg}) (9.8 \text{ m/s}^2) = \boxed{8.9 \times 10^{-30} \text{ N}}$$

The accelerating force is approximately 10^{11} times the weight of the electron.

---

33. The parachute on a race car of weight 8820 N opens at the end of a quarter-mile run when the car is traveling at 55 m/s. What is the total retarding force required to stop the car in a distance of 1000 m in the event of a brake failure?

Solution

$W = 8820 \text{ N}, \quad g = 9.80 \text{ m/s}^2, \quad v_o = 55 \text{ m/s}, \quad v_f = 0, \quad x_f - x_o = 1000 \text{ m}$

$$m = \frac{W}{g} = \frac{8820 \text{ N}}{9.80 \text{ m/s}^2} = \boxed{900 \text{ kg}}$$

$v_f^2 = v_o^2 + 2a(x - x_o), \quad 0 = 55^2 + 2a(1000), \quad \text{giving } a = -1.51 \text{ m/s}^2$

$$\Sigma F = ma = (900 \text{ kg})(-1.51 \text{ m/s}^2) = \boxed{-1.36 \times 10^3 \text{ N}}$$

The minus sign means that the force is a retarding force.

---

39. Two masses of 3 kg and 5 kg are connected by a light string that passes over a smooth pulley as in Figure 5.12. Determine (a) the tension in the string, (b) the acceleration of each mass, and (c) the distance each mass moves in the first second of motion if they start from rest.

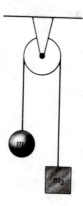

Figure 5.12

Solution

$$m_1 a = T - m_1 g \quad (1) \qquad\qquad m_2 a = m_2 g - T \quad (2)$$

Add (1) and (2) $\qquad\qquad (m_1 + m_2)a = (m_2 - m_1)g$

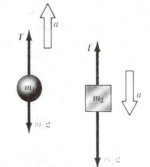

$$a = \frac{m_2 - m_1}{(m_2 + m_1)g} = \frac{5 - 3}{(5 + 3)(9.80)} = \boxed{2.45 \text{ m/s}^2}$$

$$T = m_2(g - a) = 5(9.80 - 2.45) = \boxed{36.8 \text{ N}}$$

Substitute a into (1) $\qquad \Rightarrow T = m_1(a + g) = \dfrac{2m_1 m_2 g}{m_1 + m_2}$

$$s = \frac{at^2}{2} \qquad (v_o = 0),$$

At $\ t = 1$ s, $\quad s = \dfrac{(2.45)(1^2)}{2} = \boxed{1.23 \text{ m}}$

---

50. A racing car accelerates uniformly from 0 to 80 mi/h in 8 s. The external force that accelerates the car is the frictional force between the tires and the road. If the tires do not spin, determine the *minimum* coefficient of friction between the tires and the road.

Solution

$F = \mu N = ma$ and in this case the normal force $N = mg$; therefore $F = \mu mg = ma$ or $\mu = \dfrac{a}{g}$

The acceleration is found from $\qquad a = \dfrac{v - v_o}{t} = \dfrac{(80 \text{ mi/h})(0.447 \text{ m/s/mi/h})}{8 \text{ s}} = 4.47 \text{ m/s}^2$

Substituting this value into the expression for μ, we find $\qquad \mu = \dfrac{4.47 \text{ m/s}^2}{9.80 \text{ m/s}^2} = \boxed{0.456}$

---

55. Two blocks connected by a light rope are being dragged by a horizontal force F (Fig. 5.29). Suppose that $F = 50$ N, $m_1 = 10$ kg, $m_2 = 20$ kg, and the coefficient of kinetic friction between each block and the surface is 0.1.
(a) Draw a free-body diagram for each block.
(b) Determine the tension, T, and the acceleration of the system.

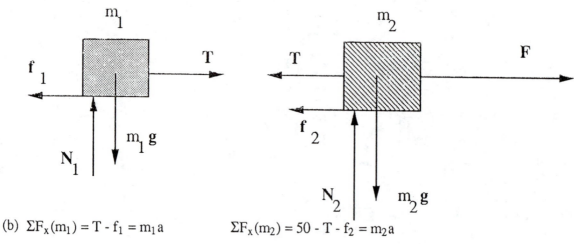

(b) $\Sigma F_x(m_1) = T - f_1 = m_1 a$ $\qquad$ $\Sigma F_x(m_2) = 50 - T - f_2 = m_2 a$

$\Sigma F_y(m_1) = N_1 - m_1 g = 0$ $\qquad$ $\Sigma F_y(m_2) = N_2 - m_2 g = 0$

$f_1 = \mu_1 N_1 = (0.1)(10 \text{ kg})(9.80 \text{ m/s}^2) = 9.80$ N

$f_2 = \mu_2 N_2 = (0.1)(20 \text{ kg})(9.80 \text{ m/s}^2) = 19.6$ N

$T - 9.8 = 10a,$ $\quad$ $50 - T - 19.6 = 20a$

Adding the expression above gives $50 - 29.4 = 30a$,

$$a = \boxed{0.687 \text{ m/s}^2} \qquad T = 10a + 9.80 = \boxed{16.7 \text{ N}}$$

---

57. A 3-kg block starts from rest at the top of a 30° incline and slides a distance of 2 m down the incline in 1.5 s. Find (a) the acceleration of the block, (b) the coefficient of kinetic friction between the block and the plane, (c) the frictional force acting on the block, and (d) the speed of the block after it has slid 2 m.

Solution

$m = 3$ kg, $\quad$ $\theta = 30°$, $\quad$ $x = 2$ m, $\quad$ $t = 1.5$ s

$$x = \frac{1}{2} at^2 \qquad 2 \text{ m} = \frac{1}{2} a (1.5 \text{ s})^2 \rightarrow a = \frac{4}{1.5^2} = \boxed{1.78 \text{ m/s}^2}$$

$N + f + mg = ma$

x: $0 - f + mg \sin 30° = ma \rightarrow f = m(g \sin 30° - a)$

y: $N + 0 - mg \cos 30° = 0 \rightarrow N = mg \cos 30°$

$f = m(g \sin 30° - a) = (3)(9.80 \sin 30° - 1.78) = \boxed{9.37 \text{ N}}$

$\mu_k = \dfrac{f}{N} = \dfrac{m(g \sin 30° - a)}{mg \cos 30°} = \tan 30° - \dfrac{a}{g(\cos 30°)} = \boxed{0.368}$

$v^2 = v_0^2 + 2a(x - x_0) \qquad x - x_0 = 2 \text{ m}$

$v^2 = 0 + 2(1.78)(2) = 7.11 \rightarrow v = (7.11)^{1/2} = \boxed{2.67 \text{ m/s}}$

---

65. A mass M is held in place by an applied force F_A and a pulley system as shown in Figure 5.33. The pulleys are massless and frictionless. Find (a) the tension in each section of rope, T_1, T_2, T_3, T_4, and T_5; and (b) the applied force F_A.

Solution

Draw free body diagrams and apply Newton's 2nd law.

$\Sigma F = 0 = T_5 - mg, \qquad T_5 = mg$

Assume frictionless pulleys

$T_1 = T_2 = T_3$

$\Sigma F = 0 = T_2 + T_3 - T_5$

$2T_2 = T_5 \qquad T_2 = \dfrac{T_5}{2}$

$T_1 = T_2 = T_3 = \dfrac{mg}{2}$

$F_A = T_1 = \dfrac{mg}{2}$

$\Sigma F = 0 = T_4 - T_1 - T_2 - T_3$

$T_4 = T_1 + T_2 + T_3 = \dfrac{3mg}{2}$

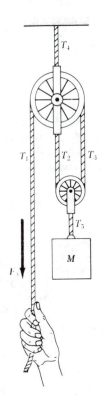

Figure 5.33

---

69. Three blocks are in contact with each other on a frictionless, horizontal surface as in Figure 5.36. A horizontal force F is applied to m_1. If $m_1 = 2$ kg, $m_2 = 3$ kg, $m_3 = 4$ kg, and F = 18 N, find (a) the acceleration of the blocks, (b) the *resultant* force on each block, and (c) the magnitude of the contact forces between the blocks.

Solution

(a) F = ma;   18 = (2 + 3 + 4)a;   a = $\boxed{2 \text{ m/s}^2}$

(b) The force on each block can be found by knowing mass and acceleration:

$F_1 = m_1 a = 2(2) = \boxed{4 \text{ N}}$

$F_2 = m_2 a = 3(2) = \boxed{6 \text{ N}}$

$F_3 = m_3 a = 4(2) = \boxed{8 \text{ N}}$

(c) The force on each block is the resultant of all contact forces. Therefore,

$F_1 = 4$ N = F - P, where P is the contact force between m_1 and m_2.

P = $\boxed{14 \text{ N}}$

$F_2 = 6$ N = P - Q, where Q is the contact force between m_2 and m_3.

Q = $\boxed{8 \text{ N}}$

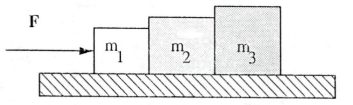

Figure 5.36

73. A 2-kg block is placed on top of a 5-kg block as in Figure 5.39. The coefficient of kinetic friction between the 5-kg block and the surface is 0.2. A horizontal force F is applied to the 5-kg block. (a) Draw a free-body diagram for each block. What force accelerates the 2-kg block? (b) Calculate the force necessary to pull both blocks to the right with an acceleration of 3 m/s². (c) Find the minimum coefficient of static friction between the blocks such that the 2-kg block does not slip under an acceleration of 3 m/s².

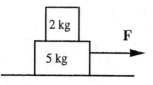

Figure 5.39

Solution

(a)   The force of static friction between the
      blocks accelerates the 2-kg block.

(b) $\Sigma F = ma$, $F - \mu N_2 = ma$,

      $F - (0.2)[(5 + 2)(9.80)] = (5 + 2)3$

      Therefore, $F = \boxed{34.7 \text{ N}}$

(c) $f = \mu_1 N_1 = \mu_1(2)(9.80) = m_1 a = 2(3)$

      Therefore, $\mu_1 = \boxed{0.306}$

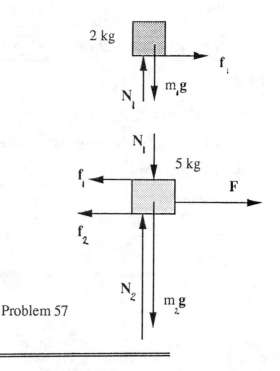

Problem 57

---

74. A 5-kg block is placed on top of a 10-kg block
(Fig. 5.40). A horizontal force of 45 N is
applied to the 10-kg block, while the 5-kg
block is tied to the wall. The coefficient of
kinetic friction between the moving surfaces is
0.2. (a) Draw a free-body diagram for each
block and identify the action-reaction forces
between the blocks. (b) Determine the tension
in the string and the acceleration of the 10-kg
block.

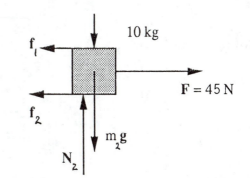

Figure 5.40

Solution

(a)

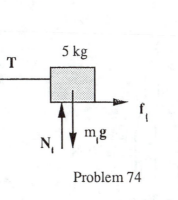

Problem 74

(b) 5 kg: $\Sigma F_x = ma$ $N_1 = 5g = 5(9.80) = 49$ N

     $f_1 - T = 0$

     $T = f_1 = \mu mg = (0.2)(5)(9.80) = \boxed{9.8 \text{ N}}$

     10 kg: $\Sigma F_x = ma$ $\Sigma F_y = 0$

       $45 - f_1 - f_2 = 10a$ $N_2 - N_1 - 10g = 0$

       $f_2 = \mu N_2 = \mu(N_1 + 10g) = (0.2)(49 + 98) = 29.4$ N

       $45 - 9.8 - 29.4 = 10a$

       $a = \boxed{0.58 \text{ m/s}^2}$

---

75. An inventive child named Pat wants to reach an apple in a tree without climbing the tree. Sitting in a chair connected to a rope that passes over a frictionless pulley (Fig. 5.41), Pat pulls on the loose end of the rope with such a force that the spring scale reads 250 N. Pat's true weight is 320 N and the chair weighs 160 N. (a) Draw free-body diagrams for Pat and the chair considered as separate systems, and another diagram for Pat and the chair considered as one system. (b) Show that the acceleration of the system is *upward* and find its magnitude. (c) Find the force that Pat exerts on the chair.

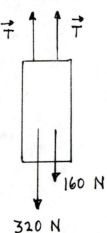

Figure 5.41

Solution

(a)

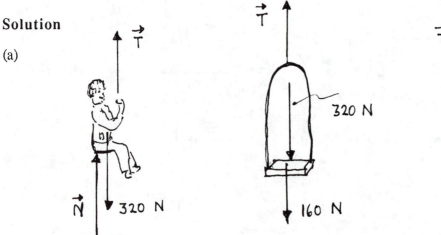

(b) First consider Pat and the chair as the system. Note that *two* ropes support the system, and $T = 60$ lb. in each rope.

$$\Sigma F = ma$$

$2T - 480 = ma$ where $m = \dfrac{480}{9.80} = 49.0$ kg

Solving for a gives $a = \dfrac{(500 - 480)}{49} = \boxed{0.408 \text{ m/s}^2}$

(c) ΣF (on Pat) $= N + T - 320 = ma$ where $m = \dfrac{320}{9.80} = 32.7$ kg

$N = ma + 320 - T = 32.7(0.408) + 320 - 250 = \boxed{83.3 \text{ N}}$

---

83. What horizontal force must be applied to the cart shown in Figure 5.46 in order that the blocks remain *stationary* relative to the cart? Assume all surfaces, wheels, and pulley are frictionless. (*Hint:* Note that the tension in the string accelerates m_1.)

Solution

Note that m_2 should be in *contact* with the cart.

$$\Sigma F = ma$$

For m_1: $T = m_1 a$

$a = \dfrac{m_2 g}{m_1}$

For m_2: $T - m_2 g = 0$

For all 3 blocks: $F = (M + m_1 + m_2)a$

$F = (M + m_1 + m_2)\left(\dfrac{m_2 g}{m_1}\right)$

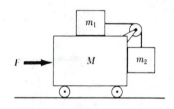

Figure 5.46

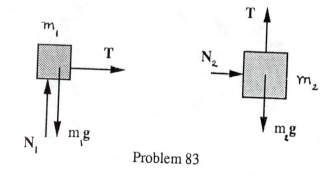

Problem 83

87. Two blocks of mass 2 kg and 7 kg are connected by a light string that passes over a frictionless pulley (Fig. 5.49). The inclines are smooth. Find (a) the acceleration of each block and (b) the tension in the string.

Solution

Since it has a larger mass, we expect the 7-kg block to move down the plane. The acceleration for both blocks should have the same magnitude since they are joined together by a nonstretching string.

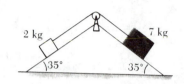

Figure 5.49

$$\Sigma F_1 = m_1 a_1 \qquad -m_1 g \sin 35° + T = m_1 a$$

$$\Sigma F_2 = m_2 a_2 \qquad -m_2 g \sin 35° + T = -m_2 a$$

and $\quad -(2)(9.80) \sin 35° + T = 2a$

$\quad -(7)(9.80) \sin 35° + T = -7a$

$T = \boxed{17.5 \text{ N}}, \qquad a = \boxed{3.12 \text{ m/s}^2}$

6

Circular Motion and Other Applications of Newton's Laws

OBJECTIVES

1. Discuss Newton's universal law of gravity (the inverse-square law), and understand that it is an *attractive* force between two *particles* separated by a distance r.

2. Discuss the nature of the fundamental forces in nature (gravitational, electromagnetic, and nuclear) and characterize the properties and relative strengths of these forces.

3. Apply Newton's second law to uniform and nonuniform circular motion.

4. Remember that Newton's laws are only valid in inertial frames of reference. When motion is described by an observer in a noninertial frame, the observer must invent fictitious forces which arise due to the acceleration of the reference frame.

5. Recognize that motion of an object through a liquid or gas can involve resistive forces which have a complicated velocity dependence.

SKILLS

Section 6.10 deals with the motion of a body through a gas or liquid. If you covered this section in class, the following solution to Eq. 6.6 (when the resistive force $\mathbf{R} = -b\mathbf{v}$) may be useful to know:

$$\frac{dv}{dt} = g - \frac{b}{m} v \qquad (6.6)$$

In order to solve this equation, it is convenient to change variables. If we let $x = g - \frac{b}{m} v$, it follows that $dx = -\frac{b}{m} dv$. With these substitutions, Eq. 6.6 becomes

$$-\left(\frac{m}{b}\right)\frac{dx}{dt} = x \quad \text{or} \quad \frac{dx}{x} = -\frac{b}{m} dt$$

Integrating this expression (now that the variables are separated) gives

$$\int \frac{dx}{x} = -\frac{b}{m} \int dt \quad \text{or} \quad ln \; x = -\frac{b}{m} t + \text{const.}$$

This is equivalent to $x = \text{const } e^{-bt/m} = g - \frac{b}{m} v$. Taking $v = 0$ at $t = 0$, we see that const $= g$, so

$$v = \frac{mg}{b} (1 - e^{-bt/m}) = v_t (1 - e^{-t/\tau}) \qquad (6.7)$$

NOTES FROM SELECTED CHAPTER SECTIONS_____

6.1 NEWTON'S SECOND LAW APPLIED TO UNIFORM CIRCULAR MOTION

If a particle moves in a circle of radius r with *constant speed*, it undergoes a centripetal acceleration v^2/r directed towards the center of rotation. (Recall that in this case, the centripetal acceleration arises from the change in *direction* of **v**.) Newton's second law applied to the motion says that the centripetal acceleration arises from some external, centripetal force acting towards the center of rotation (Fig. 6.1). That is

$$\Sigma \mathbf{F} \text{ (along } \hat{\mathbf{r}}) = m\mathbf{a}_r = -\frac{mv^2}{r}\hat{\mathbf{r}}$$

where $\hat{\mathbf{r}}$ is a unit vector pointing radially outward from the center.

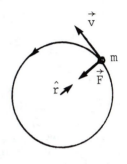

$|\mathbf{v}|$ = constant

Figure 6.1

The *universal gravitational constant,* G, is not to be confused with the acceleration due to gravity, the attractive nature of the force, and the fact that the force on m_1 due to m_2 is equal and opposite the force on m_2 due to m_1. (Newton's third law)

Figure 6.2

Earth

Figure 6.3

The gravitational force exerted by a spherically symmetric mass distribution on a particle outside the sphere is the *same* as if the entire mass of the sphere were concentrated at its center. The gravitational force of attraction on a mass m at the surface of the earth is shown in Fig. 6.3.

6.2 NONUNIFORM CIRCULAR MOTION

If the particle moves in a circular path such that the *speed changes in time,* the particle also has a tangential component of acceleration a_t, whose magnitude is dv/dt (Figure 6.4). In this case, the *total* acceleration **a** is the vector sum of a_r and a_t .

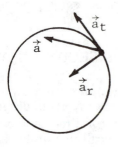

$|v| = \neq$ constant

Figure 6.4

6.3 MOTION IN ACCELERATED FRAMES

An observer in a noninertial frame (accelerated frame) making observations of the motion of a particle must introduce fictitious forces to make Newton's laws work in that frame. Such fictitious forces are sometimes referred to as *inertial forces*.

6.4 MOTION IN THE PRESENCE OF RESISTIVE FORCES

A body moving through a gas or liquid experiences a resistive force (opposing its motion) which can have a complicated velocity dependence. The body reaches a terminal velocity (maximum velocity) when the downward force of gravity is balanced by the upward resistive force. That is, when $\Sigma F = 0$, **a** = 0 and **v** = constant.

EXAMPLE PROBLEM SOLUTION

Example 6.1 A curve on a flat road surface has a radius of curvature of 80 m as in Figure 6.5. (This is an *end* view and the center of curvature is to the left in this figure.) What is the maximum speed a car can have around this curve before it skids if the coefficient of friction between the tires and road is 0.6?

Solution

In this example, the friction force f provides the centripetal force. Taking $f = f_{max} = \mu N = \mu mg$ when the car is on the verge of skidding, we get

$$f_{max} = \mu mg = \frac{mv^2}{r}$$

or

$$v = \sqrt{\mu rg} = \sqrt{0.6 \times 80 \text{ m} \times 9.8 \text{ m/s}^2} = 21.7 \text{ m/s}$$

This corresponds to about 49 miles/hr.

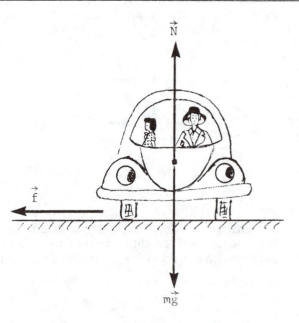

Figure 6.5

ANSWERS TO SELECTED QUESTIONS

6. Why is it that an astronaut in a space capsule orbiting the earth experiences a feeling of weightlessness?

Answer: Both the astronaut and his space capsule are actually in free-fall. Since they both have the same acceleration toward the earth, the astronaut can float around in the capsule as though he were weightless.

7. Why does mud fly off a rapidly turning wheel?

Answer: The mud flies off the wheel because the force which holds it to the wheel is not great enough to provide the necessary centripetal force to make it follow the circular path of the turning wheel.

8. A pail of water can be whirled in a vertical path such that none is spilled. Why does the water stay in, even when the pail is above one's head?

Answer: The tendency of the water at the top of the circular path is to move in a straight line path tangent to the circular path followed by the container. As a result, it is forced against the bottom of the bucket, and the normal force exerted on the water provides the centripetal force to move it in a circular path.

9. Imagine that you attach a heavy object to one end of a spring and then whirl the spring and object in a horizontal circle (by holding the free end of the spring). Does the spring stretch? If so, why? Discuss in terms of centripetal force.

Answer: The spring will stretch until the force it exerts on the mass is large enough to provide the necessary centripetal force to cause the mass to follow a circular path.

10. It has been suggested that rotating cylinders about ten miles in length and five miles in diameter be placed in space for colonies. The purpose of the rotation is to simulate gravity for the inhabitants. Explain this concept for producing an effective gravity.

Answer: The centripetal force on the inhabitants is provided by the normal force exerted on them by the cylinder wall. If the rotation rate is adjusted to such a speed that this normal force is equal to their weight on earth, the inhabitants would not be able to distinguish between this artificial gravity and normal gravity.

11. Why does a pilot tend to black out when he pulls out of a steep dive?

Answer: When pulling out of a dive, blood leaves the pilot's head because there is not a large enough centripetal force to cause it to follow the circular path of the plane. This loss of blood from the brain can cause the pilot to black out.

12. Cite an example of a situation in which an automobile driver can have a centripetal acceleration but no tangential acceleration.

Answer: An automobile driver moving in a circular path with constant tangential velocity has a centripetal acceleration but no tangential acceleration.

13. Is it possible for a car to move in a circular path in such a way that it has a tangential acceleration but no centripetal acceleration?

Answer: Any object that moves such that the direction of its velocity changes has an acceleration. A car moving in a circular path will always have a centripetal acceleration.

16. Centrifuges are often used in dairies to separate the cream from the milk. Which remains on the bottom?

Answer: The heavier particles require a larger centripetal force to make them follow the circular path of the centrifuge. Thus, they are forced to the bottom of the container where the normal force exerted by the container wall can be exerted on them. The lighter particles stay on top because the liquid below them can supply the smaller normal force required.

SOLUTIONS TO SELECTED END-OF-CHAPTER PROBLEMS

5. A 3-kg mass attached to a light string rotates in circular motion on a horizontal, frictionless table. The radius of the circle is 0.8 m, and the string can support a mass of 25 kg before breaking. What range of speeds can the mass have before the string breaks?

Solution

The string will break if the tension T exceeds the weight $mg = (25 \text{ kg})(9.80 \text{ m/s}^2) = 245$ N. As the 3-kg mass rotates in a horizontal circle, the tension provides the centripetal force, so

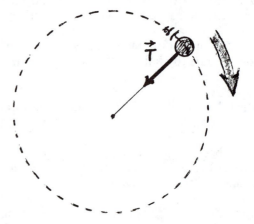

$$T = \frac{mv^2}{r} = \frac{(3 \text{ kg}) v^2}{(0.8 \text{ m})}$$

Then

$$v^2 = \frac{rT}{m} = \frac{(0.8 \text{ m})T}{(3 \text{ kg})} \leq \frac{(0.8 \text{ m})}{(3 \text{ kg})} T_{max}$$

Substituting $T_{max} = 245$ N, we find

$$v^2 \leq 65.3 \frac{m^2}{s^2}$$

or

$$\boxed{0 < v < 8.08 \text{ m/s}}$$

13. A coin is placed 30 cm from the center of a rotating, horizontal turntable. The coin is observed to slip when its speed is 50 cm/s. (a) What provides the centripetal force when the coin is stationary relative to the turntable? (b) What is the coefficient of static friction between the coin and the turntable?

Solution

(a) The centripetal force is provided by the force of static friction.

(b) The forces on the coin shown in the free-body diagram are the normal force, the weight, and the force of static friction. The only force in the radial direction is **f**, therefore

$$f = m \frac{v^2}{r} \qquad (1)$$

Since the normal force balances the weight, $N - mg = 0$, or

$$N = mg \qquad (2)$$

Dividing (1) by (2) gives

$$\frac{f}{N} = \frac{v^2}{rg} < \mu_s$$

The coin slips when $\frac{v^2}{rg} = \mu_s$. Taking r = 30 cm, v = 50 cm/s, and g = 980 cm/s² gives

$$\mu_s = \frac{(50 \text{ cm/s})^2}{(30 \text{ cm})(980 \text{ cm/s}^2)} = \boxed{0.085}$$

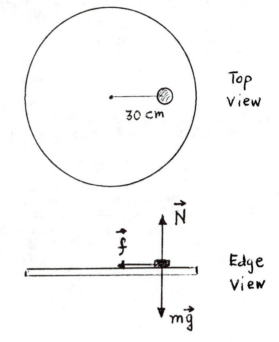

21. Tarzan (m = 85 kg) tries to cross a river by swinging from a vine. The vine is 10 m long, and his speed at the bottom of the swing is 8 m/s. Tarzan doesn't know that the vine has a breaking strength of 1000 N. Does he make it safely across the river?

Solution

The forces acting on Tarzan are the force of gravity Mg and the tension T. At the lowest point in his motion, the force **T** is upward and M**g** is downward as in the free-body diagram. Thus, Newton's second law gives

$$T - Mg = M \frac{v^2}{r}$$

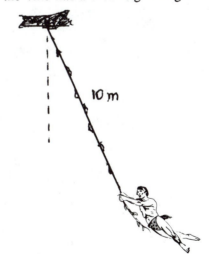

Solving for T, with v = 8 m/s, r = 10 m, and M = 85 kg gives

$$T = M\left(g + \frac{v^2}{r}\right) = (85 \text{ kg})\left[9.80 \text{ m/s}^2 + \frac{(8 \text{ m/s})^2}{10 \text{ m}}\right] = 1.38 \times 10^3 \text{ N}$$

Since T *exceeds* the breaking strength of the vine (1000 N), Tarzan *doesn't make it!* The vine breaks *before* he reaches the bottom of the swing.

==========

25. A 0.5-kg object is suspended from the ceiling of an accelerating boxcar as in Figure 6.11. If a = 3 m/s², find (a) the angle that the string makes with the vertical and (b) the tension in the string.

Solution

The only forces acting on the suspended object are the force of gravity mg and the tension **T** as shown in the free-body diagram. Applying Newton's second law in the x and y directions gives

x: $T \sin \theta = ma$ (1)

y: $T \cos \theta - mg = 0$

or

$T \cos \theta = mg$ (2)

Dividing (1) by (2) gives

(a) $\tan \theta = \frac{a}{g} = \frac{3 \text{ m/s}^2}{9.80 \text{ m/s}^2} = 0.306$

$\theta = \boxed{17.0°}$

(b) We can find T using (1).

$$T = \frac{ma}{\sin \theta} = \frac{(0.5 \text{ kg})(3 \text{ m/s}^2)}{\sin (17.0°)} = \boxed{5.13 \text{ N}}$$

Free-Body Diagram

29. A skydiver of mass 80 kg jumps from a slow-moving aircraft and reaches a terminal speed of 50 m/s. (a) What is the acceleration of the skydiver when her speed is 30 m/s? What is the drag force on the diver when her speed is (b) 50 m/s and (c) 30 m/s?

Solution

(a) For an object moving at high speeds, we can take the drag force R to be proportional to v^2 (Eq. 6.8), so Newton's second law gives

$$mg - B'v^2 = ma$$

or

$$a = g - Bv^2 \qquad (1)$$

When the skydiver reaches terminal velocity,

$$a = \frac{dv}{dt} = 0 \qquad \text{so} \qquad g - Bv_t^2 = 0$$

or

$$B = \frac{g}{v_t^2} = \frac{9.80 \text{ m/s}^2}{(50 \text{ m/s})^2} = 3.92 \times 10^{-3} \text{ m}^{-1}$$

Hence, when v = 30 m/s, (1) gives

$$a = (9.80 \text{ m/s}^2) - (3.92 \times 10^{-3} \text{ m}^{-1})(30 \text{ m/s})^2 = \boxed{6.27 \text{ m/s}^2}$$

(b) The drag force, which is *upward*, has a magnitude of ma = mBv^2. Hence, when v = 50 m/s,

$$R = mBv^2 = (80 \text{ kg})(3.92 \times 10^{-3} \text{ m}^{-1})(50 \text{ m/s})^2 = \boxed{784 \text{ N}}$$

(c) When v = 30 m/s, the drag force is

$$R = mBv^2 = (80 \text{ kg})(3.92 \times 10^{-3} \text{ m}^{-1})(30 \text{ m/s})^2 = \boxed{282 \text{ N}}$$

---

33. A small, spherical bead of mass 3 g is released from rest at t = 0 in a bottle of liquid shampoo. The terminal velocity, v_t, is observed to be 2 cm/s. Find (a) the value of the constant b in Equation 6.7, (b) the time, τ, it takes to reach $0.63v_t$, and (c) the value of the retarding force when the bead reaches terminal velocity.

Solution

(a) The velocity v varies with time according to Equation 6.7,

$$v = \frac{mg}{b}\left(1 - e^{-bt/m}\right) = v_t\left(1 - e^{-bt/m}\right) \qquad (1)$$

where $v_t = \frac{mg}{b}$ is the terminal velocity that is reached when $a = \frac{dv}{dt} = 0$. Hence,

$$b = \frac{mg}{v_t} = \frac{(3 \times 10^{-3} \text{ kg})(9.80 \text{ m/s}^2)}{(2 \times 10^{-2} \text{ m/s})} = \boxed{1.47 \text{ N·s/m}}$$

(b) To find the time it takes v to reach $0.63v_t$, we set $v = 0.63v_t$ into (1) and solve for t, to give

$$0.63v_t = v_t(1 - e^{-bt/m})$$

$$1 - e^{-bt/m} = 0.63$$

or

$$t = \boxed{2.03 \times 10^{-3} \text{ s}}$$

(c) From (1) we find

$$\frac{dv}{dt} = \frac{v_t b}{m} e^{-bt/m}$$

At $t = 0$, $\quad a = \frac{dv}{dt} = \frac{v_t b}{m}$, $\quad$ hence Newton's second law gives

$$F = ma = m\left(\frac{v_t b}{m}\right) = v_t b = (2 \times 10^{-2} \text{ m/s})\left(1.47 \frac{\text{N·s}}{\text{m}}\right) = \boxed{2.94 \times 10^{-2} \text{ N}}$$

39. Because of the earth's rotation about its axis, a point on the equator experiences a centripetal acceleration of 0.034 m/s^2, while a point at the poles experiences no centripetal acceleration. (a) Show that at the equator the gravitational force on an object (the true weight) must *exceed* the object's apparent weight. (b) What is the apparent weight at the equator and at the poles of a person having a mass of 75 kg? (Assume the earth is a uniform sphere and take $g = 9.800 \text{ m/s}^2$.)

Solution

(a) Let **N** represent the force exerted on the object by the scale, which is the "apparent weight." The true weight is mg. Summing up forces on the object in the direction towards the earth's center gives

$$mg - N = ma_c \qquad (1)$$

where $a_c = \frac{v^2}{R_e}$ is the centripetal acceleration directed toward the center of the earth. Thus, we see that

$$N = m(g - a_c) < mg$$

or

$$mg = N + ma_c > N \qquad (2)$$

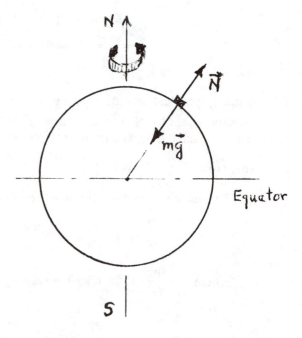

76

(b) Taking m = 75 kg, a_c = 0.034 m/s^2, and g = 9.800 m/s^2 gives

at Equator: $\quad$ N = m(g – a_c) = (75 kg)(9.800 m/s^2 – 0.034 m/s^2) = $\boxed{732.45 \text{ N}}$

at Poles: $\quad$ N = mg = (75 kg)(9.800 m/s^2) = $\boxed{735.00 \text{ N}}$
(a_c = 0)

---

45. The pilot of an airplane executes a constant-speed loop-the-loop maneuver in a vertical plane. The speed of the airplane is 300 mi/h, and the radius of the circle is 1200 ft. (a) What is the pilot's apparent weight at the lowest point if his true weight is 160 lb? (b) What is his apparent weight at the highest point? (c) Describe how the pilot could experience weightlessness if both the radius and velocity can be varied. (Note: His apparent weight is equal to the force of the seat on his body.)

Solution

(a) $\quad v = (300 \text{ mph}) \left(\dfrac{88 \text{ ft/s}}{60 \text{ mph}} \right) = 440 \text{ ft/s}$

At the lowest point, his seat pushes up at him and his weight seems to increase. Since N – W = ma, and N = W' = "apparent weight,"

$$W' = W + ma = 160 \text{ lb} + \left(\frac{160 \text{ lb}}{32 \text{ ft/s}^2} \right) \left(\frac{440 \text{ ft/s}}{1200 \text{ ft}} \right)^2 = \boxed{967 \text{ lb}}$$

(b) At the highest point, the force of the seat on the pilot is directed away from the pilot (downward), and we find

$$W' = W - ma = \boxed{-647 \text{ lb}}$$

Since W' < 0, the pilot must be strapped down. Note that W' is *upward* in this case.

(c) When W' = 0, then $W = ma = \dfrac{mv^2}{R}$. If we can vary R and v such that the above is true, then the pilot feels weightless.

7
Work and Energy

OBJECTIVES

1. Define the work done by a constant force, and realize that work is a scalar.

2. Take the scalar or dot product of any two vectors **A** and **B** using the definition **A·B** ≡ AB cos θ, or by writing **A** and **B** in unit vector form and using the multiplication table for unit vectors.

3. Recognize that the work done by a force can be positive, negative, or zero, and describe at least one example of each case.

4. Describe the work done by a force which *varies* with position. In the one-dimensional case, note that the work done equals the area under the F_x versus x curve.

5. Define the kinetic energy of an object of mass m moving with a speed v.

6. Relate the work done by the net force on an object to the *change* in kinetic energy. The relation $W = \Delta K = K_f - K_i$ is called the work-energy theorem, and is valid whether or not the (resultant) force is constant. That is, if we know the net work done on a particle as it undergoes a displacement, we also know the *change* in its kinetic energy. This is the most important concept in this chapter, so you must understand it thoroughly.

7. Define the concepts of average power and instantaneous power (the time rate of doing work).

SKILLS

There are two new mathematical skills you must learn. The first is the definition of the scalar (or dot) product, **A·B** ≡ AB cos θ, where θ is the angle between **A** and **B**. Since **A·B** is a scalar, then the order of product can be interchanged. That is, **A·B** = **B·A**. Furthermore, **A·B** can be positive, negative, or zero depending on the value of θ. (That is, cos θ, varies from -1 to +1.) If vectors are expressed in unit vector form, then the dot product is conveniently carried out using the multiplication table for unit vectors:

$$\mathbf{i \cdot i} = \mathbf{j \cdot j} = \mathbf{k \cdot k} = 1; \qquad \mathbf{i \cdot j} = \mathbf{i \cdot k} = \mathbf{k \cdot j} = 0$$

The second operation introduced in this chapter is the definite integral. In Section 7.4, it is shown that the work done by a *variable* force F_x in displacing a particle a small distance Δx is given by

$$\Delta W \approx F_x \Delta x$$

(ΔW equals the area of the shaded rectangle in Fig. 7.1.) The total work done by F_x as the particle is displaced from x_i to x_f is given approximately by the *sum* of such terms.

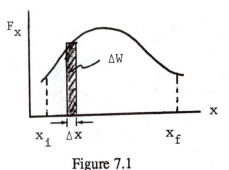

Figure 7.1

If we take such a sum, letting the widths of the displacements approach dx, the number of terms in the sum becomes very large and we get the actual work done:

$$W = \lim_{\Delta x \to 0} \sum_{x_i}^{x_f} F_x \Delta x = \int_{x_i}^{x_f} F_x dx$$

The quantity on the right is a definite integral, which graphically represents the *area under the F_x versus x curve*, as in Fig.7.1. Appendix B6 of the text represents a brief review of integration operations, with some examples.

NOTES FROM SELECTED CHAPTER SECTIONS

7.2 WORK DONE BY A CONSTANT FORCE

Work done by a constant force is defined as the product of the component of the force in the direction of the displacement and the magnitude of the displacement.

Unit of work in the SI system is the newton·meter, N·m.   1 newton·meter = 1 joule (J).

7.3 SCALAR PRODUCT OF TWO VECTORS

The *scalar product* or dot product of any two vectors is a scalar quantity equal to the product of the magnitudes of the two vectors and the cosine of the angle included between the directions of the two vectors.

7.4 WORK DONE BY A VARYING FORCE

Work done by a varying force is equal to the area under the force-displacement curve.

7.5 WORK AND KINETIC ENERGY

The work done by a force F in displacing a particle equals the change in the kinetic energy of the particle. This is known as the *work energy theorem.*

7.6 POWER

Power is the time rate of doing work or expending energy.   The SI unit of power is the watt, W. 1 W = 1 J/s.

EQUATIONS AND CONCEPTS

The work done on a body by a *constant* force **F** is defined to be the product of the component of force in the direction of the displacement and the displacement.

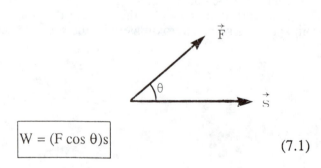

$$\boxed{W = (F \cos \theta)s}$$

(7.1)

It is convenient to express the work done by a constant force as the *dot product* (scalar product) **F·s**.

$$W = \mathbf{F}\cdot\mathbf{s} = Fs\cos\theta \tag{7.4}$$

The dot product of any two vectors **A** and **B** is defined to be a scalar quantity whose magnitude is $AB\cos\theta$.

$$\mathbf{A}\cdot\mathbf{B} = AB\cos\theta \tag{7.5}$$

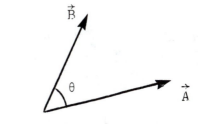

Note that work done by a force can be positive, negative, or zero depending on the value of θ. Work done by a force is positive if **F** has a component in the direction of **s** $(0 \le \theta < 90°)$; W is negative if the projection of **F** onto **s** is opposite to **s** $(90° < \theta < 180°)$. Finally, W is zero if **F** is perpendicular to **s** $(\theta = 90°)$. Work is a *scalar* quantity which has SI units of joules (J), where $1\,\text{J} = 1\,\text{N·m}$.

The work done by the force of sliding friction **f** is always negative since **f** opposes the displacement.

$$W_f = -fs \tag{7.2}$$

If a force acting along x *varies* with position, and the body is displaced from x_i to x_f, the *work done by that force* is given by an integral expression. Graphically, the work done equals the area under the F_x versus x curve.

$$W = \int_{x_i}^{x_f} F_x\,dx \tag{7.11}$$

If a mass is connected to a spring of force constant k, *the work done by the spring force* (-kx) as the mass moves from x_i to x_f is given by

$$W_s = \frac{1}{2}kx_i^2 - \frac{1}{2}kx_f^2 \tag{7.14b}$$

The *kinetic energy* K of a particle of mass m moving with a speed v is defined to be $\frac{1}{2}mv^2$.

$$K \equiv \frac{1}{2}mv^2 \tag{7.17}$$

The *work-energy theorem* states that the work done by the resultant force on a body equals the *change* in kinetic energy of the body.

$$W = K_f - K_i = \Delta K \tag{7.19}$$

Note that if W is positive, the kinetic energy increases; if W is negative, the kinetic energy decreases. Therefore, the speed of a body will only change if there is net work done on it. (When W = 0, the resultant work is zero, and $\Delta K = 0$.)

The *average power* supplied by a force is the ratio of the work done by that force to the time interval over which it acts.

$$\overline{P} = \frac{\Delta W}{\Delta t}$$ (7.22)

The *instantaneous power* is equal to the limit of the average power as the time interval approaches zero.

$$P = \frac{dW}{dt}$$ (7.23)

The SI unit of power is J/s, which is called a watt (W); 1 W = 1 J/s.

EXAMPLE PROBLEM SOLUTIONS

Example 7.1 A force of 10 N is applied to a crate at an angle of 37° with the horizontal (Figure 7.2). The crate moves a total distance of 5 m.

Figure 7.2

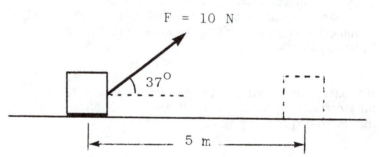

(a) What is the work done by the 10 N force?

Solution

$$W = Fs \cos \theta = (10 \text{ N}) (5 \text{ m}) \cos (37°)$$

$$W = 40 \text{ N·m} = 40 \text{ J}$$

Alternate Solution

We could also write $\mathbf{F} = 10 \cos (37°) \mathbf{i} + 10 \sin (37°) \mathbf{j}$ and $\mathbf{s} = 5\mathbf{i}$, and apply Equation 7.4. The results are the same. Try it!

Comments

We see that the calculation of the work done by a constant force **F** requires only a knowledge of **F** and the displacement **s**. However, in this problem there are other forces acting on the crate. If the surface is frictionless, there are two other forces, namely the weight **W** and the normal force **N**. However, the work done by these forces is *zero* since they are both *perpendicular* to **s**. In other words, for **N**, $\theta = 90°$, and since $\cos(90°) = 0$, $W = 0$. Similarly for **W**, $\theta = 90°$ and the work done by **W** is zero.

(b) Suppose that in this example a frictional force of 5 N is acting on the crate in the negative x direction. What is the work done by this force?

Solution

In this case, the frictional force is in the opposite direction to **s** since the sliding friction force opposes the motion of an object. Using Equation 7.1 gives

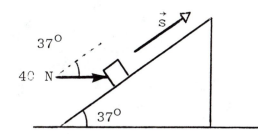

$$W = fs \cos(180°) = -fs = -(5 \text{ N})(5 \text{ m})$$

$$W = -25 \text{ N·m} = -25 \text{ J}$$

Alternate Solution

$\mathbf{f} = -5\mathbf{i}, \quad \mathbf{s} = 5\mathbf{i}, \quad W = \mathbf{f}\cdot\mathbf{s} = -5\mathbf{i}\cdot5\mathbf{i} = -25 \text{ J}.$ That is, the frictional force does *negative* work on the crate. This is equivalent to saying the crate does work on the environment, since the surface heats up during the motion, and so forth. Whenever $\theta < 90°$, W is positive or **F** has a component along **s**, whereas when $\theta > 90°$, W is negative, meaning **F** has a component in the direction -**s**.

Example 7.2 A 3-kg block is pushed up an inclined plane a distance of 2 m by a horizontal force of 40 N (Figure 7.3). The coefficient of sliding friction is 0.1 and the angle of inclination is 37°.

(a) What is the work done by the 40 N force?

Solution

The displacement **s** is up the plane and equals 2 m. The horizontal 40 N force has a component 40 cos (37°) along the incline, in the direction of **s**, therefore

$$W_F = Fs \cos (37°) = (40 \text{ N}) (2 \text{ m}) (0.8) = 64 \text{ J}$$

Figure 7.3

(b) What is the work done by gravity?

Solution

The weight vector **W** = mg can be resolved into components perpendicular and parallel to the incline. The component perpendicular to the plane does no work, since it is also perpendicular to s, so W = 0. The component 3g sin (37°) points down the incline and opposes s; hence $\theta = 180°$, and the work done by gravity is

$$W_g = Fs \cos \theta = 3g \sin (37)\,(2 \text{ m}) \cos (180°)$$

$$W_g = (3 \text{ kg})(9.8 \text{ m/s}^2)(0.6)(2 \text{ m})(-1)$$

$$W_g = -35 \text{ J}$$

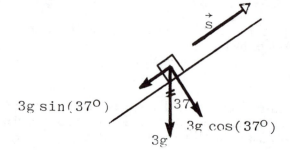

(c) What is the work done by friction?

Solution

The maximum frictional force is given by $f_k = \mu_k N$. But from the free body diagram, we see that

$$N = 3g \cos (37°) + 40 \sin (37°) = 48 \text{ N}$$

Therefore, f_k becomes

$$f_k = \mu_k N = 0.1(48 \text{ N}) = 4.8 \text{ N}$$

Since f_k is opposite to s, $\theta = 180°$, and the work done by f_k is

$$W_f = f_k s \cos (180°) = -f_k s$$

$$W_f = - (4.8 \text{ N})(2 \text{ m}) = -9.6 \text{ J}$$

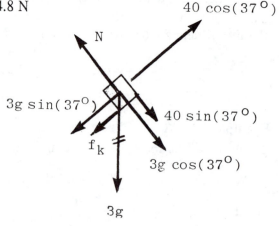

Note that we have calculated the work done by each force separately, even though all forces act on the block at once. The *total* work done by *all* forces is the sum of parts (a), (b), and (c); or $W_{tot} = 19$ J.

ANSWERS TO SELECTED QUESTIONS

6. Can the kinetic energy of an object have a negative value?

Answer: No. Kinetic energy = $mv^2/2$. Since v^2 is always positive, KE is always positive.

8. What can be said about the speed of an object if the net work done on that object is zero?

Answer: Its speed remains unchanged. This can be seen from the work-energy theorem. Since $W = \Delta KE = 0$, it follows that $v_f = v_i$.

9. Using the work-energy theorem, explain why the force of kinetic friction always has the effect of reducing the kinetic energy of a particle.

Answer: The work done by the force of sliding friction is always negative. Since the work done is equal to the change in kinetic energy, it follows that the final kinetic energy is less than the initial kinetic energy.

13. One bullet has twice the mass of a second bullet. If both are fired such that they have the same velocity, which has more kinetic energy? What is the ratio of kinetic energies of the two bullets?

Answer: The kinetic energy of the more massive bullet is twice that of the lower mass bullet.

14. When a punter kicks a football, is he doing any work on the ball while his toe is in contact with the ball? Is he doing any work on the ball after it loses contact with his toe? Are there any forces doing work on the ball while it is in flight?

Answer: As the punter kicks the football, he exerts a force on the ball and also moves the ball. Thus, he is doing work on the ball. After his toe loses contact with the ball, the punter no longer exerts a force on it, and thus, he can no longer do work on it. While in flight, the only force doing any work on the ball, in the absence of air resistance, is the weight of the ball.

15. Discuss the work done by a pitcher throwing a baseball. What is the approximate distance through which the force acts as the ball is thrown?

Answer: A pitcher does work on a baseball only while the ball is in his hand. The ball moves a distance of about one meter while in his hand.

16. Estimate the time it takes you to climb a flight of stairs; then approximate the power required to perform this task. Express your value in horsepower.

Answer: Let us guess that the time is equal to 10 s, and the height of the stairs is 3 m. If you assume that the force you must overcome to climb the stairs is your weight, which we will take to be 150 N, the approximate work done is (150 N) (3 m) = 450 J. Hence, the power is (450 J)/(10 s) = 45 W.

17. Do frictional forces always reduce the kinetic energy of a body? If your answer is no, give examples which illustrate the effect.

Answer: No. For example, if you place a crate on the bed of a truck and the truck accelerates, the friction force acting on the crate is what causes it to accelerate, assuming it doesn't slip. Another example is a car which gets its acceleration because of the friction force between the road and its tires. This force is in the direction of motion of the car and produces an increase in the car's kinetic energy. Of course, the source of the energy is the combustion of fuel in the car's engine.

SOLUTIONS TO SELECTED END-OF-CHAPTER PROBLEMS

7. A horizontal force of 150 N is used to push a 40-kg box on a rough, horizontal surface through a distance of 6 m. If the box moves at constant speed, find (a) the work done by the 150-N force, (b) the work done by friction, and (c) the coefficient of kinetic friction.

Solution

$s = 6$ m

Since $a = 0$, ($v =$ const)

$\Sigma F_x = 150 - f = 0$

$f = 150$ N

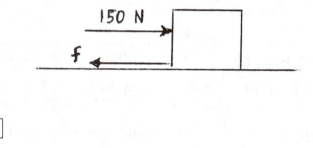

(a) $W_F = (150 \text{ N})(6 \text{ m}) = \boxed{900 \text{ J}}$

(b) $W_f = -fs = -(150 \text{ N})(6 \text{ m}) = \boxed{-900 \text{ J}}$

(c) $f = \mu N = \mu Mg$

$\mu = \dfrac{f}{Mg} = \dfrac{150 \text{ N}}{(40 \text{ kg})(9.80 \text{ m/s}^2)} = \boxed{0.383}$

---

11. A cart loaded with bricks has a total mass of 18 kg and is pulled at constant speed by a rope. The rope is inclined at 20° above the horizontal and the cart moves on a horizontal plane. The coefficient of kinetic friction between the ground and the cart is 0.5. (a) What is the tension in the rope? When the cart is moved 20 m, (b) how much work is done on the cart by the rope? (c) How much work is done by the friction force?

Solution

(a) Since $v =$ constant, $a = 0$, so $\Sigma F = 0$ gives

(1) $T \cos \theta - f = 0$

(2) $N + T \sin \theta - mg = 0$

(3) $f = \mu N = \mu(mg - T \sin \theta)$

Substituting (3) into (1) gives

$\quad T \cos \theta - \mu mg + \mu T \sin \theta = 0$

or,

$$T = \frac{\mu mg}{\cos \theta + \mu \sin \theta} = \frac{(0.5)(18 \text{ kg})(9.80 \text{ m/s}^2)}{\cos 20° + (0.5) \sin 20°} = \boxed{79.4 \text{ N}}$$

(b) For s = 20 m, W_T = (T cos θ)s = (79.4 N)(cos 20°)(20 m) = $\boxed{1.49 \text{ kJ}}$

(c) $W_{net} = W_T + W_f = 0$, so

$$W_f = -W_T = \boxed{-1.49 \text{ kJ}}$$

17. A force **F** = (6**i** – 2**j**) N acts on a particle that undergoes a displacement **s** = (3**i** + **j**) m. Find (a) the work done by the force on the particle and (b) the angle between **F** and **s**.

Solution

(a) W = **F·s** = (6**i** – 2**j**)·(3**i** + **j**) = (6)(3) + (–2)(1) = 18 – 2 = $\boxed{16.0 \text{ J}}$

(b) $F = \sqrt{F_x^2 + F_y^2} = \sqrt{6^2 + (-2)^2} = 6.32$ N

$s = \sqrt{s_x^2 + s_y^2} = \sqrt{3^2 + 1^2} = 3.16$ m

W = Fs cos θ

$\cos \theta = \dfrac{W}{Fs} = \dfrac{16 \text{ J}}{(6.32 \text{ N})(3.16 \text{ m})} = 0.8012$

$\theta = \cos^{-1}(0.8012) = \boxed{36.8°}$

29. If an applied force varies with position according to $F_x = 3x^3 - 5$, where x is in m, how much work is done by this force on an object that moves from x = 4 m to x = 7 m?

Solution

F = $3x^3 - 5$

$$W = \int_{x_1}^{x_2} F dx = \int_4^7 (3x^3 - 5) dx = \frac{3}{4}x^4 - 5x \Big|_4^7$$

$$= \frac{3}{4}(7^4 - 4^4) - 5(7 - 4) = \boxed{1.59 \text{ kJ}}$$

37. A 40-kg box initially at rest is pushed a distance of 5 m along a rough, horizontal floor with a constant applied force of 130 N. If the coefficient of friction between the box and floor is 0.3, find (a) the work done by the applied force, (b) the work done by friction, (c) the change in kinetic energy of the box, and (d) the final speed of the box.

Solution

(a) Assuming that the applied force is *horizontal*, we have

$$W_F = \mathbf{F} \cdot \mathbf{s} = Fs = (130\ N)(5\ m) = \boxed{650\ J}$$

(b) $f_k = \mu_k N = \mu_k mg = 0.3(40\ kg)(9.80\ m/s^2) = 117.6\ N$

$$W_{f_k} = \mathbf{f_k} \cdot \mathbf{s} = -f_k s = -(117.6\ N)(5\ m) = \boxed{-588\ J}$$

(c) $\Delta K = W_{Net} = W_F + W_{f_k} = 650\ J + (-588\ J) = \boxed{62.0\ J}$

(d) $\frac{1}{2} m v_f^2 - \frac{1}{2} m v_o^2 = W_{Net}$

$$\frac{1}{2}(40\ kg)v_f^2 - 0 = 62.0\ J$$

$$v_f^2 = \frac{124\ J}{40\ kg} \Rightarrow \boxed{v_f = 1.76\ m/s}$$

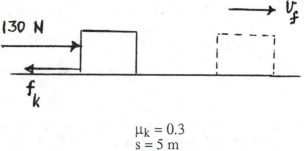

$\mu_k = 0.3$
$s = 5\ m$
$v_o = 0$

---

49. A 4-kg block is given an initial speed of 8 m/s at the bottom of a 20° incline. The frictional force that retards its motion is 15 N. (a) If the block is directed *up* the incline, how far will it move before it stops? (b) Will it slide back down the incline?

Solution

(a) If s is the displacement along the incline, the work done by friction is $W_f = -fs$, while the work done by the force of gravity is

$$W_g = -mgs(\sin 20°)$$

Since the normal force and the component of weight mg cos 20° are both perpendicular to the displacement, they do no work. Hence, from the work-energy theorem,

$f = 15\ N$
$v_o = 8\ m/s$
$m = 4\ kg$

$$W_f + W_g = K_f - K_i$$

$$-fs - mgs(\sin 20°) = 0 - \frac{1}{2} mv_0^2$$

$$-15s - (4 \text{ kg})(9.80 \text{ m/s}^2)(0.342)s = -\frac{1}{2}(4 \text{ kg})(8 \text{ m/s})^2$$

$$28.4s = 128$$

$$s = \boxed{4.51 \text{ m}}$$

(b) Because $f > mg \sin \theta$, the block will *not* slide back down the incline.

---

61. A 65-kg athlete runs a distance of 600 m up a mountain inclined at 20° to the horizontal. He performs this feat in 80 s. Assuming that air resistance is negligible, (a) how much work does he perform and (b) what is his power output during the run?

Solution

(a) Assuming the athlete runs at constant speed, we have

$$W_A + W_g = 0$$

where W_A is the work done by the athlete and W_g is the work done by gravity. In this case,

$$W_g = -mgs(\sin \theta)$$

so

$$W_A = -W_g = + mgs(\sin \theta) = (65 \text{ kg})(9.80 \text{ m/s}^2)(600 \text{ m}) \sin 20°$$

$$= \boxed{1.31 \times 10^5 \text{ J}}$$

m = 65 kg
Δt = 80 s
s = 600 m

(b) His power output is given by

$$P_A = \frac{W_A}{\Delta t} = \frac{1.31 \times 10^5 \text{ J}}{80 \text{ s}} = \boxed{1.63 \text{ kW}} \quad \text{(or 2.19 h.p.)}$$

---

75. A 4-kg particle moves along the x-axis. Its position varies with time according to $x = t + 2t^3$, where x is in m and t is in s. Find (a) the kinetic energy at any time t, (b) the acceleration of the particle and the force acting on it at time t, (c) the power being delivered to the particle at time t, and (d) the work done on the particle in the interval t = 0 to t = 2 s. (Note: P = dW/dt.)

Solution

Given $m = 4$ kg and $x = t + 2t^3$, we find

(a) $v = \dfrac{dx}{dt} = \dfrac{d}{dt}(t + 2t^3) = 1 + 6t^2$

$\quad\quad K = \dfrac{1}{2}mv^2 = \dfrac{1}{2}(4)(1 + 6t^2)^2 = \boxed{(2 + 24t^2 + 72t^4)\ J}$

(b) $a = \dfrac{dv}{dt} = \dfrac{d}{dt}(1 + 6t^2) = \boxed{12t\ m/s^2}$

$\quad\quad F = ma = 4(12t) = \boxed{48t\ N}$

(c) $P = \dfrac{dW}{dt} = \dfrac{dK}{dt} = \dfrac{d}{dt}(2 + 24t^2 + 72t^4) = \boxed{(48t + 288t^3)\ W}$ $\quad \left[or\ use\ P = Fv = 48t\left(1 + 6t^2\right)\right]$

(d) $W = K_f - K_i$ where $t_i = 0$ and $t_f = 2$ s.

At $t_i = 0$, $K_i = 2$ J

At $t_f = 2$ s, $K_f = [2 + 24(2)^2 + 72(2)^4] = 1250$ J

Therefore, $W = \boxed{1.25 \times 10^3\ J}$ $\quad \left(or\ use\ W = \displaystyle\int_{t_i}^{t_f} P\,dt = \int_{o}^{2}(48t + 288t^3)dt\ etc.\right)$

81. A 0.4-kg particle slides on a horizontal, circular track 1.5 m in radius. It is given an initial speed of 8 m/s. After one revolution, its speed drops to 6 m/s because of friction. (a) Find the work done by the force of friction in one revolution. (b) Calculate the coefficient of kinetic friction. (c) What is the total number of revolutions the particle will make before coming to rest?

Solution

(a) $W_f = \Delta K = \dfrac{1}{2}m\left(v_f^2 - v_i^2\right) = \dfrac{1}{2}(0.4\ kg)[(6\ m/s)^2 - (8\ m/s)^2] = \boxed{-5.60\ J}$

(b) $W_f = -5.60\ J = f\Delta s = -\mu mg(2\pi r)$

$$\mu = \frac{5.60 \text{ J}}{(0.4 \text{ kg})(9.80 \text{ m/s}^2)(2\pi)(1.5 \text{ m})} = \boxed{0.152}$$

(c) After n revolutions, v = 0 and all the initial kinetic energy K_i is lost due to friction.

$$K_i = \frac{1}{2} m v_i^2 = \frac{1}{2} (0.4 \text{ kg})((8 \text{ m/s})^2 = 12.8 \text{ J}$$

For one revolution, $W_f = -5.60$ J, so for n revolutions

$$W_f = -5.60n \text{ J}$$

Since $W_f = \Delta K = -K_i$, we find

$$-5.60n = -12.8 \text{ J}$$

$$n = \frac{12.8}{5.60} = \boxed{2.29 \text{ revolutions}}$$

8

Potential Energy
and Conservation of Energy

OBJECTIVES

1. Recognize the properties of conservative and nonconservative forces.

2. Recognize that a potential energy function U can only be defined when dealing with a conservative force.

3. Understand the distinction between kinetic energy (energy associated with motion), potential energy (energy associated with the position or coordinates of a system), and the total mechanical energy of a system.

4. State the law of conservation of mechanical energy, noting that mechanical energy is conserved only when conservative forces act on a system. This extremely powerful concept is most important in all areas of physics.

5. Compute the potential energy function associated with a conservative force such as the force of gravity and the spring force.

6. Recognize that the gravitational potential energy function, $U_g = mgy$, can be positive, negative, or zero, depending on the location of the coordinate system used to measure y.

7. Recognize that the spring potential energy function, $U_s = \frac{1}{2} kx^2$, is either positive or zero, where x is the elongation (or compression) of the spring measured from equilibrium.

8. Be aware of the fact that although U depends on the origin of the coordinate system, the *change* in potential energy, ΔU, is *independent* of the coordinate system used to define U.

9. Account for nonconservative forces acting on a system using the work-energy theorem. In this case, the work done by all nonconservative forces equals the change in total mechanical energy of the system.

SKILLS

As we have seen, many problems in physics can be solved using the principle of conservation of mechanical energy. Other examples involving springs as part of the system will follow. The following steps should be followed in applying this principle:

1 Define your system, which may consist of more than one object.

2. Select a reference position for the zero point of potential energy (both gravitational and spring), and use this throughout your analysis. If there is more than one conservative force, then remember to write expressions for the potential energy associated with each force.

3. Determine whether or not friction forces are present. Remember that if friction or air resistance are present, mechanical energy *is not conserved*.

4. If there are no friction forces present, then mechanical energy is conserved, and you can proceed to write the total initial energy E_i at some point as the sum of the kinetic and potential energy at that point. Then, write an expression for the total final energy $E_f = K_f + U_f$ at the final point that is of interest. Since mechanical energy is conserved, you can equate the two total energies and solve for the quantity that is unknown.

5. If friction forces are present, then you should first write expressions for the total initial and total final energies. In this case, however, the total final energy differs from the total initial energy, the difference being the work done by the nonconservative forces. That is, you should apply $W_{nc} = E_f - E_i$.

NOTES FROM SELECTED CHAPTER SECTIONS_____

8.1 CONSERVATIVE AND NONCONSERVATIVE FORCES

A force is said to be *conservative* if the work done by that force on a body moving between any two points is independent of the path taken. In addition, the work done by a conservative force is zero when the body moves through any closed path and returns to its initial position. *Nonconservative* forces are those for which the work done on a particle moving between two points depends on the path. Furthermore, the work done by a nonconservative force for any closed path is not necessarily zero.

8.2 POTENTIAL ENERGY

It is possible to define a *potential energy function* associated with a conservative force such that the work done by the conservative force equals the negative of the change in the potential energy associated with the force.

8.3 CONSERVATION OF MECHANICAL ENERGY

The law of *conservation of mechanical energy* states that the total mechanical energy of a system remains constant if the only force that does work on the system is a conservative force.

8.4 GRAVITATIONAL POTENTIAL ENERGY NEAR THE EARTH'S SURFACE

The *work done by the force of gravity* is equal to the initial value of the potential energy minus the final value of the potential energy.

8.5 NONCONSERVATIVE FORCES AND THE WORK-ENERGY THEOREM

The *total work* done on a system by all forces (conservative and nonconservative) equals the *change in the kinetic energy* of the system. The work done by all *nonconservative* forces equals the change in the total *mechanical energy* of the system.

8.8 ENERGY DIAGRAMS AND STABILITY OF EQUILIBRIUM

Positions of stable equilibrium correspond to points for which U(x) has a minimum value; positions of unstable equilibrium correspond to those points for which U(x) has a maximum value; finally, a position of neutral equilibrium corresponds to a region over which U remains constant.

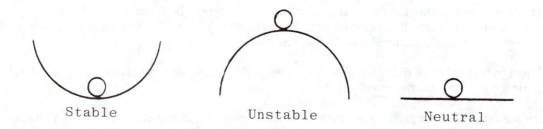

8.9 CONSERVATION OF ENERGY IN GENERAL

The energy conservation principle can be generalized to include all forms of energies and energy transformations. Energy may be transformed from one form to another, but *the total energy of an isolated system remains constant.* That is, *energy cannot be created or destroyed.*

EQUATIONS AND CONCEPTS

A potential energy function U can be defined for a conservative force.

$$U_f - U_i = -\int_{x_i}^{x_f} F_x dx \qquad (8.2)$$

The law of *conservation of mechanical energy* says that if only conservative forces act on a system, the sum of the kinetic and potential energies remains constant--or is conserved. That is,

$$K_i + U_i = K_f + U_f \qquad (8.5)$$

It is useful to define the *total mechanical energy* E of the system as the sum of the kinetic and potential energies. According to this important conservation law, if the kinetic energy of the system increases by some amount, the potential energy must decrease by the same amount--and vice versa.

$$E_i = E_f \qquad (8.6)$$

The force of gravity mg is one example of a conservative force. The work done by the gravitational force in moving a particle between two points whose y coordinates are y_i and y_f is given by Eq. 8.8.

$$W_g = -mg(y_f - y_i) \qquad (8.8)$$

It is useful to define the *gravitational potential energy* in terms of the coordinate y measured from an arbitrary reference level.

$$U_g = mgy \qquad (8.9)$$

Another common conservative force is the spring force, $F_s = -kx$, sometimes called the linear restoring force. *The work done by the spring* in moving an attached mass from position x_i to position x_f depends on the initial and final positions and the elastic constant of the spring.

$$W_s = \frac{1}{2} kx_i^2 - \frac{1}{2} x_f^2$$

The quantity $\frac{1}{2} kx^2$ is referred to as the *elastic potential energy* stored in the spring.

$$U_s = \frac{1}{2} kx^2 \qquad (8.14)$$

Note that both the gravitational force and the spring force satisfy the required properties of a conservative force. That is, the work done is path independent and is zero for any closed path.

If nonconservative forces, as well as conservative, act on a system, we can use the work-energy theorem to show that *the work done by all nonconservative forces equals the change in the total mechanical energy of the system.*

$$\boxed{W_{nc} = (K_f + U_f) - (K_i + U_i) = E_f - E_i} \qquad (8.13)$$

Note that if W_{nc} is negative (for example, when sliding friction is present), the mechanical energy decreases; if W_{nc} is positive, the mechanical energy increases; if $W_{nc} = 0$, the mechanical energy remains constant or is conserved.

If the potential energy function associated with a system moving along the x direction is known, *the conservative force associated with that function is given by Eq. 8.18.*

$$\boxed{F_x = - \frac{dU(x)}{dx}} \qquad (8.18)$$

EXAMPLE PROBLEM SOLUTIONS

Example 8.1   A 0.2 kg bead is forced to slide on a frictionless wire as in Fig. 8.1. The bead starts from rest at A and ends up at B after colliding with a light spring of force constant k. If the spring compresses a distance of 0.1 m, what is the force constant of the spring?

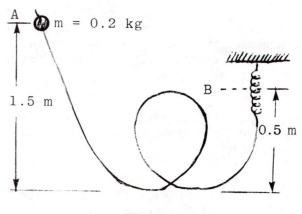

Figure 8.1

Solution

The gravitational potential energy of the bead at A *with respect to the lowest point* is

$$U_i = mgh_i = (0.2 \text{ kg}) (9.8 \text{ m/s}^2) (1.5 \text{ m}) = 2.94 \text{ J}$$

The kinetic energy of the bead at A is zero since it starts from rest. The gravitational potential energy of the bead at B is

$$U_f = mgh_f = (0.2 \text{ kg}) (9.8 \text{ m/s}^2) (0.5 \text{ m}) = 0.98 \text{ J}$$

Since the spring is part of the system, we must also take into account the energy stored in the spring at B. Since the spring compresses a distance $x_m = 0.1$ m, we have

$$U_s = \frac{1}{2} k x_m^2 = \frac{1}{2} k (0.1)^2$$

Using the principle of energy conservation gives

$$U_i = U_f + U_s$$

$$2.94 \text{ J} = 0.98 \text{ J} + \frac{1}{2} k \, (0.1)^2$$

$$k = 392 \text{ N/m}$$

Example 8.2   A block of mass 0.2 kg is given an initial speed $v_0 = 5$ m/s on a horizontal, rough surface of length 2 m as in Fig. 8.2. The coefficient of kinetic friction on the horizontal surface is 0.30. If the curved part of the track is frictionless, how high does the block rise before coming to rest at B?

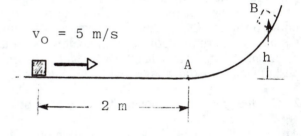

Figure 8.2

Solution

The initial kinetic energy of the block is

$$K_0 = \frac{1}{2} m v_0{}^2 = \frac{1}{2} (0.2 \text{ kg}) \, (5 \text{ m/s})^2 = 2.50 \text{ J}$$

The work done by friction along the horizontal track is

$$W_f = -fd = -\mu mgd = -(0.30) \, (0.2) \, (9.8) \, (2) = -1.18 \text{ J}$$

Using the work-energy theorem, we can find the kinetic energy at A

$$W_f = K_A - K_0 = K_A - 2.50$$

$$K_A = 2.50 + W_f = 2.50 - 1.18 = 1.32 \text{ J}$$

Since the curved track is frictionless, we can equate the kinetic energy of the block at A to its gravitational potential energy at B.

$$mgh = K_A = 1.32 \text{ J}$$

$$h = \frac{1.32 \text{ J}}{(0.2 \text{ kg}) \, (9.8 \text{ ms}^2)} = 0.673 \text{ m}$$

You should be able to show that the block finally comes to rest 2.24 m to the left of A, after sliding down from B.

ANSWERS TO SELECTED QUESTIONS

1. A bowling ball is suspended from the ceiling of a lecture hall by a strong cord. The bowling ball is drawn from its equilibrium position and released from rest at the tip of the demonstrator's nose. If the demonstrator remains stationary, explain why she will not be struck by the ball on its return swing. Would the demonstrator be safe if the ball were given a push from this position?

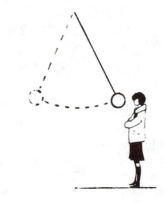

Answer: The total energy of the system (bowling ball) must be conserved. Since the ball initially has a potential energy mgh, and no kinetic energy, it cannot have any kinetic energy when returning to its initial position. Of course, air resistance will cause the ball to return to a point slightly below its initial position. On the other hand, if the ball is given a push, the demonstrator's nose will be in big trouble.

2. Can the gravitational potential energy of an object ever have a negative value? Explain.

Answer: The potential energy mgy of an object depends on the position of the reference frame. If the origin is below the object, we call the potential energy positive (positive y value). If the object is below the origin, the potential energy is negative (negative y value). This is the convention used in the text.

3. A ball is dropped by a person from the top of a building, while another person at the bottom observes its motion. Will these two people agree on the value of the ball's potential energy? On the *change* in potential energy of the ball? On the kinetic energy of the ball?

Answer: The two will not necessarily agree on the potential energy, since this depends on the origin--which may be chosen differently for the two observers. However, the two *must* agree on the value of the *change* in potential energy, which is independent of the choice of the reference frames. The two will also agree on the kinetic energy of the ball, assuming both observers are at rest with respect to each other, and hence measure the same v.

6. When nonconservative forces act on a system, does the total mechanical energy remain constant?

Answer: No. Nonconservative forces such as friction will either decrease or increase the total mechanical energy of a system. The mechanical energy remains constant when conservative forces *only* act on the system, or when nonconservative forces do zero work.

7. If three different conservative forces and one nonconservative force act on a system, how many potential energy terms will appear in the work-energy theorem?

Answer: Three. One for each conservative force.

8. A block is connected to a spring that is suspended from the ceiling. If the block is set in motion and air resistance is neglected, will the total energy of the system be conserved? How many forms of potential energy are there for this situation?

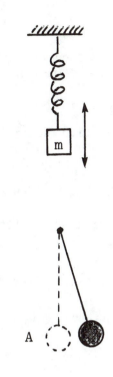

Answer: Yes, the total mechanical energy is conserved since there are only conservative forces acting: gravity and the spring force. Hence, there are two forms of potential energy: gravitational potential energy (mgy) and spring potential energy $\left(\frac{1}{2}ky^2\right)$.

9. Consider a ball fixed to one end of a rigid rod with the other end pivoted on a horizontal axis so that the rod can rotate in a vertical plane. What are the positions of stable and unstable equilibrium?

Answer: Only one position (A), the lowest point, is stable. All other positions are ones of unstable equilibrium.

SOLUTIONS TO SELECTED END-OF-CHAPTER PROBLEMS

7. A single conservative force $F_x = (2x + 4)$ N acts on a 5-kg particle, where x is in m. As the particle moves along the x axis from x = 1 m to x = 5 m, calculate (a) the work done by this force, (b) the change in the potential energy of the particle, and (c) its kinetic energy at x = 5 m if its speed at x = 1 m is 3 m/s.

Solution

$$F_x = (2x + 4)\text{ N} \quad x_i = 1\text{ m} \quad x_f = 5\text{ m} \quad v_i = 3\text{ m/s}$$

(a) $\displaystyle W_F = \int_{x_i}^{x_f} F_x\,dx = \int_{x_i}^{x_f} (2x + 4)\,dx = x^2 + 4x\Big]_1^5 = 5^2 + 4(5) - \left[1^2 + 4(1)\right] = \boxed{40.0\text{ J}}$

(b) $\Delta U = -W_F = \boxed{-40.0\text{ J}}$

(c) $\Delta K + \Delta U = 0$ $\qquad\qquad K_i = \frac{1}{2}mv_i^2 = \frac{1}{2}(5\text{ kg})(3\text{ m/s})^2$

$\quad\;\; \Delta K = -\Delta U = 40.0\text{ J}$ $\qquad\quad K_i = 22.5\text{ J}$

$\quad\;\; K_f - \frac{1}{2}mv_i^2 = 40.0\text{ J}$

$\quad\;\; K_f = 40.0\text{ J} + 22.5\text{ J} = \boxed{62.5\text{ J}}$

11. A bead slides without friction around a loop-the-loop (Fig. 8.17). If the bead is released from a height h = 3.5R, what is its speed at point A? How large is the normal force on it if its mass is 5.0 g?

Solution

It is convenient to choose the reference point of potential energy to be at the lowest point of the bead's motion. Since $v_i = 0$ at the start,

$$E_i = K_i + U_i = 0 + mgh = mg(3.5R)$$

The total energy of the bead at point A can be written as

$$E_A = K_A + U_A = \frac{1}{2}mv_A^2 + mg(2R)$$

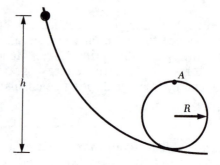

Figure 8.17

Since mechanical energy is conserved, $E_i = E_A$, and we get

$$\frac{1}{2}mv_A^2 + mg(2R) = mg(3.5R)$$

$$v_A^2 = 3gR \quad \text{or} \quad v_A = \boxed{\sqrt{3gR}}$$

To find the normal force at the top, it is useful to construct a free-body diagram as shown, where both **N** and **mg** are downward. Newton's second law gives

$$N + mg = \frac{mv_A^2}{R} = \frac{m(3gR)}{R} = 3mg$$

$$N = 3mg - mg = 2mg$$

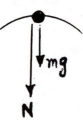

$$N = 2(5 \times 10^{-3} \text{ kg})(9.80 \text{ m/s}^2) = \boxed{0.098 \text{ N}} \text{ downward}$$

23. A 25-kg child on a swing 2 m long is released from rest when the swing supports make an angle of 30° with the vertical. (a) Neglecting friction, find the child's speed at the lowest position. (b) If the speed of the child at the lowest position is 2 m/s, what is the energy loss due to friction?

Solution

(a) First, note that the child falls through a vertical distance of

$$h = (2\ m) - (2\ m) \cos 30° = 0.268\ m$$

Taking $U = 0$ at the bottom, and using conservation of energy gives

$$K_i + U_i = K_f + U_f$$

$$0 + mgh = \frac{1}{2}mv^2 + 0$$

$$v = \sqrt{2gh} = \sqrt{2(9.80\ m/s^2)(0.268\ m)} = \boxed{2.29\ m/s}$$

(b) If $v_f = 2$ m/s, and friction is present, then

$$W_f = \Delta K + \Delta U = \left(\frac{1}{2}mv_f{}^2 - \frac{1}{2}mv_i{}^2\right) + 0 - mgh$$

$$W_f = \frac{1}{2}(25\ kg)\left[(2\ m/s)^2 - 0\right] + 0 - (25\ kg)(9.80\ m/s^2)(0.268\ m) = \boxed{-15.6\ J}$$

---

33. A block of mass 0.25 kg is placed on a vertical spring of constant k = 5000 N/m, and is pushed downward compressing the spring a distance of 0.1 m. As the block is released, it leaves the spring and continues to travel upward. To what maximum height above the point of release does the block rise?

Solution

Taking $U_g = 0$ to be at the point of release, and noting that $v_i = 0$, gives

$$E_i = K_i + U_i = 0 + (U_s + U_g)_i$$

$$E_i = \frac{1}{2}kx^2 + 0 = \frac{1}{2}(5000\ N/m)(0.1\ m)^2 = 25.0\ J$$

When the mass reaches its maximum height h, $v_f = 0$, and the spring is unstretched, so $U_s = 0$.

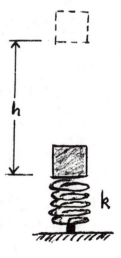

$$E_f = K_f + U_f = 0 + mgh = (0.25\ kg)(9.80\ m/s^2)h$$

Since mechanical energy is conserved, we have $E_f = E_i$, or

$$(0.25\ kg)(9.80\ m/s^2)h = 25.0\ J \quad \text{or} \quad h = \boxed{10.2\ m}$$

---

39. A potential energy function for a two-dimensional force is of the form $U = 3x^3y - 7x$. Find the force that acts at the point (x, y).

Solution

$$U = 3x^3y - 7x$$

$$F_x = -\frac{\partial U}{\partial x} = -\frac{\partial}{\partial x}(3x^3y - 7x) = -9x^2y + 7$$

$$F_y = -\frac{\partial U}{\partial y} = -\frac{\partial}{\partial y}(3x^3y - 7x) = -3x^3$$

Therefore,

$$\mathbf{F} = F_x\mathbf{i} + F_y\mathbf{j} = \boxed{(7 - 9x^2y)\mathbf{i} - 3x^3\mathbf{j}}$$

47. The particle described in Problem 46 (Fig. 8.32) is released from point A at rest. The speed of the particle at point B is 1.5 m/s. (a) What is its kinetic energy at B? (b) How much energy is lost as a result of friction as the particle goes from A to B? (c) Is it possible to determine μ from these results in any simple manner? Explain.

Solution

Let us take U = 0 at B. Since $v_i = 0$ at A, $K_A = 0$ and $U_A = mgR$.

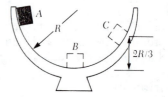

(a) Since $v_B = 1.5$ m/s,

$$K_B = \frac{1}{2}mv_B^2 = \frac{1}{2}(0.2 \text{ kg})(1.5 \text{ m/s})^2 = \boxed{0.225 \text{ J}}$$

Figure 8.32

(b) At A, $E_i = K_A + U_A = 0 + mgR = (0.2 \text{ kg})(9.80 \text{ m/s}^2)(0.3 \text{ m}) = 0.588$ J

At B, $E_f = K_B + U_B = 0.225$ J

Hence, the energy lost = $E_i - E_f = 0.588$ J $- 0.225$ J $= \boxed{0.363 \text{ J}}$

(c) Even though the energy lost is known, both the normal force and the friction force change with position as the block slides on the inside of the bowl. Therefore, there is no easy way to find μ.

57. A ball whirls around in a vertical circle at the end of a of a string. If the ball's total energy remains constant, show that the tension in the string at the bottom is greater than the tension at the top by six times the weight of the ball.

Solution

Applying Newton's second law at the bottom (b) and top (t) of the circular path gives

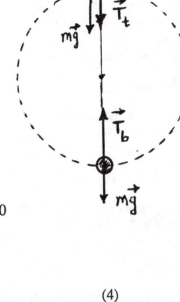

$$T_b - mg = \frac{mv_b^2}{R} \qquad (1)$$

$$T_t + mg = \frac{mv_t^2}{R} \qquad (2)$$

Subtracting (1) and (2) gives

$$T_b = T_t + 2mg + \frac{m(v_b^2 - v_t^2)}{R} \qquad (3)$$

Also, energy must be conserved; that is, $\Delta K + \Delta U = 0$. So,

$$\frac{1}{2} mv_b^2 - \frac{1}{2} mv_t^2 + (0 - 2mgR) = 0$$

or

$$m\frac{\left(v_b^2 - v_t^2\right)}{R} = 4mg \qquad (4)$$

Substituting (4) into (3) gives $\boxed{T_b = T_t + 6mg}$, which ends the proof.

59. A 20-kg block is connected to a 30-kg block by a light string that passes over a frictionless pulley. The 30-kg block is connected to a light spring of force constant 250 N/m, as in Figure 8.38. The spring is unstretched when the system is as shown in the figure, and the incline is smooth. The 20-kg block is pulled a distance of 20 cm down the incline (so that the 30-kg block is 40 cm above the floor) and is released from rest. Find the speed of each block when the 30-kg block is 20 cm above the floor (that is, when the spring is unstretched).

Solution

Let x be the distance the spring is stretched from equilibrium (x = 0.2 m), which corresponds to the upward displacement of the 30-kg mass. Also let $U_g = 0$ be measured from the lowest position of the 20-kg mass when the system is released from rest. If v is the speed of each block as they pass through the original unstretched position, then $\Delta K + \Delta U_s + \Delta U_g = 0$ gives

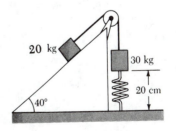

Figure 8.38

$$\left[\frac{1}{2}(m_1 + m_2)v^2 - 0\right] + \left(0 - \frac{1}{2}kx^2\right) + (m_2gx\sin\theta - m_1gx) = 0$$

$$\frac{1}{2}(50 \text{ kg})v^2 - \frac{1}{2}(250 \text{ N/m})(0.2 \text{ m})^2 + (20 \text{ kg})(9.80 \text{ m/s}^2)(0.2 \text{ m})\sin 40° - (30 \text{ kg})(9.80 \text{ m/s}^2)(0.2 \text{ m}) = 0$$

Solving for v gives

$$v = \boxed{1.24 \text{ m/s}}$$

61. A crane is to lift 2000 kg of material to a height of 150 m in 1 min at a uniform rate. What electric power is required to drive the crane motor if 35 percent of the electric power is converted to mechanical power? (Is it reasonable to neglect the kinetic energy in this process?)

Solution

The crane must move the material at a uniform velocity (upward) of

$$v = \frac{d}{t} = \frac{150 \text{ m}}{60 \text{ s}} = \frac{15}{6}\text{ m/s}$$

The mechanical power that must be delivered to lift this material at this velocity is

$$P_m = Fv = (mg)v = (2000 \text{ kg})(9.80 \text{ m/s}^2)\left(\frac{15}{6}\text{ m/s}\right) = 49 \text{ kW}$$

Since 35% of the electric power is converted to mechanical power, we have

$$P_m = 0.35P_e$$

or

$$P_e = \frac{P_m}{0.35} = \frac{49 \text{ kW}}{0.35} = \boxed{140 \text{ kW}}$$

Linear Momentum and Collisions

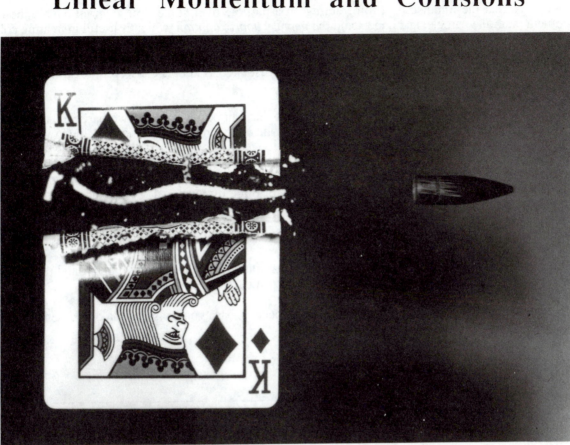

OBJECTIVES

1. Understand the concept of linear momentum of a particle and the relation between the resultant force on a particle and the time rate of change of its momentum (Newton's second law).

2. Recognize that the impulse of a force acting on a particle over some time interval equals the *change* in momentum of the particle, and understand the *impulse approximation* which is useful in treating collisions.

3. Derive the law of conservation of linear momentum for a two-particle system from Newton's second and third laws, and recognize that the linear momentum of any isolated system is conserved, regardless of the nature of the force between the particles.

4. Describe and distinguish the two types of collisions that can occur between two particles, namely elastic and inelastic collisions. Recognize that a *perfectly* inelastic collision is an inelastic collision in which the colliding particles stick together after the collision, and hence move as a composite particle.

5. Understand the fact that conservation of linear momentum applies not only to head-on collisions (one-dimensional), but also to glancing collisions (two- or three-dimensional). For example, in two-dimensional collisions, the total momentum in the x and y directions is conserved.

6. Understand and describe the concept of center of mass as applied to a collection of particles or a rigid body.

7. Describe the motion of the center of mass of a collection of particles or a rigid body, and recognize that total momentum of the center of mass remains constant in time for an *isolated* system.

8. Understand that the principle of the operation of a rocket is based on conservation of momentum of the system (rocket plus its ejected fuel).

SKILLS

The following procedure is recommended when dealing with problems involving collisions between two objects.

1. Set up a coordinate system and define your velocities with respect to that system. It is convenient to have the x axis coincide with one of the initial velocities.

2. In your sketch of the coordinate system, draw all velocity vectors with labels and include all the given information.

3. Write expressions for the x and y components of the momentum of each object before and after the collision. Remember to include the appropriate signs for the components of the velocity vectors. For example, if an object is moving in the negative x direction, its x component of velocity must be taken to be negative. It is essential that you pay careful attention to signs.

4. Now write expressions for the total momentum in the x direction before and after the collision and equate the two. Repeat this procedure for the total momentum in the y direction. These steps follow from the fact that because the momentum of the *system* is conserved in any collision, the total momentum along any direction must be conserved. It is important to emphasize that it is the momentum of the *system* (the two colliding objects) that is conserved, not the momentum of the individual objects.

5. If the collision is inelastic, kinetic energy is *not* conserved, and you should then proceed to solve the momentum equations for the unknown quantities.

6. If the collision is elastic, kinetic energy is also conserved, so you can equate the total kinetic energy before the collision to the total kinetic energy after the collision. This gives an additional relationship between the various velocities.

NOTES FROM SELECTED CHAPTER SECTIONS_____

9.1 LINEAR MOMENTUM AND IMPULSE

The *time rate of change of the momentum* of a particle is equal to the resultant force on the particle. The *impulse* of a force equals the change in momentum of the particle on which the force acts. Under the *impulse approximation,* it is assumed that one of the forces acting on a particle is of short time duration but of much greater magnitude than any of the other forces.

9.2 CONSERVATION OF LINEAR MOMENTUM FOR A TWO-PARTICLE SYSTEM

If two particles of masses m_1 and m_2 form an *isolated system*, then the total momentum of the system remains constant.

9.3 COLLISIONS

For *any type of collision,* the total momentum before the collision equals the total momentum just after the collision.

In an *inelastic collision*, the total momentum is conserved; however, the total kinetic energy is not conserved.

In a *perfectly inelastic* collision, the two colliding objects stick together following the collision.

In an *elastic collision,* both momentum and kinetic energy are conserved.

9.5 TWO-DIMENSIONAL COLLISIONS

The law of conservation of momentum is not restricted to one-dimensional collisions. If two masses undergo a *two-dimensional* (glancing) *collision* and there are no external forces acting, the total momentum in each of the x, y, and z directions is conserved.

9.7 MOTION OF A SYSTEM OF PARTICLES

The center of mass of a system of particles moves like an imaginary particle of mass M (equal to the total mass of the system) under the influence of the resultant external force on the system.

EQUATIONS AND CONCEPTS

The *linear momentum* **p** of a particle is defined as the product of its mass m with its velocity **v**.

$$\boxed{\mathbf{p} = m\mathbf{v}}$$

(9.1)

The *impulse* of a force **F** acting on a particle *equals the change in momentum of the particle*.

$$\boxed{\mathbf{I} = \Delta\mathbf{p} = \int_{t_i}^{t_f} \mathbf{F}\,dt}$$

(9.6)

To calculate impulse, we usually define a *time-averaged force* **F** which would give the same impulse to the particle as the actual time-varying force over the time interval Δt.

$$\boxed{\overline{\mathbf{F}} = \frac{1}{\Delta t}\int_{t_i}^{t_f} \mathbf{F}\,dt}$$

(9.7)

In the impulse approximation, the force **F** appearing in Eq. 9.6 acts for a short time and is much larger than any other force present. This approximation is usually assumed in problems involving collisions, where the force is the contact force between the particles during the collision.

When two particles interact with each other (but are otherwise isolated from their surroundings), *Newton's third law tells us that the force of particle 1 on particle 2 is equal and opposite to the force of particle 2 on particle 1*. Since force is the time rate of change of momentum (Newton's second law), one finds that the total momentum of the isolated pair of particles is conserved.

$$\boxed{\mathbf{p}_1 + \mathbf{p}_2 = \text{constant}}$$

(9.10)

In general, when the *external force* acting on a system of particles is zero, the *total linear momentum of the system is conserved*. This important statement is known as the *law of conservation of momentum*. It is especially useful in treating problems involving collisions between two bodies.

$$\boxed{\mathbf{p}_{1i} + \mathbf{p}_{2i} = \mathbf{p}_{1f} + \mathbf{p}_{2f}}$$

(9.12)

There are two kinds of collisions that can occur between two bodies. *An elastic collision is one in which both linear momentum and kinetic energy are conserved.* An *inelastic collision is one in which only linear momentum is conserved.* A perfectly inelastic collision is an inelastic collision in which the two bodies stick together after the collision. Note that momentum is conserved in any type of collision. Furthermore, note that when we say that the momentum is conserved, we are speaking about the momentum of the *entire system*. That is, the momentum of each particle may change as the result of the collision, but the momentum of the system remains unchanged.

$$\mathbf{p}_1 + \mathbf{p}_2 = \text{constant}$$
$$K_1 + K_2 = \text{constant}$$
(Elastic)

$$\mathbf{p}_1 + \mathbf{p}_2 = \text{constant}$$ (Inelastic)

The *x coordinate of the center of mass of n particles* whose individual coordinates are x_1, x_2, x_3, . . . and whose masses are m_1, m_2, m_3, . . . is given by Eq. 9.27. The y and z coordinates of the center of mass are defined by similar expressions.

$$x_c \equiv \frac{\Sigma m_i x_i}{\Sigma m_i}$$
(9.27)

The *x coordinate of the center of mass of a rigid body* of mass M can be calculated by dividing the body into elements of mass dm: the y and z coordinates of the center of mass are given by similar expressions.

$$x_c = \frac{1}{m} \int x \, dm$$
(9.31)

The *velocity of the center of mass of a system of particles* is given by Eq. 9.34, where v_i is the velocity of the ith particle and M is the total mass of the system.

$$v_c = \frac{\Sigma m_i v_i}{M}$$
(9.34)

The *total momentum of a system of particles* is equal to the product of the total mass M with the velocity of the center of mass.

$$\mathbf{P} = M v_c$$
(9.35)

The *acceleration of the center of mass of a system of particles* is given by Eq. 9.36.

$$\mathbf{a}_c = \frac{\Sigma m_i \mathbf{a}_i}{M}$$
(9.36)

Newton's second law applied to a system of particles says that the *resultant external force* acting on the system *equals the time rate of change of the total momentum.*

$$\Sigma \mathbf{F}_{ext} = \frac{d\mathbf{P}}{dt} = M \mathbf{a}_c$$
(9.38)

The principle behind the operation of a rocket is the law of conservation of momentum as applied to the rocket and its ejected fuel. If a rocket moves in the absence of gravity and ejects fuel with an exhaust velocity v_e, *its change in velocity* is given by Eq. 9.41, where M_i and M_f refer to its initial and final masses.

$$\Delta v = v_f - v_i = v_e \, ln\left(\frac{M_i}{M_f}\right)$$
(9.41)

EXAMPLE PROBLEM SOLUTIONS

Example 9.1 Locate the center of mass of the four particles located at the corners of a rectangle as shown in Fig. 9.1.

Solution

Since the 1 kg mass is at the origin, its coordinates are (0,0). The x-coordinate of the center of mass is given by Eq. 9.27:

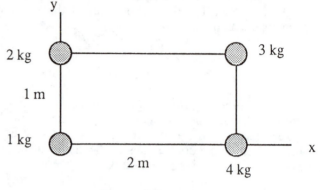

Figure 9.1

$$x_c = \frac{\Sigma m_i x_i}{\Sigma m_i} = \frac{1(0) + 2(0) + 3(2) + 4(2)}{1 + 2 + 3 + 4} = \frac{14}{10} = 1.4 \text{ m}$$

Likewise, the y-coordinate of center of mass is

$$y_c = \frac{\Sigma m_i y_i}{\Sigma m_i} = \frac{1(0) + 2(1) + 3(1) + 4(0)}{1 + 2 + 3 + 4} = \frac{5}{10} = 0.5 \text{ m}$$

Example 9.2 An object, mass M, is in the shape of a right triangle whose dimensions are shown in Fig. 9.2. Locate the coordinates of the center of mass, assuming the object has a uniform mass per unit area.

Solution

To evaluate the x-coordinate of the center of mass, we divide the triangle into narrow strips of width dx and height y as in Fig. 9.2. The mass of this strip dm can be expressed as

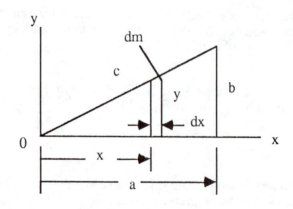

Figure 9.2

$$dm = \frac{\text{Total Mass}}{\text{Total Area}} \times \text{Area of Strip}$$

or

$$dm = \frac{M}{\frac{1}{2}ab} (ydx) = \left(\frac{2M}{ab}\right) ydx$$

Therefore, the x-coordinate of the center of mass is

$$x_c = \frac{1}{M} \int x\, dm = \frac{1}{M} \int_o^a x \left(\frac{2M}{ab}\right) y\, dx = \frac{2}{ab} \int_o^a xy\, dx$$

In order to evaluate this integral, we must express the variable y in terms of the variable x. From similar triangles in Fig. 9.2, we see that

$$\frac{y}{x} = \frac{b}{a} \quad \text{or} \quad y = \frac{b}{a} x$$

With this substitution, x_c becomes

$$x_c = \frac{2}{ab} \int_o^a x\left(\frac{b}{a} x\right) dx = \frac{2}{a^2} \int_o^a x^2 dx = \frac{2}{a^2}\left[\frac{x^3}{3}\right]_o^a = \frac{2}{3}a$$

By a similar calculation, you can easily show that the y-coordinate of the center of mass is given by $y_c = \frac{1}{3} b$. Thus, the results are

$$x_c = \frac{2}{3}a \quad \text{and} \quad y_c = \frac{1}{3}b$$

Example 9.3 A block of mass m slides down a smooth, curved track and collides head-on with an identical block at the bottom of the track as in Fig. 9.3.

(a) If the collision is assumed to be *perfectly elastic*, find the speed of each block after the collision, and the speed of block B when it reaches point C.

Solution

The speed of block A just before the collision can be obtained using conservation of energy. The potential energy of the block at point A (relative to the bottom of the track) is transferred into kinetic energy at the bottom. That is,

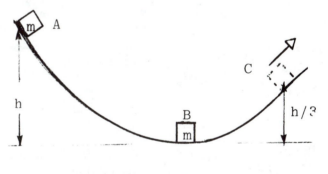

Figure 9.3

$$\tfrac{1}{2} mv^2 = mgh \quad \text{or} \quad v = \sqrt{2gh}$$

After the elastic collision, the two blocks *exchange* velocities. This can be seen from Eq. 9.16 of the text, with $m_1 = m_2$, $v_{1i} = v$, and $v_{2i} = 0$. Hence, the velocity of block B after the collision is $\sqrt{2gh}$, while block A comes to rest.

To find the speed of block B at point C, we again use conservation of energy, noting that the kinetic energy of block B at the bottom must equal the sum of its kinetic and potential energies at C. That is,

$$\frac{1}{2} mv^2 = \frac{1}{2} mv_c{}^2 + mg\left(\frac{h}{3}\right)$$

$$\frac{1}{2} m(2gh) = \frac{1}{2} mv_c{}^2 + mg\left(\frac{h}{3}\right)$$

$$v_c{}^2 = 2gh - \frac{2}{3} gh = \frac{4}{3} gh$$

or

$$v_c = 2\sqrt{\frac{gh}{3}}$$

(b) If the collision is perfectly inelastic (the two blocks stick together), find the speed of the blocks right after the collision and the maximum distance they move above point B.

Solution

In this case, *energy is not conserved* as a result of the inelastic collision. However, momentum is conserved. Equating the momentum just after the collision to the momentum just before the collision gives

$$mv = (m + m)v'$$

$$v' = \frac{1}{2} v = \frac{1}{2} \sqrt{2gh}$$

To find the maximum height h' the blocks will reach, we can use conservation of energy *after* the collision. To do this, we equate the final potential energy to the initial kinetic energy at B. This gives

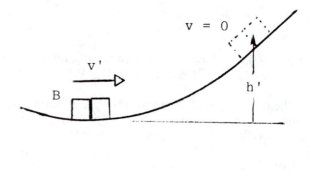

$$(m + m)gh' = \frac{1}{2} (m + m)v'^2$$

$$h' = \frac{1}{2g} v'^2 = \frac{1}{2g}\left(\frac{1}{2} \sqrt{2gh}\right)^2 = \frac{1}{4} h$$

Example 9.4   A block of mass M moving with a speed v_0, on a frictionless horizontal surface, collides with a spring attached to a block of mass 2M initially at rest, as in Figure 9.4. Find the maximum distance the spring will be compressed.

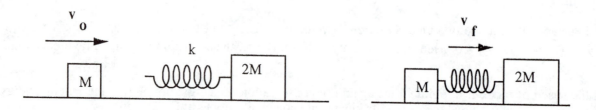

Figure 9.4

Solution

Since momentum is conserved, we can equate the initial momentum of the block of mass M to the final momentum of the composite system:

$$Mv_o = (M + 2M) v_f$$

$$v_f = \left(\frac{M}{M + 2M}\right) v_o = \frac{1}{3} v_o$$

Note that v_f is also the velocity of the center of mass of the system, where it is assumed that the spring has reached its maximum compression, so that both blocks have the same speed.

Since the surface is smooth, we can apply conservation of energy to find the distance x the spring is compressed.

$$KE_i \text{ (block M)} = KE_f \text{ (system)} + PE_f \text{ (spring)}$$

$$\frac{1}{2} Mv_o^2 = \frac{1}{2} (M + 2M) v_f^2 + \frac{1}{2} kx^2$$

$$Mv_o^2 = 3M \left(\frac{v_o}{3}\right)^2 + kx^2$$

$$x^2 = \frac{M}{k} v_o^2 - \frac{M}{k} \frac{v_o^2}{3} = \frac{2Mv_o^2}{3k}$$

$$x = \sqrt{\frac{2Mv_o^2}{3k}}$$

ANSWERS TO SELECTED QUESTIONS

1. If the kinetic energy of a particle is zero, what is its linear momentum? If the total energy of a particle is zero, is its linear momentum necessarily zero? Explain.

Answer: If $KE = mv^2/2 = 0$, then $v = 0$. Therefore, it follows that the linear momentum $= mv = 0$. Although the *total* energy of a particle may be zero, its linear momentum is not necessarily zero. A reference frame can be chosen such as the total energy $= KE + PE = 0$. That is, the particle has kinetic energy, and hence a velocity which satisfies the condition that $mv^2/2 + PE = 0$. (In this case, the PE must be negative.)

2. If the velocity of a particle is doubled, by what factor is its momentum changed? What happens to its kinetic energy?

Answer: Since $p = mv$, doubling v will double the momentum. On the other hand, since $KE = mv^2/2$, doubling v would quadruple the kinetic energy.

4. Does a large force always produce a larger impulse on a body than a smaller force? Explain.

Answer: No, not necessarily. The impulse of a force depends on the (average) force and the time over which the force acts. The statement is only true if the times over which the forces act are equal.

6. If two objects collide and one is initially at rest, is it possible for both to be at rest after the collision? Is it possible for one to be at rest after the collision? Explain.

Answer: Because momentum must be conserved, and the initial momentum is nonzero, the final momentum must be nonzero. Hence, they cannot both be at rest after the collision. However, it is possible for one to be at rest after the collision. This, in fact, occurs when two equal masses undergo a head-on, *elastic* collision (say, two billiard balls).

8. Is it possible to have a collision in which all of the kinetic energy is lost? If so, cite an example.

Answer: Yes. If two equal masses moving in opposite directions with equal speeds collide inelastically (they stick together), they are at rest after the collision. For example, two carts on a frictionless surface (air tract) can be made to stick together with sticky tape or putty. What do you suppose happens to the kinetic energy?

9. In a perfectly elastic collision between two particles, does the kinetic energy of each particle change as a result of the collision?

Answer: No, not necessarily. The kinetic energies after the collision depend on the masses of the particles and their initial velocities. If the particles have equal mass, they exchange velocities.

17. A sharpshooter fires a rifle while standing with the butt of the gun against his shoulder. If the forward momentum of a bullet is the same as the backward momentum of the gun, why isn't it as dangerous to be hit by the gun as by the bullet?

Answer: It is the product mv which is the same for both the bullet and the gun. The bullet has a large v and a small m, while the gun has a small v and a large m. Furthermore, the bullet carries much more kinetic energy than the gun.

18. A piece of mud is thrown against a brick wall and sticks to the wall. What happens to the momentum of the mud? Is momentum conserved? Explain.

Answer: Initially the mud had momentum toward the wall, but when it sticks to the wall nothing appears to have any momentum. However, this "lost" momentum is actually given to the wall and earth, causing both to move. Because of the enormous mass of the earth, its recoil speed is imperceptible.

19. Early in this century, Dr. Robert Goddard proposed sending a rocket to the moon. Critics took the position that in a vacuum, such as exists between the earth and the moon, the gases emitted by the rocket would have nothing to push against to propel the rocket. According to *Scientific American* (Jan., 1975), Goddard placed a gun in a vacuum and fired a blank cartridge from it. (A blank cartridge fires only the wadding and hot gases of the burning gunpowder.) What happened when the gun was fired?

Answer: The gun recoiled in the direction opposite the motion of the hot gases, because of momentum conservation.

5. The force F_x acting on a 2-kg particle varies in time as shown in Figure 9.27. Find (a) the impulse of the force, (b) the final velocity of the particle if it is initially at rest, and (c) the final velocity of the particle if it is initially moving along the x axis with a velocity of –2 m/s.

Solution

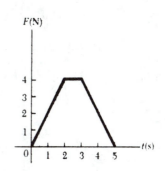

Figure 9.27

(a) $I = \int_{t_i}^{t_f} F\,dt = \int_0^5 F\,dt$ = area under graph

$$= \int_0^2 F\,dt + \int_2^3 F\,dt + \int_3^5 F\,dt$$

$$= \frac{1}{2}(4\text{ N})(2\text{ s}) + (4\text{ N})(1\text{ s}) + \frac{1}{2}(4\text{ N})(2\text{ s}) = \boxed{12\ \text{N·s}}$$

(b) $I = p_f - p_i = mv_f - mv_i$ (where $v_i = 0$)

$$v_f = v_i + \frac{I}{m} = 0 + \frac{12\text{ N·m}}{2\text{ kg}} = \boxed{6.00\text{ m/s}}$$

(c) If $v_i = -2$ m/s,

$$v_f = v_i + \frac{I}{m} = -2\text{ m/s} + 6\text{ m/s} = \boxed{4.00\text{ m/s}}$$

11. A 0.5-kg football is thrown with a speed of 15 m/s. A stationary receiver catches the ball and brings it to rest in 0.02 s. (a) What is the impulse delivered to the ball? (b) What is the average force exerted on the receiver?

Solution

$$v_i = 15\text{ m/s} \qquad v_f = 0 \qquad \Delta t = 0.02\text{ s} \qquad m = 0.5\text{ kg}$$

(a) $I = \Delta p = mv_f - mv_i = (0.5\text{ kg})(0) - (0.5\text{ kg})(15\text{ m/s}) = \boxed{-7.50\ \text{kg·m/s}}$

(b) $F_{av} = \frac{|I|}{\Delta t} = \frac{7.50\text{ kg·m/s}}{0.02\text{ s}} = \boxed{375\text{ N}}$

19. A 60-kg boy and a 40-kg girl, both wearing skates, face each other at rest. The boy pushes the girl, sending her eastward with a speed of 4 m/s. Describe the subsequent motion of the boy. (Neglect friction.)

Solution

$$m_B = 60 \text{ kg} \qquad m_G = 40 \text{ kg}$$

$$v_{Gf} = 4 \text{ m/s} \qquad v = B_{Bf} = ?$$

Since they are both at rest initially, $v_{Bi} = v_{Gi} = 0$. Momentum is conserved, and we have

$$m_B v_{Bf} = m_G v_{Gf} = m_B v_{Bi} + m_G v_{Gi} = 0$$

$$v_{Bf} = -\frac{m_G}{m_B} v_{Gf} = -\frac{(40 \text{ kg})}{(60 \text{ kg})} (4 \text{ m/s}) = \boxed{-2.67 \text{ m/s}}$$

That is, the boy moves westward with a speed of 2.67 m/s.

---

35. A 12-g bullet is fired into a 100-g wooden block initially at rest on a horizontal surface. After impact, the block slides 7.5 m before coming to rest. If the coefficient of friction between the block and the surface is 0.65, what was the speed of the bullet immediately before impact?

Solution

Since the collision is *totally inelastic*, and momentum is conserved,

$$m_1 v_0 = (m_1 + m_2)v_i$$

$$v_i = \left(\frac{12 \text{ g}}{12 \text{ g} + 100 \text{ g}}\right) v_0 = (0.10714)v_0 \quad (1)$$

The initial kinetic energy of the block and bullet just after the collision is *lost* due to the non-conservative friction force. Since

$$W_f = \Delta K = -\frac{1}{2}(m_1 + m_2)v_i^2, \quad \text{and}$$

$$W_f = -fs = -\mu mgs, \quad \text{we get}$$

$$\frac{1}{2}(m_1 + m_2)v_i^2 = \mu(m_1 + m_2)gs \quad (2)$$

Using (1) gives

$$v_i^2 = \mu gs$$

$$(0.10714)^2 v_0^2 = 2(0.65)(9.80 \text{ m/s}^2)(7.5 \text{ m})$$

$$v_0 = \boxed{91.2 \text{ m/s}}$$

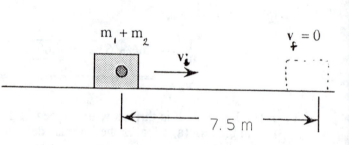

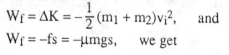

Problem 35

37. Consider a frictionless track ABC as shown in Figure 9.35. A block of mass $m_1 = 5$ kg is released from A. It makes a head-on elastic collision with a block of mass $m_2 = 10$ kg at B, initially at rest. Calculate the maximum height to which m_1 will rise after the collision.

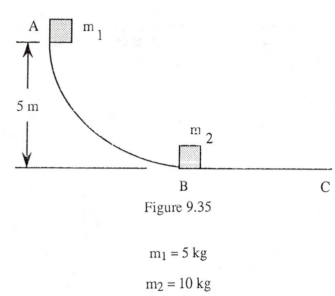

Figure 9.35

$m_1 = 5$ kg

$m_2 = 10$ kg

Solution

First, let us find the velocity of m_1 at B just *before* the collision. From conservation of energy, and the fact that $v_A = 0$, we get

$$K_A + U_A = K_B + U_B$$

$$0 + mgh = \frac{1}{2} mv_B^2 + 0$$

$$v_B = \sqrt{2gh} = \sqrt{2(9.80 \text{ m/s}^2)(5 \text{ m})} = 9.90 \text{ m/s}$$

Now use Equations 9.22 and 9.23 to get the velocities of m_1 and m_2 just *after* the collision:

$$v_{1f} = \left(\frac{m_1 - m_2}{m_1 + m_2}\right) v_{1i} = \left(\frac{5 - 10}{5 + 10}\right) v_{1i} = -\frac{1}{3} (9.90 \text{ m/s}) = -3.30 \text{ m/s}$$

$$v_{2f} = \left(\frac{2m_1}{m_1 + m_2}\right) v_{1i} = \left(\frac{2(5)}{5 + 10}\right) v_{1i} = \frac{2}{3} (9.90 \text{ m/s}) = 6.60 \text{ m/s}$$

Thus, the 5-kg mass (m_1) moves to the *left* after the collision, while the 10-kg mass (m_2) moves to the *right*. To find the maximum height, we again apply conservation of energy to m_1 and find

$$m_1 gh' = \frac{1}{2} m_1 v_{1f}^2$$

$$h' = \frac{v_{1f}^2}{2g} = \frac{(-3.30 \text{ m/s})^2}{2(9.80 \text{ m/s}^2)} = \boxed{0.556 \text{ m}}$$

55. A shell is fired from a cannon at level ground to hit an enemy bunker situated R away from the cannon. At the highest point of the trajectory, the shell explodes into two equal parts. One part of the shell is found to hit the ground at R/2. At what distance from the cannon would the second fragment fall?

Solution

The center of mass of the fragments must follow the parabolic trajectory that the shell would follow if it had not exploded. Since the shell explodes into two equal parts, each of equal mass, and one part lands at R/2, the other part *must* land at 3R/2. This is because the center of mass of these two parts is at x = R.

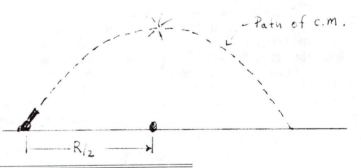

117

59. A 2-kg particle has a velocity of $(2\mathbf{i} - 3\mathbf{j})$ m/s, and a 3-kg particle has a velocity of $(\mathbf{i} + 6\mathbf{j})$ m/s. Find (a) the velocity of the center of mass and (b) the total momentum of the system.

Solution

(a) $\mathbf{v}_c = \dfrac{m_1\mathbf{v}_1 + m_2\mathbf{v}_2}{m_1 + m_2} = \dfrac{(2 \text{ kg})[(2\mathbf{i} - 3\mathbf{j}) \text{ m/s}] + (3 \text{ kg})[(\mathbf{i} + 6\mathbf{j}) \text{ m/s}]}{(2 \text{ kg} + 3 \text{ kg})} = \boxed{(1.40\mathbf{i} + 2.40\mathbf{j}) \text{ m/s}}$

(b) $\mathbf{p}_c = (m_1 + m_2)\mathbf{v}_c = (2 \text{ kg} + 3 \text{ kg})((1.40\mathbf{i} + 2.40\mathbf{j}) \text{ m/s} = \boxed{(7.00\mathbf{i} + 12.0\mathbf{j}) \text{ kg·m/s}}$

---

67. Fuel aboard a rocket has a density of 1.4×10^3 kg/m^3 and is ejected with a speed of 3.0×10^3 m/s. If the engine is to provide a thrust of 2.5×10^6 N, what volume of fuel must be burned per second?

Solution

$$\text{Thrust} = v_e \left|\frac{dM}{dt}\right| = 2.5 \times 10^6 \text{ N}$$

$$\left|\frac{dM}{dt}\right| = \frac{2.5 \times 10^6 \text{ N}}{3.0 \times 10^3 \text{ m/s}} = 833 \text{ kg/s}$$

Since $\rho = 1.4 \times 10^3$ kg/m^3, and $\rho = M/V$, then it follows that

$$\frac{dV}{dt} = \frac{1}{\rho}\left|\frac{dM}{dt}\right| = \frac{833 \text{ kg/s}}{1.4 \times 10^3 \text{ kg/m}^3} = \boxed{0.595 \text{ m}^3/\text{s}}$$

---

79. An 80-kg astronaut is working on the engines of his ship, which is drifting through space with a constant velocity. The astronaut, wishing to get a better view of the universe, pushes against the ship and later finds himself 30 m behind the ship. Without a thruster, the only way to return to the ship is to throw his 0.5-kg wrench directly away from the ship. If he throws the wrench with a speed of 20 m/s, how long does it take the astronaut to reach the ship?

Solution

The total momentum of the astronaut and wrench is *zero* relative to a frame moving with the astronaut. Since momentum is conserved, then after the wrench is thrown to the left, the astronaut acquires a velocity v to the *right*:

$$m_1v_1 + m_2v_2 = 0$$

$$(80 \text{ kg})v + (0.5 \text{ kg})(-20 \text{ m/s}) = 0$$

$$v = 0.125 \text{ m/s}$$

The time it takes the astronaut to travel 30 m at this velocity is

$$t = \frac{d}{v} = \frac{30 \text{ m}}{0.125 \text{ m/s}} = \boxed{240 \text{ s}}$$

81. A chain of length L and total mass M is released from rest with its lower end just touching the top of a table, as in Figure 9.41a. Find the force of the table on the chain after the chain has fallen through a distance x, as in Figure 9.41b. (Assume each link comes to rest the instant it reaches the table.)

Solution

Since the mass per unit length is uniform, we can express an element of mass dm having a length dx as

$$dm = \left(\frac{M}{L}\right) dx$$

The magnitude of the force due to the falling chain is given by

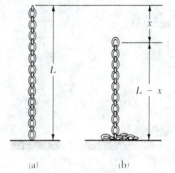

Figure 9.41

$$F_1 = \frac{dp}{dt} = v \frac{dm}{dt} = v\left(\frac{M}{L}\right)\frac{dx}{dt} = \left(\frac{M}{L}\right) v^2$$

when we have used v = dx/dt. After falling a distance x, the square of the velocity of each link $v^2 = 2gx$ (from kinematics), hence

$$F_1 = \frac{2 M g x}{L}$$

Now the links of length x already on the table exert a force F_2 equal their weight, given by

$$F_2 = \frac{M g x}{L}$$

Hence, the *total* force on the table is

$$F_{total} = F_1 + F_2 = \frac{2 M g x}{L} + \frac{M g x}{L} = \boxed{\frac{3 M g x}{L}}$$

That is, *the total force is three times the weight of the chain on the table at that instant.*

10

Rotation of a Rigid Body
About a Fixed Axis

OBJECTIVES

1. Define the angular velocity and angular acceleration of a particle or body rotating about a fixed axis.

2. Recognize that if a body rotates about a fixed axis, every particle on the body has the same angular velocity and angular acceleration. For this reason, rotational motion can be simply described using these quantities.

3. Note the similarity between the equations of rotational kinematics (constant α) and those of linear kinematics (constant a).

4. Describe and understand the relationships between linear speed and angular speed ($v = r\omega$), and between linear acceleration and angular acceleration ($a_t = r\alpha$).

5. Calculate the moment of inertia I of a system of particles or a rigid body about a specific axis. Note that the value of I depends on (a) the mass distribution and (b) the axis about which the rotation occurs. The parallel-axis theorem is useful for calculating I about an axis parallel to one that goes through the center of mass.

6. Describe the rotational kinetic energy of a body rotating about a fixed axis, ($K = \frac{1}{2}I\omega^2$), and recognize that this represents the sum of the kinetic energies of the various segments of the body as they move about the axis of rotation.

7. Understand the concept of torque associated with a force, noting that the torque associated with a force has a magnitude equal to the force times the moment arm. Furthermore, note that the value of the torque depends on the origin about which it is evaluated.

8. Show that the net torque on a rigid body about some axis is proportional to the angular acceleration; that is, $\tau = I\alpha$, where I is the moment of inertia about the axis about which the net torque is evaluated.

9. Recognize the fact that the work-energy theorem can be applied to a rotating rigid body. That is, the net work done on a rigid body rotating about a fixed axis equals the change in its rotational kinetic energy.

SKILLS

You should know how to calculate the moment of inertia of a system of particles about a specified axis. The technique is straightforward, and consists of applying $I = \Sigma m_i r_i^2$, where m_i is the mass of the ith particle and r_i is the distance from the axis of rotation to the particle.

The calculation of the moment of inertia for rigid bodies requires some knowledge of integral calculus. Your instructor may skip over this material if you are not familiar with the operation of integration. When using the basic definition given by Eq. 10.16, namely, $I = \int r^2 dm$, you must find an expression relating the element of mass, dm, to the coordinates of the body and its density.

Once the moment of inertia about an axis through the center of mass I_c is known, you can easily evaluate the moment of inertia about any axis parallel to the axis through the center of mass using the *parallel axis theorem*:

$$I = I_c + Md^2 \qquad (10.17)$$

where d is the distance between the two axes. For example, the moment of inertia of a solid cylinder about an axis through its center (the z axis in Fig. 10.1) is given by $I_z = \frac{1}{2}MR^2$. Hence, the moment of inertia about the z' axis located a distance d = R from the z axis is

$$I_{z'} = I_z + MR^2 = \frac{1}{2}MR^2 + MR^2 = \frac{3}{2}MR^2$$

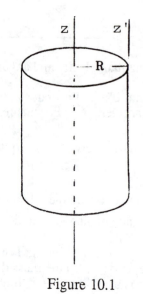

Figure 10.1

NOTES FROM SELECTED CHAPTER SECTIONS

10.1 ANGULAR VELOCITY AND ANGULAR ACCELERATION

Pure rotational motion refers to the motion of a rigid body about a fixed axis.

One *radian* (rad) is the angle subtended by an arc length equal to the radius of the arc.

In the case of *rotation about a fixed axis*, every particle on the rigid body has the same angular velocity and the same angular acceleration.

10.2 ROTATIONAL KINEMATICS: ROTATIONAL MOTION WITH CONSTANT ANGULAR ACCELERATION

The *kinematic expressions* for rotational motion under constant angular acceleration are of the *same form* as those for linear motion under constant linear acceleration with the substitutions $x \to \theta$, $v \to \omega$, and $a \to \alpha$.

10.6 TORQUE

Torque is the physical quantity which is a measure of the tendency of a force to cause rotation of a body about a specified axis. It is important to remember that torque must be defined with respect to a *specific axis* of rotation. Torque which has the *SI units* of N·m must not be confused with force.

10.8 WORK AND ENERGY IN ROTATIONAL MOTION

The *work-energy theorem in rotational motion* states that the net work done by external forces in rotating a symmetric rigid body about a fixed axis equals the change in the body's rotational kinetic energy.

EQUATIONS AND CONCEPTS

The *average angular velocity* $\overline{\omega}$ of a particle or body rotating about a fixed axis equals the ratio of the angular displacement $\Delta\theta$ to the time interval Δt, where θ is measured in radians.

$$\overline{\omega} \equiv \frac{\Delta\theta}{\Delta t}$$ (10.2)

The *instantaneous angular velocity* ω is defined as the limit of the average angular velocity as Δt approaches zero.

$$\omega = \frac{d\theta}{dt}$$ (10.3)

The *average angular acceleration* $\overline{\alpha}$ of a rotating body is defined as the ratio of the change in angular velocity to the time interval Δt.

$$\overline{\alpha} \equiv \frac{\Delta\omega}{\Delta t}$$ (10.4)

The *instantaneous angular acceleration* equals the limit of the average angular acceleration as Δt approaches zero.

$$\alpha = \frac{d\omega}{dt}$$ (10.5)

If a particle or body rotates about a fixed axis with *constant* angular acceleration, we can apply the *equations of rotational kinematics*.

$$\omega = \omega_o + \alpha t$$ (10.6)

$$\theta = \theta_o + \omega_o t + \frac{1}{2}\alpha t^2$$ (10.7)

$$\omega^2 = \omega_o^2 + 2\alpha(\theta - \theta_o)$$ (10.8)

If a rigid body rotates about a fixed axis, the linear velocity of any point on the body a distance r from the axis of rotation is related to the angular velocity through the relation $v = r\omega$. Similarly, the tangential acceleration of any point on the body is related to the angular acceleration through the relation $a_t = r\alpha$. Note that *every point on the body has the same ω and α, but not every point has the same v and a_t.*

$$v = r\omega$$ (10.9)

$$a_t = r\alpha$$ (10.10)

The *moment of inertia of a system of particles* is defined by Eq. 10.14, where m_i is the mass of the ith particle and r_i is its distance from a specified axis. Note that I has SI units of $kg \cdot m^2$.

$$I = \Sigma m_i r_i^2 \qquad (10.14)$$

The *moment of inertia of a rigid body* is given by an integral expression, where the body is divided into elements of mass dm and r is the distance from that element to the axis of rotation.

$$I = \int r^2 dm \qquad (10.16)$$

The *kinetic energy* of a rigid body rotating with an angular velocity ω about some axis can be expressed in the form of Eq. 10.15. Note that it does not represent a new form of energy. It is simply a convenient form for representing rotational kinetic energy.

$$K = \frac{1}{2} I \omega^2 \qquad (10.15)$$

The *torque* τ due to an applied force has a magnitude given by the product of the force and its moment arm d, where d equals the *perpendicular distance* from the rotation axis to the line of action of **F**. Torque is a measure of the ability of a force to rotate a body about a specified axis. Note that the torque depends on the axis of rotation, which must be specified when τ is evaluated.

$$\tau = Fd \qquad (10.18)$$

The *net torque* acting on a rigid body about some axis is equal to the product of the moment of inertia and angular acceleration, where I is the moment of inertia about the axis of rotation. This is only true for a plane laminar body or for the case when the axis of rotation is a principal axis.

$$\tau_{net} = I\alpha \qquad (10.20)$$

If a net torque τ acts on a rigid body, the *power supplied to the body* at the instant its angular velocity is ω is given by Eq. 10.23.

$$P = \tau \omega \qquad (10.23)$$

The *work-energy theorem* says that the net work done by external forces in rotating a rigid body about a fixed axis equals the change in the body's rotational kinetic energy.

$$W = \frac{1}{2} I \omega^2 - \frac{1}{2} I \omega_o^2 \qquad (10.24)$$

EXAMPLE PROBLEM SOLUTIONS_____

Example 10.1 A uniform solid disc of radius R = 0.5 m and mass M = 4 kg is mounted on a frictionless axle as shown in Fig. 10.2. A second smaller disc of radius r = 0.1 m and mass m = 1 kg is fastened to the large disk and a rope is wrapped around the smaller disk. A constant tension T = 20 N is maintained on the rope.

(a) What is the angular acceleration of the system?

Solution

The torque on the system about the axle is Tr. Since $\tau = I\alpha$, we have

$$\tau = Tr = I\alpha$$

or

$$\alpha = \frac{Tr}{I} \qquad (1)$$

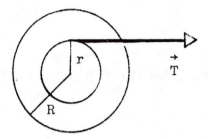

Figure 10.2

But the moment of inertia of the system is the sum of the moments of inertia for both disks. That is,

$$I = \frac{1}{2}\,MR^2 + \frac{1}{2}\,mr^2 = \frac{1}{2}\,(4\text{ kg})(0.5\text{ m})^2 + \frac{1}{2}\,(1\text{ kg})(0.1\text{ m})^2 = 0.50\text{ kg·m}^2 \qquad (2)$$

Substituting this value and the given values of T and r into Eq. (1) gives

$$\alpha = \frac{Tr}{I} = \frac{(20\text{ N})(0.10\text{ m})}{0.50\text{ kg·m}^2} = 4.0\text{ rad/s}^2$$

(b) If the system starts from rest, find the angular velocity after 5 s have elapsed.

Solution

Since the tension is constant, the torque and angular acceleration remain constant and we can use the equations of kinematics.

$$\omega = \omega_0 + \alpha t = 0 + \left(4\,\frac{\text{rad}}{\text{s}^2}\right)(5\text{ s}) = 20\text{ rad/s}$$

(c) What is the angular displacement of the wheel in the first 5 seconds?

Solution

$$\theta - \theta_0 = \omega_0 t + \frac{1}{2}\,\alpha t^2 = \frac{1}{2}\left(4\,\frac{\text{rad}}{\text{s}^2}\right)(5\text{ s})^2 = 50\text{ rad}$$

(d) Show that the work done by the force T in the first 5 seconds equals the change in rotational kinetic energy. That is, verify the work-energy theorem, Eq. 10.24.

Solution

The work done by the force **T** is Ts, where s is the total distance a point on the rim of the *smaller* disk moves in 5 seconds. Using the results to part (c), we find that

$$s = r(\theta - \theta_0) = (0.10 \text{ m})(50 \text{ rad}) = 5.0 \text{ m}$$

Therefore,

$$W = Ts = (20 \text{ N})(5.0 \text{ m}) = 100 \text{ J}$$

The initial kinetic energy is zero, and the final rotational kinetic energy is $\frac{1}{2} I\omega^2$, where $\omega = 20$ rad/s. Therefore,

$$\Delta K = \frac{1}{2} I\omega^2 = \frac{1}{2}(0.5 \text{ kg·m}^2)(20 \text{ rad/s})^2 = 100 \text{ J}$$

Thus, we see that the work-energy theorem is verified.

Example 10.2   Two masses m_1 and m_2 are connected to each other by a light cord which passes over two identical pulleys, each having a moment of inertia I. (See Fig. 10.3.) Find the acceleration of each mass and the tensions T_1, T_2, and T_3 in the cord. (Assume that no slipping occurs between the cord and pulleys.)

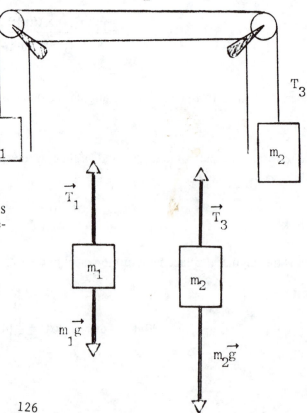

Figure 10.3

Solution

First, let us write Newton's second law of motion as applied to each block, taking $m_2 > m_1$. The free-body diagrams are shown at the right.

$$T_1 - m_1 g = m_1 a \qquad (1)$$

$$m_2 g - T_3 = m_2 a \qquad (2)$$

Next, we must include the effect of the pulleys on the motion. Free-body diagrams for the pulleys are shown below.

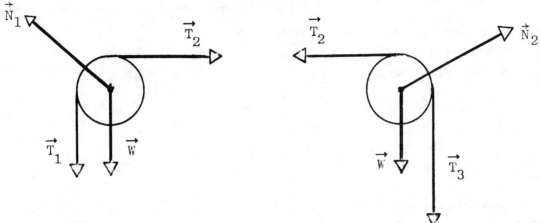

The net torque about the axle for the pulley on the left is $(T_2 - T_1)R$, while the net torque for the pulley on the right is $(T_3 - T_2)R$. Using the relation $\tau_{net} = I\alpha$ for each wheel, and noting that each wheel has the same α gives

$$(T_2 - T_1)R = I\alpha \qquad\qquad (3)$$

$$(T_3 - T_2)R = I\alpha \qquad\qquad (4)$$

We now have four equations with four unknowns; namely, a, T_1, T_2, and T_3. These can be solved simultaneously. Adding Eqs. (3) and (4) gives

$$(T_3 - T_1)R = 2I\alpha \qquad\qquad (5)$$

Adding Eqs. (1) and (2) gives

$$T_1 - T_3 + m_2g - m_1g = (m_1 + m_2)a$$

or
$$T_3 - T_1 = (m_2 - m_1)g - (m_1 + m_2)a \qquad\qquad (6)$$

Substituting Eq. (6) into Eq. (5), we have

$$[(m_2 - m_1)g - (m_1 + m_2)a]\, R = 2I\alpha$$

Since $\alpha = \dfrac{a}{R}$, this can be simplified as follows:

$$(m_2 - m_1)g - (m_1 + m_2)a = 2I\frac{a}{R^2}$$

or
$$\boxed{a = \frac{(m_2 - m_1)g}{m_1 + m_2 + 2\dfrac{I}{R^2}}} \qquad\qquad (7)$$

This value of a can then be substituted into Eqs. (1) and (2) to give T_1 and T_3. Finally, T_2 can be found from Eq. (3) or Eq. (4).

ANSWERS TO SELECTED QUESTIONS

2. A wheel rotates counterclockwise in the xy plane. What is the direction of ω? What is the direction of α if the angular velocity is decreasing in time?

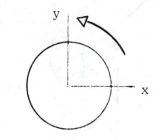

Answer: From the right-hand rule (Fig. 10.3 in the text), we see that ω is in the +z direction -- or *out* of the paper. Since ω is decreasing in time, α is *into* the paper (opposite ω).

3. Are the kinematic expressions for θ, ω, and α valid when the angular displacement is measured in degrees instead of in radians?

Answer: Yes. However, it is conventional to use radians.

4. A turntable rotates at a constant rate of 45 rev/min. What is the magnitude of its angular velocity in rad/s? What is its angular acceleration?

Answer: The frequency of rotation is 45 rotations/min $= \frac{45}{60}$ rotations/s. Since 1 rotation corresponds to an angular displacement of 2π rads, the angular frequency is $\omega = 2\pi f = 2\pi \left(\frac{45}{60}\right) = 4.71 \frac{rad}{s}$. Since ω is constant, the angular acceleration is zero.

5. When a wheel of radius R rotates about a *fixed axis*, do all points on the wheel have the same angular velocity? Do they all have the same linear velocity? If the angular velocity is constant and equal to ω_0, describe the linear velocities and linear accelerations of the points at $r = 0$, $r = \frac{R}{2}$, and $r = R$.

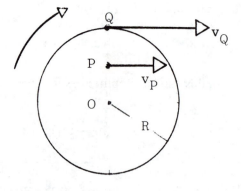

Answer: Yes. All points have the same angular velocity. This, in fact, is what makes angular quantities so useful in describing rotational motion. Not all points have the same linear velocities. The point at $r = 0$ has zero linear velocity and acceleration; the point P at $\frac{R}{2}$ has a linear velocity

$$v = \frac{R}{2}\omega_0$$

and a linear acceleration equal to the centripetal acceleration $v^2/R/2 = R\omega_0^2/2$. (The tangential acceleration is zero since ω_0 is constant.) The point Q on the rim has a linear velocity $v = R\omega_0$ and a linear acceleration equal to $R\omega_0^2$.

7. A wheel is in the shape of a hoop as in Fig. 10.8. In two separate experiments, the wheel is rotated from rest to an angular velocity ω. In one experiment, the rotation occurs about the z axis through 0; in the other, the rotation occurs about an axis parallel to z through P. Which rotation requires more work?

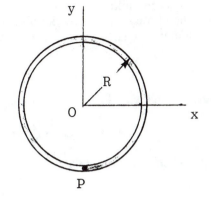

Answer: Rotation about the axis through P requires more work since the moment of inertia is twice as large about this axis. Specifically, $W = \Delta K = \frac{1}{2} I \omega^2$, and $I_p = 2I_z$; hence, it follows that it takes twice as much work to rotate the hoop about the axis through P. From the parallel axis theorem,

$$I_p = I_z + MR^2 = MR^2 + MR^2 = 2MR^2$$

9. Suppose that only two external forces act on a rigid body, and the two forces are equal in magnitude but opposite in direction. Under what conditions will the body rotate?

Answer: The body will rotate if the lines of action of the two forces do not coincide. If they do, the net torque about *any* axis would be zero, and the body would have no tendency to rotate -- assuming it has no rotational motion before the forces were applied.

12. If a small sphere of mass M were placed at the end of the rod in Fig. 10.21, would the result for ω be greater than, less than, or equal to the value obtained in Example 10.13?

Answer: The angular velocity would be less than that obtained in Example 10.13. This is due to the increase in the moment of inertia about the pivot when the mass is added. Following the procedure in Example 10.11, one finds that the angular velocity at the lowest position is

$$\omega = \frac{3}{2} \sqrt{g/L},$$

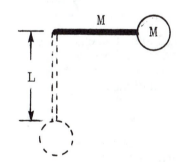

compared to the value of $1.732\sqrt{g/L}$ in the absence of the sphere.

SOLUTIONS TO SELECTED END-OF-CHAPTER PROBLEMS

7. A grinding wheel, initially at rest, is rotated with constant angular acceleration $\alpha = 5$ rad/s^2 for 8 s. The wheel is then brought to rest with uniform negative acceleration in 10 revolutions. Determine the negative acceleration required and the time needed to bring the wheel to rest.

Solution

We are given that $\omega_i = 0$, and $\alpha = 5$ rad/s^2 from t = 0 to t = 8 s. Hence, the angular velocity at t = 8 s is

$$\omega = \omega_i + \alpha t = 0 + (5 \text{ rad/s}^2)(8 \text{ s}) = 40 \text{ rad/s}$$

Since the wheel is then brought to rest after 10 revolutions, its angular displacement during this interval is

$$\theta = (10 \text{ rev})(2\pi \text{ rad/rev}) = 20\pi \text{ rad}.$$

In this second interval, $\omega = 0$, and $\omega_0 = 40$ rad/s, therefore

$$\omega^2 = \omega_0^2 + 2\alpha\theta = 0$$

$$\alpha = -\frac{\omega_0^2}{2\theta} = -\frac{(40 \text{ rad/s})^2}{2(20\pi) \text{ rad}} = \boxed{-12.7 \text{ rad/s}^2}$$

Also, since $\omega = \omega_0 + \alpha t$, taking $\omega = 0$ gives

$$t = -\frac{\omega_0}{\alpha} = -\frac{40 \text{ rad/s}}{(-12.7 \text{ rad/s}^2)} = \boxed{3.14 \text{ s}}$$

---

11. A wheel 2 m in diameter rotates with a constant angular acceleration of 4 rad/s^2. The wheel starts at rest at t = 0, and the radius vector at point P on the rim makes an angle of 57.3° with the horizontal at this time. At t = 2 s, find (a) the angular speed of the wheel, (b) the linear velocity and acceleration of the point P, and (c) the position of the point P.

Solution   Given r = 1 m, $\alpha = 4$ rad/s^2, $\omega_0 = 0$, and $\theta_0 = 57.3° = 1$ rad at t = 2 s

(a) $\omega = \omega_0 + \alpha t = 0 + \alpha t$

   At t = 2 s, $\omega = (4 \text{ rad/s}^2)(2 \text{ s}) = \boxed{8.00 \text{ rad/s}}$

(b) $v = r\omega = (1 \text{ m})(8 \text{ rad/s}) = \boxed{8.00 \text{ m/s}}$

   $a_r = r\omega^2 = (1 \text{ m})(8 \text{ rad/s})^2 = \boxed{64.0 \text{ m/s}^2}$ $a_t = r\alpha = (1 \text{ m})(4 \text{ rad/s}^2) = \boxed{4.00 \text{ m/s}^2}$

(c) $\theta = \theta_0 + \omega_0 t + \frac{1}{2}\alpha t^2 = (1 \text{ rad}) + \frac{1}{2}(4 \text{ rad/s}^2)(2 \text{ s})^2 = \boxed{9.00 \text{ rad}}$

13. A disk 8 cm in radius rotates at a constant rate of 1200 rev/min about its axis. Determine (a) the angular speed of the disk, (b) the linear speed at a point 3 cm from its center, (c) the radial acceleration of a point on the rim, and (d) the total distance a point on the rim moves in 2 s.

Solution

(a) $\omega = 2\pi f = (2\pi \text{ rad/rev})\left(\dfrac{1200}{60}\right) \text{rev/s} = 40\pi \text{ rad/s} = \boxed{126 \text{ rad/s}}$

(b) $v = \omega R = (40\pi \text{ rad/s})(0.03 \text{ m}) = \boxed{3.77 \text{ m/s}}$

(c) $a_r = \omega^2 r = (40\pi \text{ rad/s})^2(0.08 \text{ m}) = \boxed{1.26 \text{ km/s}^2}$

(d) $s = \theta R = \omega t R = (40\pi \text{ rad/s})(2 \text{ s})(0.08 \text{ m}) = \boxed{20.1 \text{ m}}$

---

17. The four particles in Figure 10.24 are connected by light, rigid rods. If the system rotates in the xy plane about the z axis with an angular velocity of 6 rad/s, calculate (a) the moment of inertia of the system about the z axis and (b) the kinetic energy of the system.

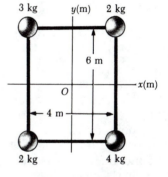

Solution

(a) All four particles are at the same distance r from the z axis:

$$r^2 = (3 \text{ m})^2 + (2 \text{ m})^2 = 13 \text{ m}^2$$

Figure 10.24

Therefore,

$$I_z = \Sigma m_i r_i^2 = (3 \text{ kg})(13 \text{ m}^2) + (2 \text{ kg})(13 \text{ m}^2) + (4 \text{ kg})(13 \text{ m}^2) + (2 \text{ kg})(13 \text{ m}^2)$$

$$= (11 \text{ kg})(13 \text{ m}^2) = \boxed{143 \text{ kg·m}^2}$$

(b) $K = \dfrac{1}{2} I_z \omega^2 = \dfrac{1}{2} (143 \text{ kg·m}^2)(6 \text{ rad/s})^2 = \boxed{2.57 \text{ kJ}}$

---

31. A wheel 1 m in diameter rotates on a fixed, frictionless, horizontal axle. Its moment of inertia about this axis is 5 kg·m². A constant tension of 20 N is maintained on a rope wrapped around the rim of the wheel, so as to cause the wheel to accelerate. If the wheel starts from rest at t = 0, find (a) the angular acceleration of the wheel, (b) the wheel's angular speed at t = 3 s, (c) the kinetic energy of the wheel at t = 3 s, and (d) the length of rope unwound in the first 3 s.

Solution

(a) $\tau = FR = (20 \text{ N})(0.5 \text{ m}) = 10 \text{ N·m}$

Also, $\alpha = \dfrac{\tau}{I} = \dfrac{10 \text{ N·m}}{5 \text{ kg·m}^2} = \boxed{2.00 \text{ rad/s}^2}$

(b) $\omega = \alpha t = (2 \text{ rad/s})(3 \text{ s}) = \boxed{6.00 \text{ rad/s}}$

(c) $K = \dfrac{1}{2} I\omega^2 = \dfrac{1}{2}(5 \text{ kg·m}^2)(6 \text{ rad/s})^2 = \boxed{90.0 \text{ J}}$

(d) $\theta = \dfrac{1}{2}\alpha t^2 = \dfrac{1}{2}(2 \text{ rad/s}^2)(3 \text{ s})^2 = 9 \text{ rad}$

$s = R\theta = (0.5 \text{ m})(9 \text{ rad}) = \boxed{4.50 \text{ m}}$

---

33. (a) A uniform solid disk of radius R and mass M is free to rotate on a frictionless pivot through a point on its rim (Fig. 10.32). If the disk is released from rest in the position shown by the solid line, what is the velocity of its center of mass when it reaches the position indicated by the broken line? (b) What is the speed of the lowest point on the disk in the dotted position? (c) Repeat part (a) if the object is a uniform hoop.

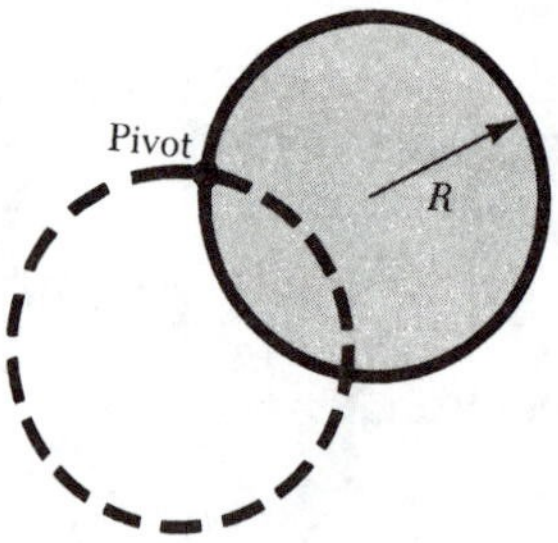

Solution

Figure 10.32

(a) Conservation of energy gives $Mgh = \dfrac{1}{2} I\omega^2$. In this case $I = \dfrac{3MR^2}{2}$ and h = R. Therefore

$$MgR = \frac{1}{2}\left(\frac{3MR^2}{2}\right)\omega^2 \quad \text{or} \quad \omega = \sqrt{\frac{4g}{3R}}$$

$$v_c = R\omega = R\sqrt{\frac{4g}{3R}} = \boxed{\sqrt{\frac{4gR}{3}}}$$

(b) $v_L = 2r\omega = \boxed{4\sqrt{\dfrac{gR}{3}}}$

(c) In this case $I = 2MR^2$, and conservation of energy gives

$$MgR = \frac{1}{2}(2MR^2)\omega^2 \quad \text{or} \quad \omega = \sqrt{\frac{g}{R}} \text{ rad/s}$$

$$v_c = r\omega = R\sqrt{\frac{g}{R}} = \boxed{\sqrt{gR}}$$

---

39. A uniform solid cylinder of mass M and radius R rotates on a horizontal, frictionless axle (Fig. 10.33). Two equal masses hang from light cords wrapped around the cylinder. If the system is released from rest, find (a) the tension in each cord, (b) the acceleration of each mass, and (c) the angular velocity of the cylinder after the masses have descended a distance h.

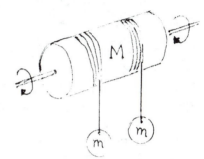

Solution

For the cylinder,

Figure 10.33

$$\Sigma\tau = (2T)R = \frac{1}{2}MR^2\frac{a}{R} \qquad (1)$$
$$T = \tfrac{1}{4}Ma$$

where a is the acceleration of the falling masses. For each of the falling masses, Newton's second law gives

$$\Sigma F = ma = mg - T \qquad (2)$$

(a) Combining Equations (1) and (2) we find

$$T = \boxed{\dfrac{Mmg}{M + 4m}} \qquad (3)$$

(b) Substituting (3) into (2) gives

$$a = g - \frac{T}{m} = \boxed{\dfrac{4mg}{M + 4m}} \qquad (4)$$

(c) After the masses have fallen a distance h, we have $v^2 = 2ah$. Also, $\omega R = v$; hence we find

$$\omega = \boxed{\left[\dfrac{8mgh}{R^2(M + 4m)}\right]^{1/2}}$$

---

41. A 4-m length of light nylon cord is wound around a uniform cylindrical spool of radius 0.5 m and 1-kg mass. The spool is mounted on a frictionless axle and is initially at rest. The cord is pulled from the spool with a constant acceleration of 2.5 m/s². (a) How much work has been done on the spool when it reaches an angular speed $\omega = 8$ rad/s? (b) Assuming there is enough cord on the spool, how long will it take the spool to reach an angular speed of 8 rad/s? (c) Is there enough cord on the spool to enable the spool to reach this angular speed of 8 rad/s?

Solution

(a) $W = \frac{1}{2} I\omega^2 - \frac{1}{2} I\omega_o^2;$ $\omega_o = 0;$ $\omega = 8$ rad/s

$I = \frac{1}{2} mR^2 = \frac{1}{2} (1 \text{ kg})(0.5 \text{ m})^2 = 0.125 \text{ kg·m}^2$

$W = \frac{1}{2} (0.125 \text{ kg·m}^2)(8 \text{ rad/s})^2 = \boxed{4.00 \text{ J}}$

(b) $\alpha = \frac{a_t}{r} = \frac{2.5 \text{ m/s}^2}{0.5 \text{ m}} = 5.00 \text{ rad/s}^2$

$\alpha = \frac{\omega - \omega_o}{t}$ and $\omega_o = 0,$ therefore

$t = \frac{\omega}{\alpha} = \frac{8 \text{ rad/s}}{5.00 \text{ rad/s}^2} = \boxed{1.60 \text{ s}}$

(c) $\theta = \theta_o + \omega_o t + \frac{1}{2} \alpha t^2,$ $\theta_o = 0,$ and $\omega_o = 0$ so that

$\theta = \frac{1}{2} \alpha t^2 = \frac{1}{2} (5.00 \text{ rad/s}^2)(1.60 \text{ s})^2 = 6.40 \text{ rad}$

$s = r\theta = (0.5 \text{ m})(6.40 \text{ rad}) = 3.20 \text{ m}$

When the spool reaches an angular velocity of 8 rad/s, 1.60 s will have elapsed and 3.20 m of cord will have been removed from the spool. YES!

43. A long uniform rod of length L and mass M is pivoted about a horizontal, frictionless pin through one end. The rod is released from rest in a vertical position as in Figure 10.34. At the instant the rod is horizontal, find (a) the angular velocity of the rod, (b) its angular acceleration, (c) the x and y components of the acceleration of its center of mass, and (d) the components of the reaction force at the pivot.

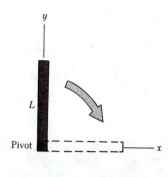

Figure 10.34

Solution

(a) Since no nonconservative forces act, $\Delta E = 0$

$$(K_f - K_i) + (U_f - U_i) = 0 \qquad \text{or} \qquad \left(\tfrac{1}{2}I\omega^2 - 0\right) + \left(0 - \tfrac{1}{2}mgL\right) = 0$$

so

$$\omega = \boxed{\left(\frac{3g}{L}\right)^{1/2}}$$

(b) $\tau = I\alpha$ so that in the horizontal position, $-\tfrac{1}{2}mgL = \left(\tfrac{1}{3}mL^2\right)\alpha$ or

$$\alpha = \boxed{-\frac{3g}{2L}}$$

(c) $a_x = a_r = -\dfrac{L}{2}\omega^2 = \boxed{-\dfrac{3g}{2}}$ (to the left)

$a_y = a_t = \alpha\dfrac{L}{2} = \boxed{-\dfrac{3g}{4}}$ (downward)

(d) Using F = ma, we have

$$R_x = ma_x = \boxed{-\frac{3mg}{2}}$$

$$R_y - mg = -ma \qquad \text{or} \qquad R_y = \boxed{-\frac{mg}{4}}$$

Angular Momentum and Torque
As Vector Quantities

OBJECTIVES

1. Define the cross product (magnitude and direction) of any two vectors, **A** and **B**, and state the various properties of the cross product.

2. Define the angular momentum **L** of a particle moving with a velocity **v** relative to a specified point, and the torque τ acting on the particle relative to that point. Note that both **L** and τ are quantities which depend on the choice of the origin since they involve the vector position **r** of the particle. (That is, $\mathbf{L} = \mathbf{r} \times \mathbf{p}$ and $\tau = \mathbf{r} \times \mathbf{F}$.)

3. Derive the relationship between the net torque on a particle and the time rate of change of its angular momentum. Note that the relation $\tau = \dfrac{d\mathbf{L}}{dt}$ is the rotational analog of Newton's second law, $\mathbf{F} = \dfrac{d\mathbf{p}}{dt}$.

4. Describe the total angular momentum of a system of particles and a rigid body rotating about a fixed axis.

5. Apply the conservation of angular momentum principle to a body rotating about a fixed axis, in which the moment of inertia changes due to a change in the mass distribution.

6. Give a qualitative description of the motion of a spinning top and gyroscope.

7. Describe the center of mass motion of a rigid body which undergoes both rotation about some axis and translation through space. Note that for pure rolling motion of an object such as a sphere or cylinder, the total kinetic energy can be expressed as the sum of a rotational kinetic energy about the center of mass plus the translational kinetic energy of the center of mass.

SKILLS

The operation of the vector or cross product is used for the first time in this chapter. (Recall that the angular momentum **L** of a particle is defined as $\mathbf{L} \equiv \mathbf{r} \times \mathbf{p}$, while torque is defined by the expression $\tau = \mathbf{r} \times \mathbf{F}$.) Let us briefly review the cross-product operation and some of its properties.

If **A** and **B** are any two vectors, their cross product, written as $\mathbf{A} \times \mathbf{B}$, is also a vector **C**. That is,

$$C = A \times B$$

where the magnitude of **C** is given by

$$C = |\mathbf{C}| = AB \sin \theta$$

and θ is the smaller angle between **A** and **B** as in Fig. 11.1 on the following page.

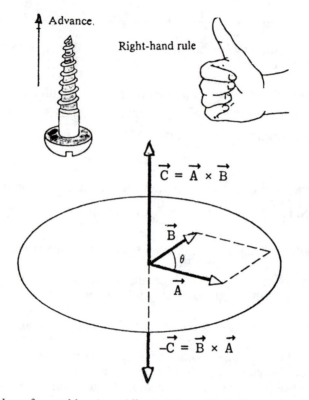

Figure 11.1

The direction of **C** is perpendicular to the plane formed by **A** and **B**, and its sense is determined by the right-hand rule. You should practice this rule for various choices of vector pairs. Note that **B** × **A** is directed *opposite* to **A** × **B**. That is, **A** × **B** = − **B** × **A**.

This follows from the right-hand rule. You should not confuse the cross product of two vectors, which is a vector quantity, with the dot product of two vectors, which is a scalar quantity. (Recall that the dot product is defined as **A·B** = AB cos θ.

Note that the cross product of any vector with itself is zero. That is, **A** × **A** = 0 since in this case θ = 0°, so sin (0) = 0, and from Eq. 11.8, it follows that **A** × **A** = 0.

Very often, vectors will be expressed in unit vector form, and it is convenient to make use of the multiplication table for unit vectors. These follow directly from Eq. 11.8 and the fact that **i**, **j**, and **k** represent a set of mutually orthogonal vectors as shown in Fig. 11.2.

i × **i** = **j** × **j** = **k** × **k** = 0

i × **j** = - **j** × **i** = **k**

j × **k** = - **k** × **j** = **i**

k × **i** = - **i** × **k** = **j**   (11.7)

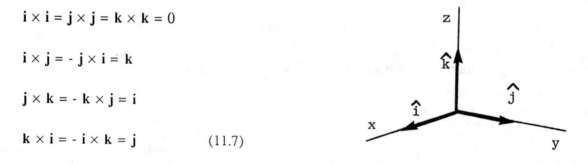

Figure 11.2

For example, if **A** = 3**i** + 5**j** and **B** = 4**j**, then

138

$$\mathbf{A} \times \mathbf{B} = (3\mathbf{i} + 5\mathbf{j}) \times (4\mathbf{j}) = 3\mathbf{i} \times 4\mathbf{j} + 5\mathbf{j} \times 4\mathbf{j} = 12\mathbf{k}$$

That is, the magnitude of $\mathbf{A} \times \mathbf{B}$ is 12 units and the direction of $\mathbf{A} \times \mathbf{B}$ is along the +z axis.

NOTES FROM SELECTED CHAPTER SECTIONS

11.1 ROLLING MOTION OF A RIGID BODY

The *total kinetic energy* of a body undergoing rolling motion is the sum of the rotational kinetic energy about the center of mass and the translational kinetic energy of the center of mass.

11.3 ANGULAR MOMENTUM OF A PARTICLE

The *torque* acting on a particle is equal to the time rate of change of its angular momentum.

11.5 CONSERVATION OF ANGULAR MOMENTUM

The *total angular momentum* of a system is constant if the resultant external torque acting on the system is zero. The resultant torque acting about the center of mass of a body equals the time rate of change of angular momentum, regardless of the motion of the center of mass.

EQUATIONS AND CONCEPTS

The *cross product* of any two vectors $\mathbf{A}$ and $\mathbf{B}$ is a vector $\mathbf{C}$ whose magnitude is given by $AB \sin \theta$ and whose direction is perpendicular to the plane formed by $\mathbf{A}$ and $\mathbf{B}$. The sense of $\mathbf{C}$ can be determined from the right-hand rule.

$$\mathbf{C} = \mathbf{A} \times \mathbf{B} \tag{11.8}$$

$$|\mathbf{C}| = AB \sin \theta \tag{11.9}$$

The *angular momentum* of a particle whose linear momentum is $\mathbf{p}$ and whose vector position is $\mathbf{r}$ is defined as $\mathbf{L} \equiv \mathbf{r} \times \mathbf{p}$. The SI unit of angular momentum is $kg \cdot m/s^2$. Note that both the magnitude and direction of $\mathbf{L}$ depend on the choice of origin.

$$\mathbf{L} \equiv \mathbf{r} \times \mathbf{p} \tag{11.15}$$

The *torque* acting on a particle whose vector position is $\mathbf{r}$ can be expressed as $\mathbf{r} \times \mathbf{F}$, where $\mathbf{F}$ is the external force acting on the particle. Torque also depends on the choice of the origin and has the SI unit of $N \cdot m$.

$$\tau = \mathbf{r} \times \mathbf{F} \tag{11.17}$$

If the same origin is used to define $\mathbf{L}$ and τ, then the *torque* on the particle *equals the time rate of change of its angular momentum.* This expression is the rotational analog of Newton's second law, $\mathbf{F} = \dfrac{d\mathbf{p}}{dt}$, and is the basic equation for treating rotating rigid bodies and rotating particles.

$$\boxed{\tau = \frac{d\mathbf{L}}{dt}} \qquad (11.19)$$

The angular momentum of a system of particles is obtained by taking the vector sum of the individual angular momenta about some point in an inertial frame. The individual momenta may change with time, which can change the total angular momentum. However, the total angular momentum of the system will only change if a net *external* torque acts on the system. In fact, *the net torque acting on a system of particles equals the time rate of change of the total angular momentum.*

$$\boxed{\Sigma\tau_{ext} = \frac{d\mathbf{L}}{dt}} \qquad (11.20)$$

The *angular momentum of a rigid body* in the form of a plane lamina rotating about a *fixed axis* is given by the product $I\omega$, where I is the moment of inertia about the axis of rotation and ω is the angular velocity.

$$\boxed{L = I\omega} \qquad (11.21)$$

Taking the time derivative of Eq. 11.21, and using Eq. 11.20, we find that *the net external torque acting on a rigid body is proportional to the angular acceleration* α. Hence, if the external torque is zero, the body has no angular acceleration. In this case, the body either rotates with constant angular velocity or is at rest.

$$\boxed{\Sigma\tau_{ext} = I\alpha} \qquad (11.23)$$

The law of conservation of angular momentum states that if the resultant external torque acting on a system is zero, the total angular momentum is constant. This follows from Eq. 11.20.

$$\boxed{\begin{array}{l} \text{If } \Sigma\tau_{ext} = \dfrac{d\mathbf{L}}{dt} = 0 \\[2mm] \mathbf{L} = \text{Constant} \end{array}}$$

$$(11.24)$$
$$(11.25)$$

If we apply this result to a body rotating about a fixed axis, and the moment of inertia changes from I_i to I_f, then the conservation of angular momentum can be used to find the final angular velocity in terms of the initial angular velocity.

$$\boxed{I_i\omega_i = I_f\omega_f = \text{Constant}} \qquad (11.27)$$

If a uniform body of circular cross section rolls on a rough surface without slipping, the *velocity and acceleration of the center of mass* are simply related to the angular velocity and angular acceleration according to Eqs. 11.1 and 11.2.

$$v_c = R\omega \qquad (11.1)$$
$$a_c = R\alpha \qquad (11.2)$$

The total kinetic energy of a rigid body rolling on a rough surface can be expressed as the sum of the rotational kinetic energy about the center of mass and the translational kinetic energy of the center of mass.

$$K = \frac{1}{2} I_c \omega^2 + \frac{1}{2} M v_c^2 \qquad (11.4)$$

If a body rolls down an incline *without slipping,* one can use conservation of energy to find the *velocity of the center of mass* as the body falls through a vertical distance h, starting from rest. The result is given by Eq. 11.6. From this expression, one can also find the acceleration of the center of mass.

$$v_c = \left(\frac{2gh}{1 + \dfrac{I_c}{MR^2}} \right)^{\frac{1}{2}} \qquad (11.6)$$

EXAMPLE PROBLEM SOLUTIONS

Example 11.1 A disk of moment of inertia I_0 rotates with angular velocity ω_0 about a vertical axle as in Fig. 11.3a. Two identical pieces of putty, each of mass m, are dropped onto the disk at points located distances a and b from the axis. Find the final angular velocity of the system.

(a)   (b)

Figure 11.3

Solution

The problem is easily analyzed using conservation of angular momentum. The initial angular momentum of the disk is given by

$$L_i = I_0 \omega_0 \qquad (1)$$

and is directed along the z axis (upwards). The final angular momentum of the system is given by

$$L_f = I \omega_f \qquad (2)$$

where I is the moment of inertia of the disk plus that of the added pieces of putty about the axis of rotation:

$$I = I_0 + I_{putty} = I_0 + ma^2 + mb^2 = I_0 + m(a^2 + b^2) \tag{3}$$

Since angular momentum is conserved, then $L_i = L_f$. Using this condition, together with Eqs. (1), (2), and (3) gives

$$I\omega_f = I_0\omega_0$$

$$\omega_f = \frac{I_0\omega_0}{I} = \frac{I_0\omega_0}{I_0 + m(a^2 + b^2)}$$

Note that $\omega_f < \omega_0$, as you would expect, since the inertia of the system has increased due to the added mass.

As an extension of this problem, you should be able to show that the final energy of the system, E_f, is *less* than the initial energy $E_0 = \frac{1}{2} I_0\omega_0^2$. In fact, the ratio of these two quantities is given by

$$\frac{E_f}{E_0} = \frac{I_0}{I_0 + m(a^2 + b^2)}$$

What do you suppose accounts for the decrease in energy?

Example 11.2   A projectile of mass m moves with a speed v_0 to the right as in Fig. 11.4a. The projectile strikes and sticks to the end of a stationary bar of mass M, length L, pivoted about a frictionless axle through its center. Find the angular velocity of the system right after the collision.

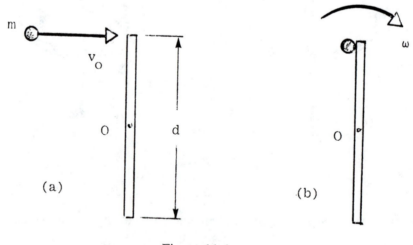

Figure 11.4

Solution

The initial angular momentum of the system about an axis through the pivot is that of the mass m:

$$L_i = mv_0\left(\frac{d}{2}\right) \tag{1}$$

After the collision, the mass and bar move with a common angular velocity ω about 0 (Fig. 11.4b on the previous page). The final angular momentum about 0 is given by

$$L_f = I\omega \tag{2}$$

where I is the moment of inertia of the system (bar + mass):

$$I = I_{bar} + I_{mass} + \frac{1}{12}Md^2 + m\left(\frac{d}{2}\right)^2$$

$$= \frac{1}{12}Md^2 + m\frac{d^2}{4} = \left(\frac{M+3m}{12}\right)d^2 \tag{3}$$

Since angular momentum is conserved, then $L_i = L_f$. Using this condition, together with Eqs. (1), (2), and (3) gives

$$\left(\frac{M+3m}{12}\right)d^2\omega = mv_0\left(\frac{d}{2}\right)$$

or

$$\omega = \frac{6mv_0}{(M+3m)d}$$

As an extra exercise, you should show that the fractional loss in energy due to the collision is given by

$$\text{Fractional Energy Loss} = \frac{M}{3m}$$

ANSWERS TO SELECTED QUESTIONS

1. Is it possible to calculate the torque acting on a rigid body without specifying the origin? Is the torque independent of the location of the origin?

Answer: The answer is *no* to both questions. The value of τ depends on the choice of the origin, since $\tau = r \times F$, and the value of r changes if the origin is shifted.

2. Is the triple product defined by $A \cdot (B \times C)$ equal to a scalar or vector quantity? Note that the operation $(A \cdot B) \times C$ has no meaning. Explain.

Answer: $A \cdot (B \times C)$ is a scalar since $B \times C$ is a vector which we can call D, and $A \cdot D$ by definition is a scalar. $(A \cdot B) \times C$ has no meaning since the brackets around $(A \cdot B)$ imply this operation comes first (a scalar quantity) and you cannot cross a scalar with a vector.

3. In the expression for torque, $\tau = \mathbf{r} \times \mathbf{F}$, is **r** equal to the moment arm? Explain.

Answer: No. The vector **r** is the position vector of the applied force relative to some origin. The moment arm is the projection of **r** along a direction *perpendicular* to **F** -- whose magnitude is the perpendicular distance to the line of action of **F** from the axis of rotation.

4. Can a particle moving in a straight line have nonzero angular momentum?

Answer: No. A particle has angular momentum about an origin when moving in a straight line, as long as the line of motion does not pass through the origin. Its angular momentum is zero if the line of motion passes through the origin, in which case **r** is parallel to **v**, and $\mathbf{L} = \mathbf{r} \times \mathbf{p} = \mathbf{r} \times m\mathbf{v} = 0$.

6. If the torque acting on a particle about an *arbitrary* origin is zero, what can you say about its angular momentum about that origin?

Answer: Since $\tau = \dfrac{d\mathbf{L}}{dt} = 0$, we conclude that **L** is a constant of the motion.

7. A particle moves in a straight line, and you are told that the torque acting is zero about some unspecified origin. Does this necessarily imply that the net force on the particle is zero? Can you conclude that its velocity is constant? Explain.

Answer: No, the net force is not necessarily zero. If the line of action of the net force passes through the origin, the net torque about that origin is zero, yet $F_{net} \neq 0$. Therefore, you cannot conclude that its velocity is constant.

8. Suppose that the velocity vector of a particle is completely specified. What can you conclude about the *direction* of its angular momentum vector with respect to the direction of motion?

Answer: Since $\mathbf{L} = \mathbf{r} \times \mathbf{p}$, **L** must be *perpendicular* to the direction of motion.

9. If the net torque acting on a rigid body is nonzero about some origin, is there any other origin about which the net torque is zero?

Answer: Yes. If the origin is one for which the moment arm is zero, the net torque will be zero about that origin. In fact, any origin for which $\mathbf{r} \times \mathbf{F} = 0$ (that is, **r** parallel to **F**), will satisfy this condition.

10. If a system of particles is in motion, is it possible for the total angular momentum to be zero about some origin? Explain.

Answer: Yes. The vector sum of the individual angular momentum can be zero. For example, the two identical particles shown at the right moving in circular orbits, but in the opposite sense, have angular momentum which cancel each other. (L_1 points *out* of the page while L_2 points *into* the page. Also, $L_1 = L_2 = mvr$.)

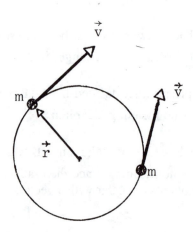

144

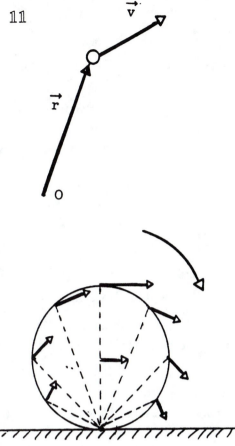

11. A ball is thrown in such a way that it does not spin about its own axis. Does this mean that the angular momentum is zero about an arbitrary origin? Explain.

Answer: No. Angular momentum is defined as $\mathbf{L} = \mathbf{r} \times \mathbf{p} = \mathbf{r} \times m\mathbf{v}$. As you can see from the diagram, the particle has nonzero angular momentum about an arbitrary origin even when the particle moves in a straight line. ($\mathbf{L} = 0$ only when $\mathbf{r}$ is parallel to $\mathbf{v}$, or when $\mathbf{v} = 0$.)

14. When a cylinder rolls on a horizontal surface, as in Fig. 11.4, are there any points on the cylinder that have only a vertical component of velocity at some instant? If so, where are they?

Answer: No. The diagram shows that every point has a horizontal component equal to $R\omega$, where R is the radius of the cylinder. However, points along a vertical line through the center of mass have only a horizontal component of velocity.

SOLUTIONS TO SELECTED END-OF-CHAPTER PROBLEMS

3. (a) Determine the acceleration of the center of mass of a uniform solid disk rolling down an incline and compare this acceleration with that of a uniform hoop. (b) What is the minimum coefficient of friction required to maintain pure rolling motion for the disk?

Solution

$$\Sigma F_x = mg \sin \theta - f = ma_c \qquad (1)$$

$$\Sigma F_y = N - mg \cos \theta = 0 \qquad (2)$$

$$\tau = fr = I_c \alpha = \frac{I_c a_c}{r} \qquad (3)$$

(a) For a disk, $I_c = \frac{1}{2}mr^2$. From (3) we find $f = \frac{1}{2}ma_c$. Substituting this into (1) gives

$$mg \sin \theta - \frac{1}{2}ma_c = ma_c$$

$$a_c = \boxed{\frac{2}{3} g \sin \theta}$$

For a hoop, $I_c = mr^2$. From (3) we find $f = \dfrac{mr^2 a_c}{r^2} = ma_c$. Substituting this into (1) gives

$$mg\sin\theta - ma_c = ma_c \qquad \text{or} \qquad \boxed{a_c = \frac{1}{2}g\sin\theta}$$

(b) $f = \mu N$ and from (2) we find $N = mg\cos\theta$. Thus

$$f = \mu mg\cos\theta = \frac{1}{2}ma_c = \frac{1}{2}m\left(\frac{2}{3}\right)g\sin\theta = \frac{1}{3}mg\sin\theta$$

so

$$\mu = \frac{1}{3}\left(\frac{\sin\theta}{\cos\theta}\right) = \boxed{\frac{1}{3}\tan\theta}$$

---

11. If $\left|\mathbf{A}\times\mathbf{B}\right| = \mathbf{A}\cdot\mathbf{B}$, what is the angle between $\mathbf{A}$ and $\mathbf{B}$?

Solution

We are given the condition $\left|\mathbf{A}\times\mathbf{B}\right| = \mathbf{A}\cdot\mathbf{B}$, This says that

$$AB\sin\theta = AB\cos\theta \quad \Rightarrow \quad \tan\theta = 1 \qquad \text{or} \qquad \theta = \boxed{45°}$$

---

25. A particle of mass 0.4 kg is attached to the 100-cm mark of a meter stick of mass 0.1 kg. The meter stick rotates on a horizontal, smooth table with an angular velocity of 4 rad/s. Calculate the angular momentum of the system if the stick is pivoted about an axis (a) perpendicular to the table through the 50-cm mark and (b) perpendicular to the table through the 0-cm mark.

$$\textbf{Solution} \qquad M = 0.1\text{ kg} \qquad m = 0.4\text{ kg} \qquad L = 1\text{ m} \qquad \omega = 4\text{ rad/s}$$

(a) From the parallel axis theorem, we have

$$I_B = \frac{1}{2}ML^2 + mL^2\left(\frac{1}{2}\right)^2 = \left(\frac{M}{12}+\frac{m}{4}\right)L^2 = \left(\frac{0.1\text{ kg}}{12}+\frac{0.4\text{ kg}}{4}\right)(1\text{ m})^2 = 0.1083\text{ kg·m}^2$$

$$L_B = I_B\omega = (0.1083\text{ kg·m}^2)(4\text{ rad/s}) = \boxed{0.433\ \text{kg·m}^2\text{/s}}$$

(b) $$I_A = \frac{1}{3}ML^2 + mL^2 = \left(\frac{M}{3}+m\right)L^2 = 0.433\text{ kg·m}^2$$

$$L_A = I_A\omega = (0.433\text{ kg·m}^2)(4\text{ rad/s}) = \boxed{1.73\ \text{kg·m}^2\text{/s}}$$

39. A string is wound around a uniform disk of radius R and mass M. The disk is released from rest with the string vertical and its top end tied to a fixed support (Fig. 11.31). As the disk descends, show that (a) the tension in the string is one-third the weight of the disk, (b) the acceleration of the center of mass is 2g/3, and (c) the velocity of the center of mass is $(4gh/3)^{1/2}$. Verify your result to (c) using the energy approach.

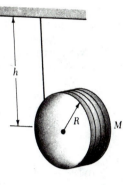

Figure 11.31

Solution

$$\Sigma F = T - Mg = -Ma; \quad \Sigma \tau = TR = I\alpha = \frac{1}{2} MR^2 \left(\frac{a}{R}\right)$$

(a) Combining the above two equations, we find

$$T = M(g - a) \quad \text{and} \quad a = \frac{2T}{M}, \quad \text{thus} \quad T = \frac{Mg}{3}$$

(b) $\quad a = \frac{2T}{M} = \left(\frac{2}{M}\right)\left(\frac{Mg}{3}\right) \quad \text{or} \quad a = \frac{2}{3} g$

(c) Requiring conservation of mechanical energy, we have

$$\Delta U + \Delta K_{rot} + \Delta K_{trans} = 0, \quad \text{or}$$

$$(0 - mgh) + \frac{1}{2}\left(\frac{1}{2} MR^2\right) \omega^2 - 0 + \left(\frac{1}{2} Mv^2 - 0\right) = 0$$

When there is no slipping, $\quad \omega = \frac{v}{R} \quad \text{and} \quad v = \sqrt{\frac{4gh}{3}}$

---

49. A mass m is attached to a cord passing through a small hole in a frictionless, horizontal surface (Fig. 11.36). The mass is initially orbiting in a circle of radius r_0 with velocity v_0. The cord is then slowly pulled from below, decreasing the radius of the circle to r. (a) What is the velocity of the mass when the radius is r? (b) Find the tension in the cord as a function of r. (c) How much work is done in moving m from r_0 to r? (Note: The tension depends on r.) (d) Obtain numerical values for v, T, and W when r = 0.1 m, if m = 50 g, r_0 = 0.3 m, and v_0 = 1.5 m/s.

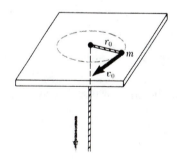

Figure 11.36

Solution

(a) Although an external force (tension of rope) acts on the mass, no external torques act. Therefore $L =$ constant and $mv_0 r_0 = mvr$ and $v = \dfrac{v_0 r_0}{r}$.

(b) $T = mv^2/r$. Substituting for v from (a), we find $T = \dfrac{mv_0{}^2 r_0{}^2}{r^3}$

(c) $W = \Delta K = \dfrac{1}{2}m\left(v^2 - v_0{}^2\right) = \dfrac{mv_0{}^2}{r^2}\left(\dfrac{r_0{}^2}{r^2} - 1\right)$

(d) Using the data given we find

$$v = \boxed{4.50 \text{ m/s}} \qquad T = \boxed{10.1 \text{ N}} \qquad W = \boxed{0.450 \text{ J}}$$

---

59. Suppose a solid disk of radius R is given an angular velocity ω_0 about an axis through its center and is then lowered to a rough, horizontal surface and released, as in Problem 58 (Fig. 11.41). Furthermore, assume that the coefficient of friction between the disk and surface is μ. (a) Show that the *time* it takes pure rolling motion to occur is given by $R\omega_0/3\mu g$. (b) Show that the *distance* the disk travels before pure rolling occurs is given by $R^2\omega_0{}^2/18\mu g$. (See hint in Problem 58.)

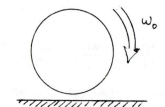

Figure 11.41

Solution

(a) After pure rolling occurs, $v - v_0 = at$ when $v_0 = 0$, and from Prob. 58, $\omega = \dfrac{1}{3}\omega_0$ so that

$v = \dfrac{1}{3}R\omega_0$. Using these expressions for v and v_0 in the first equation, we find

$$\dfrac{1}{3}R\omega_0 = at \quad \text{where} \quad a = \dfrac{F}{m} = \dfrac{-\mu mg}{m} = -\mu g$$

Therefore, $\qquad\qquad\qquad t = \dfrac{R\omega_0}{3\mu g}$

(b) The distance of travel is $\Delta x = v_0 t + \dfrac{1}{2}at^2$. Using the result from part (a), we find

$$\Delta x = \dfrac{1}{2}(\mu g)\left(\dfrac{\frac{1}{3}R\omega_0}{\mu g}\right)^2 = \dfrac{R^2\omega_0{}^2}{18\mu g}$$

12

Static Equilibrium of a Rigid Body

OBJECTIVES

1. Describe the two necessary conditions of equilibrium for a rigid body.

2. Locate the center of gravity of a system of particles or a rigid body and understand the subtle difference between center of gravity and center of mass.

3. Analyze problems of rigid bodies in static equilibrium using the procedures presented in Section 12.3 of the text.

SKILLS

Since this chapter represents application of Newton's laws to a special situation, namely, rigid bodies in static equilibrium, it is important that you understand and follow the procedures for analyzing such problems. The following skills must be mastered in this regard.

1. The need to recognize all *external* forces acting on the body, and the construction of an accurate free-body diagram.

2. Resolving the external forces into their rectangular components, and applying the first condition of equilibrium $\Sigma F_x = 0$ and $\Sigma F_y = 0$.

3. You must choose a convenient origin for calculating the net torque on the body. The choice of this origin is arbitrary. (The torque equation gives information which is not offered by applying $\Sigma F = 0$.)

4. Solving the set of simultaneous equations obtained from the two conditions of equilibrium.

The following procedure is recommended when analyzing a body in equilibrium under the action of several external forces:

1. Make a sketch of the object under consideration.

2. Draw a free-body diagram and label all external forces acting on the object. Try to guess the correct direction for each force. If you select an incorrect direction that leads to a negative sign in your solution for a force, do not be alarmed; this merely means that the direction of the force is the opposite of what you assumed.

3. Resolve all forces into rectangular components, choosing a convenient coordinate system. Then apply the first condition for equilibrium, which balances forces. Remember to keep track of the signs of the various force components.

4. Choose a convenient axis for calculating the net torque on the object. Remember that the choice of the origin for the torque equation is *arbitrary;* therefore, choose an origin that will simplify your calculation as much as possible. Becoming adept at this is a matter of practice.

5. The first and second conditions of equilibrium give a set of linear equations with several unknowns. All that is left is to solve the simultaneous equations for the unknowns in terms of the known quantities.

12.1 THE CONDITIONS OF EQUILIBRIUM OF A RIGID BODY

A *rigid body* is defined as one that does not deform under the application of external forces. There are two necessary conditions for *equilibrium of a rigid body*: (1) the resultant external force must be zero and (2) the resultant external torque must be zero about any axis. Two forces are equal if, and only if, they have equal magnitudes and they have equal torques about any specified axis.

12.2 THE CENTER OF GRAVITY

In order to compute the torque due to the weight force, all of the weight of a body can be considered to be concentrated at a point called the *center of gravity*. At the point of the center of gravity, a force of magnitude mg and directed opposite the force of gravity will balance the body if no other external forces are acting. The center of gravity of an object coincides with its center of mass if the object is in a uniform gravitational field.

12.4 ELASTIC PROPERTIES OF SOLIDS

The elastic properties of solids are described in terms of *stress* and *strain*. Stress is a quantity that is proportional to the force causing a deformation of the object. Strain is a measure of the degree of the resulting deformation.

The *elastic modulus* of a material is the ratio of stress to strain for that material. There is an elastic modulus for each of three types of deformation: *Young's modulus* which measures resistance to change in length, *Shear modulus* which measures resistance to relative motion of the planes of a solid, and *Bulk modulus* which measures the resistance to a change in volume.

EQUATIONS AND CONCEPTS_____

In general, *an object at rest or one moving with constant velocity will only do so if the resultant force on it is zero*. This is a statement of the *first condition of equilibrium* -- and corresponds to the condition of translational equilibrium.

$$\Sigma F = 0$$

(12.2)

Since Eq. 12.2 is a *vector sum of* all *external forces* acting on the body, this necessarily implies that the sum of the x, y, and z components separately must be zero.

$$\Sigma F_x = 0$$
$$\Sigma F_y = 0$$
$$\Sigma F_z = 0$$

The *second condition of equilibrium* of a rigid body requires that the *vector sum of the torques relative to any origin must be zero*. This is the condition of rotational equilibrium.

$$\Sigma \tau = 0$$

(12.3)

If all the forces acting on a rigid body lie in a common plane, say the xy plane, then there is no z component of force. In this case, we only have to deal with three equations--two of which correspond to the first condition of equilibrium, the third coming from the second condition. In this case, the torque vector lies along the z axis. All problems in this chapter fall into this category.

$$\Sigma F_x = 0$$
$$\Sigma F_y = 0$$
$$\Sigma \tau_z = 0$$

(12.4)

Young's modulus Y is a measure of the resistance of a body to elongation, and is equal to the ratio of the tensile stress (the force per unit area) to the tensile strain (the change in length over the original length).

$$Y = \frac{F/A}{\Delta L/L_0}$$

(12.7)

The *Shear modulus* S is a measure of the deformation which occurs when a force is applied parallel to one of the body's surface. It equals the ratio of the shearing stress to the shearing strain.

$$S = \frac{F/A}{\Delta x/h}$$

(12.8)

The *Bulk modulus* B is a parameter which characterizes the response of a body to uniform pressure on all sides. It is defined as the ratio of the volume stress (the pressure) to the volume strain ($\Delta V/V$).

$$B = -\frac{\Delta P}{\Delta V/V}$$

(12.9)

EXAMPLE PROBLEM SOLUTION

Example 12.1   A uniform beam of weight 300 N and length 5 m is supported by a pivot at one end and a rope connected to a 900 N weight through a frictionless pulley as in Fig. 12.1. A man weighing 600 N stands at the position a distance x from the pivot such that the beam remains horizontal and the rope makes an angle of 37° with the horizontal. Obtain a numerical value for x and find the force exerted by the pivot on the beam.

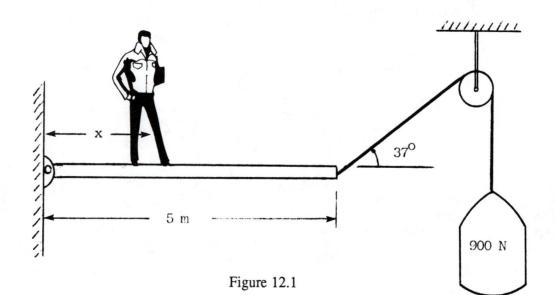

Figure 12.1

Solution

First, let us construct a free-body diagram for the beam. In the diagram below, **w** is the weight of the beam, **W** is the weight of the man, **T** is the force of tension, and **R** is the reaction force at the pivot.

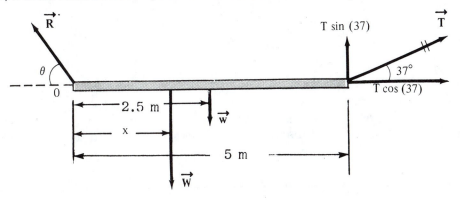

First note that the tension **T** in the rope is uniform and equal in magnitude to 900 N, that of the suspended weight. (That is, if we "isolate" the 900 N weight, the rope must exert a force of 900 N upwards to keep the weight in equilibrium.) The first condition of equilibrium requires that $\mathbf{R} + \mathbf{W} + \mathbf{w} + \mathbf{T} = 0$. In component form, this becomes

$$\Sigma F_x = T \cos (37°) - R \cos \theta = 0$$

$$\Sigma F_y = R \sin \theta + T \sin (37°) - W - w = 0$$

Substituting T = 900 N, W = 600 N, and w = 300 N gives

$$720 - R \cos \theta = 0 \qquad\qquad (1)$$

$$R \sin \theta - 360 = 0 \qquad\qquad (2)$$

We can solve these two equations for R and θ by noting that from Eq. (1), $R = 720/\cos \theta$. Substituting this into Eq. (2) gives

$$\left(\frac{720}{\cos \theta}\right) \sin \theta - 360 = 0$$

$$\tan \theta = \frac{360}{720} = 0.5$$

$$\theta = 26.6°$$

Substituting this value into Eq. (1) gives

$$R = \frac{720}{\cos \theta} = \frac{720}{\cos (26.6°)} = 805 \text{ N}$$

To find the value of x such that the beam remains horizontal, we must apply the second condition of equilibrium. In order to do this, it is convenient to use the pivot as the origin of coordinates and resolve the force **T** into components parallel and perpendicular to the beam. Note that the moment arm of the parallel component, T cos (37°) is *zero* about the pivot. The moment arm of the perpendicular component, T sin (37°), is 5 m, the length of the beam. Likewise, the moment arms of the forces **w**, **W**, and **R** are 2.5 m, x and 0, respectively. Therefore,

$$\Sigma\tau_o = 5T \sin(37°) + (0)T \cos(37°) - 2.5w - xW + (0)R = 0$$

or

$$2700 - 750 - 600\,x = 0$$

$$x = \frac{1950}{600} = 3.25 \text{ m}$$

It is important to recognize that we could not have obtained x without using the second condition of equilibrium. It is often useful to construct a table of forces, moment arms and torques "to keep the books straight." The table below gives this information for the present example, with the origin at the pivot.

Force Component	Moment Arm, m	Torque Component, N·m
T cos (37°)	0	0
T sin (37°)	5	5T sin (37°)
w	2.5	-2.5w
W	x	-xW
R	0	0

You should also find x using a *different* origin for the torques, say the point at the right end of the beam. For this choice of origin, the moment arms of the forces will differ from those in the table above, but the value you obtain for x *must* be the same as that calculated using the pivot as the origin. In other words, the solution must be *independent* of the choice of the origin. This is a good method for checking your solution to a given problem. Sometimes the direction of **R** is not obvious from inspection. However, if you choose the wrong direction for R_x or R_y (or both), your solution will yield negative values for these components, but the magnitudes will be correct. Working a number of problems will help clarify this point.

ANSWERS TO SELECTED QUESTIONS

1. Can a body be in equilibrium if only one external force acts on it? Explain.

Answer: No. The first condition requires that the resultant force on the body must be zero. If only one force acts, the resultant force cannot be zero.

2. Can a body be in equilibrium if it is in motion? Explain.

Answer: Yes. A body moving with constant velocity is in equilibrium. For example, a rocket coasting in free space experiences no external forces.

7. Give an example in which the net torque acting on an object is zero and yet the net force is nonzero.

Answer: Imagine two forces applied to an object at its center of gravity. One force is 10 N to the right; the other is 6 N to the left. There is no torque about the center of gravity, yet there is a net force of 4 N to the right.

9. Can an object be in equilibrium if the only torques acting on it produce clockwise rotation?

Answer: No. For an object to be in equilibrium, the clockwise and counterclockwise torques must be equal.

10. A tall crate and a short crate of equal mass are placed side-by-side on an incline (without touching each other). As the incline angle is increased, which crate will topple first? Explain.

Answer: The center of gravity of the taller crate is higher; therefore, the torque produced by its weight is greater. Thus, it will tip first.

11. A male and a female student are asked to do the following task. Face a wall, step three foot lengths away from the wall, and then lean over and touch the wall with your nose, keeping your hands behind your back. The male usually fails, but the female succeeds. How would you explain this?

Answer: The higher center of gravity of the male causes the torque produced by his weight to tip him forward.

12. When lifting a heavy object, why is it recommended to straighten your back as much as possible rather than bend over and lift mainly with the arms?

Answer: Compressional forces in the spine are relatively safe, but torques on the spine can cause discs to slip out of alignment.

SOLUTIONS TO SELECTED END-OF-CHAPTER PROBLEMS_____

3. A uniform beam of weight W and length L has weights W_1 and W_2 at two positions, as in Figure 12.21. The beam is resting at two points. For what value of x will the beam be balanced at P such that the normal force at 0 is zero?

Solution Take torques about P.

$$\Sigma \tau_p = N_0 \left[\frac{L}{2} + d\right] - W_1 \left[\frac{L}{2} + d\right] - Wd + W_2x = 0$$

We want to find x for which $N_0 = 0$. Let $N_0 = 0$ and solve for x.

$$-W_1 \left[\frac{L}{2} + d\right] - Wd + W_2x = 0$$

$$\boxed{x = \frac{W_1 \left(\frac{L}{2} + d\right) + Wd}{W_2}}$$

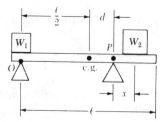

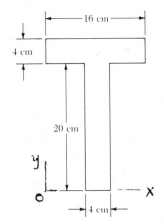

Figure 12.21

---

7. A flat plate in the shape of a letter T is cut with the dimensions shown in Figure 12.23. Locate the center of gravity. (Hint: Note that the weights of the two rectangular parts are proportional to their volumes.)

Solution

Note that the center of gravity is located on the lateral line of symmetry of the T. Let m_1 = the mass of the vertical section and m_2 = the mass of the horizontal section of T. Then

$$y_{cm} = \frac{m_1y_1 + m_2y_2}{m_1 + m_2} = \frac{A_1y_1 + A_2y_2}{A_1 + A_2}$$

Measuring from an origin at the bottom of the T, $y_1 = 10$ cm, $y_2 = 22$ cm, $A_1 = 80$ cm^2, $A_2 = 64$ cm^2 and

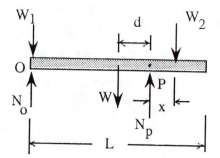

Figure 12.23

$$y_{cm} = \frac{(80)(10) \text{ cm}^3 + (64)(22) \text{ cm}^3}{144 \text{ cm}^2} = \boxed{15.3 \text{ cm}}$$

Likewise, $x_{cm} = 8.00$ cm from the left side of the "tee."

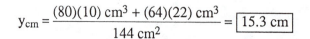

33. A bridge of length 50 m and mass 8×10^4 kg is supported at each end as in Figure 12.29. A truck of mass 3×10^4 kg is located 15 m from one end. What are the forces on the bridge at the points of support?

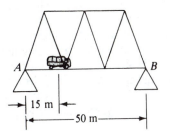

Solution

Let N_A and N_B be the normal forces at the points of support. Choosing the origin at point A and using $\Sigma F_y = \Sigma \tau = 0$, we find

Figure 12.29

$$N_A + N_B - (8 \times 10^4 \text{ kg})g - (3 \times 10^4 \text{ kg})g = 0$$

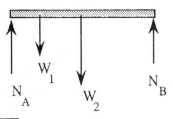

and $\quad -(3 \times 10^4 \text{ kg})(15 \text{ m}) - (8 \times 10^4 \text{ kg})g(25 \text{ m}) + N_B(50 \text{ m}) = 0$

The equations combine to give

$$N_A = \boxed{5.98 \times 10^5 \text{ N}} \quad \text{and} \quad N_B = \boxed{4.80 \times 10^5 \text{ N}}$$

---

37. A uniform beam of length 4 m and mass 10 kg supports a 20-kg mass as in Figure 12.33. (a) Draw a free-body diagram for the beam. (b) Determine the tension in the supporting wire and the components of the reaction force at the pivot.

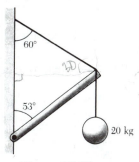

Solution

Choosing the origin at the pivot point and applying Equations 12.2 and 12.3, we have

Figure 12.33

$$\Sigma F_x = R_x - T \cos 30° = 0$$

$$\Sigma F_y = R_y + T \sin 30° - (10 + 20)g = 0$$

$$\Sigma \tau = LT \sin 67° - \frac{L}{2}(10)g \sin 53° - L(20)g \sin 53° = 0$$

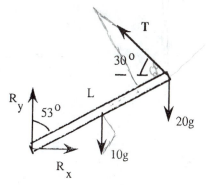

Solving the above equations give

$$R_x = \boxed{184 \text{ N}} \qquad R_y = \boxed{188 \text{ N}}$$

$$\text{and} \qquad T = \boxed{213 \text{ N}}$$

41. A 15-m uniform ladder weighing 500 N rests against a frictionless wall. The ladder makes a 60° angle with the horizontal. (a) Find the horizontal and vertical forces that the earth exerts on the base of the ladder when an 800-N firefighter is 4 m from the bottom. (b) If the ladder is just on the verge of slipping when the firefighter is 9 m up, what is the coefficient of static friction between ladder and ground?

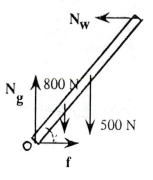

Solution

(a) $\Sigma F_x = f - N_w = 0$

$\Sigma F_y = N_g - 800 - 500 = 0$ and

$\Sigma \tau_O = 4(800) \sin 30° + 7.5(500) \sin 30° - 15(N_w) \sin 60° = 0$

Solve the torque equation to find the force exerted by the wall.

$$N_w = \frac{[4(800) + 7.5(500)] \tan 30°}{15} = 267.5 \text{ N}$$

Next substitute this value into the F_x equation to find $f = N_w = \boxed{268 \text{ N}}$ in the positive x-direction.

Solve the ΣF_y equation to find $N_g = \boxed{1300 \text{ N}}$ in the positive y-direction.

(b) In this case, the torque equation gives

$$\Sigma \tau_O = 9(800)\sin 30° + 7.5(500)\sin 30° - 15(N_w)\sin 60° = 0$$

$$N_w = 421 \text{ N}$$

Since $\quad f = N_w = 421 \text{ N}$ and $f = f_{max} = \mu N_g,$

we find

$$\mu = \frac{f_{max}}{N_g} = \frac{421 \text{ N}}{1300 \text{ N}} = \boxed{0.324}$$

43. A 10,000-N shark is supported by a cable attached to a 4-m rod that can pivot at the base. Calculate the cable tension needed to hold the system in the position shown in Figure 12.38. Find the horizontal and vertical forces exerted on the base of the rod. (Neglect the weight of the rod.)

Solution

$$\Sigma F_x = F_H - T \cos 20° = 0$$

$$\Sigma F_y = F_v + T \sin 20° - 10000 = 0 \quad \text{and}$$

$$\Sigma \tau = 4(10000) \cos 60° - 4\,T \sin 80° = 0$$

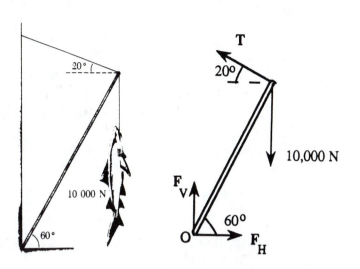

Figure 12.38

Solve the torque equation to find $\quad T = \dfrac{10\,000 \cos 60°}{\sin 80°} = \boxed{5.08 \times 10^3 \text{ N}}$

Solve the ΣF_x equation to find $\quad F_H = T \cos 20° = \boxed{4.77 \times 10^3 \text{ N}}$

Solve the ΣF_y equation to find $\quad F_v = 10\,000 - T \sin 20° = \boxed{8.26 \times 10^3 \text{ N}}$

49. A uniform beam of weight w is inclined at an angle θ to the horizontal with its upper end supported by a horizontal rope tied to a wall and its lower end resting on a rough floor (Fig. 12.43). (a) If the coefficient of static friction between the beam and floor is μ_s, determine an expression for the *maximum* weight W that can be suspended from the top before the beam slips. (b) Determine the magnitude of the reaction force at the floor and the magnitude of the force of the beam on the rope at P in terms of w, W, and μ_s.

Solution

(a) We can use $\Sigma F_x = \Sigma F_y = \Sigma \tau = 0$ with origin at the point of contact on the floor.

Figure 12.43

Then

$$\Sigma F_x = T - \mu N = 0,$$

$$\Sigma F_y = N - W - w = 0 \qquad \text{and}$$

$$\Sigma \tau = W(\cos\theta)L + w(\cos\theta)\frac{L}{2} - T(\sin\theta)L = 0$$

Solving the above equations gives

$$W = \frac{w}{2}\left[\frac{2\mu_s \sin\theta - \cos\theta}{\cos\theta - \mu_s \sin\theta}\right]$$

(b)   At the floor, we see that the normal force is in the y-direction and frictional force is in the x-direction. The reaction force then is

$$R = \sqrt{N^2 + (\mu_s N)^2} = (W + w)\sqrt{1 + \mu_s^2}$$

At point P, the force of the beam on the rope is

$$F = \sqrt{T^2 + W^2} = \sqrt{W^2 + \mu_s^2(W + w)^2}$$

---

56. A wire of length L, Young's modulus Y, and cross-sectional area A is stretched elastically by an amount ΔL. By Hooke's law, the restoring force is given by $-k\,\Delta L$. (a) Show that the constant k is given by $k = \dfrac{YA}{L}$. (b) Show that the work done in stretching the wire by an amount ΔL is given by

$$\text{Work} = \frac{1}{2}\frac{YA}{L}(\Delta L)^2$$

Solution

(a)   According to Hooke's law $|F| = k\,\Delta L$. Young's modulus is defined as

$$Y = \frac{F/A}{\Delta L/L} = k\frac{L}{A} \qquad \text{or} \qquad k = \frac{YA}{L}$$

(b) $W = -\displaystyle\int_0^{\Delta L} F dx = -\int_0^{\Delta L}(-kx)dx = \frac{YA}{L}\int_0^{\Delta L} x dx = \frac{1}{2}\frac{YA}{L}(\Delta L)^2$

13

Oscillatory Motion

OBJECTIVES

1. Describe the general characteristics of simple harmonic motion, and the significance of the various parameters which appear in the expression for the displacement versus time, $x = A \cos(\omega t + \delta)$.

2. Start with the expression for the displacement versus time for the simple harmonic oscillator, and obtain equations for the velocity and acceleration as functions of time.

3. Understand the phase relations between displacement, velocity, and acceleration for simple harmonic motion, noting that acceleration is proportional to the displacement, but in the opposite direction.

4. Obtain a value for the phase constant δ, given the initial displacement and initial velocity of the body undergoing simple harmonic motion.

5. Describe and understand the conditions of simple harmonic motions executed by the mass-spring system (where the frequency depends on k and m) and the simple pendulum (where the frequency depends on L and g).

6. Apply energy principles to the simple harmonic oscillator, noting that the total energy is conserved if one assumes there are no nonconservative forces acting on the system.

7. Discuss the relationship between simple harmonic motion and the motion of a point on a circle moving with uniform angular velocity.

8. Give a qualitative description of damped oscillations and forced oscillations.

SKILLS

Most of this chapter deals with simple harmonic motion, and the properties of the displacement expression

$$x(t) = A \cos(\omega t + \delta) \tag{13.1}$$

In order to obtain the velocity v(t) and acceleration a(t) of the system, one must be familiar with the derivative operation as applied to trigonometric functions. In particular, note that

$$\frac{d}{dt} \cos(\omega t + \delta) = -\omega \sin(\omega t + \delta)$$

$$\frac{d}{dt} \sin(\omega t + \delta) = \omega \cos(\omega t + \delta)$$

Using these results, and x(t) from Eq. 13.1, we see that

$$v(t) = \frac{dx(t)}{dt} = -A\omega \sin(\omega t + \delta) \tag{13.5}$$

and

$$a(t) = \frac{dv(t)}{dt} = -A\omega^2 \cos(\omega t + \delta) \tag{13.6}$$

By direct substitution, you should be able to show that Eq. 13.1 represents a general solution to the equation of motion for the mass-spring system (a second-order homogeneous differential equation) given by

$$\frac{d^2x}{dt^2} + \frac{k}{m}x = 0$$

where

$$\omega = \sqrt{\frac{k}{m}}$$

In treating the motion of the simple pendulum, we made use of the small angle approximation $\sin\theta \cong \theta$. This approximation enables us to reduce the equation of motion to that of the simple harmonic oscillator. The small angle approximation for $\sin\theta$ follows from inspecting the series expansion for $\sin\theta$, where θ is in *radians*:

$$\sin\theta = \theta - \frac{\theta^3}{3!} + \frac{\theta^5}{5!} - \ldots$$

For small values of θ, the higher order terms in θ^3, θ^5 . . . are *small* compared to θ, so it follows that $\sin\theta \cong \theta$. The difference between $\sin\theta$ and θ is less than 1% for $0 < \theta < 15°$ (where $15° \cong 0.26$ rad).

NOTES FROM SELECTED CHAPTER SECTIONS_____

13.1 SIMPLE HARMONIC MOTION

Oscillatory motions are exhibited by many physical systems such as a mass attached to a spring, a pendulum, atoms in a solid, stringed musical instruments, and electrical circuits driven by a source of alternating current. *Simple harmonic motion* of a mechanical system corresponds to the oscillation of an object between two points for an indefinite period of time, with no loss in mechanical energy.

An object exhibits simple harmonic motion if the net external force acting on it is a *linear restoring force*.

The value of the phase constant δ depends on the initial displacement and initial velocity of the body. Two special cases are discussed in Section 13.1. Fig. 13.1 on the following page represents plots of the displacement, velocity, and acceleration versus time assuming that at $t = 0$, $x_0 = A$ and $v_0 = 0$. In this case, one finds that $\delta = 0$. Note that the velocity is 90° out of phase with the displacement. That is, when $|x|$ is a maximum, v is zero; while $|v|$ is a maximum when x is zero. Furthermore, note that the acceleration is 180° out of phase with the displacement. That is, when x is a maximum and positive, $|a|$ is a maximum, but a is negative. In other words, a is proportional to x, but in the *opposite* direction.

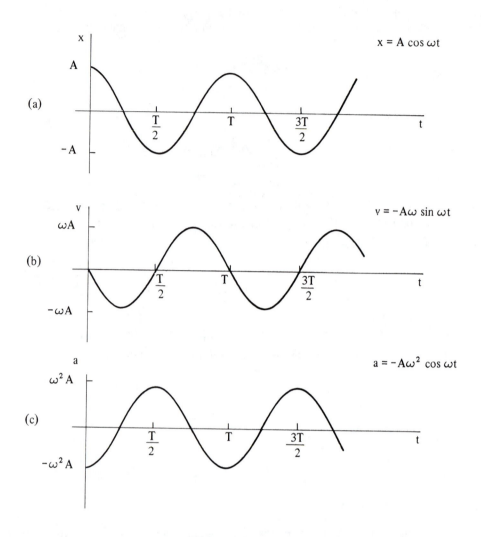

$x = A \cos \omega t$

(a)

$v = -A\omega \sin \omega t$

(b)

$a = -A\omega^2 \cos \omega t$

(c)

Figure 13.1

Representation of (a) the displacement, (b) the velocity, and (c) the acceleration as a function of time for an object moving with simple harmonic motion.

13.2 MASS ATTACHED TO A SPRING

The most common system which undergoes simple harmonic motion is the mass-spring system shown in Fig. 13.2 on the following page. The mass is assumed to move on a horizontal, *frictionless* surface. The point x = 0 is the equilibrium position of the mass; that is, the point where the mass would reside if left undisturbed. In this position, there is no horizontal force on the mass. When the mass is displaced a distance x from its equilibrium position, the spring produces a linear restoring force given by Hooke's law, F = –kx, where k is the force constant of the spring, and has SI units of N/m. The minus sign means that F is to the *left* when the displacement x is positive, whereas F is to the *right* when x is negative. In other words, the direction of the force F is *always* towards the equilibrium position.

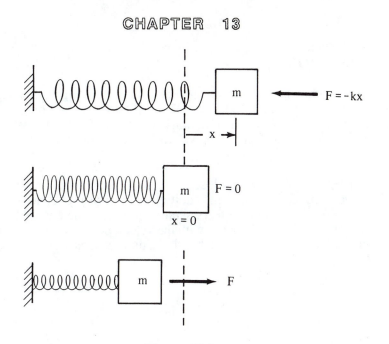

Figure 13.2

Oscillating motion of a mass on the end of a spring.

13.3 ENERGY OF THE SIMPLE HARMONIC OSCILLATOR

You should study carefully the comparison between the motion of the mass-spring system and that of the simple pendulum. In particular, notice that when the displacement is a maximum, the energy of the system is entirely potential energy; whereas, when the displacement is zero, the energy is entirely kinetic energy. This is consistent with the fact that $v = 0$ when $|x| = A$, while $v = v_{max}$ when $x = 0$. For an arbitrary value of x, the energy is the sum of K and U.

13.4 THE PENDULUM

A *simple pendulum consists* of a mass M attached to a light string of length L as shown in Fig. 13.3. When the angular displacement θ is small during the entire motion (less than about 15°), the pendulum exhibits simple harmonic motion. In this case, the resultant force acting on the mass m equals the component of weight *tangent* to the circle, and has a magnitude mgsin θ. Since this force is always directed towards $\theta = 0$, it corresponds to a restoring force. For small θ, we use the small angle approximation sin $\theta \cong \theta$. In this approximation, the equation of motion reduces to Eq. 13.22.

This equation is *identical* in form to Eq. 13.13 for the mass-spring system. Its solution is therefore of the general form $\theta = \theta_0 \cos (\omega t + \delta)$, where ω is given by Eq. 13.23. The period of motion is given by Eq. 13.24. In other words, the period depends only on the length of the pendulum and the acceleration of gravity. The period *does not* depend on mass, so we conclude that *all* simple pendula of equal length oscillate with the same frequency and period.

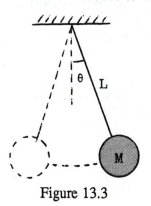

Figure 13.3

13.6 DAMPED OSCILLATIONS

Damped oscillations occur in realistic systems in which retarding forces such as friction are present. These forces will reduce the amplitudes of the oscillations with time, since mechanical energy is continually lost by the system. When the retarding force is assumed to be proportional to the velocity, but small compared to the restoring force, the system will still oscillate, but the amplitude will decrease exponentially with time.

It is possible to compensate for the energy lost in a damped oscillator by adding an additional driving force that does positive work on the system. This additional energy supplied to the system must at least equal the energy lost due to friction to maintain constant amplitude. The energy transferred to the system is a maximum when the driving force is in phase with the velocity of the system. The amplitude is a maximum when the frequency of the driving force matches the natural (resonance) frequency of the system.

EQUATIONS AND CONCEPTS

Applying *Newton's second law* to the motion in the x direction gives $F = ma_x = -kx$. Since $a_x = \dfrac{d^2x}{dt^2}$, this is equivalent to Eq. 13.13.

$$\frac{d^2x}{dt^2} = -\frac{k}{m}x \qquad (13.13)$$

The general solution to Eq. 13.13 represents the *displacement versus time*, x(t), provided that $\omega^2 = \dfrac{k}{m}$. In this expression, A represents the *amplitude* of the motion, $\omega t + \delta$ is the *phase*, ω is the *angular frequency* (rad/s), and δ is the *phase constant*.

$$x(t) = A \cos(\omega t + \delta) \qquad (13.1)$$

$$\omega^2 = \frac{k}{m} \qquad (13.14)$$

The *period of motion* T equals the time it takes the mass to complete *one* oscillation; that is, the time it takes the mass to return to its original position for the first time. The *frequency* of the motion, f, numerically equals the inverse of the period and represents the number of oscillations per unit time. T is measured in seconds, while f is measured in s^{-1} or Hertz (Hz).

$$T = \frac{2\pi}{\omega} = 2\pi\sqrt{\frac{m}{k}} \qquad (13.16)$$

$$f = \frac{1}{T} = \frac{\omega}{2\pi} \qquad (13.3)$$

Taking the first derivative of x with respect to time gives the *velocity of the oscillator as a function of time*.

$$v = \frac{dx}{dt} = -\omega A \sin(\omega t + \delta) \qquad (13.5)$$

The acceleration as a function of time is equal to the time derivative of the velocity (or the second derivative of the displacement). Note from Eq. 13.7 that the *acceleration* (and hence the force) *is always proportional to and opposite the displacement.*

$$a = \frac{dv}{dt} = -\omega^2 A \cos(\omega t + \delta) \quad (13.6)$$

or

$$a = -\omega^2 x \quad (13.7)$$

The kinetic energy of a simple harmonic oscillator is given by $\frac{1}{2} mv^2$, while the potential energy is equal to $\frac{1}{2} kx^2$. Using Eqs. 13.1 and 13.5, together with the fact that $\omega^2 = k/m$, gives the *total* energy E of the oscillator. Note that E remains constant since we have assumed there are no nonconservative forces acting on the system. The total energy of the simple harmonic oscillator is a constant of the motion and is proportional to the amplitude.

$$E = \frac{1}{2} mv^2 + \frac{1}{2} kx^2$$

or

$$E = \frac{1}{2} kA^2 \quad (13.20)$$

Energy conservation can be used to obtain an expression for velocity as a function of position.

$$v = \pm \sqrt{\frac{k}{m}(A^2 - x^2)} \quad (13.21)$$

The equation of motion for the simple pendulum assumes a small displacement so that $\sin\theta \approx \theta$.

$$\frac{d^2\theta}{dt^2} = -\frac{g}{L}\theta \quad (13.22)$$

The period and frequency of a simple pendulum depend only on the length of the supporting string and the value of the acceleration due to gravity.

$$\omega = \sqrt{\frac{g}{L}} \quad (13.23)$$

$$T = 2\pi\sqrt{\frac{L}{g}} \quad (13.24)$$

ANSWERS TO SELECTED QUESTIONS

1. What is the total distance traveled by a body executing simple harmonic motion in a time equal to its period if its amplitude is A?

Answer: It travels a distance of 2A.

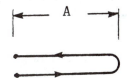

2. If the coordinate of a particle varies as x = –A cos ωt, what is the phase constant δ in Eq. 13.1? At what position does the particle begin its motion?

Answer: $\delta = \pm \pi/2$; At t = 0, x = –A

3. Does the displacement of an oscillating particle between t = 0 and a later time t necessarily equal the position of the particle at time t? Explain.

Answer: The two will be equal if the origin of coordinates coincides with the position of the particle at t = 0.

5. Can the amplitude A and phase constant δ be determined for an oscillator if only the position is specified at t = 0? Explain.

Answer: No. It is necessary to know both the position and velocity at t = 0.

7. If a mass-spring system is hung vertically and set into oscillation, why does the motion eventually stop?

Answer: There will always be some friction present, such as air resistance.

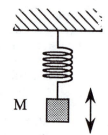

8. Explain why the kinetic and potential energies of a mass-spring system can never be negative.

Answer: The kinetic energy is proportional to the square of the speed, while the potential energy is proportional to the square of the displacement. Therefore, both must be positive quantities.

9. A mass-spring system undergoes simple harmonic motion with an amplitude A. Does the total energy change if the mass is doubled but the amplitude is not changed? Do the kinetic and potential energies depend on the mass? Explain.

Answer: No. Since $E = \frac{1}{2} kA^2$, changing the mass has no effect on the total energy. However, the kinetic energy depends on the mass.

10. What happens to the period of a simple pendulum if its length is doubled? What happens to the period if the mass that is suspended is doubled?

Answer: Since $T = 2\pi \sqrt{\frac{L}{g}}$, doubling L will increase T by a factor of $\sqrt{2}$. Doubling the mass will not change the period.

11. A simple pendulum is suspended from the ceiling of a stationary elevator, and the period is determined. Describe the changes, if any, in the period if the elevator (a) accelerates upward, (b) accelerates downward, and (c) moves with constant velocity.

Answer: If it accelerates upwards, the effective "g" is greater than the acceleration of gravity, so the period decreases. If it accelerates downward, the effective "g" is less than the acceleration of gravity, so the period increases. If it moves with constant velocity, the period does not change. (If the pendulum is in free fall, it does not oscillate.)

12. A simple pendulum undergoes simple harmonic motion when θ is small. Will the motion be *periodic* if θ is large? How does the period of motion change as θ increases?

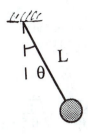

Answer: Yes. The period will increase as the amplitude of motion increases.

14. Will damped oscillations occur for any values of b and k? Explain.

Answer: No. If the resistive force is greater than the restoring force of the spring, (that is, if $b^2 > 4mk$), the system will be overdamped and will not oscillate.

15. Is it possible to have damped oscillations when a system is at resonance? Explain.

Answer: Yes. At resonance, the amplitude of a damped oscillator will remain constant. If the system were not damped, the amplitude would increase without limit at resonance.

17. A platoon of soldiers marches in step along a road. Why are they ordered to break step when crossing a bridge?

Answer: There is a chance that the periodic driving force associated with their marching will be close to one of the natural frequencies of vibration of the bridge and thus cause it to collapse.

SOLUTIONS TO SELECTED END-OF-CHAPTER PROBLEMS

5. The displacement of a body is given by the expression $x = (8.0 \text{ cm})\cos(2t + \pi/3)$, where x is in cm and t is in s. Calculate (a) the velocity and acceleration at $t = \pi/2$ s, (b) the maximum speed and the earliest time $(t > 0)$ at which the particle has this speed, and (c) the maximum acceleration and the earliest time $(t > 0)$ at which the particle has this acceleration.

Solution

$$x = (8.0 \text{ cm})\cos\left(2t + \frac{\pi}{3}\right)$$

(a) $\quad v = -(16.0 \text{ cm/s})\sin\left(2t + \frac{\pi}{3}\right)$ $\qquad$ at $t = \dfrac{\pi}{2}$ s, $v = \boxed{13.9 \text{ cm/s}}$

$\quad a = -(32.0 \text{ cm/s}^2)\cos\left(2t + \frac{\pi}{3}\right)$ $\qquad$ at $t = \dfrac{\pi}{2}$ s, $a = \boxed{16.0 \text{ cm/s}^2}$

(b) $\quad v_{max} = \boxed{16.0 \text{ cm/s}}$ $\qquad$ This occurs when $\quad t = \dfrac{1}{2}\left[\sin^{-1}(1) - \frac{\pi}{3}\right] = \boxed{0.262 \text{ s}}$

(c) $\quad a_{max} = \boxed{32.0 \text{ cm/s}^2}$ $\qquad$ This occurs when $\quad t = \dfrac{1}{2}\left[\cos^{-1}(-1) - \frac{\pi}{3}\right] = \boxed{1.05 \text{ s}}$

15. A 0.5-kg mass attached to a spring of force constant 8 N/m vibrates with simple harmonic motion with an amplitude of 10 cm. Calculate (a) the maximum value of its speed and acceleration, (b) the speed and acceleration when the mass is at x = 6 cm from the equilibrium position, and (c) the time it takes the mass to move from x = 0 to x = 8 cm.

Solution

(a) $\omega = \sqrt{\dfrac{k}{m}} = \sqrt{\dfrac{8 \text{ N/m}}{0.5 \text{ kg}}} = 4 \text{ s}^{-1}$

Therefore, position is given by x = (10 cm)sin(4t). From this we find that

$v = (40 \text{ cm/s})\cos(4t)$ $\qquad$ $v_{max} = \boxed{40 \text{ cm/s}}$

$a = -(160 \text{ cm/s}^2)\sin(4t)$ $\qquad$ $a_{max} = \boxed{160 \text{ cm/s}^2}$

(b) $t = \dfrac{1}{4}\sin^{-1}\left(\dfrac{x}{10}\right)$

When x = 6 cm, t = 0.161 s and we find

$v = (40 \text{ cm/s})\cos[4(0.161)] = \boxed{32 \text{ cm/s}}$

$a = -(160 \text{ cm/s}^2)\sin[4(0.161)] = \boxed{-96 \text{ cm/s}^2}$

(c) Using $t = \dfrac{1}{4}\sin^{-1}\left(\dfrac{x}{10}\right)$

When x = 0, t = 0 and when x = 8 cm, t = 0.232 s. Therefore $\Delta t = \boxed{0.232 \text{ s}}$

23. A particle executes simple harmonic motion with an amplitude of 3.0 cm. At what displacement from the midpoint of its motion will its speed equal one half of its maximum speed?

Solution

From energy considerations, $v^2 + \omega^2 x^2 = \omega^2 A^2$

$v_{max} = \omega A$ and $v = \dfrac{v_{max}}{2} = \dfrac{\omega A}{2}$ so $\dfrac{1}{2}\omega^2 A^2 + \omega^2 x^2 = \omega^2 A^2$

From this we find $x^2 = \dfrac{3A^2}{4}$ and $x = \pm \dfrac{A\sqrt{3}}{2} = \pm \dfrac{3\sqrt{3}}{2} = \pm \boxed{2.60 \text{ cm}}$ where A = 3.0 cm

33. A physical pendulum in the form of a planar body exhibits simple harmonic motion with a frequency of 1.5 Hz. If the pendulum has a mass of 2.2 kg and the pivot is located 0.35 m from the center of mass, determine the moment of inertia of the pendulum.

Solution

$f = 1.5 \text{ Hz}, \quad d = 0.35 \text{ m}, \quad \text{and} \quad m = 2.2 \text{ kg}$

$$T = \frac{1}{f} \qquad T = 2\pi\sqrt{\frac{I}{mgd}} \qquad T^2 = \frac{4\pi^2 I}{mgd}$$

$$I = \frac{T^2 mgd}{4\pi^2} = \left(\frac{1}{f}\right)^2 \frac{mgd}{4\pi^2} = \frac{(2.2 \text{ kg})(9.80 \text{ m/s}^2)(0.35 \text{ m})}{(1.5 \text{ s}^{-1})^2(4\pi^2)}$$

$$I = \boxed{8.50 \times 10^{-2} \text{ kg·m}^2}$$

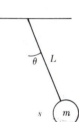

47. When the simple pendulum illustrated in Figure 13.20 makes an angle θ with the vertical, its speed is v. (a) Calculate the total mechanical energy of the pendulum as a function of v and θ. (b) Show that when θ is small, the potential energy can be expressed as

$$\frac{1}{2} mgL\theta^2 = \frac{1}{2} m\omega^2 s^2$$

Hint: In part (b), approximate cos θ by

$$\cos \theta \approx 1 - \frac{\theta^2}{2}$$

Figure 13.20

Solution

(a) $E = \frac{1}{2} I\omega^2 + mgh$ where $I = mL^2$ and $\omega = \frac{v}{L}$

When the pendulum makes an angle θ with the vertical, the mass is a distance h above the lowest point, where $h = L(1 - \cos \theta)$.

$$E = \frac{1}{2} mv^2 + mgL(1 - \cos \theta)$$

(b) $U = mgL(1 - \cos \theta)$ and for small θ, $U \approx mgL\left[1 - \left(1 - \frac{\theta^2}{2}\right)\right] = \frac{mg \cdot L\theta^2}{2}$

Also since $\theta L = s$ and $\omega^2 = \frac{g}{L}$, we have $U = \frac{m\omega^2 s^2}{2}$

51. A mass M is attached to the end of a uniform rod of mass m and length L, which is pivoted at the top (Fig. 13.22). (a) Determine the tensions in the rod at the pivot and at the point P when the system is stationary. (b) Calculate the period of oscillation for small displacements from equilibrium, and determine this period for L = 2 m. (Hint: Assume the mass M is a point mass, and make use of Eq. 13.26.)

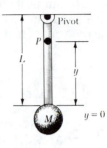

Figure 13.22

Solution

(a)   At the pivot, $T = Mg + Mg = 2Mg$. At P, a fraction of the rod's mass $\left(\dfrac{y}{L}\right)$ pulls down as well as does the ball. At this point, $T = Mg\left(\dfrac{y}{L}\right) + Mg = Mg\left(1 + \dfrac{y}{L}\right)$

(b)   Relative to the pivot, $I = \dfrac{1}{3}ML^2 + ML^2$. For a physical pendulum, $T = 2\pi\sqrt{\dfrac{I}{mgd}}$ where $m = 2M$ and d is the distance from the pivot to the center of mass. Therefore,

$$d = \frac{\dfrac{ML}{2} + ML}{M + M} = \frac{3L}{4} \qquad \text{and we have} \qquad T = \frac{4\pi}{3}\sqrt{\frac{2L}{g}}$$

For L = 2 m, $\qquad T = \dfrac{4\pi}{3}\sqrt{\dfrac{2(2\text{ m})}{9.80\text{ m/s}^2}} = \boxed{2.68\text{ s}}$

---

53. A small, thin disk of radius r and mass m is attached rigidly to the face of a second thin disk of radius R and mass M as shown in Figure 13.24. The center of the small disk is located at the edge of the large disk. The large disk is mounted at its center on a frictionless axle. The assembly is rotated through an angle θ from its equilibrium position and released. (a) Show that the speed of the center of the small disk as it passes through the equilibrium position is

$$v = 2\sqrt{\frac{Rg(1 - \cos\theta)}{(M/m) + (r/R)^2 + 2}}$$

(b) Show that the period of the motion is

$$T = 2\pi\sqrt{\frac{(M + 2m)R^2 + mr^2}{2mgR}}$$

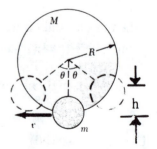

Figure 13.24

Solution

(a) $\Delta K + \Delta U = 0$, thus $K_{top} + U_{top} = K_{bot} + U_{bot}$, where $K_{top} = U_{bot} = 0$. Therefore $mgh = \frac{1}{2} I\omega^2$,

but $h = R - R\cos\theta = R(1 - \cos\theta)$, $\omega = \frac{v}{R}$, and $I = \frac{1}{2} MR^2 + \frac{1}{2} mr^2 + mR^2$. Substituting we find

$$mgR(1 - \cos\theta) = \frac{1}{2}\left(\frac{1}{2} MR^2 + \frac{1}{2} mr^2 + mR^2\right)\frac{v^2}{R^2}$$

$$mgR(1 - \cos\theta) = \left(\frac{1}{4} M + \frac{1}{4}\frac{r^2}{R^2} m + \frac{1}{2} m\right)v^2$$

and $$v^2 = \frac{4gR(1 - \cos\theta)}{\frac{M}{m} + \frac{r^2}{R^2} + 2}$$

so $$v = 2\sqrt{\frac{Rg(1 - \cos\theta)}{\frac{M}{m} + \frac{r^2}{R^2} + 2}}$$

(b) $T = 2\pi\sqrt{\dfrac{I}{M_T g d_{cm}}}$ $\qquad M_T = M + m$ $\qquad d_{cm} = \dfrac{mR + M(0)}{m + M}$

$$T = 2\pi\sqrt{\frac{\frac{1}{2} MR^2 + \frac{1}{2} mr^2 + mR^2}{mgR}} = 2\pi\sqrt{\frac{(M + 2m)R^2 + mr^2}{2mgR}}$$

---

65. A mass m is connected to two rubber bands of length L, each under tension T, as in Figure 13.31. The mass is displaced by a *small* distance y vertically. Assuming the tension does not change appreciably, show that (a) the restoring force is $-(2T/L)y$ and (b) the system exhibits simple harmonic motion with an angular frequency given by $\omega = \sqrt{2T/mL}$.

Solution

Figure 13.31

(a) $\Sigma F = -2T\sin\theta \mathbf{j}$ where $\theta = \tan^{-1}\left(\frac{y}{L}\right)$

Therefore, for a small displacement $\sin\theta \approx \tan\theta = \frac{y}{L}$ and $F = \left(-\frac{2Ty}{L}\right)\mathbf{j}$

(b) For a spring system, $F = -kx$ becomes $F = -\left(\frac{2T}{L}\right)y$

Therefore, $\omega = \sqrt{\frac{k}{m}} = \sqrt{\frac{2T}{mL}}$

The Law of Universal Gravitation

OBJECTIVES

1. State Kepler's three laws of planetary motion and recognize that the laws are empirical in nature; that is, they are based on astronomical data.

2. Describe the nature of Newton's universal law of gravity, and the method of deriving Kepler's third law ($T^2 \sim r^3$) from this law for circular orbits.

3. Recognize that Kepler's second law is a consequence of conservation of angular momentum and the central nature of the gravitational force.

4. Understand the concepts of the gravitational field and the gravitational potential energy, and know how to derive the expression for the potential energy for a pair of particles separated by a distance r.

5. Describe the total energy of a planet or earth satellite moving in a circular orbit about a large body located at the center of motion. Note that the total energy is negative, as it must be for any closed orbit.

6. Understand the meaning of escape velocity, and know how to obtain the expression for v_{esc} using the principle of conservation of energy.

7. Learn the method for calculating the gravitational force between a particle and an extended object. In particular, you should be familiar with the force between a particle and a spherical body when the particle is located outside and inside the spherical body.

SKILLS

In this chapter we made use of the definite integral in evaluating the potential energy function associated with the conservative gravitational force. You should be familiar with the following type of definite integral:

$$\int_{x_1}^{x_2} x^n dx = \frac{x^{n+1}}{n+1}\Bigg]_{x_1}^{x_2} = \frac{x_2^{n+1} - x_1^{n+1}}{n+1} \ (n \neq -1)$$

For example, in deriving Eq. 14.10, we used the above expression as follows:

$$\int_{r_i}^{r_f} \frac{dr}{r^2} = \int_{r_i}^{r_f} r^{-2} dr = \frac{r_f^{-1} - r_i^{-1}}{-2+1} = -\left(\frac{1}{r_f} - \frac{1}{r_i}\right)$$

NOTES FROM SELECTED CHAPTER SECTIONS

14.1 NEWTON'S UNIVERSAL LAW OF GRAVITY

Newton's law of gravitation states that every particle in the universe attracts every other particle with a force that is directly proportional to the product of their masses and inversely proportional to the square of their distance of separation. The gravitational force is an *action-at-a-distance force* which always exists between two particles regardless of the medium which separates them.

14.4   KEPLER'S LAWS

Using astronomical data provided by Brahe, Kepler deduced the following empirical laws as they apply to our solar system:

1. All planets move in elliptical orbits with the sun at one of the focal points.

2. The radius vector drawn from the sun to any planet sweeps out equal areas in equal times.

3. The square of the orbital period of any planet is proportional to the cube of the semi-major axis for the elliptical orbit.

Kepler's second law is a consequence of the central nature of the gravitational force, which leads to conservation of angular momentum. Kepler's third law follows from the inverse square nature of the gravitational force.

14.7   GRAVITATIONAL POTENTIAL ENERGY

The *gravitational potential energy* for any pair of particles varies as $1/\pi$. The potential energy is negative since the force is one of attraction and the potential energy is taken to be zero when the distance of separation between the two particles is infinity. The absolute value of the potential energy is the *binding energy* of the system.

14.8   ENERGY CONSIDERATIONS IN PLANETARY AND SATELLITE MOTION

Both the total energy and the total angular momentum of a planet-sun system are *constants of the motion*.

The *escape velocity* of an object is independent of the mass of the object and is independent of the direction of the velocity.

14.10   GRAVITATIONAL FORCE BETWEEN A PARTICLE AND A SPHERICAL MASS

You should be aware of some interesting special cases of gravitational force between a particle and a spherically symmetric mass distribution.

For the case of a *spherical shell*:

1. If a particle of mass m is located *outside* a spherical shell of mass M, the spherical shell attracts the particle as though the mass of the shell were concentrated at its center.

2. If the particle is located *inside* the spherical shell, the force on it is zero.

Note that the shell of mass does *not* act as a gravitational shield. The particle may experience forces due to other masses outside the shell.

For the case of a *solid sphere*:

1. If a particle of mass m is located *outside* a homogeneous solid sphere of mass M, the sphere attracts the particle as though the mass of the sphere were concentrated at its center. This follows from case 1 above, since a solid sphere can be considered a collection of concentric spherical shells.

2. If a particle of mass m is located (at a distance r from the center) *inside* a homogeneous solid sphere of mass M and radius R, the force on m is due *only* to the mass M' contained within the sphere of radius r < R.

$$F = -\frac{GmM}{R^3} r\,\hat{r} \quad \text{for } r < R$$

That is, the force goes to zero at the center of the sphere.

3. If a particle is located *inside* a solid sphere having a density ρ that is spherically symmetric but *not* uniform, then M' is given by an integral of the form $M' = \int \rho \, dV$, where the integration is taken over the volume contained *within* the dotted surface. This integral can be evaluated if the radial variation of ρ is given. The integral is easily evaluated if the mass distribution has spherical symmetry; that is, if ρ is a function of r only. In this case, we take the volume element dV as the volume of a spherical shell of radius r and thickness dr, so that $dV = 4\pi r^2 \, dr$.

EQUATIONS AND CONCEPTS

The *universal law of gravity* states that any pair of particles *attract* each other with a force that is proportional to the product of their masses and inversely proportional to the square of their separation.

$$F = G \frac{m_1 m_2}{r^2} \tag{14.1}$$

The constant G is called the *universal constant of gravity*.

$$G = 6.673 \times 10^{-11} \frac{N \cdot m^2}{kg^2} \tag{14.2}$$

The acceleration due to gravity, g', decreases with increasing altitude, h, measured from the earth's surface.

$$g' = \frac{GM_e}{(R_e + h)^2} \tag{14.5}$$

A *gravitational field* **g** exists at some point in space if a particle of mass m experiences a gravitational force **F** = m**g** at that point. That is, the gravitational field represents the ratio of the gravitational force experienced by the mass divided by that mass.

$$\mathbf{g} = \frac{\mathbf{F}}{m} \tag{14.8}$$

Since the gravitational force is conservative, we can define a gravitational energy function corresponding to that force. As a mass m moves from one position to another in the presence of the earth's gravity, its potential changes by an amount given by Eq. 14.10, where r_i and r_f are the initial and final distances of the mass from the center of the earth.

$$\Delta U = -GM_e m \left(\frac{1}{r_i} - \frac{1}{r_f} \right) \tag{14.10}$$

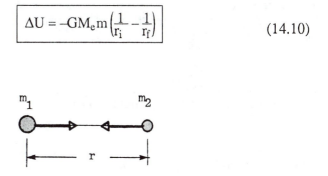

Figure 14.1

The *gravitational potential energy* associated with *any pair* of particles of masses m_1 and m_2 separated by a distance r is given by Eq. 14.12. The negative sign in this expression corresponds to the attractive nature of the gravitational force. An external agent must do positive work to increase the separation of the particles.

$$U = -G\frac{m_1 m_2}{r}$$

(14.12)

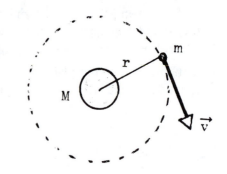

As a body of mass m moves in a circular orbit around a very massive body of mass M (where M >> m), the *total energy* of the system is the sum of the kinetic energy of m (taking the massive body to be at rest) and the potential energy of the system. When the two contributions are evaluated, one finds that the *total energy* E is negative, and given by Eq. 14.17. This arises from the fact that the (positive) kinetic energy is equal to one-half the magnitude of the (negative) potential energy.

Figure 14.1

$$E = -\frac{GMm}{2r}$$

(14.17)

The *escape velocity* is defined as the *minimum* velocity a body must have, when projected from the earth whose mass is M_e and radius is R_e, in order to escape the earth's gravitational field (that is, to just reach r = ∞ with zero speed). Note that v_{esc} does not depend on the mass of the projected body.

$$v_{esc} = \sqrt{\frac{2GM_e}{R_e}}$$

(14.19)

The *potential energy* associated with a particle of mass m and an *extended* body of mass M can be evaluated using Eq. 14.20, where the extended body is divided into segments of mass dM, and r is the distance from dM to the particle.

$$U = -Gm\int \frac{dM}{r}$$

(14.20)

If a particle of mass m is *outside* a *uniform* solid sphere or spherical shell of radius R, the sphere attracts the particle as though the mass of the sphere were concentrated at its center.

$$F = -\frac{GMm}{r^2}\hat{r} \qquad (r > R)$$

(14.22a)

If a particle is located *inside a uniform spherical shell*, the force acting on the particle is *zero*.

$$F = 0 \qquad (r < R)$$

(14.22b)

If a particle is *inside* a *homogeneous solid sphere* of radius R, the force acting on the particle acts toward the center of the sphere and is proportional to the distance r from the center to the particle.

$$F = -\frac{GmM}{R^3}r\hat{r} \qquad (r < R)$$

(14.24)

EXAMPLE PROBLEM SOLUTIONS

Example 14.1 Three point masses of 3 kg, 4 kg, and 5 kg are placed as shown in Figure 14.2. Calculate the resultant gravitational force on the 4-kg mass. (Assume they are isolated from the rest of the universe.)

Solution

First, we will calculate the individual forces on the 4-kg mass, and then take a vector sum to get the resultant force on it.

The force on the 4-kg mass due to the 3-kg mass is *upwards*, and is given by

$$\mathbf{F}_{43} = G\,\frac{m_3 m_4}{r^2}\,\mathbf{j} = 6.67 \times 10^{-11}\,\frac{Nm^2}{kg^2} \times \frac{3\ kg \times 4\ kg}{(2\ m)^2}\,\mathbf{j}$$

or

$$\mathbf{F}_{43} = 2.0 \times 10^{-10}\,\mathbf{j}\ N$$

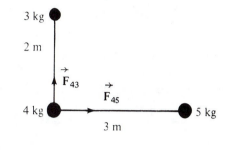

Figure 14.2

The force on the 4-kg mass due to the 5-kg mass is to the *right*, and is given by

$$\mathbf{F}_{45} = 6.67 \times 10^{-11}\,\frac{Nm^2}{kg^2} \times \frac{4\ kg \times 5\ kg}{(3\ m)^2}\,\mathbf{i} = 1.5 \times 10^{-10}\,\mathbf{i}\ N$$

Therefore, the resultant force on the 4-kg mass is the *vector* sum of $\mathbf{F}_{43}$ and $\mathbf{F}_{45}$.

$$\mathbf{F} = \mathbf{F}_{43} + \mathbf{F}_{45} = (1.5 \times 10^{-10}\,\mathbf{i} + 2.0 \times 10^{-10}\,\mathbf{j})\ N$$

The magnitude of this force is 2.5×10^{-10} N, which is equivalent to only 5.6×10^{-11} lb! The direction of **F** is given by arc tan (2/1.5) = 53°. That is, **F** makes an angle of 53° with the x axis.

Example 14.2 Calculate the total potential energy of the system of 3 particles shown in Fig. 14.2.

Solution

For a system of three particles, summing over all *pairs* according to Eq. 14.9 gives

$$U = U_{12} + U_{13} + U_{23}$$

$$U = -G\left(\frac{m_1 m_2}{r_{12}} + \frac{m_1 m_3}{r_{13}} + \frac{m_2 m_3}{r_{23}}\right)$$

Note that in Fig. 14.2, $m_1 = 3$ kg, $m_2 = 4$ kg, $m_3 = 5$ kg, $r_{12} = 2$ m, $r_{13} = \sqrt{13}$ m and $r_{23} = 3$ m. Therefore

$$U = -6.67 \times 10^{-11}\left(\frac{3 \times 4}{2} + \frac{3 \times 5}{\sqrt{13}} + \frac{4 \times 5}{3}\right) J = -1.12 \times 10^{-9}\ J$$

Example 14.3   Assume the moon orbits the earth in a circular path with a period of 27 days, at a distance of 3.8×10^8 m. Using this information, calculate the mass of the earth.

Solution

The centripetal force acting on the moon is provided by the gravitational attraction between the moon and the earth. Therefore, Newton's second law applied to the moon, together with Eq. 14.1, gives

$$F_m = G \frac{M_e M_m}{r^2} = M_m \frac{v^2}{r} \qquad (1)$$

where v is the orbital velocity of the moon. Since the moon travels a distance of $2\pi r$ in a time of 27 days, where $r = 3.8 \times 10^8$ m, we get

$$v = \frac{2\pi r}{T} = \frac{2\pi \times 3.8 \times 10^8 \text{ m}}{27 \text{ days} \times 8.64 \times 10^4 \frac{s}{day}} = 1.02 \times 10^3 \text{ m/s} \qquad (2)$$

Therefore, simplifying (1) for the mass of the earth gives

$$M_e = \frac{v^2 r}{G} = \frac{\left(1.02 \times 10^3 \frac{m}{s}\right)^2 \times 3.8 \times 10^8 \text{ m}}{6.67 \times 10^{-11} \frac{N \cdot m^2}{kg^2}}$$

or

$$M_e \approx 5.93 \times 10^{24} \text{ kg}$$

This is to be compared with the more accurate value of 5.98×10^{24} kg. Using this value of M_e, and the fact that the moon's mass is about 7.34×10^{22} kg, you should show that the force of attraction between the earth and the moon is equal to about 2.03×10^{20} N. This provides a centripetal acceleration for the moon equal to 2.8×10^{-8} m/s^2.

═══════════════════════════

ANSWERS TO SELECTED QUESTIONS

4. If a system consists of five distinct particles, how many terms appear in the expression for the total potential energy?

Answer: Ten. For five particles, there are ten distinct pairs of particles, as shown in the diagram.

5. Is it possible to calculate the potential energy function associated with a particle and an extended body without knowing the geometry or mass distribution of the extended body?

Answer: No. Both are required as discussed in Section 14.6.

6. Does the escape velocity of a rocket depend on its mass? Explain.

Answer: No. Eq. 14.16 shows that the escape velocity is independent of the mass of the rocket.

9. Is the magnitude of the potential energy associated with the earth-moon system greater than, less than, or equal to the kinetic energy of the moon relative to the earth?

Answer: The potential energy is greater than the kinetic energy by a factor of two.

10. Explain carefully why there is no work done on a planet as it moves in a circular orbit around the sun, even though a gravitational force is acting on the planet. What is the *net* work done on a planet during each revolution as it moves around the sun in an elliptical orbit?

Answer: The displacement is always perpendicular to the central force as shown in the figure, therefore the work done by the gravitational force must be zero. Furthermore, the net work done is also zero for the same reason.

11. A particle is projected through a small hole into the interior of a large spherical shell. Describe the motion of the particle in the interior of the shell.

Answer: The particle will move with constant velocity in the direction of its original motion until it hits the opposite side of the shell. Remember that the gravitational force (and field) is zero inside the shell.

13. Neglecting the density variation of the earth, what would be the period of a particle moving in a smooth hole dug through the earth's center?

Answer: 84.3 min. The answer is the same as that obtained in Example 14.7. The period T is 84.3 min. for *any* tunnel through the earth, including the one that passes through its center.

14. With reference to Fig. 14.8, consider the area swept out by the radius vector in the time intervals $t_2 - t_1$ and $t_4 - t_3$. Under what condition is A_1 equal to A_2?

Answer: The areas will be equal if the time intervals are equal.

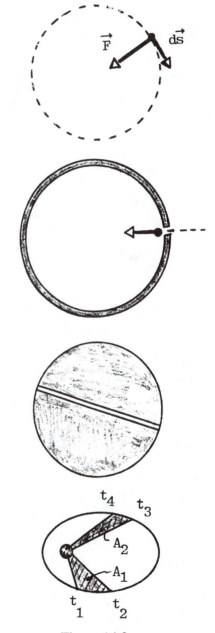

Figure 14.8

15. If A_1 equals A_2 in Fig. 14.8, is the average speed of the planet in the time interval $t_2 - t_1$ less than, equal to, or greater than its average speed in the time interval $t_4 - t_3$?

Answer: The average speed in the interval $t_2 - t_1$ is greater than the average speed in the interval $t_4 - t_3$.

16. At what position in its elliptical orbit is the speed of a planet a maximum? At what position is the speed a minimum?

Answer: The speed is a maximum at the perihelion, p, and is a minimum at the aphelion, a.

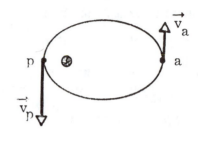

SOLUTIONS TO SELECTED END-OF-CHAPTER PROBLEMS

3. Three 5-kg masses are located at the corners of an equilateral triangle having sides 0.25 m in length. Determine the magnitude and direction of the resultant gravitational force on one of the masses due to the other two masses.

Solution

The magnitude of either force is given by

$$F = (6.672 \times 10^{-11} \ \text{N·m}^2/\text{kg}^2)(5 \ \text{kg})(5 \ \text{kg})/(0.25 \ \text{m})^2 = 2.67 \times 10^{-8} \ \text{N}$$

$$\Sigma F_x = (2.67 \times 10^{-8} \ \text{N}) + (2.67 \times 10^{-8} \ \text{N})\cos(60°) = \boxed{4.01 \times 10^{-8} \ \text{N}}$$

$$\Sigma F_y = (2.67 \times 10^{-8} \ \text{N})\sin(60°) = \boxed{2.31 \times 10^{-8} \ \text{N}}$$

$$R = \sqrt{\Sigma F_x{}^2 + \Sigma F_y{}^2} = \boxed{4.62 \times 10^{-8} \ \text{N}}$$

Problem 3

$$\theta = \tan^{-1}\sqrt{\frac{\Sigma F_x}{\Sigma F_y}} = \boxed{30°}$$ from the side of the triangle or the force is directed along the bisector

toward the other masses.

17. At its aphelion, the planet Mercury is 6.99×10^{10} m from the sun, and at its perihelion, it is 4.60×10^{10} m from the sun. If its orbital speed is 3.88×10^4 m/s at the aphelion, what is its orbital speed at the perihelion?

Solution

$\Sigma \tau = 0$ and L = constant; therefore $mv_a r_a = mv_p r_p$ and

$$v_p = \left(\frac{r_a}{r_p}\right) v_a = \left(\frac{6.99}{4.60} \times 10^{10} \text{ m}\right)(3.88 \times 10^4 \text{ m/s}) = \boxed{5.90 \times 10^4 \text{ m/s}}$$

25. A system consists of three particles, each of mass 5 g, located at the corners of an equilateral triangle with sides of 30 cm. (a) Calculate the potential energy of the system. (b) If the particles are released simultaneously, where will they collide?

Solution

(a) $U_{Tot} = U_{12} + U_{13} + U_{23} = -\dfrac{3GM^2}{r}$

$$= -\frac{3(6.67 \times 10^{-11} \text{ N·m}^2/\text{kg}^2)(0.005 \text{ kg})^2}{(0.3 \text{ m})} = \boxed{-1.67 \times 10^{-14} \text{ J}}$$

(b) At the center of the equilateral triangle

Problem 25

31. (a) Calculate the minimum energy required to send a 3000-kg spacecraft from the earth to a distant point in space where earth's gravity is negligible. (b) If the journey is to take three weeks, what *average* power would the engines have to supply?

Solution

(a) $v_{esc} = \sqrt{\dfrac{2GM_e}{R_e}} = 1.12 \times 10^4$ m/s

$$K = \frac{1}{2} mv_{esc}^2 = \frac{1}{2}(3000 \text{ kg})(1.12 \times 10^4 \text{ m/s})^2 = \boxed{1.88 \times 10^{11} \text{ J}}$$

(b) $P_{av} = \dfrac{K}{\Delta t} = \dfrac{1.88 \times 10^{11} \text{ J}}{\left(21 \text{ days} \times 8.64 \times 10^4 \text{ s/day}\right)} = \boxed{103 \text{ kW}}$

Problem 31

39. A *nonuniform* rod of length L is placed along the x axis at a distance h from the origin, as in Figure 14.16. The mass per unit length, λ, varies according to the expression $\lambda = \lambda_o + Ax^2$, where λ_o and A are constants. Find the force on a particle of mass m placed at the origin. (Hint: An element of the rod has a mass $dM = \lambda\, dx$.)

Solution

The force exerted on the mass by an element of length of the rod is given by $dF = Gm\, dm/x^2$ where x is the distance between the mass m and the increment of length dx. Using

$$dm = \lambda\, dx = (\lambda_o + Ax^2)\, dx$$

we have

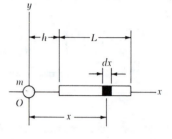

Figure 14.16

$$F = \int dF = Gm \int_d^{L+d} \frac{(\lambda_o + Ax^2)}{x^2}\, dx = \boxed{\frac{GML\lambda_o}{d(L+d)} + GmAL}$$

---

41. A uniform solid sphere has a radius of 0.4 m and a mass of 500 kg. Find the magnitude of the force on a particle of mass 50 g located (a) 1.5 m from the center of the sphere, (b) at the surface of the sphere, and (c) 0.2 m from the center of the sphere.

Solution

(a) $F = \dfrac{GmM}{r^2} = \dfrac{(6.67 \times 10^{-11}\ \text{N·m}^2/\text{kg}^2)(0.05\ \text{kg})(500\ \text{kg})}{(1.5\ \text{m})^2}$

$F = \boxed{7.41 \times 10^{-10}\ \text{N}}$

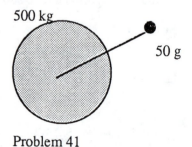

Problem 41

(b) $F = \dfrac{(6.67 \times 10^{-11}\ \text{N·m}^2/\text{kg}^2)(0.05\ \text{kg})(500\ \text{kg})}{(0.4\ \text{m})^2}$

$F = \boxed{1.04 \times 10^{-8}\ \text{N}}$

(c) In this case the mass m is a distance r from a sphere of mass,

$M = (500\ \text{kg})(0.2\ \text{m}/0.4\ \text{m})^3 = 62.5\ \text{kg}$ and

$F = \dfrac{(6.67 \times 10^{-11}\ \text{N·m}^2/\text{kg}^2)(0.05\ \text{kg})(62.5\ \text{kg})}{(0.2\ \text{m})^2} = \boxed{5.21 \times 10^{-9}\ \text{N}}$

---

55. Two hypothetical planets of masses m_1 and m_2 and radii r_1 and r_2, respectively, are at rest when they are an infinite distance apart. Because of their gravitational attraction, they head toward each other on a collision course. (a) When their center-to-center separation is d, find the speed of each planet and their *relative* velocity. (b) Find the kinetic energy of each planet *just* before they collide if $m_1 = 2 \times 10^{24}$ kg, $m_2 = 8 \times 10^{24}$ kg, $r_1 = 3 \times 10^6$ m, and $r_2 = 5 \times 10^6$ m. (Hint: Note that both energy and momentum are conserved.)

Solution

(a) At infinite separation $U = 0$ and at rest $K = 0$. Since energy is conserved, we have

$$0 = \frac{1}{2} m_1 v_1^2 + \frac{1}{2} m_2 v_2^2 + - \frac{Gm_1 m_2}{d} \tag{1}$$

The initial momentum is zero and momentum is conserved.

Therefore $\qquad\qquad 0 = m_1 v_1 - m_2 v_2 \tag{2}$

Combine (1) and (2) to find

$$v_1 = m_2 \sqrt{\frac{2G}{d(m_1 + m_2)}} \qquad \text{and} \qquad v_2 = m_1 \sqrt{\frac{2G}{d(m_1 + m_2)}}$$

Relative velocity $\qquad v_r = v_1 - (-v_2) = \sqrt{\frac{2G(m_1 + m_2)}{d}}$

(b) Substitute the given numerical values into the equation found for v_1 and v_2 in part (a) to find $v_1 = 1.03 \times 10^4$ m/s and $v_2 = 2.58 \times 10^3$ m/s. Therefore

$$K_1 = \frac{1}{2} m_1 v_1^2 = \boxed{1.07 \times 10^{32} \text{ J}}$$

and

$$K_2 = \frac{1}{2} m v_2^2 = \boxed{2.67 \times 10^{31} \text{ J}}$$

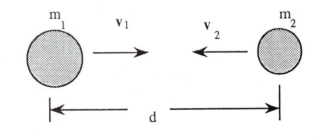

Problem 55

63. A sphere of mass M and radius R has a *nonuniform* density that varies with r, the distance from its center, according to the expression $\rho = Ar$, for $0 \le r \le R$. (a) What is the constant A in terms of M and R? (b) Determine the force on a particle of mass m placed *outside* the sphere. (c) Determine the force on the particle if it is *inside* the sphere. (Hint: See Section 14.10.)

Solution

(a) If we consider a hollow shell in the sphere with radius r and thickness dr, then $dM = \rho dV = \rho(4\pi r^2 dr)$. The total mass is then

$$M = \int \rho dV = \int_0^R (Ar)(4\pi r^2 dr) = \pi A R^4 \qquad \text{and} \qquad \boxed{A = \frac{M}{\pi R^4}}$$

(b) The total mass of the sphere acts as if it were at the center of the sphere and $F = GmM/r^2$ directed toward the center of the sphere.

(c) Inside the sphere at a distance r from the center, $dF = (Gm/r^2)dM$ where dM is just the mass of a shell enclosed within the radius r.

$$F = Gm \int_0^r \frac{dM}{r^2} = -Gm \int_0^r \frac{Ar(4\pi r^2)dr}{r^2} = \boxed{\frac{2GmMr^2}{R^4}}$$

---

67. A satellite is in a circular orbit about a planet of radius R. If the altitude of the satellite is h and its period is T, (a) show that the density of the planet is given by

$$\rho = \frac{3\pi}{GT^2}\left(1 + \frac{h}{R}\right)^3$$

(b) Calculate the average density of the planet if the period is 200 min and the satellite's orbit is close to the planet's surface.

Solution

(a) The density is just $\rho = M/V = 3M/4\pi R^3$. From Kepler's 3rd law,

$$T^2 = \left(\frac{4\pi^2}{GM}\right)r^3 \qquad \text{where} \qquad r = R + h.$$

Combining these two equations, we find

$$\rho = \frac{3\pi}{GT^2}\left(1 + \frac{h}{R}\right)^3$$

(b) If the satellite is close to the surface, then $h/R \ll 1$ and

$$\rho \cong \frac{3\pi}{GT^2} = \frac{3\pi}{(6.67 \times 10^{-11}\ \text{N·m}^2/\text{kg}^2)(1.2 \times 10^4\ \text{s})^2} = \boxed{981\ \text{kg/m}^3}$$

---

69. A particle of mass m is located *inside* a uniform solid sphere of radius R and mass M. If the particle is at a distance r from the center of the sphere, (a) show that the gravitational potential energy of the system is given by $U = (GmM/2R^3)r^2 - 3GmM/2R$. (b) How much work is done by the gravitational force in bringing the particle from the surface of the sphere to its center?

Solution

(a) $U = \int F dr$. Initially take the particle from ∞ and move it to the sphere's surface. Then

$$U = \int_{\infty}^{R} \left(\frac{GmM}{r^2}\right) dr = -\frac{GmM}{R}$$

Now move it to a position r from the center of the sphere. The force in this case is a function of the mass enclosed by r at any point. Since

$$\rho = \frac{M}{\frac{4}{3}\pi R^3}$$

we have

$$U = \int_{R}^{r} \frac{Gm 4\pi r^3 \rho \, dr}{3r^2} = \frac{GmM}{R^3}\left(\frac{r^2}{2} - \frac{R^2}{2}\right)$$

and the total gravitational potential is

$$U = \frac{GmM}{2R^3}(r^2 - 3R^2) = \left(\frac{GmM}{2R^3}\right)r^2 - \frac{3GmM}{2R}$$

(b) $U(R) = -\dfrac{GMm}{R}$ and $U(0) = -\dfrac{3GmM}{2R}$ therefore

$$W_g = -[U(0) - U(R)] = \boxed{\frac{GMm}{2R}}$$

15

Fluid Mechanics

OBJECTIVES

1. Discuss the general properties of the three states of matter.

2. Define the density of a substance and understand the concept of pressure at a point in a fluid, and the variation of pressure with depth.

3. Understand the origin of buoyant forces, state and explain Archimedes' principle, and be able to work problems involving buoyant forces.

4. State the simplifying assumptions of an ideal fluid moving with streamline flow.

5. Derive the equation of continuity and Bernoulli's equation for an ideal fluid in motion, and understand the physical significance of each equation.

6. Present a qualitative discussion of some applications of Bernoulli's equation, such as air lift and available energy from winds.

NOTES FROM SELECTED CHAPTER SECTIONS

15.1 STATES OF MATTER

Matter is generally classified as being in one of three states: solid, liquid, or gas.

In a *crystalline solid*, the atoms are arranged in an ordered periodic structure; while in an *amorphous solid* (i.e. glass), the atoms are present in a disordered fashion.

In the *liquid state*, thermal agitation is greater than in the solid state, the molecular forces are weaker, and molecules wander throughout the liquid in a random fashion.

The molecules of a *gas* are in constant random motion and exert weak forces on each other. The distances separating molecules are large compared to the dimensions of the molecules.

15.2 DENSITY AND PRESSURE

The *specific gravity* of a substance is a dimensionless quantity which is the ratio of the density of the substance to the density of water.

15.3 VARIATION OF PRESSURE WITH DEPTH

In a *fluid at rest*, all points at the same depth are at the same pressure. *Pascal's law* states that a change in pressure applied to an *enclosed* fluid is transmitted undiminished to every point in the fluid and the walls of the containing vessel.

15.4 PRESSURE MEASUREMENTS

The *absolute pressure* of a fluid is the sum of the *gauge pressure* and atmospheric pressure. The SI unit of pressure is the Pascal (Pa). Note that $1 \text{ Pa} \equiv 1 \text{ N/m}^2$.

15.5 BUOYANT FORCES AND ARCHIMEDES' PRINCIPLE

Any object partially or completely submerged in a fluid experiences a buoyant force equal in magnitude to the weight of the fluid displaced by the object and acting vertically upward through the point which was the center of gravity of the displaced fluid.

15.6 FLUID DYNAMICS

When fluid is in motion, its flow can be characterized as being one of two main types. The flow is said to be steady if each particle of the fluid follows a smooth path, and the paths of each particle do not cross each other. Above a certain critical speed, fluid flow becomes nonsteady or turbulent, characterized by small whirlpool-like regions.

The term *viscosity* is commonly used in fluid flow to characterize the degree of internal friction in the fluid. This internal friction is associated with the resistance to two adjacent layers of the fluid to move relative to each other. Because of viscosity, part of the kinetic energy of a fluid is converted to thermal energy.

In our model of an ideal fluid, we make the following four assumptions:

1. *Nonviscous fluid.* In a nonviscous fluid, internal friction is neglected. An object moving through a nonviscous fluid would experience no retarding viscous force.

2. *Steady flow.* In a steady flow, we assume that the velocity of the fluid at each point remains constant in time.

3. *Incompressible fluid.* The density of the fluid is assumed to remain constant in time.

4. *Irrotational flow.* Fluid flow is irrotational if there is no angular momentum of the fluid about any point. If a small wheel placed anywhere in the fluid does not rotate about its center of mass, the flow would be considered irrotational. (If the wheel were to rotate, as it would if turbulence were present, the flow would be rotational.)

15.7 STREAMLINES AND THE EQUATION OF CONTINUITY

The path taken by a fluid particle under *steady flow* is called a *streamline*. The velocity of the fluid particle is always tangent to the streamline at that point and no two streamlines can cross each other. A set of streamlines form a *tube of flow*. In *steady flow*, particles of the fluid cannot flow into or out of a tube of flow.

EQUATIONS AND CONCEPTS

The *density* of a homogeneous substance is defined as its mass per unit volume.

$$\rho = \frac{m}{V}$$

(15.1)

Average pressure on a surface is defined as the ratio of force to area.

$$P = \frac{F}{A}$$

(15.2)

The *absolute pressure* at a depth h below the surface of a liquid open to the atmosphere is greater than atmospheric pressure.

$$P = P_a + \rho g h$$

(15.7)

According to *Bernoulli's equation* (for a non-viscous, incompressible fluid in steady flow), the sum of pressure, kinetic energy per unit volume, and potential energy per unit volume has the same value at all points along a streamline.

$$P + \frac{1}{2}\rho v^2 + \rho gy = \text{constant} \qquad (15.13)$$

The maximum power per unit area of a wind machine will double if the wind velocity increases by 26%.

$$\frac{P_{max}}{A} = \frac{8}{27}\rho v^3 \qquad (15.18)$$

The *coefficient of viscosity* for a fluid is defined as the ratio of the shearing stress to the rate of change of the shear strain.

$$\eta = \frac{FL}{Av} \qquad (15.19)$$

The onset of turbulent flow is determined by a dimensionless parameter called the *Reynolds number*.

$$RN = \frac{\rho vd}{\eta} \qquad (15.21)$$

ANSWERS TO SELECTED QUESTIONS

1. Two glass tumblers that weigh the same but have different shapes and different cross-sectional areas are filled to the same level with water. According to the expression $P = P_a + \rho gh$, the pressure is the same at the bottom of both tumblers. In view of this, why does one tumbler weigh more than the other?

Answer: The hydrostatic pressure represents the force per unit area which would be exerted on any object placed at that depth. The weight of each tumbler depends on the density of the fluid, the volume of the fluid, and the weight of the empty tumbler.

3. When you drink a liquid through a straw, you reduce the pressure in your mouth and let the atmosphere move the liquid. Explain how this works. Could you use a straw to sip a drink on the moon?

Answer: When you reduce the pressure in your mouth, the push of the atmosphere on the surface of the liquid forces the liquid up the straw and into the mouth. Because there is no atmospheric pressure on the moon, you could not sip a drink there.

4. Indian fakirs stretch out for a nap on a bed of nails. How is this possible?

Answer: If you try to stand on a single nail, the pressure on your foot is your weight divided by the very small area of the point of the nail. This pressure is sufficiently large to penetrate the skin. But, if your weight is distributed over several hundred nails, as is the case when you lie prone on a bed of nails, the pressure at any point is your weight divided by the cross-sectional areas of *all* the nails in contact with your body. This produces a considerably smaller pressure than that produced by a single nail. It is not comfortable even then.

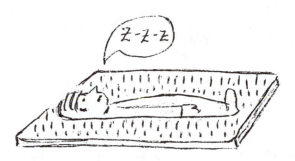

5. Pascal used a barometer with water as the working fluid. Why is it impractical to use water for a typical barometer?

Answer: Water has a density of only 10^3 kg/m^3 as compared to 13.6×10^3 kg/m^3 for mercury. Therefore, the corresponding height of the water column in a water barometer would be 13.6 times greater than that of the mercury barometer. For example, one atmosphere, which corresponds to a mercury column of 0.76 m, would be equivalent to a water barometer of height 10.3 m or about 33 ft.

6. A person sitting in a boat floating in a small pond throws a heavy anchor overboard. Does the level of the pond rise, fall, or remain the same?

Answer: The level of the pond falls. This is because the anchor displaces more water while in the boat. A floating object displaces a volume of water, whose weight is equal to the weight of the object. A submerged object displaces a volume of water equal to the volume of the object. Since the density of the anchor is greater than the density of water, a volume of water which weighs the same as the anchor will be greater than the volume of the anchor.

9. A fish rests on the bottom of a bucket of water while the bucket is being weighed. When the fish begins to swim around, does the weight change?

Answer: In either case the scale is supporting the container, the water, and the fish. Therefore, the weight reading of the scale remains the same.

10. Will a ship ride higher in the water of an island lake or in the ocean? Why?

Answer: The buoyant force on an object such as a ship is equal to the weight of the water displaced by the ship. Because of the greater density of salty ocean water, less water needs to be displaced by the boat to enable it to float. Thus, the boat floats higher in the ocean than in a freshwater inland lake.

11. If 1,000,000 N of weight were placed on the deck of the World War II battleship North Carolina, the ship would sink only 2.5 cm lower in the water. What is the cross-sectional area of the ship at water level?

Answer: The buoyant force exerted by the water on the ship must be increased by 1,000,000 N in order to support the ship. In order to do this, an amount of water weighing 1,000,000 N must be displaced. The area can be calculated by realizing that the weight equals the product: (density of water) (increase in depth) (area of cross-section) (acceleration due to gravity). Using the values given in this example, the area is found to be approximately $4.1 \times 10^3 \, m^2$.

12. Lead has a greater density than iron, and both are denser than water. Is the buoyant force on a lead object greater than, less than, or equal to the buoyant force on an iron object of the same volume?

Answer: The buoyant forces are the same, since the buoyant force equals the weight of the displaced water.

13. An ice cube is placed in a glass of water. What happens to the level of the water as the ice melts?

Answer: It stays the same.

14. A woman wearing high-heeled shoes is invited into a home in which the kitchen has a newly installed vinyl floor covering. Why should the homeowner be concerned?

Answer: The woman can exert enough pressure on the floor to puncture or dent the floor covering. The large pressure is produced because her weight is distributed over the very small cross-sectional area of her high-heels.

15. A typical silo on a farm has many bands wrapped around its perimeter as shown in the photograph. Why is the spacing between successive bands smaller at the lower regions of the silo?

Answer: If one thinks of the grain stored in the silo as a fluid, the pressure it exerts on the walls of the silo increases with depth. Thus, the bands must be closer together near the bottom to overcome the larger outward force produced.

(Photographed by Jim Lehman, James Madison University.)

16. The water supply for a city is often provided from reservoirs built on high ground. Water flows from the reservoir through pipes and into your home when you turn the tap on your faucet. Why is the water flow more rapid out of a faucet on the first floor of a building than in an apartment on a higher floor?

Answer: The water supplied to the building flows through a pipe connected to the water tower. Near the earth, the water pressure is greater because the pressure increases with increasing depth beneath the surface of the water. The penthouse apartment is not as far below the water surface; hence, the water flow will not be as rapid as in a lower floor.

17. Smoke rises in a chimney faster when a breeze is blowing. Use Bernoulli's principle to explain this phenomena.

Answer: The rapidly moving air above the chimney exerts less downward pressure on the smoke particles than would stationary air above the chimney. As a result, the smoke goes up the chimney.

18. Why do many trailer trucks use wind deflectors on the top of their cabs? (See photograph.) How do such devices reduce fuel consumption?

Answer: The wind deflectors produce a more streamline flow of air over the top of the truck, thereby decreasing air resistance, and reducing fuel consumption.

19. Consider the cross-section of the wing on an airplane. The wing is designed such that the air travels faster over the top than under the bottom. Explain why there is a net upward force (lift) on the wing due to the Bernoulli principle.

Answer: The rapidly moving air over the upper surface exerts a smaller downward force than the slower moving air beneath the wing. This follows from Bernoulli's equation, and the fact that $v_2 > v_1$. The net result is an upward force (or lift) on the wing.

20. When a fast-moving train passes a train at rest, the two tend to be drawn together. How does the Bernoulli effect explain this phenomenon?

Answer: As air is displaced by the moving train, that portion passing between the trains has a higher relative velocity than the air on the outside, which is free to expand. Thus, the air pressure is lower between the trains than on the sides of the trains away from the constriction.

21. A baseball heading for home plate is seen from above to be spinning counterclockwise. In which direction does the ball deflect?

Answer: From the viewpoint of a right-handed batter, the ball moves from left to right and spins counterclockwise. Thus, the air motion is retarded above the ball and helped along beneath the ball. This causes the air pressure above the ball to be lower than the air pressure beneath the ball, so the ball will rise.

22. A tornado or hurricane will often lift the roof of a house. Use the Bernoulli principle to explain why this occurs. Why should you keep your windows open during these conditions?

Answer: The rapidly moving air characteristic of a tornado or hurricane causes the external pressure to fall below atmospheric pressure. However, the stationary air inside the house remains at normal atmospheric pressure. The pressure difference results in a large upward force on the roof, which can be large enough to explode the roof off the house. If your windows are left open, the pressure difference between outside and inside will be greatly reduced, and help prevent such a disaster.

24. If you hold a sheet of paper and blow across the top surface, the paper rises. Explain.

Answer: The rapidly moving air over the top of the paper exerts a smaller (downward) force on the paper than does the slower moving air on the under surface of the paper, which exerts an upward force. Hence, the paper rises.

25. If air from a hair dryer is blown over the top of a Ping-Pong ball, the ball can be suspended in air. Explain how the ball can remain in equilibrium.

Answer: The rapidly moving stream of air above the ball exerts a smaller downward force than the slower moving air beneath the ball, which exerts an upward force. When the difference in these forces is equal to the weight of the ball, the ball will suspend in mid air.

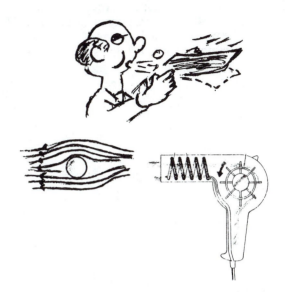

26. Two ships passing near each other in a harbor tend to be drawn together and run the risk of a sideways collision. How does the Bernoulli principle explain this?

Answer: See the answer to question 20.

27. When ski-jumpers are airborne, why do they bend their bodies forward and keep their hands at their sides?

Answer: The ski-jumper's body will be in the shape of the airfoil, which causes the pressure above the body to be lower than the pressure below the body, resulting in an airlift.

(Question 27). (© Thomas Zimmerman, FPG International)

SOLUTIONS TO SELECTED END-OF-CHAPTER PROBLEMS

3. Estimate the density of the *nucleus* of an atom. What does this result suggest concerning the structure of matter? (Use the fact that the mass of a proton is 1.67×10^{-27} kg and its radius is about 10^{-15} m.)

Solution

N = sum of number of protons and neutrons in the nucleus, $m = Nm_p$, and $V = N\left(\dfrac{4\pi r^3}{3}\right)$

$$\rho = \frac{m}{V} = \frac{Nm_p}{N\left(\frac{4\pi r^3}{3}\right)} = \frac{1.67 \times 10^{-27} \text{ kg}}{\frac{4}{3}\pi(10^{-15} \text{ m})^3} = \boxed{3.99 \times 10^{17} \text{ kg/m}^3}$$

11. The spring of the pressure gauge shown in Figure 15.4 has a force constant of 1000 N/m, and the piston has a diameter of 2 cm. Find the depth in water for which the spring compresses by 0.5 cm.

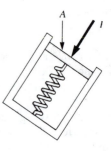

Figure 15.4

Solution

$$F_{el} = F_{fluid} \quad \text{or} \quad kx = \rho g h A; \quad \text{and}$$

$$h = \frac{kx}{\rho g A} = \frac{(1000 \text{ N/m}^2)(0.005 \text{ m})}{(10^3 \text{ kg/m}^3)(9.80 \text{ m/s}^2)\pi(0.01 \text{ m})^2} = \boxed{1.62 \text{ m}}$$

27. A cube of wood 20 cm on a side and having a density of 0.65×10^3 kg/m³ floats on water. (a) What is the distance from the top of the cube to the water level? (b) How much lead weight has to be placed on top of the cube so that its top is just level with the water?

Solution

(a) According to Archimedes principle,

$$B = \rho_w V g = (1 \text{ g/cm}^3)[20 \times 20 \times (20 - h)]g$$

but

$$B = \text{Weight of Block} = Mg = \rho_{wood} V_{wood} g = (0.65 \text{ g/cm}^3)(20 \text{ cm})^3 g$$

$$(0.65)(20)^3 g = (1)(20)(20)(20 - h)g$$

$$20 - h = 20(0.65)$$

$$h = 20(1 - 0.65) \text{ cm} = \boxed{7.00 \text{ cm}}$$

(b) $B = W + Mg$ where $M = \text{mass of lead}$

$$1(20)^3 g = (0.65)(20)^3 g + Mg$$

$$M = 20^3(1 - 0.65) = 20^3(0.35) = 2800 \text{ g} = \boxed{2.80 \text{ kg}}$$

37. A large storage tank filled with water develops a small hole in its side at a point 16 m below the water level. If the rate of flow from the leak is 2.5×10^{-3} m³/min, determine (a) the speed at which the water leaves the hole and (b) the diameter of the hole.

Solution

Assuming the top is open to the atmosphere, then $P_1 = P_a$

Flow rate = 2.5×10^{-3} m³/min = 4.167×10^{-5} m³/s

(a) $A_1 \gg A_2$,   so $v_1 \ll v_2$.   Assuming $v_1 \approx 0$,   and $P_1 = P_2 = P_a$,

$$P_1 + \frac{1}{2}\rho v_1^2 + \rho g y_1 = P_2 + \frac{1}{2}\rho v_2^2 + \rho g y_2$$

$$v_2 = \sqrt{2gy_1} = \sqrt{2(9.80 \text{ m/s}^2)(16 \text{ m})} = \boxed{17.7 \text{ m/s}}$$

(b)   Flow rate = $A_2 v_2 = \left(\frac{\pi d^2}{4}\right)(17.7 \text{ m/s}) = 4.167 \times 10^{-5}$ m³/s

Thus, $d = \boxed{1.73 \times 10^{-3} \text{ m}} = \boxed{1.73 \text{ mm}}$

---

45. Each wing of an airplane has an area of 25 m². If the speed of the air is 50 m/s over the lower wing surface and 65 m/s over the upper wing surface, determine the weight of the airplane. (Assume the plane travels in level flight at constant speed at an elevation where the density of air is 1 kg/m³. Also assume that all of the lift is provided by the wings.)

Solution

$$P_1 + \frac{1}{2}\rho v_1^2 = P_2 + \frac{1}{2}\rho_2 v_2^2$$

$$\Delta P = P_1 - P_2 = \frac{1}{2}\rho(v_2^2 - v_1^2)$$

$$W = F_{net} = \Delta P(A) = \frac{1}{2}\rho(v_2^2 - v_1^2)A$$

Therefore,

$$W = \frac{1}{2}(1 \text{ kg/m}^3)[(65 \text{ m/s})^2 - (50 \text{ m/s})^2](50 \text{ m}^2) = \boxed{4.31 \times 10^4 \text{ N}}$$

57. A sample of copper is to be subjected to a hydrostatic pressure that will increase its density by 0.1 percent. What pressure is required?

Solution

$$B = -\frac{\Delta P}{\Delta V/V} = 14 \times 10^{10} \text{ N/m}^2 \text{ for copper}$$

If $\quad \frac{\Delta \rho}{\rho} = 0.001, \quad$ then $\quad \frac{\Delta V}{V} = -0.001$

Solving for ΔP gives

$$\Delta P = -B\left(\frac{\Delta V}{V}\right) = (-14 \times 10^{10} \text{ N/m}^2)(-0.001) = \boxed{1.40 \times 10^8 \text{ P}_a}$$

61. The true weight of a body is its weight when measured in a vacuum where there are no buoyant forces. A body of volume V is weighed in air on a balance using weights of density ρ. If the density of air is ρ_a and the balance reads W', show that the true weight W is given by

$$W = W' + \left(V - \frac{W'}{\rho g}\right)\rho_a g$$

Solution

The "balanced" condition is one in which the apparent weight of the body equals the apparent weight of the weights. This condition can be written as $W - W_b = W' - W_b'$, where W_b and W_b' are the buoyant forces on the body and weights respectively. The buoyant force experienced by an object of volume V in air equals (Volume of object)($\rho_a g$), so we have

$$W_b = V\rho_a g \qquad \text{and} \qquad W_b' = \left(\frac{W'}{\rho g}\right)\rho_a g$$

Therefore

$$W = W' + \left[V - \frac{W'}{\rho g}\right]\rho_a g$$

67. With reference to Figure 15.8, show that the total torque exerted by the water behind the dam about an axis through 0 is $\frac{1}{6}\rho gwH^3$. Show that the effective line of action of the total force exerted by the water is at a distance $\frac{1}{3}H$ above 0.

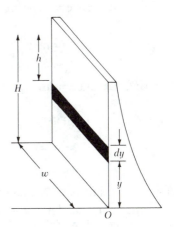

Solution

The torque is $\tau = \int d\tau = \int rdF$. From Figure 15.8, we have

$$\tau = \int_0^H y[\rho g(H - y)w \, dy] = \frac{1}{6}\rho gwH^3$$

Figure 15.8

The total force is given as $\frac{1}{2}\rho gwH^2$. If this were applied at a height y_{eff} such that the torque remains unchanged, we have

$$\frac{1}{6}\rho gwH^3 = y_{eff}\left[\frac{1}{2}\rho gwH^2\right] \quad \text{and} \quad \boxed{y_{eff} = \frac{1}{3}H}$$

---

73. Consider a composite "raft" consisting of two square slabs, each of side s, attached face to face. One slab has density ρ_1 and thickness h_1, while the other has density $\rho_2 > \rho_1$ and thickness h_2. (a) Find the average density $\overline{\rho}$ of the raft. (b) Assume that $\overline{\rho} < \rho_w$, so that the raft floats in water. The raft is placed in water with the denser slab on the bottom. Find d, the depth of the bottom surface of the raft. (c) If the raft is placed in water with the denser slab on the *top*, find d', the depth of the bottom surface of the raft. Comment on your answer. (d) For which of the orientations described in (b) and (c) is the gravitational potential energy of the entire system (consisting of the raft and the body of water in which it is floating) greater? Find the potential energy difference.

Solution

(a) $\quad \rho_{av} = \frac{M}{V} = \frac{\rho_1 s^2 h_1 + \rho_2 s^2 h_2}{s^2(h_1 + h_2)} = \frac{\rho_1 h_1 + \rho_2 h_2}{h_1 + h_2}$

(b) We need $\quad \rho_w s^2 d = M \quad$ so $\quad d = \frac{M}{\rho_w s^2} = \frac{\rho_1 h_1 + \rho_2 h_2}{\rho_w}$

(c) Same result $\quad d' = d$

(d) Gravitational energy of water is same in both cases, but gravitational potential energy of the raft differs. The potential energy is higher when slab #2 is on top.

P.E. Difference = m_2g(height diff. of C. M. of #2) – m_1g(height diff. of C. M. of #1)

$$= m_2g\left[\left(h_1 + \frac{h_2}{2}\right) - \frac{h_2}{2}\right] - m_1g\left[\left(h_2 + \frac{h_1}{2}\right) - \frac{h_1}{2}\right]$$

P.E. Difference = $(m_2h_1 - m_1h_2)g$ but $m_2 = \rho_2s^2h_2$ and $m_1 = \rho_1s^2h_1$

so   P.E. Diff. = $s^2(\rho_2h_2h_1 - \rho_1h_2h_1)g = (\rho_2 - \rho_1)s^2h_1h_2g$

---

75. A cable of mass density ρ_c and diameter d extends vertically downward a distance h through water, and a block of mass M_b and density ρ_b is hung from the bottom end of the cable. Both ρ_c and ρ_b exceed ρ_w, the density of water. Find (a) the tension T_L at the lower end of the cable, (b) the tension T_u at the upper end of the cable, and (c) the tension T_L' and T_u' that would exist at the lower and upper ends of the cable if the entire assembly were in air rather than water. (Neglect the buoyant force provided by the air.) (d) Evaluate T_L, T_u, T_L', and T_u' for the case of a 100-meter steel cable supporting a prefabricated concrete object of mass 2.00 metric tons: $\rho_c = 7.86 \times 10^3$ kg/m³, d = 2×10^{-2} m, h = 100 m, $M_b = 2.00 \times 10^3$ kg, and $\rho_b = 2.38 \times 10^3$ kg/m³.

Solution

(a)   The required tension T_L is given by

$$T_L = M_bg - \left(\frac{M_b}{\rho_b}\right)\rho_wg = \left(1 - \frac{\rho_w}{\rho_b}\right)M_bg$$

(b) $T_u = T_L + \frac{1}{4}(\rho_c - \rho_w)\pi d^2hg$

(c) $T_L' = M_bg$ $T_u' = T_L' + \frac{1}{4}\rho_c\pi d^2hg$

(d) $T_L = \left(1 - \frac{1}{2.38}\right)(2 \times 10^3)(9.80) = \boxed{1.14 \times 10^4 \text{ N}}$

$T_u = (1.138 \times 10^4) + \frac{1}{4}\left[\left(1 - \frac{1}{7.86}\right)(7.86 \times 10^3)\pi(2 \times 10^{-2})^2(100)(9.80)\right]$

$= (1.138 \times 0.211) \times 10^4 \text{ N} = \boxed{1.35 \times 10^4 \text{ N}}$

$T_L' = M_bg = \boxed{1.96 \times 10^4 \text{ N}}$

$T_u' - T_L' = \frac{1}{4}\rho_c\pi d^2hg = 2.422 \times 10^3 \text{ N}$

or $T_u' = (1.962 + 0.2422) \times 10^4 \text{ N} = \boxed{2.20 \times 10^4 \text{ N}}$

16

Wave Motion

OBJECTIVES

1. Recognize whether or not a given function is a possible description of a traveling wave.

2. Express a given harmonic wave function in several alternative forms involving different combinations of the wave parameters: wavelength, period, phase velocity, wave number, angular frequency, and harmonic frequency.

3. Given a specific wave function for a harmonic wave, obtain values for the characteristic wave parameters: A, ω, k, λ, f, and ϕ.

4. Calculate the rate at which energy is transported by harmonic waves in a string.

5. Make calculations which involve the relationships between wave speed and the inertial and elastic characteristics of a string through which the disturbance is propagating.

6. Plot a curve showing the shape of a wave form due to a specific wave function or the shape of a string due to interfering traveling waves at any stated instant of time.

NOTES FROM SELECTED CHAPTER SECTIONS

16.1 INTRODUCTION

The production of *mechanical waves* require: (1) an *elastic medium* which can be disturbed, (2) an *energy source* to provide a disturbance or deformation in the medium, and (3) a physical mechanism by way of which adjacent portions on the medium can *influence* each other. The three parameters important in characterizing waves are (1) wavelength, (2) frequency, and (3) wave velocity.

16.2 TYPES OF WAVES

Transverse waves are those in which particles of the disturbed medium move along a direction which is perpendicular to the direction of the wave velocity. For *longitudinal waves*, the particles of the medium undergo displacements which are parallel to the direction of wave motion.

16.3 ONE-DIMENSIONAL TRAVELING WAVES

A one-dimensional traveling wave can be described mathematically by its *wave function* $y(x, t) = f(x \mp vt)$. If the wave is assumed to be traveling along the x-direction, then $y(x, t)$ represents the y-coordinate of any point on the string at any time, t.

In the expression for the wave function, the *negative sign* describes a wave pulse traveling *toward the right* and the *positive sign* describes the wave function for a pulse traveling *toward the left*.

In the case of a wave on a string at a *fixed value of x*, the wave function represents the y-coordinate of a particular *point as a function of time*. On the other hand, if *t is fixed,* the wave function defines a curve showing the *shape of the wave pulse* at a given time.

16.4 SUPERPOSITION AND INTERFERENCE OF WAVES

If two or more waves are moving through a medium, the *resultant wave function* is the *algebraic sum* of the wave functions of the individual waves. Two traveling waves can pass through each other without being destroyed or altered.

16.5 THE VELOCITY OF WAVES ON STRINGS

For linear waves, the *velocity* of *mechanical waves* depends only on the physical properties of the medium through which the disturbance travels. In the case of waves on a *string*, the velocity depends on the tension in the string and the mass per unit length (linear mass density).

16.7 HARMONIC WAVES

A harmonic wave is one whose shape is sinusoidal at every instant of time. Harmonic waves which differ in phase are shown in Figure 16.1.

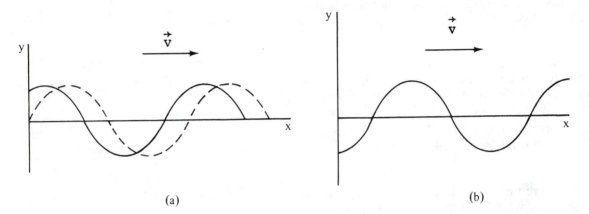

(a)   (b)

Figure 16.1   (a) A harmonic wave traveling to the right, with $\phi \neq 0$. The solid curve represents the wave at t = 0; the dotted curve represents the wave at some later time t.
(b) A traveling wave, where $\phi = \pi/2$, relative to the dotted curve of (a).

In the expression for the wave function (Eq. 16.6), the quantity A is called the *amplitude* of the wave and is the maximum possible value of the displacement, y. The constant λ, called the *wavelength*, equals the distance between any two points that are in phase (are the same displacement from y = 0 and are moving in the same direction). The amplitude A and wavelength λ are shown in Figure 16.2.

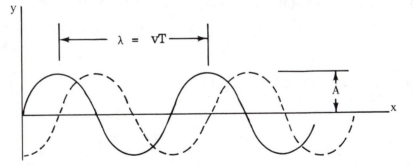

Figure 16.2   Schematic representation of a harmonic wave. The solid curve is the wave at t = 0; the dotted curve is the displaced wave at some time t.

In a time t, each point along the profile of the wave moves a distance x = vt. The time required to travel a distance equal to λ, the distance between successive points which are in phase, is called the *period*, T. This is the time after which any point in the medium (string) completes one cycle in its harmonic motion.

EQUATIONS AND CONCEPTS

The *wave speed* or *phase velocity* is in general the rate at which the profile of the disturbance moves along the direction of travel (e.g. the x-axis).

$$v = \frac{dx}{dt}$$

(16.3)

In the special case of a transverse pulse moving along a stretched string, the wave speed depends on the tension in the string F and the linear density μ of the string (mass per unit length).

$$v = \sqrt{\frac{F}{\mu}}$$

(16.4)

The *wave function* for a harmonic wave can be written for the case when y = 0 at t = 0, or for the more general case when y has some arbitrary value between 0 and A when t = 0. In the latter case, a *phase constant* ϕ must be included. The value of ϕ can be determined from initial conditions.

$$y = A \sin\left[\frac{2\pi}{\lambda}(x - vt)\right]$$

$$y = A \sin\left[\frac{2\pi}{\lambda}(x - vt) - \phi\right]$$

(16.6)

It is convenient to define three additional characteristic wave quantities: *the wave number* k, the *angular frequency* ω, and the *harmonic frequency* f. See Programmed Exercises 1E through 1H.

$$k = \frac{2\pi}{\lambda}$$

(16.9)

$$\omega = \frac{2\pi}{T}$$

(16.10)

$$f = \frac{1}{T}$$

(16.12)

The expression for the *wave function* can be written in a more compact form in terms of the parameters defined above.

$$y = A \sin(kx - \omega t)$$

(16.11)

$$y = A \sin(kx - \omega t - \phi)$$

(16.15)

The *wave speed* v or phase velocity can also be expressed in alternative forms.

$$v = \frac{\omega}{k}$$

(16.13)

$$v = \lambda f$$

(16.14)

The *transverse velocity* v_y of a point on a harmonic wave is out of phase with the *transverse acceleration* a_y of that point by $\pi/2$ radians.

$$v_y = -A\omega\cos(kx - \omega t)$$

(16.16)

$$a_y = -A\omega^2\sin(kx - \omega t)$$

(16.17)

The *power* transmitted by any harmonic wave is proportional to the square of the frequency and the square of the amplitude according to Eq. 16.21, where μ is the mass per unit length of the string.

$$\boxed{\text{Power} = \frac{1}{2}\mu v \omega^2 A^2}$$

(16.21)

Any wave function having the form $y = f(x \pm vt)$ satisfies the *linear wave equation* given by Eq. 16.26.

$$\boxed{\frac{\partial^2 y}{\partial x^2} = \frac{1}{v^2}\frac{\partial^2 y}{\partial t^2}}$$

(16.26)

EXAMPLE PROBLEM SOLUTION

Example 16.1   A harmonic wave in a stretched string is described by the function

$$y = (0.25 \text{ m}) \cos (0.1x - 20t)$$

where x and y are in m and t is in s. Determine for this wave motion (a) phase angle, (b) amplitude, (c) angular frequency, (d) harmonic frequency, (e) wave number, (f) wavelength, (g) wave speed, and (h) direction of motion.

Solution   The wave function is generally expressed in the standard form given by Eq. 16.15:

$$y = A \sin (kx - \omega t - \phi)$$

This can be written as $y = A \sin (kx - \omega t) \cos \phi - A \cos (kx - \omega t) \sin \phi$

When the equation above is compared with the given wave function, it is seen that

$$\cos \phi = 0 \text{ and } \sin \phi = -1$$

This requires that the phase angle $\phi = 3\pi/2$. Further comparison of the values of the given wave function with the constants in the standard form of the harmonic wave equation yields

Amplitude: $A = 0.25$ m

Angular frequency: $\omega = 20 \text{ s}^{-1}$

Harmonic frequency: $2\pi f = \omega = 20 \text{ s}^{-1}$; $f = 3.18$ Hz

Wave number: $k = 0.1 \text{ m}^{-1}$

Wavelength: $\dfrac{2\pi}{\lambda} = k = 0.1 \text{ m}^{-1}$; $\lambda = 20\pi$ m

Wave speed: $v = f\lambda = (3.18 \text{ s}^{-1})(20\pi \text{ m}) = 200$ m/s

Direction of travel: Since the given function is of the form $f(x - vt)$, the direction of travel is toward the right (along the + x-axis).

ANSWERS TO SELECTED QUESTIONS

2. How would you set up a longitudinal wave in a stretched spring? Would it be possible to set up a transverse wave in a spring?

Answer: A longitudinal wave can be set up in a stretched spring by compressing the coils in a small region, and releasing the compressed region. The disturbance will proceed to propagate as a longitudinal pulse. It would not be possible to set up a *pure* transverse wave in a spring. The wave will always have some longitudinal component.

3. By what factor would you have to increase the tension in a stretched string in order to double the wave speed?

Answer: Since the wave speed is proportional to the square root of the tension, you would have to quadruple the tension to double the wave speed.

7. If you were to shake one end of a stretched rope periodically three times each second, what would be the period of the harmonic waves set up in the rope?

Answer: Since the frequency of vibration is three cycles per second, the period (which equals the inverse of the frequency) is equal to 1/3 second.

9. Consider a wave traveling on a stretched rope. What is the difference, if any, between the speed of the wave and the speed of a small section of the rope?

Answer: The speed of the wave is a constant determined by the tension in the rope and its mass per unit length. The velocity of a small segment of the string is simple harmonic motion along a direction perpendicular to the motion of the wave pattern.

10. If a long rope is hung from a ceiling and waves are sent up the rope from its lower end, the waves do not ascend with constant speed. Explain.

Answer: The speed of the wave changes because the tension along the string is not constant.

11. What happens to the wavelength of a wave on a string when the frequency is doubled? Assume the tension in the string remains the same.

Answer: In order for the velocity to remain constant, doubling the frequency means that the wavelength must be reduced by a factor of two.

12. What happens to the velocity of a wave on a string when the frequency is doubled? Assume the tension in the string remains the same.

Answer: The velocity of a wave on a string depends on the tension in the string and its mass per unit length. If these remain constant, the velocity does not change.

13. How do transverse waves differ from longitudinal waves?

Answer: In a transverse wave, the vibrating material moves perpendicular to the motion of the wave pattern. The motion of the material is parallel to the motion of the wave pattern in a longitudinal wave.

14. When all the strings on a guitar are stretched to the same tension, will the velocity of a wave along the more massive bass strings be faster or slower than the velocity of a wave on the lighter strings?

Answer: The velocity of a wave on a string is decreased by increasing the mass of the string. Thus, the wave will move slower on the more massive bass strings.

SOLUTIONS TO SELECTED END-OF-CHAPTER PROBLEMS

7. Two harmonic waves in a string are defined by the following functions:

$$y_1 = (2 \text{ cm}) \sin(20x - 30t) \qquad y_2 = (2 \text{ cm}) \sin(25x - 40t)$$

where the y's and x are in cm and t is in s. (a) What is the phase difference between these two waves at the point x = 5 cm at t = 2 s? (b) What is the positive x value closest to the origin for which the two phases will differ by $\pm \pi$ at t = 2 s? (This is where the two waves will totally destroy each other.)

Solution

(a) $\phi_1 = 20(5) - 30(2) = 40$ rad

  $\phi_2 = 25(5) - 40(2) = 45$ rad

  $\Delta\phi = 5$ radians $= 286° = -74°$

(b) $\Delta\phi = |20x - 30t - (25x - 40t)| = |-5x + 10t| = |-5x + 20| = \pm 5\pi$

  for x < 4, $-5x + 20 = \pm 5\pi$

    $5x = 20 \pm 5\pi$

    $x = 4 \pm \pi = 7.14$ cm, $x = \boxed{0.858 \text{ cm}}$

17. A 30-m steel wire and a 20-m copper wire, both with 1-mm diameters, are connected end-to-end and stretched to a tension of 150 N. How long will it take a transverse wave to travel the entire length of the two wires?

Solution

The total time is the sum of the two times.

In one wire $t = \dfrac{L}{v} = L\sqrt{\dfrac{\mu}{T}}$ where $\mu = \rho A = \dfrac{1}{4}\pi\rho d^2$

Thus, $t = L\sqrt{\dfrac{\pi\rho d^2}{4T}}$

for copper $t_1 = (20)\sqrt{\dfrac{(\pi)(8920)(0.001)^2}{(4)(150)}} = 0.137$ s

for steel
$$t_2 = (30)\sqrt{\frac{(\pi)(7860)(0.001)^2}{(4)(150)}} = 0.192 \text{ s}$$

The total time is $0.137 + 0.192 = \boxed{0.329 \text{ s}}$

---

27. (a) Write the expression for y as a function of x and t for a sinusoidal wave traveling along a rope in the *negative* x direction with the following characteristics: $y_{max} = 8$ cm, $\lambda = 80$ cm, $f = 3$ Hz, and $y(0, t) = 0$ at $t = 0$. (b) Write the expression for y as a function of x for the wave in (a) assuming that $y(x, 0) = 0$ at the point $x = 10$ cm.

Solution

(a) $A = y_{max} = 8$ cm $= 0.08$ m $\qquad\qquad k = \frac{2\pi}{\lambda} = \frac{2\pi}{0.8 \text{ m}} = 7.85 \text{ m}^{-1}$

$\omega = 2\pi f = 2\pi(3) = 6\pi$ rad/s

Therefore, $y = A \sin(kx + \omega t)$,   or $y = (0.0800 \text{ m}) \sin(7.85x + 6\pi t)$,   (where $y = 0$ at $t = 0$).

(b)   In general, $y = (0.0800 \text{ m}) \sin(7.85x + 6\pi t + \phi)$. Assuming $y(x, 0) = 0$ at $x = 0.1$ m, then we require that $0 = (0.0800 \text{ m}) \sin(0.785 + \phi)$,   or $\phi = -0.785$ rad

Therefore, $y = (0.0800 \text{ m}) \sin(7.85x + 6\pi t - 0.785)$

---

39. A harmonic wave in a string is described by the equation:

$$y = (0.15 \text{ m})\sin(0.8x - 50t)$$

where x and y are in m and t is in s. If the mass per unit length of this string is 12 g/m, determine (a) the speed of the wave, (b) the wavelength, (c) the frequency, and (d) the power transmitted to the wave.

Solution

Compare $y = (0.15 \text{ m}) \sin(0.8x - 50t)$ with $y = A \sin(kx - \omega t)$

(a) $v = f\lambda = \frac{\omega}{2\pi}\frac{2\pi}{k} = \frac{\omega}{k} = \frac{50}{0.8}$ m/s $= \boxed{62.5 \text{ m/s}}$

(b) $\lambda = \frac{2\pi}{k} = \frac{2\pi}{0.8}$ m $= \boxed{7.85 \text{ m}}$

(c) $f = \frac{\omega}{2\pi} = \frac{50}{2\pi} = \boxed{7.96 \text{ Hz}}$

(d) $P = \frac{1}{2}\mu\omega^2 A^2 v = \frac{1}{2}(0.012 \text{ kg/m})(50 \text{ s}^{-1})^2(0.15 \text{ m})^2(62.5 \text{ m/s})$ W $= \boxed{21.1 \text{ W}}$

---

43. In Section 16.9 it is verified that $y_1 = A \sin(kx - \omega t)$ is a solution to the wave equation. The wave function $y_2 = B \cos(kx - \omega t)$ describes a wave $\pi/2$ radians out of phase with the first. (a) Determine whether or not $y = A \sin(kx - \omega t) + B \cos(kx - \omega t)$ is a solution to the wave equation. (b) Determine if $y = A \sin(kx) \cdot B(\cos \omega t)$ is a solution to the wave equation.

Solution

(a) $y = A \sin(kx - \omega t) + B \cos(kx - \omega t)$

$$\frac{\partial^2 y}{\partial t^2} = -\omega^2 A \sin(kx - \omega t) - \omega^2 B \cos(kx - \omega t)$$

$$\frac{\partial^2 y}{\partial x^2} = -k^2 A \sin(kx - \omega t) - k^2 B \cos(kx - \omega t)$$

Substitution into the general wave equation shows that this is a solution.

(b) $y = A(\sin kx) B(\cos \omega t)$

$$\frac{\partial^2 y}{\partial t^2} = -AB\omega^2 (\sin kx)(\cos \omega t)$$

$$\frac{\partial^2 y}{\partial x^2} = -ABk^2 (\sin kx)(\cos \omega t)$$

Substitution into the general wave equation shows that this is also a solution.

---

52. A rope of total mass m and length L is suspended vertically. Show that a transverse wave pulse will travel the length of the rope in a time $t = 2\sqrt{\dfrac{L}{g}}$. (Hint, First find an expression for the velocity at any point a distance x from the lower end of the rope, by considering the tension in the rope as resulting from the weight of the segment below that point.)

Solution

$v = \sqrt{\dfrac{F}{\mu}}$ where $F = \mu xg$, the weight of length x of rope. Therefore, $v = \sqrt{gx}$. But $v = \sqrt{\dfrac{dx}{dt}}$ so that

$dt = \dfrac{dx}{\sqrt{gx}}$ and $t = \displaystyle\int_0^L \dfrac{dx}{\sqrt{gx}} = 2\sqrt{\dfrac{L}{g}}$

---

57. It is stated in Problem 52 that a wave pulse will travel from the bottom to the top of a rope of length L in a time $t = 2\sqrt{\dfrac{L}{g}}$. Use this result to answer the following questions. (It is *not* necessary to set up any new integrations.) (a) How long does it take for a wave pulse to travel halfway up the rope of length L? (Give your answer as a fraction of the quantity $\left(2\sqrt{\dfrac{L}{g}}\right)$. (b) A pulse starts traveling up the rope. How far has the pulse traveled after a time $\sqrt{\dfrac{L}{g}}$?

Solution

(a) The speed in the lower half of a rope of length L is the same function of distance (from the bottom end) as the speed along the entire length of a rope of length $\dfrac{L}{2}$.

Thus the time required $= 2\sqrt{\dfrac{L'}{g}}$ with $L' = \dfrac{L}{2}$

Therefore the time required $= 2\sqrt{\dfrac{L}{2g}} = 0.707\left[2\sqrt{\dfrac{L}{g}}\right]$

It takes the pulse more than 70% of the total time to cover 50% of the distance.

(b) By the same reasoning applied in part (a), the distance climbed in τ is given by $\dfrac{g\tau^2}{4}$.

For $\tau = \dfrac{t}{2} = \sqrt{\dfrac{L}{g}}$, we find the *distance climbed* $= \dfrac{L}{4}$. In half the total trip time, the pulse has climbed $\dfrac{1}{4}$ of the total length.

Sound Waves

OBJECTIVES

1. Calculate the speed of sound in various media in terms of the appropriate elastic properties of the medium (these can include bulk modulus, Young's modulus, and the pressure-volume relationships of an ideal gas) and the corresponding inertial properties (usually the mass density).

2. Describe the harmonic displacement and pressure variation as functions of time and position for a harmonic sound wave.

3. Relate the displacement amplitude to the pressure amplitude for a harmonic sound wave and calculate the wave intensity from each of these parameters.

4. Understand the basis of the logarithmic intensity scale (decibel scale). Determine the intensity ratio for two sound sources whose decibel levels are known. Calculate the decibel level for some combination of sources whose individual decibel levels are known.

5. Describe the wave function for spherical and planar harmonic waves. Understand the amplitude dependence on distance from the source for spherical and plane waves.

6. Describe the various situations under which a Doppler shifted frequency is produced. Note that a Doppler shift is observed as long as there is *relative* motion between the observer and the source.

SKILLS

When making calculations using Eq. 17.11 which defines the intensity of a sound wave on the decibel scale, the properties of logarithms must be kept clearly in mind.

In order to determine the decibel level corresponding to two sources sounded simultaneously, you must first find the intensity, I, of each source in W/m^2, add these values, and then convert the resulting intensity to the decibel scale. As an illustration of this technique, note that if two sounds of intensity 40 dB and 45 dB are sounded together, the intensity level of the combined sources *is* 46.2 dB (*NOT 85 dB*). See Example 17.1 of the study guide.

The most likely error in using Eq. 17.21 to calculate the Doppler frequency shift due to relative motion between a sound source and an observer is due to using the incorrect algebraic sign for the velocity of either the observer or the source. These sign conventions are illustrated in the chart on the following page in which the directions of motion of the observer O and source S are indicated by arrows; and the correct choice of signs used in Eq. 17.21,

$$f' = f\left(\frac{v \pm v_o}{v \mp v_s}\right)$$

(17.21)

where f is the true frequency of the source and f' is the apparent frequency as measured by the observer.

Observer	Source	Equation	Remark
$\bigcirc\!\!\!\!\!\rightarrow$	(S)	$f' = f\left(\dfrac{v + v_o}{v}\right)$	Observer moving toward stationary source
$\leftarrow\!\!\bigcirc$	(S)	$f' = f\left(\dfrac{v - v_o}{v}\right)$	Observer moving away from stationary source
$\bigcirc$	$\leftarrow\!(S)$	$f' = f\left(\dfrac{v}{v - v_s}\right)$	Source moving toward stationary observer
$\bigcirc$	$(S)\!\rightarrow$	$f' = f\left(\dfrac{v}{v + v_s}\right)$	Source moving away from stationary observer
$\bigcirc\!\!\rightarrow$	$(S)\!\rightarrow$	$f' = f\left(\dfrac{v + v_o}{v + v_s}\right)$	Observer following moving source
$\leftarrow\!\!\bigcirc$	$\leftarrow\!(S)$	$f' = f\left(\dfrac{v - v_o}{v - v_s}\right)$	Source following moving observer
$\leftarrow\!\!\bigcirc$	$(S)\!\rightarrow$	$f' = f\left(\dfrac{v - v_o}{v + v_s}\right)$	Observer and source moving away from each other along opposite directions
$\bigcirc\!\!\rightarrow$	$\leftarrow\!(S)$	$f' = f\left(\dfrac{v + v_o}{v - v_s}\right)$	Observer and source moving toward each other
$\bigcirc$	(S)	$f' = f$	Observer and source both stationary

(See Example 17.2 of the study guide.)

NOTES FROM SELECTED CHAPTER SECTIONS_____

17.2 HARMONIC SOUND WAVES

The *pressure amplitude* is proportional to the *displacement amplitude*; however, the pressure wave is 90° out of phase with the displacement wave.

17.3 ENERGY AND INTENSITY IN HARMONIC SOUND WAVES

The *intensity* of a wave is the rate at which sound energy flows through a unit area perpendicular to the direction of travel of the wave.

17.5   THE DOPPLER EFFECT

In general, a Doppler effect is experienced whenever there is *relative motion* between the sound source and the observer.

EQUATIONS AND CONCEPTS

A sound wave propagates as a *compressional wave* with *a speed* v which depends on the bulk modulus B and equilibrium density ρ of the medium in which the wave is traveling. The *Bulk Modulus* is equal to the negative ratio of the pressure variation to the fractional change in volume of the medium.

$$v = \sqrt{\frac{B}{\rho}}$$
(17.1)

$$B = -\frac{\Delta P}{\Delta V/V}$$
(17.2)

The *speed of sound in a gas* depends on the characteristic parameter γ (the ratio of the specific heat at constant pressure to the specific heat at constant volume), and can be expressed either in terms of the pressure and density or the absolute temperature T and molecular weight M.

$$v = \sqrt{\frac{\gamma P}{\rho}}$$

$$v = \sqrt{\frac{\gamma RT}{M}}$$

The speed of sound in a solid depends on the value of Young's modulus Y and the density of the material. Young's modulus is equal to the ratio of the tensile stress to the tensile strain (see Chapter 15).

$$v = \sqrt{\frac{Y}{\rho}}$$

A harmonic sound wave is produced in a gas when the source is a body vibrating in simple harmonic motion (such as a vibrating guitar string). Under these conditions, both the *displacement* of the medium S and the *pressure variations* of the gas ΔP vary harmonically in time. Note that the displacement and the pressure variation are out of phase by $\pi/2$.

$$s(x,t) = s_m \cos(kx - \omega t)$$
(17.4)

$$\Delta P = \Delta P_m \sin(kx - \omega t)$$
(17.5)

A sound wave may be considered as either a displacement wave or a pressure wave. The *pressure amplitude* is *proportional* to the *displacement amplitude*.

$$\Delta P_m = \rho \, \omega v s_m$$
(17.6)

The *intensity* of a harmonic sound wave is proportional to the square of the source frequency and to the square of the displacement (or pressure) amplitude.

$$I = \frac{1}{2} \rho \, \omega^2 v s_m^2 \qquad (17.9)$$

$$I = \frac{\Delta P_m^2}{2\rho \, v} \qquad (17.10)$$

The human ear is sensitive to a wide *range of intensities*. For this reason, a *logarithmic intensity scale* is defined using a reference intensity of $I_0 = 10^{-12}$ W/m^2. On this scale, the unit of intensity is the decibel (dB).

$$\beta = 10 \log\left(\frac{I}{I_o}\right) \qquad (17.11)$$

The *amplitude* of a periodic spherical wave varies inversely with the distance r from the source.

$$\psi(r, t) = \left(\frac{s_o}{r}\right) \sin(kr - \omega t)$$

$$\text{(spherical)} \qquad (17.13)$$

At large distances from the source ($r \gg \lambda$), a spherical wave can be approximated as a *plane wave*. In this case, the *amplitude* of the wave is *constant*.

$$\psi(r, t) = s_o \sin(kx - \omega t)$$
$$\text{(plane wave)} \qquad (17.14)$$

The *intensity* I of a spherical wave decreases as the square of the distance from the source. This is so because the intensity is proportional to the *square of the amplitude*, which in turn varies inversely with distance from the source.

$$I = \frac{P_{av}}{4\pi r^2} \qquad (17.12)$$

The change in frequency heard by an observer whenever there is *relative motion between the source and the observer* is called the *Doppler effect*. The upper signs refer to motion of one toward the other and the lower signs apply to motion of one away from the other. In Eq. 17.21, v_0 and v_s are measured *relative to the medium in which the sound travels*.

$$f' = f\left(\frac{v \pm v_o}{v \mp v_s}\right) \qquad (17.21)$$

Shock waves are produced when a sound source moves through a medium with a speed v_s which is greater than the wave speed in that medium. The shock wave front has a conical shape with a half angle which depends on the *Mach number* of the source, defined as the ratio v_s/v.

$$\sin \theta = \frac{v}{v_s}$$

EXAMPLE PROBLEM SOLUTIONS

Example 17.1   Two sound sources whose intensities, when measured individually at a particular location, yield values of 40 dB and 45 dB. (a) Determine the expected intensity in dB at that location if both sources are sounded at the same time. (b) Calculate the ratio of the intensities I_2/I_1.

Solution

(a) From Eq. 17.11, $\beta = 10 \log \left(\dfrac{I}{I_o}\right)$ so that $I = I_o \, 10^{(\beta/10)}$

For the example given here, $\beta_1 = 40$ and $\beta_2 = 45$.   Therefore, $I_1 = I_o \, 10^4$ and $I_2 = I_o \, 10^{4.5}$

The combined intensity is found to be $I = I_o \, (10^4 + 10^{4.5})$

Expressed in decibels, the intensity for the combination is given by $\beta = 10 \log \left(\dfrac{I}{I_o}\right)$

$$\beta = 10 \log \left[\frac{I_o \left(10^4 + 10^{4.5}\right)}{I_o}\right] = 10 \log \left(4.16 \times 10^4\right)$$

or $\qquad\qquad\qquad\qquad\qquad\qquad \beta = 46.2 \text{ dB}$

(b) Again, directly from Eq. 17.11, for any intensity I:

$$\log \left(\frac{I}{I_o}\right) = \frac{\beta}{10}$$

Therefore, for I_1 and I_2 corresponding to $\beta_1 = 40$ dB and $\beta_2 = 45$ dB

$$\log \left(\frac{I_2}{I_o}\right) = 4.5 \quad \text{and} \quad \log \left(\frac{I_1}{I_o}\right) = 4$$

Subtracting and remembering that the difference of two logarithms is the log of the quotient yields

$$\log \left(\frac{I_2}{I_o}\right) - \log \left(\frac{I_1}{I_o}\right) = \log \left(\frac{I_2}{I_1}\right) = 0.5$$

so that $\qquad\qquad\qquad\qquad\qquad \dfrac{I_2}{I_1} = 10^{(0.5)} = \sqrt{10} = 3.16$

or source "2" has an intensity which is more than 3 times the intensity of source "1". Note that the ratio of the decibel levels of the two sounds is much smaller:

$$\frac{\beta_2}{\beta_1} = \frac{45}{40} = 1.125$$

Example 17.2   A train traveling parallel to a highway at a speed of 30 mph (44 ft/s) sounds a whistle whose "true" frequency is 300 Hz. What frequency is heard by a passenger in a car which is approaching the train at a speed of 60 mph (88 ft/s)? Take the velocity of sound in air to be 1080 ft/s.

Solution

First find a solution under the assumption that the air is at rest relative to the ground and use Eq. 17.21 for the case when observer and source are moving toward each other.

$$f' = f\left(\frac{v + v_0}{v - v_s}\right)$$

Substituting the given values

$$f' = (300 \text{ Hz})\left(\frac{1080 \text{ ft/s} + 88 \text{ ft/s}}{1080 \text{ ft/s} - 44 \text{ ft/s}}\right) = 338 \text{ Hz}$$

Now obtain a second solution under the assumption that the air is moving parallel to the direction of motion of the train with a speed (relative to ground) of 15 mph (22 ft/s). The observer and source velocities relative to the air are

$$v_0 = 88 \text{ ft/s} + 22 \text{ ft/s} = 110 \text{ ft/s}$$

$$v_s = 44 \text{ ft/s} - 22 \text{ ft/s} = 22 \text{ ft/s}$$

Substituting these values into Eq. 17.21 and using the same choice of signs as before

$$f' = (300 \text{ Hz})\left(\frac{1080 \text{ ft/s} + 110 \text{ ft/s}}{1080 \text{ ft/s} - 22 \text{ ft/s}}\right) = 337 \text{ Hz}$$

The motion of the air decreases the observed frequency by just over 0.2%.

ANSWERS TO SELECTED QUESTIONS

1. Why are sound waves characterized as longitudinal?

Answer: The molecules of the disturbed medium vibrate about some mean position in the direction *parallel to the direction of the wave motion*.

3. If an alarm clock is placed in a good vacuum and then activated, no sound will be heard. Explain.

Answer: Sound waves will only propagate through matter in the form of a gas, solid, or liquid. If there are no atoms present to be disturbed (as in a vacuum), there can be no sound wave.

4. Some sound waves are harmonic, whereas others are not. Give an example of each.

Answer: Examples of harmonic sound waves are those produced by a loudspeaker driven by a sinusoidal electrical signal, and the wave produced by a vibrating tuning fork. A sound wave which is not harmonic is a supersonic wave generated by a jet airplane.

5. In Example 17.4, we found that a point source with a power output of 80 W reduces to an intensity level of 40 dB at a distance of about 16 miles. Why do you suppose you cannot normally hear a rock concert going on 16 miles away?

Answer: There is a great deal of absorption of sound in the air, together with scattering and absorption by mountains, trees, and other obstructions.

6. If the distance from a point source is tripled, by what factor does the intensity decrease?

Answer: Since the intensity varies as $1/r^2$, tripling the value of r will decrease the intensity by a factor of 9.

7. Explain how the Doppler effect is used with microwaves to determine the speed of an automobile.

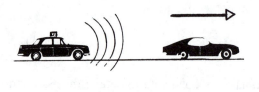

Answer: A microwave beam is sent out with a known frequency. The frequency of this signal is compared with the frequency of the signal reflected from the moving vehicle. In effect, the car acts as a moving source of microwaves. The frequency of the reflected signal depends on the speed and direction of motion of the moving vehicle.

8. If you are in a moving vehicle, explain what happens to the frequency of your echo as you move *toward* a canyon wall. What happens to the frequency as you move *away* from the wall?

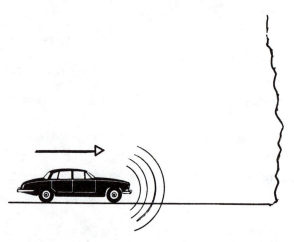

Answer: The frequency increases as you move toward the wall since there are more wavefronts reflected per unit of time. The reverse is true as you move away from the wall.

10. Of the following sounds, which is most likely to have an intensity level of 60 dB: a rock concert, the turning of a page in this text, normal conversation, a cheering crowd at a football game, or background noise at a church?

Answer: To answer this question, the student should refer to Table 15.2 in the body of the chapter. A decibel level of 60 corresponds to the sound level of normal conversation.

11. Estimate the decibel level of each of the sounds in Question 10.

Answer: Again, as in the preceding question, the student should refer to Table 15.2 in the text. A rock concert sound level is of the order of 130 dB; the turning of the pages in a text compares to the rustle of leaves which is given as 10 dB in the table. Normal conversation is at a level of 60 dB. A cheering crowd at a football game corresponds to busy street traffic approximately, and thus, has a decibel level of 70. The background noise level in a church should be about 10 dB.

12. A binary star system consists of two stars revolving about each other. If we observe the light reaching us from one of these stars as it makes one complete revolution about the other, what does the Doppler effect predict will happen to this light?

Answer: The frequency of the light from the star will be increased when it moves toward the earth and decreased when it moves away from the earth.

13. How could an object move with respect to an observer such that the sound from it is not shifted in frequency?

Answer: If the movement is perpendicular to a line drawn from source to object, there is no frequency shift.

SOLUTIONS TO SELECTED END-OF-CHAPTER PROBLEMS

13. An experimenter wishes to generate in air a sound wave that has a displacement amplitude equal to 5.5×10^{-6} m. The pressure amplitude is to be limited to 8.4×10^{-1} N/m^2. What is the minimum wavelength the sound wave can have?

Solution

We are given $s_m = 5.5 \times 10^{-6}$ m and $\Delta P_m = 0.84$ N/m^2. The pressure amplitude is given by

$$\Delta P_m = \rho v \omega s_m = \rho v \left(\frac{2\pi v}{\lambda}\right) s_m$$

or

$$\lambda = \frac{2\pi \rho v^2}{\Delta P_m} s_m = \frac{2\pi (1.2 \text{ kg/m}^3)(343 \text{ m/s})^2(5.5 \times 10^{-6} \text{ m})}{0.84 \text{ N/m}^2} = \boxed{5.81 \text{ m}}$$

23. Calculate the pressure amplitude corresponding to a sound intensity of 120 dB (a rock concert).

Solution

$$I = \frac{\Delta P_m{}^2}{2\rho v} \qquad \text{or} \qquad \Delta P_m = \sqrt{2\rho v I}$$

$$\beta = 10 \log\left(\frac{I}{I_o}\right) = 120 \text{ dB}$$

Hence,

$$I = I_o 10^{12} = (10^{-12} \text{ W/m}^2)10^{12} = 1 \text{ W/m}^2$$

$$\therefore \quad \Delta P_m = \sqrt{2\rho v I} = \sqrt{2(1.2 \text{ kg/m}^3)(343 \text{ m/s})(1 \text{ W/m}^2)} = \boxed{28.7 \text{ Pa}}$$

29. The sound level at a distance of 3 m from a source is 120 dB. At what distance will the sound level be (a) 100 dB and (b) 10 dB?

Solution

$$\beta = 10 \log\left(\frac{I}{10^{-12}}\right) \qquad I = \left[10^{\beta/10}\right]10^{-12} \text{ W/m}^2$$

$$I_{120} = 1 \text{ W/m}^2 \qquad I_{100} = 10^{-2} \text{ W/m}^2 \qquad I_{10} = 10^{-11} \text{ W/m}^2$$

(a) $P = 4\pi r^2 I$ so that $r_1{}^2 I_1 = r_2{}^2 I_2$

$$r_2 = r_1 \sqrt{\frac{I_1}{I_2}} = (3 \text{ m})\sqrt{\frac{1}{10^{-2}}} = \boxed{30.0 \text{ m}}$$

(b) $r_2 = r_1 \sqrt{\dfrac{I_1}{I_2}} = (3 \text{ m})\sqrt{\dfrac{1}{10^{-11}}} = \boxed{9.49 \times 10^5 \text{ m}}$

---

47. (a) The sound level of a jackhammer is measured as 130 dB and that of a siren as 120 dB. Find the ratio of the intensities of the two sound sources. (b) Two sources have measured intensities of $I_1 = 100 \text{ }\mu\text{W/m}^2$ and $I_2 = 200 \text{ }\mu\text{W/m}^2$. By how many dB is source 1 lower than source 2?

Solution

$$\beta = 10 \log\left(\frac{I}{I_o}\right) \qquad \text{or} \qquad I = I_o 10^{\beta/10}$$

(a) $I_1 = (10^{-12} \text{ W/m}^2)10^{130/10} = 10 \text{ W/m}^2$

$I_2 = (10^{-12} \text{ W/m}^2)10^{120/10} = 1 \text{ W/m}^2$

$$\frac{I_1}{I_2} = \boxed{10}$$

(b) $\beta_1 = 10 \log\left[\dfrac{10^{-4} \text{ W/m}^2}{10^{-12} \text{ W/m}^2}\right] = 80 \text{ dB}$

$$\beta_2 = 10 \log\left[\frac{2 \times 10^{-4} \text{ W/m}^2}{10^{-12} \text{ W/m}^2}\right] = 83 \text{ dB}$$

Difference in levels: $\beta_2 - \beta_1 = \boxed{3 \text{ dB}}$

---

49. Two ships are moving along a line due east. The trailing vessel has a speed relative to a land-based observation point of 64 km/h, and the leading ship has a speed of 45 km/h relative to that station. The two ships are in a region of the ocean where the current is moving uniformly due west at 10 km/h. The trailing ship transmits a sonar signal at a frequency of 1200 Hz. What frequency is monitored by the leading ship? (Use 1520 m/s as the speed of sound in ocean water.)

Solution

When the observer is moving in front of and in the same direction as the source,

$$f_o = f_s \left[\frac{v - v_o}{v - v_s} \right]$$

where v_o and v_s are measured relative to the *medium* in which the sound is propagated. In this case the ocean current is opposite the direction of travel of the ships and

$$v_o = 45 \text{ km/h} - (-10 \text{ km/h}) = 55 \text{ km/h} = 15.3 \text{ m/s}$$

$$v_s = 64 \text{ km/h} - (-10 \text{ km/h}) = 74 \text{ km/h} = 20.55 \text{ m/s}$$

Therefore,

$$f_o = (1200 \text{ Hz}) \left[\frac{1520 \text{ m/s} - 15.3 \text{ m/s}}{1520 \text{ m/s} - 20.55 \text{ m/s}} \right] = \boxed{1204 \text{ Hz}}$$

53. By proper excitation, it is possible to produce both longitudinal and transverse waves in a long metal rod. A particular metal rod is 150 cm long and has a radius of 0.2 cm and a mass of 50.9 g. Young's modulus for the material is 6.8×10^{10} N/m^2. What must the tension in the rod be if the ratio of the speed of longitudinal waves to the speed of transverse waves is 8?

Solution

For the longitudinal wave $v_L = \sqrt{\dfrac{Y}{\rho}}$

For the transverse wave $v_T = \sqrt{\dfrac{F}{\mu}}$

If we require $\dfrac{v_L}{v_T} = 8$ we have $F = \dfrac{\mu Y}{64 \rho}$ where $\mu = \dfrac{m}{L}$ and $\rho = \dfrac{\text{mass}}{\text{volume}} = \dfrac{m}{\pi r^2 L}$

This gives $F = \dfrac{\pi r^2 Y}{64} = \dfrac{\pi (0.002 \text{ m})^2 (6.8 \times 10^{10} \text{ N/m}^2)}{64} = \boxed{1.34 \times 10^4 \text{ N}}$

59. Three metal rods are located relative to each other as shown in Figure 17.15, where $L_1 + L_2 = L_3$. Values of density and Young's modulus for the three materials are $\rho_1 = 2.7 \times 10^3$ kg/m³, $Y_1 = 7 \times 10^{10}$ N/m², $\rho_2 = 11.3 \times 10^3$ kg/m³, $Y_2 = 1.6 \times 10^{10}$ N/m², $\rho_3 = 8.8 \times 10^3$ kg/m³, $Y_3 = 11 \times 10^{10}$ N/m². (a) If $L_3 = 1.5$ m, what must the ratio L_1/L_2 be if a sound wave is to travel the length of rods 1 *and* 2 in the same time for the wave to travel the length of rod 3? (b) If the frequency of the source is 4000 Hz, determine the phase difference between the wave traveling along rods 1 and 2 and the one traveling along rod 3.

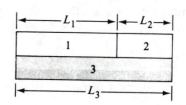

Figure 17.15

Solution

(a) The time required for a sound pulse to travel a distance L at a speed v is given by $t = \dfrac{L}{v} = \dfrac{L}{\sqrt{Y/\rho}}$.

Using this expression, we find

$$t_1 = \frac{L_1}{\sqrt{\dfrac{Y_1}{\rho_1}}} = \frac{L_1}{\sqrt{\dfrac{7 \times 10^{10}\ \text{N/m}^2}{2.7 \times 10^3\ \text{kg/m}^3}}} = 1.96 \times 10^{-4}\, L_1$$

$$t_2 = \frac{1.5 - L_1}{\sqrt{\dfrac{1.6 \times 10^{10}\ \text{N/m}^2}{11.3 \times 10^3\ \text{kg/m}^3}}} = 1.26 \times 10^{-3} - 8.40 \times 10^{-4}\, L_1$$

$$t_3 = \frac{1.5\ \text{m}}{\sqrt{\dfrac{11 \times 10^{10}\ \text{N/m}^2}{8.8 \times 10^3\ \text{kg/m}^3}}} = 4.24 \times 10^{-4}\ \text{s}$$

We require $t_1 + t_2 = t_3$, or

$$1.96 \times 10^{-4}\, L_1 + 1.26 \times 10^{-3} - 8.40 \times 10^{-4}\, L_1 = 4.24 \times 10^{-4}$$

This gives $L_1 = 1.30$ m and $L_2 = 1.50 - 1.30 = 0.20$ m and the ratio of lengths is $\dfrac{L_1}{L_2} = \boxed{6.5}$

(b) The ratio of lengths $\dfrac{L_1}{L_2}$ is adjusted in part (a) so that $t_1 + t_2 = t_3$. Therefore, sound travels the two paths in equal time intervals and the phase difference, $\Delta\phi = 0$.

18

Superposition and Standing Waves

OBJECTIVES

1. Write out the wave function which represents the superposition of the two sinusoidal waves of equal amplitude and frequency traveling in opposite directions in the same medium.

2. Identify the angular frequency, maximum amplitude, and determine the values of x which correspond to nodal and antinodal points of a standing wave, given an expression for the wave function.

3. Plot the resultant waveform due to the interference of two harmonic waves at specified times.

4. Calculate the normal mode frequencies for a string under tension, and for open and closed air columns.

5. Describe the time dependent amplitude and determine the effective frequency of vibration when two waves of slightly different frequency interfere. Also, calculate the expected beat frequency for this situation.

NOTES FROM SELECTED CHAPTER SECTIONS

18.1 SUPERPOSITION OF HARMONIC WAVES

A property of the *superposition principle* is that when two or more waves move in the same linear medium, the *net displacement* of the medium at any point (the resultant wave) equals the algebraic sum of the displacements of all the waves. If the individual waves are harmonic and of equal frequency, the resultant wave function is also harmonic and has the *same frequency* and *same wavelength* as the individual waves.

Figure 18.1 shows the resultant of two traveling harmonic waves for

(a) $\phi = 0, 2\pi, 4\pi, \ldots$ corresponding to constructive interference,

(b) $\phi = \pi, 3\pi, 5\pi, \ldots$ corresponding to destructive interference, and

(c) $0 < \phi < \pi$ for which the resultant amplitude has a value between 0 and 2A.

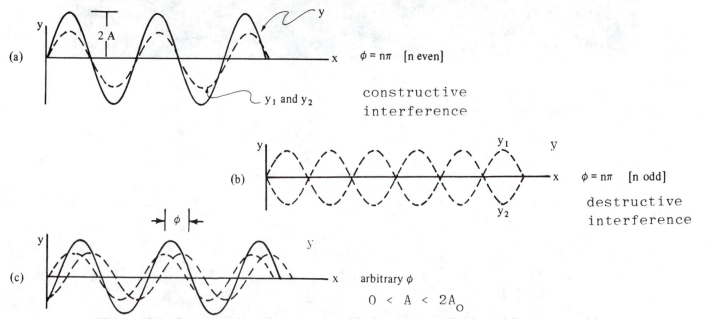

Figure 18.1 Superposition of two waves with the same amplitude and frequency, with a phase difference ϕ of (a) $n\pi$ (n even), (b) $n\pi$ (n odd), and (c) arbitrary ϕ. The dotted curves represent y_1 and y_2; the solid curves represent $y = y_1 + y_2$.

18.3 STANDING WAVES IN A STRING FIXED AT BOTH ENDS

Standing waves can be set up in a string by a continuous superposition of waves incident on and reflected from the ends of the string. The string has a number of natural patterns of vibration, called *normal modes*. Each normal mode has a *characteristic frequency*. The lowest of these frequencies is called the *fundamental frequency*, which together with the higher frequencies form a *harmonic series*.

Figure 18.2 is a schematic representation of the first three normal modes of vibration of string fixed at both ends.

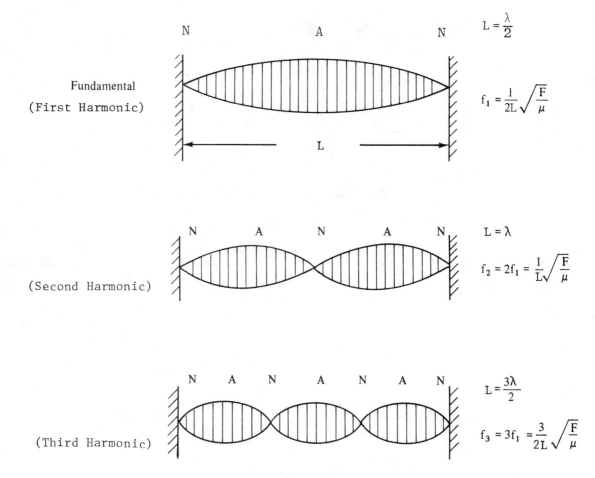

Fundamental
(First Harmonic)

$L = \frac{\lambda}{2}$

$f_1 = \frac{1}{2L}\sqrt{\frac{F}{\mu}}$

(Second Harmonic)

$L = \lambda$

$f_2 = 2f_1 = \frac{1}{L}\sqrt{\frac{F}{\mu}}$

(Third Harmonic)

$L = \frac{3\lambda}{2}$

$f_3 = 3f_1 = \frac{3}{2L}\sqrt{\frac{F}{\mu}}$

Figure 18.2 Schematic representation of standing waves on a stretched string of length L, where the envelope represents many successive vibrations. The points of zero displacement are called *nodes;* the points of maximum displacement are called *antinodes*.

18.5 STANDING WAVES IN AIR COLUMNS

Standing waves are produced in strings by interfering *transverse* waves. Sound sources can be used to produce *longitudinal* standing waves in air columns. The phase relationship between incident and reflected waves depends on whether or not the reflecting end of the air column is open or closed. This gives rise to two sets of possible standing wave conditions.

The first three natural modes of vibration for (a) an open pipe and (b) a closed pipe are shown in Figure 18.3.

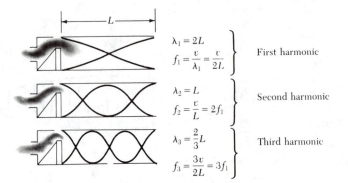

$\lambda_1 = 2L$

$f_1 = \dfrac{v}{\lambda_1} = \dfrac{v}{2L}$ } First harmonic

$\lambda_2 = L$

$f_2 = \dfrac{v}{L} = 2f_1$ } Second harmonic

$\lambda_3 = \dfrac{2}{3}L$

$f_3 = \dfrac{3v}{2L} = 3f_1$ } Third harmonic

Figure 18.3(a)   Natural modes of vibration in a hollow pipe open at each end. All harmonics are present.

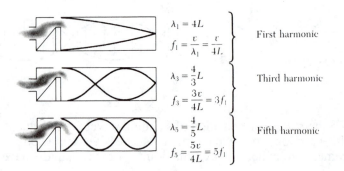

$\lambda_1 = 4L$

$f_1 = \dfrac{v}{\lambda_1} = \dfrac{v}{4L}$ } First harmonic

$\lambda_3 = \dfrac{4}{3}L$

$f_3 = \dfrac{3v}{4L} = 3f_1$ } Third harmonic

$\lambda_5 = \dfrac{4}{5}L$

$f_5 = \dfrac{5v}{4L} = 5f_1$ } Fifth harmonic

Figure 18.3(b)   Natural modes of vibration for a hollow pipe closed at one end. Only odd harmonics are present.

EQUATIONS AND CONCEPTS

The wave function which is the resultant of two traveling harmonic waves having the same direction, frequency, and amplitude is also harmonic and has the same frequency and wavelength as the individual waves.

$$y = 2A_o \cos\left(\frac{\phi}{2}\right) \sin\left(kx - \omega t - \frac{\phi}{2}\right) \qquad (18.1)$$

The amplitude of the resultant wave depends on the phase difference between the two individual waves according to:

$$y_m = 2A_o \cos\left(\frac{\phi}{2}\right)$$
(amplitude)

A *phase difference* can arise between two waves generated by the same source and arriving at a common point after having traveled along paths of *unequal path length*. In Eq. 18.2, ϕ is the phase difference and Δr is the *path difference* between the two waves.

$$\Delta r = \frac{\lambda}{2\pi} \phi \qquad (18.2)$$

A *standing wave* can be produced in a string due to the interference of two sinusoidal waves with equal amplitude and frequency traveling in opposite directions.

$$y = (2A_o \sin kx) \cos \omega t \qquad (18.3)$$

The amplitude of a standing wave is a function of the position x along the string. *Antinodes* are points of maximum displacement and *nodes* are points of zero displacement.

$$x = \begin{cases} (2n + 1)\left(\frac{\lambda}{4}\right) \text{-antinodes} \\ n\left(\frac{\lambda}{2}\right) \text{-nodes} \end{cases}$$

$$n = 0, \pm 1, \pm 2, \pm 3, \ldots$$

A series of natural patterns of vibration called *normal modes* can be excited in a string fixed at both ends. Each mode corresponds to a characteristic frequency and wavelength.

$$\lambda_n = \frac{2L}{n} \qquad (18.6)$$

$$f_n = \frac{n}{2L} v \qquad (18.7)$$

$$f_n = \frac{n}{2L} \sqrt{\frac{F}{\mu}} \qquad (18.8)$$

$$n = 1, 2, 3, \ldots$$

In an "open" pipe (open at both ends), all integral multiples of the fundamental frequency can be excited. In a "closed" pipe (closed at one end), only the odd multiples of the fundamental (odd harmonics) are possible.

$$f_n = \begin{cases} \dfrac{nv}{2L}, & n = 1, 2, 3, \ldots \text{ open pipe} \quad (18.11) \\[2ex] \dfrac{nv}{4L}, & n = 1, 3, 5, \ldots \text{ closed pipe} \quad (18.12) \end{cases}$$

Beats are formed by the combination of two waves of equal amplitude but slightly different frequencies traveling in the *same* direction.

$$y = 2A_o \cos 2\pi \left(\frac{f_1 - f_2}{2}\right) t \cos 2\pi \left(\frac{f_1 + f_2}{2}\right) t \quad (18.13)$$

The *amplitude* of the wave described by Eq. 18.13 is time dependent. Each occurrence of maximum amplitude results in a "beat" and the *beat frequency* f_b equals the difference in the frequencies of the individual waves.

$$A = 2A_o \cos 2\pi \left(\frac{f_1 - f_2}{2}\right) t \quad (18.14)$$

$$f_b = |f_1 - f_2| \quad (18.15)$$

Any *complex periodic wave form* can be represented by the combination of sinusoidal waves which form a harmonic series (combination of fundamental and various harmonics). Such a sum of sine and cosine terms is called a *Fourier series*.

$$y(t) = \sum_n \left(A_n \sin 2\pi f_n t + B_n \cos 2\pi f_n t\right) \quad (18.16)$$

ANSWERS TO SELECTED QUESTIONS

1. For certain positions of the movable section in Fig. 18.2, there is no sound detected at the receiver, corresponding to destructive interference. This suggests that perhaps energy is somehow lost! What happens to the energy transmitted by the source?

Answer: The intensity may be zero at the position of the receiver, corresponding to cancellation of the two sound waves at that point. However, the sound waves do not cancel at all points along the tube. This is equivalent to listening to two sources of sound from two speakers at different locations in a room. At certain positions in the room, the sound will have a minimum intensity, and at others it will have a maximum intensity.

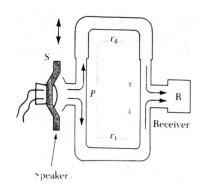

Figure 18.2

2. Does the phenomenon of wave interference apply only to harmonic waves?

Answer: No. Any waves moving in the same medium can interfere with each other. For example, two pulses moving in opposite directions on a stretched string interfere when they meet each other.

3. When two waves interfere constructively or destructively, is there any gain or loss in energy? Explain.

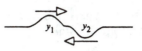

Answer: No. The energy may be transformed into other forms of energy. For example, when two pulses traveling on a stretched string in opposite directions overlap, and one is inverted, some potential energy is transferred to kinetic energy when they overlap. In fact, if they have equal and opposite amplitudes, they completely cancel each other at one point. In this case, all of the energy is transverse kinetic energy when the resultant amplitude is zero.

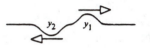

4. A standing wave is set up on a string as in Figure 18.5. Explain why no energy is transmitted along the string.

Answer: The string is fixed at both ends, and we assume that no energy is transferred to the object which supports it. Energy on the string itself is continually being transferred between potential and kinetic energy. In reality, some energy is being released in the form of sound waves and vibrations of the supporting medium.

Figure 18.5

5. What is common to *all* points (other than the nodes) on a string supporting a standing wave?

Answer: All points on the string, other than the nodes, move in the transverse direction with simple harmonic motion.

6. Some singers claim to be able to shatter a wine glass by maintaining a certain pitch in their voice over a period of several seconds (see photo). What mechanism causes the glass to break? (The glass must be very clean in order for it to break.)

Answer: The wine glass has natural frequencies of vibration. If the singer's voice has a strong component which corresponds to one of these natural frequencies, sound waves are coupled into the wine glass, setting up forced vibrations which can set up large stresses in the glass causing it to shatter.

7. What limits the amplitude of motion of a real vibrating system that is driven at one of its resonant frequencies?

Answer: The amplitude of motion will be limited by the magnitude of the dissipative (frictional) force (or forces) acting on the system.

8. If the temperature of the air in an organ pipe increases, what happens to the resonance frequencies?

Answer: The frequencies increase. Since the velocity of sound increases with increasing temperature, and the natural frequencies of vibration in the pipe are proportional to the wave velocity, it follows that the frequencies increase.

9. Explain why your voice seems to sound better than usual when you sing in the shower.

Answer: The shower acts like a resonant chamber and amplifies those frequency components of your voice which correspond to its natural frequencies of vibration.

10. What is the purpose of the slide on a trombone or the valves on a trumpet?

Answer: They change the effective length of an air column, which changes the resonance frequency of the instrument.

11. Explain why all harmonics are present in an organ pipe open at both ends, but only the odd harmonics are present in a pipe closed at one end.

Answer: The pipe which is open at both ends has antinodes at each end, which allows all possible harmonics. The pipe which is closed at one end has a node at the closed end and an antinode at the open end. These boundary conditions allow only odd harmonics.

12. Explain how a musical instrument such as a piano may be tuned using the phenomenon of beats.

Answer: A sound of the desired frequency (such as a struck tuning fork) is sounded simultaneously with the instrument. The instrument is tuned until the resulting beat frequency is minimized.

13. An airplane mechanic notices that the sound from a twin-engine aircraft rapidly varies in loudness when both engines are running. What could be causing this variation from loud to soft?

Answer: Apparently the two engines are emitting sounds having frequencies which differ only by a very small amount from each other. This results in a beat frequency, causing the variation from loud to soft.

14. At certain speeds, an automobile driven on a washboard road will vibrate disastrously and lose traction and braking effectiveness. At other speeds, either lesser or greater, the vibration is more manageable. Explain. Why are "rumble strips", which work on this same principle, often used just before stop signs?

Answer: At certain speeds the car crosses the bumps on the road at a rate which matches one of its natural frequencies of vibration. This causes the car to go into a large amplitude vibration that can be dangerous. At a speed slightly slower or faster, the natural frequency of the car and the bumps on the road are out of step, and the car moves relatively smoothly. Rumble strips are used to get your attention.

15. Why does a vibrating guitar string sound louder when placed on the instrument than it would if allowed to vibrate in the air while off the instrument?

Answer: A vibrating string is not able to set very much air into motion when vibrated alone. Thus it will not be very loud. If it is placed on the instrument, however, the string's vibration sets the sounding board of the guitar into vibration. A vibrating piece of wood is able to move a lot of air, and the note is louder.

SOLUTIONS TO SELECTED END-OF-CHAPTER PROBLEMS_____

9. Two harmonic waves are described by

$$y_1 = (3 \text{ cm}) \sin \pi(x + 0.6t)$$

$$y_2 = (3 \text{ cm}) \sin \pi(x - 0.6t)$$

Determine the *maximum* displacement of the motion at (a) x = 0.25 m, (b) x = 0.5 m, and (c) x = 1.5 m. (d) Find the three smallest values of x corresponding to antinodes.

Solution

$y_1 = (3 \text{ cm}) \sin \pi(x + 0.6t)$ $y_2 = (3 \text{ cm}) \sin \pi(x - 0.6t)$

(a)   We can take t = 0 to get the maximum y

$\qquad y = y_1 + y_2 = 3 \sin (\pi x) + 3 \sin (\pi x) = (6 \text{ cm}) \sin (\pi x)$

$\qquad$ At x = 0.25 cm, y = (6 cm) sin (π × 0.25) = $\boxed{4.24 \text{ cm}}$

(b)   At x = 0.5 cm, y = (6 cm) sin (π × 0.5) = $\boxed{6.00 \text{ cm}}$

(c)   At x = 1.5 cm, y = (6 cm) sin (π × 1.5) = $\boxed{-6.00 \text{ cm}}$

(d)   The antinodes occur when $x = \dfrac{n\lambda}{4}$ (n = 1, 3, 5, . . .)   but $k = \dfrac{2\pi}{\lambda} = \pi$, so λ = 2 cm,   and

$$x_1 = \frac{\lambda}{4} = \frac{2}{4} = \boxed{0.5 \text{ cm}}$$

$$x_2 = \frac{3\lambda}{4} = \frac{3(2)}{4} = \boxed{1.5 \text{ cm}}$$

$$x_3 = \frac{5\lambda}{4} = \frac{5(2)}{4} = \boxed{2.5 \text{ cm}}$$

13. A standing wave is formed by the interference of two traveling waves, each of which has an amplitude A = π cm, propagation number k = (π/2) cm⁻¹, and angular frequency ω = 10π rad/s. (a) Calculate the distance between the first two antinodes. (b) What is the amplitude of the standing wave at x = 0.25 cm?

Solution

(a)   Using the given parameters, the wave function is

$$y = (2\pi \text{ cm}) \sin\left(\frac{\pi x}{2}\right) \cos (10\pi t)$$

We need to find values of x for which $\sin\left(\dfrac{\pi x}{2}\right) = 1$

This condition requires that $\frac{\pi x}{2} = \pi\left(n + \frac{1}{2}\right)$; $\quad n = 0, 1, 2, \ldots$

For $n = 0$, $x = 1$ cm.   For $n = 1$, $x = 3$ cm.

Therefore the distance between antinodes, $\Delta x = \boxed{2.00 \text{ cm}}$

(b) $A = (2\pi \text{ cm}) \sin\left(\frac{\pi x}{2}\right)$;   when $x = 0.25$ cm, $A = \boxed{2.40 \text{ cm}}$

---

21. Find the fundamental frequency and the next three frequencies that could cause a standing wave pattern on a string that is 30 m long, has a mass per unit length 9×10^{-3} kg/m, and is stretched to a tension of 20 N.

Solution

$L = 30$ m, $\mu = 9 \times 10^{-3}$ kg/m, $F = 20$ N

$f_1 = \frac{v}{2L}$ where $v = \sqrt{\frac{F}{\mu}} = 47.1$ m/s,   so

$\quad f_1 = \frac{47.1 \text{ m/s}}{60 \text{ m}} = \boxed{0.786 \text{ Hz}}$

$\quad f_2 = 2f_1 = \boxed{1.57 \text{ Hz}}$

$\quad f_3 = 3f_1 = \boxed{2.36 \text{ Hz}}$

$\quad f_4 = 4f_1 = \boxed{3.14 \text{ Hz}}$

---

27. A 60-cm guitar string under a tension of 50 N has a mass per unit length of 0.1 g/cm. What is the highest resonant frequency that can be heard by a person capable of hearing frequencies up to 20,000 Hz?

Solution

$L = 60$ cm $= 0.60$ m, $F = 25$ N, $\mu = 0.1$ g/cm $= 0.001$ kg/m

$f_n = \frac{nv}{2L}$ where $v = \sqrt{\frac{F}{\mu}} = 158.1$ m/s

$f_n = n\left(\frac{158.1}{1.2}\right) = 131.8n = 20,000$

Largest $n = 151$ which corresponds to $f = \boxed{19.902 \text{ kHz}}$

---

43. A glass tube is open at one end and closed at the other (by a movable piston). The tube is filled with 30°C air, and a 384-Hz tuning fork is held at the open end. Resonance is heard when the piston is 22.8 cm from the open end and again when it is 68.3 cm from the open end. (a) What speed of sound is implied by these data? (b) Where would the piston be for the next resonance?

$f = 384$ Hz

Solution

For resonance in a half-open tube, Equation 18.12 gives

$$f = n\frac{v}{4L} \quad (n = 1, 3, 5, \ldots)$$

(a) Adding n = 1 and n = 3

$$384 \text{ s}^{-1} = \frac{v}{4(0.228 \text{ m})} \qquad \text{and} \qquad 384 \text{ s}^{-1} = \frac{3v}{4(0.683 \text{ m})}$$

In either case we find $\boxed{v = 350 \text{ m/s}}$

(b) For the next resonance, n = 5, and

$$L = \frac{5v}{4f} = \frac{5(350 \text{ m/s})}{4(384 \text{ s}^{-1})} = \boxed{1.14 \text{ m}}$$

---

45. A shower stall measures 86 cm × 86 cm × 210 cm. When you sing in the shower, which frequencies will sound the richest (resonate), assuming the shower acts as a pipe closed at both ends (nodes at both sides)? Assume also that the human voice ranges from 130 Hz to 2000 Hz (not necessarily one person's voice, however). Let the speed of sound in the hot shower stall be 355 m/s.

Solution

For a closed box, the resonant frequencies will have nodes at both sides, so the permitted wavelengths will be

$$L = \frac{n\lambda}{2}, \quad (n = 1, 2, 3, \ldots)$$

$$L = \frac{n\lambda}{2} = \frac{nv}{2f} \qquad \text{or} \qquad f = \frac{nv}{2L}$$

Therefore with L = 0.860 m and L' = 2.10 m, the resonant frequencies are

$$f_n = \boxed{n(206 \text{ Hz})} \quad \text{for} \quad L = 0.860 \text{ m}$$

and

$$f_n' = \boxed{n(84.5 \text{ Hz})} \quad \text{for} \quad L' = 2.10 \text{ m}$$

---

53. A student holds a tuning fork oscillating at 256 Hz. He walks towards a wall at a constant speed of 1.33 m/s. (a) What beat frequency does he observe between the tuning fork and its echo? (b) How fast must he walk away from the wall to observe a beat frequency of 5 Hz?

Solution

For an echo, $f = f_o \dfrac{(v + v_s)}{(v - v_s)}$ and the beat frequency $f_b = |f - f_o|$

Solving for f_b gives $f_b = f_o \dfrac{(2v_s)}{(v - v_s)}$ when approaching the wall.

(a) $f_b = \dfrac{(256 \text{ Hz})(2)(1.33 \text{ m/s})}{(343 \text{ m/s} - 1.33 \text{ m/s})} = \boxed{1.99 \text{ Hz}}$

(b) When moving away from wall, v_s changes sign. Solving for v_s gives

$$v_s = \frac{f_b v}{(2f_o - f_b)} = \frac{(5 \text{ Hz})(343 \text{ m/s})}{2(256 \text{ Hz}) - 5 \text{ Hz}} = \boxed{3.38 \text{ m/s}}$$

57. Two speakers are arranged as shown in Figure 18.3. For this problem, assume that point 0 is 12 m along the center line and the speakers are separated by a distance of 1.5 m. As the listener moves toward point P from point 0, a series of alternating minima and maxima is encountered. The distance between the first minimum and the next maximum is 0.4 m. Using 344 m/s as the speed of sound in air, determine the frequency of the speakers. (Use the approximation $\sin \theta \approx \tan \theta$.)

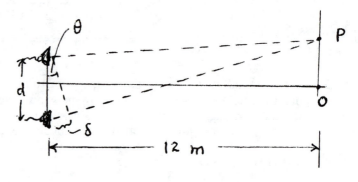

Figure 18.3

Solution

Path difference, $\delta = d \sin \theta$; and for $L \gg y$, $\delta \approx d \tan \theta = d\left(\dfrac{y}{L}\right)$

For minimum $\delta_1 = d\left(\dfrac{y_1}{L}\right) = \dfrac{\lambda}{2}$. For maximum $\delta_2 = d\left(\dfrac{y_2}{L}\right) = \lambda$

Therefore $\lambda = \dfrac{2(y_2 - y_1)d}{L} = \dfrac{2(0.4 \text{ m})(1.5 \text{ m})}{12 \text{ m}} = 0.1 \text{ m}$

and $f = \dfrac{v}{\lambda} = \dfrac{340 \text{ m/s}}{0.1 \text{ m}} = \boxed{3.40 \text{ kHz}}$

61. To maintain a length of string under tension in a horizontal position, one end of the string is connected to a vibrating blade and the other end is passed over a pulley and attached to a mass. The mass of the string is 10 g, and its *total* length is 1.25 m. (a) When the suspended mass is 10 kg, the string vibrates in three equal length segments. Determine the vibration frequency of the blade. (Assume that the point where the string passes over the pulley and the point where it is attached to the blade are both nodes. Also, ignore the contribution to the tension due to the string's mass.) (b) What mass should be attached to the string if it is to vibrate in four equal segments?

Solution

(a) When the string fastened at each end vibrates in three segments, $\lambda = \frac{2L}{3} = 0.830$ m.

The linear density of the string is $\mu = \frac{m}{L} = \frac{0.01}{1.25} = 8.00 \times 10^{-3}$ kg/m.

Since the tension is given as $F = 10g = 98$ N, the wave speed is

$$v = \sqrt{\frac{F}{\mu}} = \sqrt{\frac{98 \text{ N}}{8.00 \times 10^{-3} \text{ kg/m}}} = 111 \text{ m/s}$$

Therefore $f = \frac{v}{\lambda} = \frac{111 \text{ m/s}}{0.830 \text{ m}} = \boxed{133 \text{ Hz}}$

(b) In four segments, $\lambda = \frac{L}{2} = 0.625$ m and $v = f\lambda = 83.1$ m/s

We find the mass of string to be

$$M = \frac{\mu v^2}{g} = \frac{(8.00 \times 10^{-3} \text{ kg/m})(83.1 \text{ m/s})^2}{9.80 \text{ m/s}^2} = \boxed{5.64 \text{ kg}}$$

19

Temperature, Thermal Expansion, and Ideal Gases

OBJECTIVES_____

1. Understand the concepts of thermal equilibrium and thermal contact between two bodies, and state the zeroth law of thermodynamics.

2. Discuss some physical properties of substances which change with temperature, and the manner in which these properties are used to construct thermometers.

3. Describe the operation of the constant-volume gas thermometer and how it is used to define the ideal-gas temperature scale.

4. Convert between the various temperature scales, especially the conversion from degrees Celsius into kelvins, degrees Fahrenheit into kelvins, and degrees Celsius into degrees Fahrenheit.

5. Provide a qualitative description of the origin of thermal expansion of solids and liquids; define the linear expansion coefficient and volume expansion coefficient for an isotropic solid, and learn how to deal with these coefficients in practical situations involving expansion or contraction.

6. Understand the properties of an ideal gas and the equation of state for an ideal gas. You should also be familiar with the conditions under which a real gas behaves like an ideal gas.

NOTES FROM SELECTED CHAPTER SECTIONS_____

19.1 TEMPERATURE AND THE ZEROTH LAW OF THERMODYNAMICS

Thermal physics is the study of the behavior of solids, liquids and gases, using the concepts of heat and temperature. Two approaches are commonly used in this area of science. The first is a *macroscopic* approach, called *thermodynamics,* in which one explains the bulk thermal properties of matter. The second is a *microscopic* approach, called *statistical mechanics,* in which properties of matter are explained on an atomic scale. Both approaches require that you understand some basic concepts, such as the concepts of temperature and heat. As we will see, *all* thermal phenomena are manifestations of the laws of mechanics as we have learned them. For example, thermal energy (or heat energy) is actually a consequence of the vibrations of a large number of particles making up the system.

The concept of the *temperature* of a system can be understood in connection with a measurement, such as the reading of a thermometer. Temperature, a scalar quantity, is a property which can only be defined when the system is in thermal equilibrium with another system. Thermal equilibrium implies that two (or more) systems are at the same temperature.

The *zeroth law of thermodynamics* states that if two systems are in thermal equilibrium with a third system, they must be in thermal equilibrium with each other. The third system is usually a calibrated thermometer whose reading determines whether or not the systems are in thermal equilibrium. There are several types of thermometers which can be used.

19.3 THE CONSTANT-VOLUME GAS THERMOMETER AND THE KELVIN SCALE

The *gas thermometer* is a standard device for defining temperature. In the constant-volume gas thermometer, a low density gas is placed in a flask, and its volume is kept constant while it is heated. The pressure is measured as the gas is heated or cooled. Experimentally, one finds that the temperature is proportional to the absolute pressure.

The *thermodynamic temperature scale* is based on a scale for which b = 0 in Eq. 19.1 and the reference temperature is taken to be the *triple point of water*; that is, the temperature and pressure at which water,

water vapor, and ice coexist in equilibrium. On this scale, the SI unit of temperature is the *kelvin*, defined as the fraction $\frac{1}{273.16}$ of the temperature of the triple point of water.

The *absolute temperature scale*, or kelvin scale, is identical to the ideal-gas scale for temperatures above 1 K.

EQUATIONS AND CONCEPTS

In a constant-volume gas thermometer, the temperature T is proportional to the absolute pressure P. In Eq. 19.1, a and b are constants.

$$T = aP + b \qquad (19.1)$$

Real gases behave as ideal gases at sufficiently low gas pressures and high temperatures. The ideal-gas temperature T in the limit of low gas pressures is defined by Eq. 19.3, where P_3 is the pressure at the triple-point temperature.

$$T = (273.16 \text{ K}) \lim_{P_3 \to 0} \frac{P}{P_3}$$
$$(\text{constant } V) \qquad (19.3)$$

The *Celsius temperature* T_C is related to the absolute temperature T (in kelvins) according to Eq. 19.4, where 0°C corresponds to 273.15 K.

$$T_C = T - 273.15 \qquad (19.4)$$

The *Fahrenheit temperature* T_F can be converted to degrees Celsius using Eq. 19.5. Note that 0°C = 32°F and 100°C = 212°F.

$$T_F = \frac{9}{5} T_C + 32 \text{ F°} \qquad (19.5)$$

If a body has a length *L*, the *change in its length* ΔL due to a change in temperature is proportional to the change in temperature and the length. The proportionality constant α is called the *average coefficient of linear expansion*.

$$\Delta L = \alpha L \, \Delta T \qquad (19.6)$$

or

$$\alpha = \frac{1}{L} \frac{\Delta L}{\Delta T} \qquad (19.7)$$

If the temperature of a body of volume V changes by an amount ΔT at constant pressure, *the change in its volume* is proportional to ΔT and the original volume. The constant of proportionality β is the *average coefficient of volume expansion*. For an isotropic solid, $\beta = 3\alpha$.

$$\Delta V = \beta V \Delta T \qquad (19.8)$$

If n molecules of a dilute gas occupy a volume V, the *equation of state* which relates the variables P, V, and T at equilibrium is that of an *ideal gas*, Eq. 19.12, where R is the *universal gas constant.*

$$PV = nRT \tag{19.12}$$

$$R = 8.31 \; \frac{J}{mol \cdot K} \tag{19.13}$$

The *equation of state of an ideal gas* can also be expressed in the form of Eq. 19.14, where N is the total number of gas molecules and k is *Boltzman's constant.*

$$PV = NkT \tag{19.14}$$

$$k = 1.38 \times 10^{-23} \; \frac{J}{K} \tag{19.15}$$

ANSWERS TO SELECTED QUESTIONS

2. Is it possible for two objects to be in thermal equilibrium if they are not in contact with each other? Explain.

Answer: Yes. Two bodies are in equilibrium with each other if they are at the same temperature.

3. A piece of copper is dropped into a beaker of water. If the water's temperature rises, what happens to the temperature of the copper? Under what condition will the water and copper be in thermal equilibrium?

Answer: The temperature of the copper will decrease. The water and copper will be in thermal equilibrium when they reach a common temperature. The time this takes to occur depends on the two temperatures and the relative masses of the water and copper.

7. Explain why a column of mercury in a thermometer first descends slightly and then rises when placed in hot water.

Answer: The glass surrounding the mercury expands before the mercury. The mercury rises after it begins to heat up and approach the temperature of the hot water because its temperature coefficient of expansion is greater than that for glass.

9. A steel wheel bearing has an inside diameter which is 1 mm smaller than an axle. How can it be made to fit onto the axle without removing any material?

Answer: The bearing can be heated until its diameter has expanded to fit the axle.

10. Markings to indicate length are placed on a steel tape in a room that has a temperature of 22°C. Are measurements made with the tape on a day when the temperature is 27°C too long, too short, or accurate? Defend your answer.

Answer: Too short. At 22°C the tape would read the width of an object accurately, but an increase in temperature causes the divisions ruled on the tape to be further apart than they should be. This "too long" ruler will, then, measure objects to be shorter than they really are.

11. What would happen if the glass of a thermometer expanded more upon heating than did the liquid inside?

Answer: The liquid inside would go down in the stem as the temperature increased.

12. Determine the number of grams in one mol of the following gases: (a) hydrogen, (b) helium, and (c) carbon monoxide.

Answer: (a) 2 grams per mol, (b) 4 grams per mol, and (c) 28 grams per mol.

15. Two identical cylinders at the same temperature each contain the same kind of gas. If the volume of cylinder A is three times greater than the volume of cylinder B, what can you say about the relative pressures in the cylinders?

Answer: The pressure in cylinder A is three times greater than the pressure in cylinder B.

SOLUTIONS TO SELECTED END-OF-CHAPTER PROBLEMS

19. A structural steel I-beam is 15 m long when installed at 20°C. How much will its length change over the temperature extremes - 30°C to 50°C?

Solution

$$\Delta L = \alpha L_0 \Delta T = (11 \times 10^{-6} \ C^{\circ -1})(15 \ m)(50 + 30) \ C^{\circ} = \boxed{1.32 \times 10^{-2} \ m}$$

41. One mol of oxygen gas is at a pressure of 6 atm and a temperature of 27°C. (a) If the gas is heated at constant volume until the pressure triples, what is the final temperature? (b) If the gas is heated such that both the pressure and volume are doubled, what is the final temperature?

Solution

(a) At constant volume $\dfrac{P_1}{P_2} = \dfrac{T_1}{T_2}$, where $T_1 = 300$ K. Hence,

$$T_2 = \left(\frac{P_2}{P_1}\right) T_1 = \left(\frac{3P_1}{P_1}\right) T_1 = 3(300 \ K) = \boxed{900 \ K}$$

(b) From the ideal gas law, $\dfrac{P_1 V_1}{P_2 V_2} = \dfrac{T_1}{T_2}$ and

$$T_2 = \left(\frac{P_2}{P_1}\right)\left(\frac{V_2}{V_1}\right) T_1 = \left(\frac{2P_1}{P_1}\right)\left(\frac{2V_1}{V_1}\right) T_1 = 4T_1 = \boxed{1200 \ K}$$

53. The rectangular plate shown in Fig. 19.13 has an area A equal to Lw. If the temperature increases by ΔT, show that the increase in area is given by $\Delta A = 2\alpha A \ \Delta T$, where α is the coefficient of linear expansion. What approximation does this expression assume? (Hint: Each dimension increases according to $\Delta L = \alpha L \ \Delta T$.)

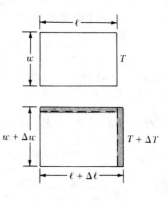

Figure 19.13

Solution

From the diagram in Fig. 19.13, we see that the *change* in area is $\Delta A = L\Delta w + w\Delta L + \Delta w\Delta L$. Since ΔL and Δw are each small quantities, the product $\Delta w\Delta L$ will be very small. Therefore, we assume $\Delta w\Delta L \approx 0$. Since $\Delta w = w\alpha\Delta T$ and $\Delta L = L\alpha\Delta T$, we then have $\Delta A = Lw\alpha\Delta T + wL\alpha\Delta T$. Finally, since $A = Lw$, we find $\Delta A = 2\alpha A\,\Delta T$.

---

60. (a) Show that the volume coefficient of thermal expansion for an ideal gas at constant pressure is given by $\beta = \frac{1}{T}$, where T is the kelvin temperature. Start with the definition of β and use the equation of state, $PV = nRT$. (b) What value does this expression predict for β at 0°C? Compare this with the experimental values for helium and air in Table 19.2.

Solution

(a)   For an ideal gas $PV = nRT$. For small change ΔV and ΔT, and at constant pressure we have $P\Delta V = nR\Delta T$. Multiplying each side of the equation by VT, we have

$$P\Delta V(VT) = nR\Delta T(VT)$$

and using $PV = nRT$, this becomes

$$T\Delta V = V\Delta T$$

But $\beta = \frac{\Delta V}{V\Delta T}$,   therefore $\beta = \frac{1}{T}$

(b)   At $T = 0°C = 273.15$ K, $\beta = \frac{1}{273.15}$ K $= 3.66 \times 10^{-3}$ °C^{-1}

Experimentally, $\beta(He) = 3.665 \times 10^{-3}$ °C^{-1} and $\beta(air) = 3.67 \times 10^{-3}$ °C^{-1}

Hence, both gases have thermal expansion coefficients very close to the values predicted by the ideal gas model.

---

61. Starting with Equation 19.12, show that the total pressure P in a container filled with a mixture of several different ideal gases is given by $P = P_1 + P_2 + P_3 + \ldots$, where P_1, P_2, etc., are the pressures that each gas would exert if it alone filled the container (or the *partial pressures* of the respective gases). This is known as *Dalton's law of partial pressures.*

Solution

For each gas alone, $P_1 = \frac{n_1kT}{V}$ and $P_2 = \frac{n_2kT}{V}$ and $P_3 = \frac{n_3kT}{V}$,   etc.

For all the gases, $P_1V_1 + P_2V_2 + P_3V_3 \ldots = (n_1 + n_2 + n_3 \ldots)kT$ and $(n_1 + n_2 + n_3 \ldots)kT = PV$

Also, $V_1 = V_2 = V_3 = \ldots = V$,   therefore $P = P_1 + P_2 + P_3 + \ldots$

---

67. An air bubble originating from a deep sea diver has a radius of 5 mm at some depth h. When the bubble reaches the surface of the water, it has a radius of 7 mm. Assuming the temperature of the air in the bubble remains constant, determine (a) the depth h of the diver, and (b) the absolute pressure at this depth.

Solution

(a) The pressure at any depth is $P = P_a + \rho gh$ where $P_a = 1.013 \times 10^5$ N/m^2 and $\rho = 10^3$ kg/m^3. Also, for an ideal gas that remains at constant temperature, $PV = P_aV_a$. This gives

$$P = \left(\frac{V_a}{V}\right)P_a = \left(\frac{r_a}{r}\right)^3 P_a = \left(\frac{7}{5}\right)^3 P_a = 2.744 P_a$$

Substituting this expression for P into the first equation and using the given numerical values, we have

$$h = \frac{P - P_a}{\rho g} = \frac{P_a}{\rho g}[2.744 - 1] = \boxed{18.0 \text{ m}}$$

(b) $P = P_a + \rho gh = 1.013 \times 10^5$ N/m^2 + $(10^3$ kg/m$^3)(9.80$ m/s$^2)(18.0$ m)

$P = 2.78 \times 10^5$ Pa = $\boxed{278 \text{ kPa}}$

---

73. A steel guitar string with a diameter of 1.00 mm is stretched between supports 80 cm apart. The temperature is 0°C. (a) Find the mass per unit length of this string. (Use the value 7.86×10^3 kg/m^3 for the density.) (b) The fundamental frequency of transverse oscillations of the string is 200 Hz. What is the tension in the string? (c) If the temperature is raised to 30°C, find the resulting values of the tension and the fundamental frequency. [Assume that both the Young's modulus (Table 12.1) and the coefficient of thermal expansion (Table 19.2) of steel have constant values between 0°C and 30°C.]

Solution

(a) $\mu = \left(\frac{\pi d^2}{4}\right)\rho = \frac{1}{4}\pi(1.00 \times 10^{-3}$ m$)^2(7.86 \times 10^3$ kg/m$^3) = \boxed{6.17 \times 10^{-3} \text{ kg/m}}$

(b) $f_1 = \frac{v}{2L}$; $v = \sqrt{\frac{F}{\mu}}$, so $f_1 = \frac{1}{2L}\sqrt{\frac{F}{\mu}}$

Therefore, $F = \mu(2Lf_1)^2 = (6.173 \times 10^{-3})(2 \times 0.80 \times 2 \times 10^2)^2 = \boxed{632 \text{ N}}$

(c) (Actual length) at 0°C = (natural length)$\left(1 + \frac{F}{AY}\right)$

$A = \left(\frac{\pi}{4}\right)(1 \times 10^{-3}$ m$)^2 = 7.854 \times 10^{-7}$ m^2 and $Y = 20 \times 10^{10}$ N/m^2

Therefore, $\frac{F}{AY} = \frac{632 \text{ N}}{(7.854 \times 10^{-7} \text{ m}^2)(20 \times 20^{10} \text{ N/m}^2)} = 4.024 \times 10^{-3}$

The natural length at 0°C = $\frac{0.800 \text{ m}}{1 + 4.024 \times 10^{-3}} = 0.79679$ m

The unstressed length at 30°C = (0.79679 m)[1 + 30(11 × 10⁻⁶)] = 0.79706 m

Then $\quad 0.800 = (0.79706)\left[1 + \dfrac{F'}{(A'Y)}\right]$

From this we find $\quad \dfrac{F'}{A'Y} = \dfrac{0.800}{0.79706} - 1 = 3.689 \times 10^{-3}$

and $\quad$ F'= A'Y(3.689 × 10⁻³) = (7.854 × 10⁻⁷)(20 × 10¹⁰)(3.689 × 10⁻³)(1 + αΔT)²

$\quad$ F' = (579.5 N)(1 + 3.3 × 10⁻⁴)² ≈ $\boxed{580 \text{ N}}$

Also, $\quad \dfrac{f_1'}{f_1} = \sqrt{\dfrac{F'}{F}},\quad$ therefore $\quad f_1' = \sqrt{\dfrac{580}{632}} \times 200 = \boxed{192 \text{ Hz}}$

20

Heat and the First Law
of Thermodynamics

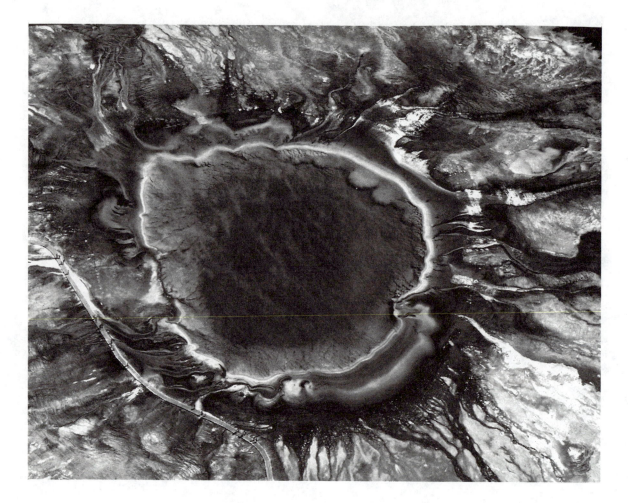

OBJECTIVES

1. Understand the concepts of heat, internal energy, and thermodynamic processes.

2. Define and discuss the calorie, heat capacity, specific heat, and latent heat.

3. Provide a qualitative description of different types of phase changes which a substance may undergo, and the changes in energy which accompany such processes.

4. Discuss the possible mechanisms which can give rise to heat transfer between a system and its surroundings; that is, heat conduction, convection, and radiation. You should also be able to state the basic law of heat conduction, and give a realistic example of each heat transfer mechanism.

5. Describe Joule's experiment for measuring the mechanical equivalent of heat, 1 cal = 4.186 J.

6. Understand how work is defined when a system undergoes a change in state, and the fact that work (like heat) depends on the path taken by the system. You should also know how to sketch processes on a PV diagram, and calculate work using these diagrams.

7. State the first law of thermodynamics ($\Delta U = Q - W$), and explain the meaning of the three forms of energy contained in this statement.

8. Discuss the implications of the first law of thermodynamics as applied to (a) an isolated system, (b) a cyclic process, (c) an adiabatic process, and (d) an isothermal process.

9. Calculate the work done when an ideal gas expands during an isothermal process.

SKILLS

Many applications of the first law of thermodynamics deal with the work done by (or on) a system which undergoes a change in state. For example, as a gas is taken from a state whose initial pressure and volume are P_i, V_i, and whose final pressure and volume are P_f, V_f, the work can be calculated if the process can be drawn on a PV diagram as in Figure 20.1. The work done during the expansion is given by the integral expression

$$W = \int_{V_i}^{V_f} P \, dV \, ,$$

which numerically represents the *area* under the PV curve (the shaded region) shown in Figure 20.1. It is important to recognize that the work *depends on the path taken as the gas goes from i to f.* That is, W depends on the specific manner in which the pressure P changes during the process.

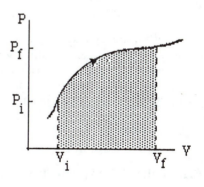

Figure 20.1

NOTES FROM SELECTED CHAPTER SECTIONS_____

20.1 HEAT AND THERMAL ENERGY

When two systems at different temperatures are in contact with each other, energy will transfer between them until they reach the same temperature (that is, when they are in thermal equilibrium with each other). This energy is called heat, or thermal energy, and the term "heat flow" refers to an energy transfer as a consequence of a temperature difference.

The unit of heat is the *calorie* (cal), defined as the amount of heat necessary to increase the temperature of 1 g of water from 14.5°C to 15.5°C. The *mechanical equivalent of heat*, first measured by Joule, is given by 1 cal = 4.186 J.

20.2 HEAT CAPACITY AND SPECIFIC HEAT

The *heat capacity*, C', of any substance is defined as the amount of heat required to increase the temperature of that substance by one Celsius degree. Its units are cal/°C.

20.3 LATENT HEAT

The *latent heat of fusion* is a parameter used to characterize a solid-to-liquid phase change; the *latent heat of vaporization* characterizes the liquid-to-gas phase change.

20.4 WORK AND HEAT IN THERMODYNAMIC PROCESSES

The work done in the expansion from the initial state to the final state is the area under the curve in a PV diagram. See Figure 20.2.

If the gas is compressed, $V_f < V_i$, and the work is negative. That is, work is done *on* the gas. If the gas expands, $V_f > V_i$, the work is positive, and the gas does work on the piston. If the gas expands at *constant pressure*, called an *isobaric process*, then $W = P(V_f - V_i)$.

The work done by a system depends on the process by which the system goes from the initial to the final state. In other words, the work done depends on the initial, final, and intermediate states of the system.

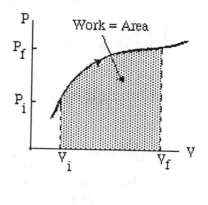

Figure 20.2

20.5 THE FIRST LAW OF THERMODYNAMICS

In the first law of thermodynamics, $\Delta U = Q - W$, Q is the heat added to the system and W is the work done by the system. Note that by convention, Q is *positive* when heat enters the system and *negative* when heat is removed from the system. Likewise, W can be positive or negative as mentioned earlier. The initial and final states must be *equilibrium* states; however, the intermediate states are, in general, nonequilibrium states since the thermodynamic coordinates undergo finite changes during the thermodynamic process. For an *infinitesimal change of the system*, we can express the *first law of thermodynamics* in the form dU = dQ - dW. It is important to note that dQ and dW *are not exact differentials*, since both Q and W are not functions of the system's coordinates. That is, both Q and W depend on the *path* taken between the initial and final equilibrium states, during which time the system interacts with its environment. On the other hand, dU is an *exact differential* and the internal energy U is a *state variable*. The function U is analogous to the potential energy function used in mechanics when dealing with conservative forces.

20.6 SOME APPLICATIONS OF THE FIRST LAW OF THERMODYNAMICS

An *isolated system* is one which does not interact with its surroundings. In such a system, Q = W = 0, so it follows from the first law that $\Delta U = 0$. That is, the internal energy of an isolated system cannot change.

A *cyclic process* is one that originates and ends up at the same state. In this situation, $\Delta U = 0$, so from the first law we see that Q = W. That is, the work done per cycle equals the heat added to the system per cycle. This is important to remember when dealing with heat engines in the next chapter.

An *adiabatic process* is a process in which no heat enters or leaves the system; that is, Q = 0. The first law applied to this process gives $\Delta U = -W$. A system may undergo an adiabatic process if it is thermally insulated from its surroundings.

An *isobaric process* is a process which occurs at constant pressure. For such a process, the heat transferred and the work done are nonzero.

An *isovolumetric process* is one which occurs at constant volume. By definition, W = 0 for such a process (since dV = 0), so from the first law it follows that $\Delta U = Q$. That is, all of the heat added to the system kept at constant volume goes into increasing the internal energy of the system.

A process that occurs at constant temperature is called an *isothermal process*, and a plot of P versus V at constant temperature for an ideal gas yields a hyperbolic curve called an *isotherm*. The internal energy of an ideal gas is a function of temperature only. Hence, in an isothermal process of an ideal gas, $\Delta U = 0$.

20.7 HEAT TRANSFER

There are three basic processes of heat transfer. These are (1) conduction, (2) convection, and (3) radiation.

Conduction is a heat transfer process which occurs when there is a *temperature gradient* across the body. That is, conduction of heat occurs only when the body's temperature is *not* uniform. For example, if you heat a metal rod at one end with a flame, heat will flow from the hot end to the colder end. If the heat flow is along x (that is, along the rod), and we define the *temperature gradient* as dT/dx, then a quantity of heat dQ will flow in a time dt along the rod. The rate of flow of heat along the rod, sometimes called the *heat current*, is proportional to the cross-sectional area of the rod, the temperature gradient, and k, the thermal conductivity of the material of which the rod is made.

When heat transfer occurs as the result of the motion of material, such as the mixing of hot and cold fluids, the process is referred to as *convection*. Convection heating is used in conventional hot-air and hot-water heating systems. Convection currents produce changes in weather conditions when warm and cold air masses mix in the atmosphere.

Heat transfer by *radiation* is the result of the continuous emission of electromagnetic radiation by all bodies.

EQUATIONS AND CONCEPTS

The *specific heat* c of a substance of mass m equals its heat capacity per unit mass.

$$c = \frac{C}{m}$$

(20.3)

The *heat energy* Q transferred between a system of mass m and its surroundings for a temperature change ΔT is given by Eq. 20.4.

$$Q = C\Delta T = mc\Delta T$$

(20.4)

A substance may undergo a phase change when heat is transferred between the substance and its surroundings. The heat Q required to change the phase of a mass m is given by Eq. 20.6, where L is called the *latent heat*.

$$Q = mL$$

(20.6)

The *work done by a gas* which undergoes an expansion or compression from initial volume V_i to final volume V_f is given by Eq. 20.8. The pressure is generally not constant, so you must exercise care in evaluating W from this equation. In general, the work done equals the area under the PV curve bounded by V_i and V_f, and the function P, as in Fig. 20.2.

$$W = \int_{V_i}^{V_f} PdV$$

(20.8)

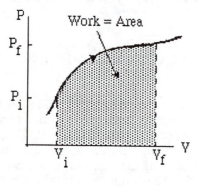

Figure 20.2

The *first law of thermodynamics* is a generalization of the law of conservation of energy that includes possible changes in internal energy. It states that the *change* in internal energy of a system, ΔU, equals the quantity Q - W, where Q is the heat added to the system.

$$\Delta U = Q - W$$

(20.9)

In an *adiabatic process*, Q = 0, and the change in internal energy equals the negative of the work done *by* the gas.

$$\boxed{\Delta U = -W} \qquad (20.11)$$

In a constant volume (isovolumetric) process, the work done is zero and all heat added to the system goes into increasing the internal energy.

$$\boxed{\Delta U = Q} \qquad (20.12)$$

An *isothermal process* is one which occurs at constant temperature. During such a process, the change in internal energy results from both heat transfer and work done. The *work done during the isothermal expansion of an ideal gas* can be calculated from Eq. 20.8 and the equation of state PV = nRT. The result is Eq. 20.13.

$$\boxed{W = nRT \, ln\left(\frac{V_f}{V_i}\right)} \qquad (20.13)$$

In the law of heat conduction, k is the *thermal conductivity* of the material and $\frac{dT}{dx}$ is the *temperature gradient*.

$$\boxed{H = -kA\left(\frac{dT}{dx}\right)} \qquad (20.15)$$

If a rod of length *L* has a *uniform* cross-sectional area and its ends are at temperatures T_1 and T_2, the *rate of heat flow* can be expressed in the form of Eq. 20.16.

$$\boxed{H = kA\frac{(T_2 - T_1)}{L}} \qquad (T_2 > T_1) \qquad (20.16)$$

The rate at which radiant energy is emitted by a body (or the power) is proportional to the fourth power of the absolute temperature T. This is known as *Stefan's law* (Eq. 20.19). In this expression, A is the surface area, σ is a universal constant equal to 5.6696×10^{-8} W/m^2·K, e is emissivity which can have a value between 0 and 1, depending on the nature of the surface.

$$\boxed{P = \sigma AeT^4} \qquad (20.19)$$

ANSWERS TO SELECTED QUESTIONS

1. Ethyl alcohol has about one half the specific heat of water. If equal masses of alcohol and water in separate beakers are supplied with the same amount of heat, compare the temperature increases of the two liquids.

Answer: The increase in the temperature of the alcohol will be twice that of the water.

3. A small crucible is taken from a 200°C oven and immersed in a tub full of water at room temperature (often referred to as *quenching*). What is the approximate final equilibrium temperature?

Answer: Room temperature. The tub of water has a much larger heat capacity than the crucible, so the increase in its temperature is negligible.

5. In a daring lecture demonstration, an instructor dips his wetted fingers into molten lead (327°C) and withdraws them quickly without getting burned. How is this possible? *(Note that this is a dangerous experiment which you should not attempt.)*

Answer: The fingers are wetted in order to create a layer of steam between the hand and the molten lead. The vapor acts as an insulator and prevents serious burns. This is similar to the technique of testing the temperature of a hot steam iron with a wetted finger.

8. Why is it possible to hold a lighted match even when it is burned to within a few millimeters of your fingertips?

Answer: Wood is a poor conductor of heat, so your fingers are protected.

9. The photograph at the right shows the pattern formed by snow on the roof of a barn. What causes the alternating pattern of snow-covered and exposed roof?

Answer: The snow-covered region is formed by wooden beams (good insulators) underneath the roof structure. The exposed regions between the beams conduct the heat to the exterior more readily (because they lack insulation), causing the snow to melt in these areas.

11. A tile floor in a bathroom may feel uncomfortably cold to your bare feet, but a carpeted floor in an adjoining room at the same temperature will feel warm. Why?

Answer: The tile is a better conductor of heat than is the carpet. Thus, heat is conducted away from your feet by the tile more rapidly than by the carpeted floor.

12. Why can potatoes be baked more quickly when a skewer has been inserted through them?

Answer: A potato is not a good conductor of heat. Thus, quite some time is required for heat to be conducted into its interior. A skewer through it, however, quickly conducts heat to the inside of the potato.

14. A piece of paper is wrapped around a rod made half of wood and half of copper. When held over a flame, the paper in contact with the wood burns but the half in contact with the metal does not. Explain.

Answer: Because copper is a good conductor of heat, the heat added to the paper is quickly conducted into the copper. Wood, however, is a poor conductor and cannot carry the added heat away fast enough to prevent the temperature of the paper from rising above its kindling point.

20. Pioneers stored fruits and vegetables in underground cellars. Discuss as fully as possible this choice for a storage site.

Answer: The factors to be considered are the insulating properties of soil, the absence of a path for heat to be radiated away from or to the vegetables, and the hindrance of the formation of convection currents in the small, enclosed space.

21. Why can you get a more severe burn from steam at 100°C than from water at 100°C?

Answer: The heated steam can liberate 540 cal/g, its heat of vaporization, when it encounters your hand before it reaches the same state of water at 100°C.

22. Concrete has a higher specific heat than soil. Use this fact to explain (partially) why cities have a higher average night-time temperature than the surrounding countryside. If a city is hotter than the surrounding countryside, would you expect breezes to blow from city to country or from country to city? Explain.

Answer: The large amounts of heat stored in the concrete during the day as the sun falls on it is released at night, resulting in an overall higher average temperature than the countryside. The heated air in a city rises to be replaced by cooler air drawn in from the countryside. Thus, breezes blow from country to city.

25. If water is a poor conductor of heat, why can it be heated quickly when placed over a flame?

Answer: The heat entering the water at the point in contact with the flame is transferred through the fluid by convection currents.

27. If you hold water in a paper cup over a flame, you can bring the water to a boil without burning the cup. How is this possible?

Answer: Because of the small thickness of the bottom of the cup, heat is rapidly conducted through the paper. Convection currents rapidly move this heat away from the surface of the paper.

SOLUTIONS TO SELECTED END-OF-CHAPTER PROBLEMS_____

7. What is the final equilibrium temperature when 10 g of milk at 10°C is added to 160 g of coffee at 90°C? (Assume the heat capacities of the two liquids are the same as water, and neglect the heat capacity of the container.)

Solution

$(mc\Delta T)_{milk} = - (mc\Delta T)_{coffee}$

$(10\ g)(1\ cal/g\cdot°C)(T - 10°C) = (160\ g)(1\ cal/g\cdot°C)(90 - T)°C$

$T = \boxed{85.3°C}$

41. Five mols of an ideal gas expand isothermally at 127°C to four times its initial volume. Find (a) the work done by the gas and (b) the heat flow into the system, both in joules.

Solution

(a) $W = nRT \, Ln\left(\dfrac{V_f}{V_i}\right) = (5 \text{ mol})(8.314 \text{ J/mol·K})(400 \text{ K})Ln(4)$ or $W = \boxed{23.1 \text{ kJ}}$

(b) Since T = Const, $\Delta U = 0$ so $Q = W = \boxed{23.1 \text{ kJ}}$

---

45. An ideal gas initially at 300 K undergoes an isobaric expansion at a pressure of 2.5 kPa. If the volume increases from 1 m³ to 3 m³ and 12 500 J of heat is added to the gas, find (a) the change in internal energy of the gas and (b) its final temperature.

Solution

(a) $\Delta U = Q - W$ where $W = P\Delta V$ so that

$\Delta U = Q - P\Delta V = 1.25 \times 10^3 \text{ J} - (2.5 \times 10^3 \text{ N/m}^2)[(3-1) \text{ m}^3] = \boxed{7.50 \text{ kJ}}$

(b) $\dfrac{V_1}{T_1} = \dfrac{V_2}{T_2}$ and $T_2 = \left(\dfrac{V_2}{V_1}\right)T_1 = \left(\dfrac{3}{1}\right)(300 \text{ K}) = \boxed{900 \text{ K}}$

---

67. One mol of an ideal gas is contained in a cylinder with a movable piston. The initial pressure, temperature, and volume are P_0, V_0, and T_0, respectively. Find the work done by the gas for the following processes and show each process on a PV diagram: (a) an isobaric compression in which the final volume is one half the initial volume, (b) an isothermal compression in which the final pressure is four times the initial pressure, (c) an isovolumetric process in which the final pressure is triple the initial pressure.

Solution

(a) The work done by the gas is the area under the PV curve, and

$$W = P_0\left(\dfrac{V_0}{2} - V_0\right) = -0.500 P_0 V_0$$

(b) In this case the area under the curve is $W = \int P dV$. Since the process is isothermal,

$$PV = P_0 V_0 = 4P_0\left(\dfrac{V_0}{4}\right) = nRT_0$$

and

$$W = \int_{V_0}^{V_0/4} P_0 V_0\left(\dfrac{dV}{V}\right) = P_0 V_0 \, Ln\left(\dfrac{V_0/4}{V_0}\right) = -P_0 V_0 \, Ln4 = -1.39 \, P_0 V_0$$

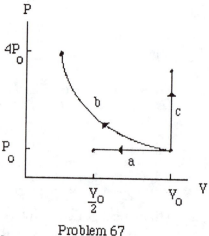

Problem 67

(c) The area under the curve is 0   and $W = 0$.

252

73. Using the data in Example 20.5 and Table 20.2, calculate the change in internal energy when 2 cm^3 of liquid helium gas at 4.2 K is converted to helium gas at 273.15 K and atmospheric pressure. (Assume that the molar heat capacity of helium gas is 24.9 J/mol·K, and note that 1 cm^3 of liquid helium is equivalent to 3.1×10^{-2} mols.)

Solution

The heat added to the system is used to boil the liquid He at a constant temperature of 4.2 K and then increase the temperature of the gas to 273.15 K. Therefore,

$$Q = mL + nC_p\Delta T$$

$$m = \rho V = (0.125 \text{ g/cm}^3)(2 \text{ cm}^3) = 0.250 \text{ g} \quad \text{and} \quad n = 6.2 \times 10^{-2} \text{ mol}$$

Therefore,

$$Q = (0.250 \text{ g})(4.99 \text{ cal/g})(4.186 \text{ J/cal}) + (6.2 \times 10^{-2} \text{ mol})(24.9 \text{ J/mol·K})(273.15 - 4.2) \text{ K}$$

or $\quad Q = 420 \text{ J}$

$W = \int p dV$ and at constant pressure, $W = p\Delta V = nR\Delta T$. Therefore,

$$W = (6.2 \times 10^{-2} \text{ mol})(8.314 \text{ J/mol·K})(273.15 - 4.2) \text{ K} = 139 \text{ J}$$

$$\Delta U = Q - W = 420 \text{ J} - 139 \text{ J} = \boxed{281 \text{ J}}$$

---

77. A vessel in the shape of a spherical shell has an inner radius a and outer radius b. The wall has a thermal conductivity k. If the inside is maintained at a temperature T_1 and the outside is at a temperature T_2, show that the rate of heat flow between the surfaces is given by

$$\frac{dQ}{dt} = \left(\frac{4\pi kab}{b-a}\right)(T_1 - T_2)$$

Solution

$$\frac{dQ}{dt} = -kA\frac{dT}{dx}$$

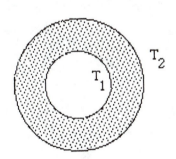

Cross-section of spherical shell

For a spherical shell of radius r and thickness dr,

$$\frac{dQ}{dt} = -k(4\pi r^2)\frac{dT}{dr}$$

Since the rate of heat transfer, $\frac{dQ}{dt}$, is constant, we can integrate over r to find the difference in temperature across the wall of the shell:

$$-4\pi k(T_2 - T_1) = \left(\frac{dQ}{dt}\right)\int_a^b \frac{dr}{r^2} = \left(\frac{dQ}{dt}\right)\left(\frac{1}{a} - \frac{1}{b}\right)$$

and

$$\frac{dQ}{dt} = \frac{4\pi k ab}{b-a}(T_1 - T_2)$$

79. The passenger section of a jet airliner is in the shape of a cylindrical tube of length 35 m and inner radius 2.5 m. Its walls are lined with a 6 cm thickness of insulating material of thermal conductivity 4×10^{-5} cal/s·cm·C°. The inside is to be maintained at 25°C while the outside is at −35°C. What heating rate is required to maintain this temperature difference? (Use the result from Problem 78.)

Solution

From problem 78, the rate of heat flow through the wall is

$$\frac{dQ}{dt} = \frac{2\pi k L(T_1 - T_2)}{\ln\left(\frac{b}{a}\right)} = \frac{2\pi(3500 \text{ cm})(4 \times 10^{-5} \text{ cal/s·cm·°C})(60°C)}{\ln\left(\frac{2.56}{2.50}\right)}$$

$$\frac{dQ}{dt} = 2.23 \times 10^3 \text{ cal/s} = \boxed{9.32 \text{ kW}}$$

This is the rate of heat loss from the plane, and consequently the rate at which energy must be supplied in order to maintain an equilibrium temperature.

81. A "solar cooker" consists of a curved reflecting mirror that focuses sunlight onto the object to be heated (Fig. 20.26). The solar power per unit area reaching the earth at some location is 600 W/m², and a small solar cooker has a diameter of 0.6 m. Assuming that 40% of the incident energy is converted into heat energy, how long would it take to completely boil off 0.5 liters of water initially at 20°C? (Neglect the heat capacity of the container.)

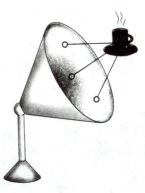

Figure 20.26

Solution

The power incident on the solar collector is

$$P_i = IA = (600 \text{ W/m}^2)\pi(0.30 \text{ m})^2 = 169.6 \text{ W}$$

or

$$P_i = \frac{169.6 \text{ W}}{4.186 \text{ J/cal}} = 40.53 \text{ cal/s}$$

For a 40% reflector, the collected power is

$$P_c = 16.21 \text{ cal/s}$$

The total energy required to increase the temperature of the water to the boiling point and to evaporate it is

$$Q = mc\Delta T + mL_v = (500 \text{ g})(1 \text{ cal/g·°C})(80°C) + (500 \text{ g})(540 \text{ cal/g}) = 3.10 \times 10^5 \text{ cal}$$

The time required is

$$\Delta t = \frac{Q}{P_c} = \frac{3.10 \times 10^5 \text{ cal}}{16.21 \text{ cal/s}} = \boxed{1.91 \times 10^4 \text{ s}} = \boxed{5.31 \text{ h}}$$

The Kinetic Theory of Gases

OBJECTIVES

1. State and understand the assumptions made in developing the molecular model of an ideal gas.

2. Recognize that the temperature of an ideal gas is proportional to the average molecular kinetic energy.

3. State the theorem of equipartition of energy, noting that each degree of freedom of a molecule contributes an equal amount of energy, of magnitude $\frac{1}{2}kT$.

4. Recognize that the internal energy of an ideal gas is proportional to the absolute temperature, and be able to derive the specific heat of an ideal gas at constant volume from the first law of thermodynamics.

5. Define an adiabatic process, and be able to derive the expression $PV^{\gamma} =$ constant, which applies to a quasi-static, adiabatic process.

6. Present a qualitative discussion of the possible degrees of freedom associated with a molecule (translational, rotational, and vibrational motions) and their contributions to the total energy and specific heats.

7. Describe the total energy and heat capacity of a solid at high temperatures using the equipartition theorem.

8. Understand the meaning of the Maxwell speed distribution function, and recognize the differences between rms speed, average speed, and most probable speed.

9. Know the meaning of the mean free path concept, and understand the nature of the Van der Waals equation of state.

NOTES FROM SELECTED CHAPTER SECTIONS

21.1 MOLECULAR MODEL FOR THE PRESSURE OF AN IDEAL GAS

A microscopic *model of an ideal gas* is based on the following assumptions:

1. *The number of molecules is large, and the average separation between them is large* compared with their dimensions. Therefore, the molecules occupy a negligible volume compared with the volume of the container.

2. *The molecules obey Newton's laws of motion, but the individual molecules move in a random fashion.* By random fashion, we mean that the molecules move in all directions with equal probability and with various speeds. This distribution of velocities does not change in time, despite the collisions between molecules.

3. *The molecules undergo elastic collisions with each other.* Thus, the molecules are considered to be structureless (that is, point masses), and in the collisions both kinetic energy and momentum are conserved.

4. *The forces between molecules are negligible except during a collision.* The forces between molecules are short-range, so that the only time the molecules interact with each other is during a collision.

5. *The gas under consideration is a pure gas.* That is, all molecules are identical.

6. *The gas is in thermal equilibrium with the walls of the container.* Hence, the wall will eject as many molecules as it absorbs, and the ejected molecules will have the same average kinetic energy as the absorbed molecules.

21.2 MOLECULAR INTERPRETATION OF TEMPERATURE

The *theorem of equipartition of energy* states that the energy of a system in thermal equilibrium is equally divided among all degrees of freedom.

21.4 ADIABATIC PROCESS FOR AN IDEAL GAS

An *adiabatic process* is one in which there is no heat transfer between a system and its surroundings.

A *quasi-static* adiabatic process is one which is slow enough to allow the system to be always near equilibrium, but fast compared with the time required for the system to exchange heat energy with its surroundings.

21.6 THE EQUIPARTITION OF ENERGY

The heat capacities of gases containing complex molecules can sometimes be explained if one includes other degrees of freedom, namely, those associated with vibrational and rotational motions. The contribution from vibrations is especially important at high temperatures, while rotational effects are significant above about 50 K. The classical model does not provide an adequate description of molecular systems in all situations.

The heat capacities of solids generally increase nonlinearly with increasing temperature, and approach a value of about 3R at sufficiently high temperatures. This result is known as the *DuLong-Petit law*. The heat capacities approach zero at $T \rightarrow 0$, but the general variation with temperature cannot be explained using classical concepts.

EQUATIONS AND CONCEPTS

In the kinetic theory of an ideal gas, one finds that the pressure of the gas is proportional to the number of molecules per unit volume and the average translational kinetic energy per molecule.

$$P = \frac{2}{3} \frac{N}{V} \left(\overline{\frac{1}{2} m v^2} \right) \qquad (21.5)$$

From Eq. 21.5, and the equation of state for an ideal gas, $PV = NkT$, we find that *the absolute temperature of an ideal gas is a direct measure of the average molecular kinetic energy.*

$$T = \frac{2}{3k} \left(\overline{\frac{1}{2} m v^2} \right) \qquad (21.6)$$

The *total internal energy* U of N molecules (or n mols) of a monatomic ideal gas is proportional to the absolute temperature.

$$U = \frac{3}{2} NkT = \frac{3}{2} nRT \qquad (21.11)$$

If we apply the first law of thermodynamics to a monatomic *ideal gas* in which heat is transferred at *constant volume,* we find that the *specific heat at constant volume* is equal to

$$C_V = \frac{3}{2} R \qquad (21.13)$$

$$\frac{3}{2} R = 2.99 \text{ cal/mol} \cdot \text{K}$$

If heat is transferred to an ideal gas at *constant pressure*, the first law of thermodynamics shows that the *specific heat at constant pressure* is greater than the specific heat at constant volume by an amount R.

$$\boxed{C_P - C_V = R}$$ (21.16)

or

$$C_P = \frac{5}{2}R$$

The *ratio of specific heats* is a dimensionless quantity γ which, for an *ideal gas*, is equal to 1.67.

$$\boxed{\gamma = \frac{C_P}{C_V} = 1.67}$$ (21.17)

If an *ideal gas* undergoes a *quasi-static, adiabatic expansion,* and we assume $PV = nRT$ is valid at any time, then the pressure and volume of the gas obey Eq. 21.18.

$$\boxed{PV^\gamma = \text{constant}}$$ (21.18)

In deriving an expression for the *speed of a sound wave propagating through a gas*, it is assumed that the variations in pressure and volume occur adiabatically.

$$v = \sqrt{\frac{\gamma P}{\rho}}$$ (21.21)

$$v = \sqrt{\frac{\gamma RT}{M}}$$ (21.22)

The Maxwell speed distribution function describes the most probable distribution of speeds of N gas molecules at temperature T (where k is the Boltzman constant).

$$\boxed{N_v = 4\pi N \sqrt{\left(\frac{m}{2\pi kT}\right)}\, v^2 e^{-\left(\frac{mv^2}{2kT}\right)}}$$ (21.24)

Specific expressions can be derived for the rms, average, and most probable speeds.

$$v_{rms} = 1.73\sqrt{\frac{kT}{m}}$$ (21.25)

$$\overline{v} = 1.60\sqrt{\frac{kT}{m}}$$ (21.26)

$$v_{mp} = 1.41\sqrt{\frac{kT}{m}}$$ (21.27)

The average distance L between collisions is called the *mean free path* (where d is the molecular "diameter" and n_v is the density of molecules).

$$L = \frac{1}{\sqrt{2}\,\pi d^2 n_v}$$ (21.28)

The Van der Waals' equation of state is a modification of the ideal gas equation of state to take into account the volume of the gas molecules and the intermolecular forces when the molecules are close together. (The constants a and b are empirical and are chosen to provide best agreement on a particular *ideal gas*.)

$$\left(P + \frac{a}{v^2}\right)(V - b) = RT \qquad\qquad (21.30)$$

EXAMPLE PROBLEM SOLUTION

Example 21.1   One mol of an ideal gas occupies a volume of 3.50×10^{-2} m³ and exerts a pressure of 3 atm on the walls of the container. Find the average translational kinetic energy per molecule and the absolute temperature of the gas.

Solution

One gram-mol of the gas contains Avogadro's number of molecules. Therefore, the number of molecules per unit volume is given by

$$\frac{N_A}{V} = \frac{6.02 \times 10^{23} \text{ molecules}}{3.50 \times 10^{-2} \text{ m}^3} = 1.72 \times 10^{25} \frac{\text{molecules}}{\text{m}^3}$$

The pressure of the gas in N/m² is

$$P = 3 \text{ atm} \times 1.013 \times 10^5 \frac{\text{N/m}^2}{\text{atm}} = 3.039 \times 10^5 \frac{\text{N}}{\text{m}^2}$$

Substituting these values into Eq. 21.5 gives

$$\frac{1}{2} m\overline{v^2} = \frac{3}{2} \frac{P}{N_A/V} = \frac{3}{2} \frac{(3.039 \times 10^5 \text{ N/m}^2)}{(1.72 \times 10^{25} \text{ molecules/m}^3)} = 2.65 \times 10^{-20} \text{ J}$$

Substituting this result into Eq. 21.6, we can find the absolute temperature T:

$$T = \frac{2}{3k}\left(\frac{1}{2} m\overline{v^2}\right) = \frac{2(2.65 \times 10^{-20} \text{ J})}{3(1.38 \times 10^{-23} \text{ J/K})} = 1280 \text{ K}$$

You should be able to show that if the temperature of the gas were reduced to 300 K (room temperature), the pressure of the gas would drop to 0.703 atm.

ANSWERS TO SELECTED QUESTIONS

2. One container is filled with helium gas and another with argon gas. If both containers are at the same temperature, which molecules have the higher rms speed?

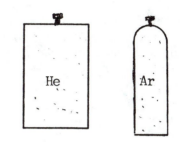

Answer: The helium molecules have the higher rms speed since the helium molecule is lighter than argon. Note that both molecules have the same average kinetic energy since they are at the same temperature.

4. A gas consists of a mixture of He and N_2 molecules. Do the lighter He molecules travel faster than the N_2 molecules? Explain.

Answer: Yes. Both molecules have the same average kinetic energy but the lighter molecules have a higher rms speed.

5. Although the average speed of gas molecules in thermal equilibrium at some temperature is greater than zero, the average velocity is zero. Explain.

Answer: The gas consists of molecules moving in various directions with a distribution of speeds. There are as many molecules traveling in one direction as in the opposite direction. When all possible directions and speeds are taken into account, the average velocity is found to be zero.

6. Why does a fan make you feel cooler on a hot day?

Answer: The fan increases the rate of evaporation of moisture from the surface of the skin. This reduces the temperature of the remaining moisture on the body. That is, evaporation is a cooling process.

7. Alcohol taken internally makes you feel warmer. Yet when it is rubbed on your body, it lowers body temperature. Explain the latter effect.

Answer: Alcohol evaporates from the body surface and removes heat from the body.

9. A vessel containing a fixed volume of gas is cooled. Does the mean free path increase, decrease, or remain constant in the cooling process? What about the collision frequency?

Answer: The mean free path does not change. However, the collision frequency decreases since the average speed of the gas molecules decreases as the gas is cooled.

10. A gas is compressed at a constant temperature. What happens to the mean free path of the molecules in this process?

Answer: The mean free path decreases. This can be understood by noting that the density of molecules increases in the process. The density of molecules is given by $n_v = P/kT$. As the gas is compressed at constant temperature, the pressure increases, which increases n_v. Since the mean free path is inversely proportional to the density, we conclude that the mean free path decreases.

SOLUTIONS TO SELECTED END-OF-CHAPTER PROBLEMS

25. Two molecules of an ideal gas ($\gamma = 1.40$) expand quasi-statically and adiabatically from a pressure of 5 atm and a volume of 12 liters to a final volume of 30 liters. (a) What is the final pressure of the gas? (b) What are the initial and final temperatures?

Solution

(a) $P_1V_1^\gamma = P_2V_2^\gamma$

$$P_2 = \left(\frac{V_1}{V_2}\right)^\gamma P_1 = \left(\frac{12}{30}\right)^{1.40}(5 \text{ atm}) = \boxed{1.39 \text{ atm}}$$

(b) $T_1 = \dfrac{P_1V_1}{nR} = \dfrac{(5)(1.01 \times 10^5)(12 \times 10^{-3})}{(2)(8.31)} = \boxed{366 \text{ K}}$

$T_2 = \dfrac{P_2V_2}{nR} = \boxed{254 \text{ K}}$

---

53. In an ultrahigh vacuum system, the pressure is measured to be 10^{-10} torr (where 1 torr = 133 N/m^2). If the gas molecules have a molecular diameter of 3 Å = 3×10^{-10} m and the temperature is 300 K, find (a) the number of molecules in a volume of 1 m^3, (b) the mean free path of the molecules, and (c) the collision frequency, assuming an average speed of 500 m/s.

Solution

(a) $PV = \left(\dfrac{N}{N_A}\right)RT$ and $N = \dfrac{PVN_A}{RT}$, so that

$$N = \frac{(10^{-10})(133)(6.02 \times 10^{23})}{(8.31)(300)} = \boxed{3.21 \times 10^{12} \text{ molecules}}$$

(b) $L = \dfrac{1}{n\sigma\sqrt{2}} = \dfrac{V}{N\sigma\sqrt{2}} = \dfrac{1 \text{ m}^3}{(3.21 \times 10^{12} \text{ molecules})\pi(3 \times 10^{-10} \text{ m})^2\sqrt{2}}$

$L = 7.78 \times 10^5 \text{ m} = \boxed{778 \text{ km}}$

(c) $f = \dfrac{v}{L} = \dfrac{500 \text{ m/s}}{7.78 \times 10^5 \text{ m}} = \boxed{6.42 \times 10^{-4} \text{ s}^{-1}}$

---

62. A cylinder containing n moles of an ideal gas undergoes a quasi-static, adiabatic expansion.

(a) Starting with the expression $W = \displaystyle\int PdV$ and using $PV^\gamma = \text{constant}$, show that the work done is given by

$$W = \frac{P_iV_i - P_fV_f}{\gamma - 1}$$

CHAPTER 21

(b) Starting with the first law in differential form, prove that the work done is also equal to $nC_v(T_i - T_f)$. Show that this result is consistent with the equation in (a).

Solution: (a) $PV^\gamma = k$ so $W = \int_i^f PdV = k\int_{V_i}^{V_f} \frac{dV}{V^\gamma} = \frac{P_iV_i - P_fV_f}{\gamma - 1}$

(b) $dU = dQ - dW$ and $dQ = 0$ for an adiabatic process. Therefore,

$$W = -\Delta U = -\frac{3}{2} nR\Delta T = nC_v(T_i - T_f)$$

To show consistency between these two equations, consider that $\gamma = \frac{C_p}{C_v}$ and $C_p - C_v = R$.

Therefore $\frac{1}{\gamma - 1} = \frac{C_v}{R}$. Using this, the result found in part (a) becomes

$$W = (P_iV_i - P_fV_f)\frac{C_v}{R}$$

Also, for an ideal gas $\frac{PV}{R} = nT$ so that $W = nC_v(T_i - T_f)$.

---

63. Twenty particles, each of mass m and confined to a volume V, have the following speeds: two have speed v; three have speed 2v; five have speed 3v; four have speed 4v; three have speed 5v; two have speed 6v; one has speed 7v. Find (a) the average speed, (b) the rms speed, (c) the most probable speed, (d) the pressure they exert on the walls of the vessel, and (e) the average kinetic energy per particle.

Solution

(a) The average speed v_{av} is just the weighted average of all the speeds.

$$v_{av} = \frac{2(v) + 3(2v) + 5(3v) + 4(4v) + 3(5v) + 2(6v) + 1(7v)}{2 + 3 + 5 + 4 + 3 + 2 + 1} = \boxed{3.65v}$$

(b) First find the average of the square of the speeds,

$$v_{av}^2 = \frac{2(v)^2 + 3(2v)^2 + 5(3v)^2 + 4(4v)^2 + 3(5v)^2 + 2(6v)^2 + 1(7v)^2}{2 + 3 + 5 + 4 + 3 + 2 + 1} = 15.95v^2$$

The root-mean square speed is then

$$v_{rms} = \sqrt{v_{av}^2} = \boxed{3.99v}$$

(c) The most probable speed is the one that most of the particles have.

That is, five particles have speed $\boxed{3.00v}$

(d) $PV = \frac{1}{2}Nmv_{av}^2$. Therefore,

$$P = \left(\frac{20}{3}\right)\frac{m(15.95)v^2}{V} = \boxed{106\,\frac{mv^2}{V}}$$

(e) The average kinetic energy for each particle is

$$K = \frac{1}{2}\,mv_{av}^2 = \frac{1}{2}\,m(15.95v^2) = \boxed{7.98mv^2}$$

---

65. A vessel contains 1 mol of helium gas at a temperature of 300 K. Calculate the approximate number of molecules having speeds in the range from 400 m/s to 410 m/s. (Hint: This number is approximately equal to $N_v\Delta v$, where Δv is the range of speeds.)

Solution

The Maxwell speed distribution for molecules of varying velocity is

$$N_v = 4\pi N\left[\frac{m}{2\pi kT}\right]^{3/2} v^2\,e^{-mv^2/2kT} \tag{1}$$

For 1 mol of gas, $N = N_A = 6.02 \times 10^{23}$. The mass m of a helium atom is approximately 4 u, or $m = 4(1.67 \times 10^{-27}\,kg)$. The number of molecules with a speed in the range dv is $N_v dv$ which is $\approx N_v\Delta v$. In this case, N_v is calculated for the average of the two velocity extremes in the interval. That is, $N_v \approx N_{405}$. Substituting these values into (1) gives

$$(N_{405})(10\,m/s) = \boxed{1.40 \times 10^{21}\ \text{molecules}}$$

---

68. The compressibility, κ, of a substance is defined as the fractional change in volume of that substance for a given change in pressure:

$$\kappa = -\frac{1}{V}\frac{dV}{dP}$$

(a) Explain why the negative sign in this expression ensures that κ will always be positive. (b) Show that if an ideal gas is compressed *isothermally*, its compressibility is given by $\kappa_1 = 1/P$. (c) Show that if an ideal gas is compressed *adiabatically*, its compressibility is given by $\kappa_2 = 1/\gamma P$. (d) Determine values for κ_1 and κ_2 for a monatomic ideal gas at a pressure of 2 atm.

Solution

(a) Since pressure increases as volume decreases (and vice versa),

$$\frac{dV}{dP} < 0 \qquad \text{and} \qquad -\left(\frac{1}{V}\right)\frac{dV}{dP} > 0$$

264

(b)   For an ideal gas, $V = \dfrac{nRT}{P}$ and $\kappa_1 = -\dfrac{1}{V}\dfrac{d}{dP}\left(\dfrac{nRT}{P}\right)$

If the compression is isothermal, T is constant and

$$\kappa_1 = -\frac{nRT}{V}\left(\frac{-1}{P^2}\right) = \frac{1}{P}$$

(c)   For an adiabatic compression, $PV^\gamma = C$ (where C is a constant) and

$$\kappa_2 = -\left(\frac{1}{V}\right)\frac{d}{dP}\left(\frac{C}{P}\right)^{1/\gamma} = \left(\frac{1}{V\gamma}\right)\frac{C^{1/\gamma}}{P^{1/\gamma + 1}} = \frac{V}{V\gamma P} = \frac{1}{\gamma P}$$

(d) $\kappa_1 = \dfrac{1}{P} = \dfrac{1}{2\ \text{atm}} = \boxed{0.5\ \text{atm}^{-1}}$

$\gamma = \dfrac{C_p}{C_v}$ and for a monatomic ideal gas, $\gamma = \dfrac{5}{3}$,   so that

$$\kappa_2 = \frac{1}{\gamma P} = \frac{1}{\left(\frac{5}{3}\right)(2\ \text{atm})} = \boxed{0.3\ \text{atm}^{-1}}$$

---

69. One mol of a gas obeying van der Waals' equation of state is compressed isothermally. At some critical temperature, T_c, the isotherm has a point of zero slope, as in Figure 21.16. That is, at $T = T_c$,

$$\frac{\partial P}{\partial V} = 0 \quad \text{and} \quad \frac{\partial^2 P}{\partial V^2} = 0$$

Using Equation 21.30 and these conditions, show that at the critical point, the pressure, volume, and temperature are given by

$$P_c = \frac{a}{27b^2}, \quad V_c = 3b, \quad \text{and} \quad T_c = \frac{8a}{27Rb}$$

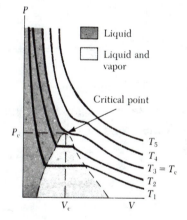

Figure 21.16

Solution

Van der Waals' equation is

$$\left(P + \frac{a}{V^2}\right)(V - b) = RT \tag{1}$$

Holding T constant, take the partial derivative $\frac{\partial}{\partial V}$ of each side to get

$$\left(\frac{\partial P}{\partial V} - \frac{2a}{V^3}\right)(V - b) + \left(P + \frac{a}{V^2}\right) = 0$$

Setting $\frac{\partial P}{\partial V} = 0$ gives

$$P + \frac{a}{V^2} = \frac{2a}{V^3}(V - b) \tag{2}$$

Taking the second derivative $\frac{\partial^2}{\partial V^2}$ of (1) gives

$$\frac{\partial^2 P}{\partial V^2} + \frac{6a}{V^4}(V - b) + 2\left(\frac{\partial P}{\partial V} - \frac{2a}{V^3}\right) = 0$$

Setting $\frac{\partial^2 P}{\partial V^2} = \frac{\partial P}{\partial V} = 0$ gives $\frac{2a}{V^3} - \frac{6ab}{V^4} = 0$ and

$$\frac{3b}{V} = 1 \tag{3}$$

Solving equations (1), (2), and (3) gives

$$P_c = \frac{a}{27b^2}, \qquad V_c = 3b, \qquad \text{and} \qquad T_c = \frac{8a}{27Rb}$$

---

71. In Equation 21.22 the temperature T must be in degrees kelvin. (a) Starting with this equation, show that the speed of sound in a gas can be expressed in the form $v = v_0 + \left(\frac{v_0}{546}\right)t$, where t is the temperature in °C and v_0 is the speed of sound in the gas at 0°C. (Hint: Assume that t « 273°C and use the expansion $\sqrt{1 + x} = 1 + \frac{1}{2}x - \frac{1}{8}x^2 + \ldots$ (b) In the case of air, show that this result leads to $v = (331 + 0.61t)$ m/s.

Solution

(a) $v = \sqrt{\frac{\gamma R T_k}{m}} = \sqrt{\frac{\gamma R(273)}{m}}\sqrt{\frac{T_k}{273}} = v_0 \sqrt{\frac{T_k}{273}}$

where $T_k \equiv$ temperature in kelvin and v_0 is the speed of sound at 0°C = 273 K. If $t \equiv$ temperature in celsius, then

$$v = v_0 \sqrt{\left(\frac{t}{273}\right) + 1} \approx v_0\left(1 + \frac{t}{546}\right)$$

(b) For air, $v_0 = 331$ m/s and $v = (331 + 0.61t)$ m/s.

Heat Engines, Entropy,
and the Second Law of Thermodynamics

OBJECTIVES

1. Understand the basic principle of the operation of a heat engine, and be able to define and discuss the *thermal efficiency* of a heat engine.

2. State the second law of thermodynamics in both the Kelvin-Planck form and the Clausius form.

3. Describe the principle of a refrigerator and define the coefficient of performance of a refrigerator.

4. Discuss the difference between reversible and irreversible processes.

5. Describe the processes which take place in an ideal heat engine taken through a *Carnot cycle*.

6. Derive the efficiency of a Carnot engine, and note that the efficiency of real heat engines is always less than the Carnot efficiency.

7. Describe the absolute temperature scale using the Carnot cycle as the basis for your definition.

8. Give a qualitative description of the gasoline engine, and discuss the four processes which occur in the Otto cycle--which approximates the processes occurring in four-stroke cycle of a conventional gasoline engine.

9. Discuss the concept of entropy, and give a thermodynamic definition of energy.

10. Calculate entropy changes for reversible processes (such as one involving an ideal gas).

11. Calculate entropy changes for irreversible processes, recognizing that the entropy change for an irreversible process is equivalent to that of a reversible process between the same two equilibrium states.

12. Discuss the importance of the first and second laws of thermodynamics as they apply to various forms of energy conversion and thermal pollution.

NOTES FROM SELECTED CHAPTER SECTIONS

22.1 HEAT ENGINES AND THE SECOND LAW OF THERMODYNAMICS

The *second law of thermodynamics* can be stated in several ways:

Clausius Statement: No thermodynamic process can occur whose only result is to transfer heat from a colder to a hotter body. Such a process is only possible if work is done on the system.

Kelvin-Planck Statement: It is impossible for a thermodynamic process to occur whose only final result is the complete conversion of heat extracted from a hot reservoir into work.

The Kelvin-Planck statement is equivalent to stating that *it is impossible to construct a perpetual motion machine of the second kind* (that is, a machine which violates the second law). Perpetual motion machines of the *first kind* are those which violate the first law of thermodynamics, which requires that energy be conserved.

22.2 REVERSIBLE AND IRREVERSIBLE PROCESSES

A process is *irreversible* if the system and its surroundings cannot be returned to their initial states. A process is *reversible* if the system passes from the initial to the final state through a succession of equilibrium states.

22.3 THE CARNOT ENGINE

The most efficient cyclic process is called the *Carnot cycle,* described in the P-V diagram of Figure 22.1. The Carnot cycle consists of two adiabatic and two isothermal processes, all being reversible.

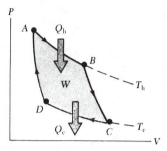

Figure 22.1   The Carnot Cycle

1.   The process A → B is an isotherm (constant T), during which time the gas expands at constant temperature T_h, and absorbs heat Q_h from the hot reservoir.

2.   The process B → C is an adiabatic expansion (Q = O), during which time the gas expands and cools to a temperature T_c.

3.   The process C → D is a second isotherm, during which time the gas is compressed at constant temperature T_c, and gives up heat Q_c to the cold reservoir.

4.   The final process D → A is an adiabatic compression in which the gas temperature increases to a final temperature of T_h.

In practice, no working engine is 100% efficient, even when losses such as friction are neglected. One can obtain some theoretical limits on the efficiency of a real engine by comparison with the ideal Carnot engine. A *reversible engine* is one which will operate with the same efficiency in the forward and reverse directions. The Carnot engine is one example of a reversible engine. *Carnot's theorems*, which are consistent with the first and second laws of thermodynamics, can be stated as follows:

Theorem I. All reversible engines operating between T_h and T_c have the *same* efficiency given by Eq. 22.2.

Theorem II. No real (irreversible) engine can have an efficiency greater than that of a reversible engine operating between the same two temperatures.

A *heat engine* is a device that converts thermal energy into other forms of energy such as mechanical and electrical energy. During its operation, a heat engine carries some working substance through a *cyclic process*, which is a process which begins and ends at the same state.

A schematic diagram of a heat engine is shown in Fig. 22.2a on the following page, where Q_h is the heat extracted from the hot reservoir at temperature T_h, Q_c is the heat rejected to the cold reservoir at temperature T_c, and W is the work done by the engine.

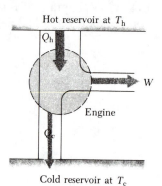

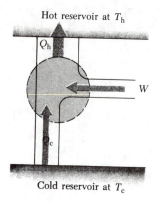

(a) Heat engine

(b) Refrigerator (heat pump)

Figure 22.2   (a) Schematic diagram of a heat engine
                (b) Schematic diagram of a refrigerator (heat pump)

22.6 HEAT PUMPS AND REFRIGERATORS

A refrigerator is a heat engine operating in reverse, as described in Figure 22.2b. During one cycle of operation, the refrigerator absorbs heat Q_c from the cold reservoir, expels heat Q_h to the hot reservoir, and the work done on the system is $W = Q_h - Q_c$.

22.7 ENTROPY
22.8 ENTROPY CHANGES IN IRREVERSIBLE PROCESSES

Entropy is a quantity used to measure the degree of *disorder* in a system. For example, the molecules of a gas in a container at a high temperature are in a more disordered state (higher entropy) than the same molecules at a lower temperature.

When heat is added to a system, dQ_r is *positive* and the entropy *increases*. When heat is removed, dQ_r is *negative* and the entropy *decreases*. Note that only *changes* in entropy are defined by Eq. 22.8; therefore, the concept of entropy is most useful when a system undergoes a *change in its state*.

When using Eq. 22.9 to calculate entropy changes, note that ΔS may be obtained even if the process is irreversible, since ΔS depends only on the initial and final equilibrium states, not on the path. In order to calculate ΔS for an irreversible process, you must devise a reversible process (or sequence of reversible processes) between the initial and final states, and compute dQ_r/T for the reversible process. The entropy change for the irreversible process is the *same* as that of the reversible process between the same initial and final equilibrium states.

The second law of thermodynamics can be stated in terms of entropy as follows: *The total entropy of an isolated system always increases in time if the system undergoes an irreversible process*. If an isolated system undergoes a *reversible* process, the total entropy *remains constant*.

EQUATIONS AND CONCEPTS

The net *work done* by a heat engine during one cycle equals the net heat flowing into the engine.

$$W = Q_h - Q_c \qquad (22.1)$$

The *thermal efficiency*, e, of a heat engine is defined as the ratio of the net work done to the heat absorbed during one cycle of the process.

$$e \equiv \frac{W}{Q_h} = 1 - \frac{Q_c}{Q_h} \qquad (22.2)$$

The *coefficient of performance* of a refrigerator is defined by the ratio of the heat absorbed, Q_c, to the work done. A good refrigerator has a high coefficient of performance.

$$\text{coefficient of performance} = \frac{Q_c}{W} \qquad (22.7)$$

Since the ratio of heats for a Carnot cycle is given by $Q_c/Q_h = T_c/T_h$, the *thermal efficiency of a Carnot engine* is e_c given by Eq. 22.4.

$$e_c = 1 - \frac{T_c}{T_h} \qquad (22.4)$$

If a system changes from one equilibrium state to another, under a reversible, quasi-static process, and a quantity of heat dQ_r is added (or removed) at the absolute temperature T, the *change in entropy* is defined by the ratio dQ_r/T.

$$dS \equiv \frac{dQ_r}{T} \qquad (22.8)$$

The *change in entropy* of a system which undergoes a reversible process between the states i and f is given by Eq. 22.9. Note that ΔS must be evaluated along a reversible path, and its value depends only on the properties of the initial and final equilibrium states.

$$\Delta S = \int_i^f \frac{dQ_r}{T} \qquad (22.9)$$

The change in entropy of a system for any arbitrary *reversible cycle* is identically *zero*.

$$\oint \frac{dQ_r}{T} = 0 \qquad (22.10)$$

During a *free expansion* (irreversible, adiabatic process), the entropy of a gas *increases*.

$$\Delta S = nR \, ln \left(\frac{V_f}{V_i} \right) \qquad (22.15)$$

During the process of irreversible heat transfer between two masses, the entropy of the system will increase.

$$\Delta S = m_1 c_1 \, ln \left(\frac{T_f}{T_1} \right) + m_2 c_2 \, ln \left(\frac{T_f}{T_2} \right) \qquad (22.17)$$

All isolated systems tend toward disorder and the increase in entropy is proportional to the probability, W, of occurrence of the event.

$$\boxed{S = k \, Ln \, W} \qquad (22.18)$$

EXAMPLE PROBLEM SOLUTION

Example 22.1   One mol of an ideal monatomic gas ($\gamma = 1.67$) is taken through the cycle described in Figure 22.3, starting at point 1 where its pressure, volume and temperature are P_0, V_0, and T_0, respectively. Determine, in terms of P_0, V_0, T_0, and R, (a) the pressures and temperatures at points 2 and 3; and (b) the work done, the heat transferred, and the change in internal energy for each of the three processes.

Solution

(a) Since the process $1 \rightarrow 2$ is an adiabatic expansion, we can apply the condition that

$$P_i V_i^{\gamma} = P_f V_f^{\gamma}, \quad \text{or}$$

$$P_0 V_0^{\gamma} = P_2 V_2^{\gamma} = P_2 (2V_0)^{\gamma}$$

Solving for P_2 gives

$$P_2 = \frac{P_0}{2^{\gamma}} \qquad (1)$$

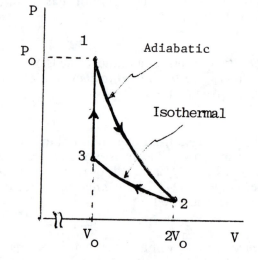

Figure 22.3

Since the gas is assumed to be ideal, the equation of state $PV = RT$ is valid at any point. Therefore, $P_2 V_2 = RT_2$ or

$$T_2 = \frac{P_2 V_2}{R} = \frac{1}{R} \left(\frac{P_0}{2^{\gamma}} \right)(2V_0) = (2)^{1-\gamma} \frac{P_0 V_0}{R}$$

Since $P_0 V_0 = RT_0$, we see that

$$T_2 = (2)^{1-\gamma} T_0 \qquad (2)$$

Furthermore, $T_3 = T_2$ since the process $2 \rightarrow 3$ is an isotherm. To find the pressure P_3, note that $P_2 V_2 = P_3 V_3$, so

$$P_3 = \frac{P_2 V_2}{V_3} = \frac{P_0}{2^{\gamma}} \left(\frac{V_2}{V_3} \right) = \frac{P_0}{2^{\gamma}} (2) = \frac{P_0}{2^{\gamma-1}} \qquad (3)$$

(b) The work done during the adiabatic expansion process $1 \to 2$ can be calculated using the fact that $PV^\gamma = P_0 V_0^\gamma = \text{const}$, so that

$$W_{12} = \int_{V_0}^{2V_0} P dV = \int_{V_0}^{2V_0} (P_0 V_0^\gamma) \frac{dV}{V^\gamma} = (P_0 V_0^\gamma) \frac{V^{1-\gamma}}{1-\gamma} \Bigg]_{V_0}^{2V_0}$$

$$= \frac{P_0 V_0^\gamma}{1-\gamma} \left[(2V_0)^{1-\gamma} - V_0^{1-\gamma} \right] = \frac{P_0 V_0}{1-\gamma} \left[2^{1-\gamma} - 1 \right]$$

The work done during the isothermal compression $2 \to 3$ (where $T_2 = T_3 = (2)^{1-\gamma} T_0$), is given by

$$W_{23} = \int_{2V_0}^{V_0} P dV = \int_{2V_0}^{V_0} RT_2 \frac{dV}{V}$$

$$= R(2)^{1-\gamma} T_0 \int_{2V_0}^{V_0} \frac{dV}{V} = R(2)^{1-\gamma} T_0 \, Ln\left(\frac{1}{2}\right) = P_0 V_0 (2)^{1-\gamma} Ln\left(\frac{1}{2}\right)$$

Finally, the work done during the process $3 \to 1$ is zero since the volume remains constant.

The heat transferred during the process $1 \to 2$ is *zero*, since the process is adiabatic. The heat transferred during the process $2 \to 3$ is equal to the work done during this process, since the gas is ideal and there is no change in internal energy [that is, $U = U(T)$ only for an ideal gas]. The heat transferred during the constant volume process $3 \to 1$ is given by

$$Q_{31} = n \, C_v \Delta T = \frac{3R}{2} \left[T_1 - T_3 \right] = \frac{3R}{2} \left[T_0 - (2)^{1-\gamma} T_0 \right] = \frac{3RT_0}{2} \left[1 - (2)^{1-\gamma} \right]$$

Applying the first law of thermodynamics, $\Delta U = Q - W$, to each of the three processes gives

$$\Delta U_{12} = \cancel{Q}_{12} - W_{12} = -W_{12} = \frac{P_0 V_0}{\gamma-1} \left[2^{1-\gamma} - 1 \right]$$

$$\Delta U_{23} = 0 \quad (\text{since } T = \text{constant})$$

$$\Delta U_{31} = Q_{31} - \cancel{W}_{31} = Q_{31} = \frac{3RT_0}{2} \left[1 - (2)^{1-\gamma} \right]$$

ANSWERS TO SELECTED QUESTIONS

1. Distinguish clearly among temperature, heat, and internal energy.

Answer: Temperature is a measure of the kinetic energy of the molecules of a substance. Heat is an energy transfer occurring between two objects because of a difference in temperature of the objects. The internal energy of a substance is a measure of the kinetic and potential energies of the atoms and molecules of the substance.

2. When a sealed Thermos bottle full of hot coffee is shaken, what are the changes, if any, in (a) the temperature of the coffee and (b) its internal energy?

Answer: Although no heat is transferred into or out of the system, work is done as the result of the agitation. Consequently, both the temperature and internal energy increase.

3. Use the first law of thermodynamics to explain why the total energy of an isolated system is always conserved.

Answer: By definition, no heat is transferred into or out of an isolated system. Therefore, from the first law, any work that is done on or by the system must appear as an increase or decrease in its internal energy.

4. Is it possible to convert internal energy to mechanical energy?

Answer: Yes. For example, the chemical energy stored in a battery can set up a current in a circuit which can be used to power a motor.

5. What are some factors that affect the efficiency of automobile engines?

Answer: Heat losses due to friction in the cylinder walls; incomplete combustion of the fuel.

6. The statement was made in this chapter that the first law says we cannot get more out of a process than we put in but the second law says that we cannot break even. Explain.

Answer: The second law says that during the operation of a heat engine some heat must be rejected to the environment. As a result, it is theoretically impossible to construct an engine that will work with 100% efficiency.

7. Is it possible to cool a room by leaving the door of a refrigerator open? What happens to the temperature of a room in which an air conditioner is left running on a table in the middle of the room?

Answer: No. A refrigerator extracts heat from its interior and rejects it to its surroundings. Under normal operation (that is, with the door closed), the room gets warmer. If the door is left open, nothing will be cooled. The same argument applies to the air conditioner which is not vented outside.

8. In practical heat engines, which do we have more control of, the temperature of the hot reservoir or the temperature of the cold reservoir? Explain.

Answer: Temperature of the cold reservoir. The temperature of the hot reservoir is set by the type of fuel that is used, the condition of the combustion and friction in the system. The temperature of the cold reservoir can be controlled by heat sinks, such as the cooling system of an automobile engine.

9. A steam-driven turbine is one major component of an electric power plant. Why is it advantageous to increase the temperature of the steam as much as possible?

Answer: The efficiency is given by $1 - (T_c/T_h)$. Therefore, increasing T_h will increase the efficiency.

10. Is it possible to construct a heat engine that creates no thermal pollution?

Answer: No. It is impossible to construct an engine which does work without rejecting some heat to a cold reservoir.

11. Electrical energy can be converted to heat energy with an efficiency of 100%. Why is this number misleading with regard to heating a home? That is, what other factors must be considered in comparing the cost of electric heating with the cost of hot air or hot water heating?

Answer: Most electric power plants produce their electricity by burning fossil fuels, with an efficiency of about 30%. When you pay your electric bill, you are indirectly paying for the oil or coal used to generate the electricity, in addition to the other services provided. In conventional hot-air or hot-water heating systems, oil or gas is used to directly heat the air or water. You must compare the lower cost of this direct heating with the cost of maintenance, initial installation, etc.

12. Discuss three common examples of natural processes that involve an increase in entropy. Be sure to account for all parts of each system under consideration.

Answer: (1) The mixing process of a hot and cold liquid. (2) The increase in entropy of an object after it collides with the ground. (3) The gas which is heated at constant volume.

13. Discuss the change in entropy of a gas that expands (a) at constant temperature and (b) adiabatically.

Answer: (a) During a constant temperature expansion of a gas, the initial energy of the gas does not change. Thus, the heat transferred is equal to the work done by the gas. For an expansion, this means that Q is positive, and thus, the entropy increases. (b) The heat transferred during an adiabatic expansion is zero. Therefore, the change in entropy during the process is also zero.

SOLUTIONS TO SELECTED END-OF-CHAPTER PROBLEMS

13. One of the most efficient engines ever built operates between 430°C and 1870°C. Its actual efficiency is 42%. (a) What is its maximum theoretical efficiency? (b) How much power does the engine deliver if it absorbs 1.4×10^5 J of heat each second?

Solution

$T_c = 430\ °C = 703\ K$ and $T_h = 1870\ °C = 2143\ K$

(a) $e = \dfrac{\Delta T}{T_h} = \dfrac{1440\ K}{2143\ K} = 0.672$ or $\boxed{67.2\%}$

(b) $Q_h = 1.40 \times 10^5\ J$ $W = 0.42 Q_h = 5.88 \times 10^4\ J$

$P = \dfrac{W}{\Delta t} = \dfrac{5.88 \times 10^4\ J}{1\ s} = \boxed{58.8\ kW}$

24. An ideal refrigerator (or heat pump) is equivalent to a Carnot engine running in reverse. That is, heat Q_c is absorbed from a cold reservoir and heat Q_h is rejected to a hot reservoir. (a) Show that the work that must be supplied to run the refrigerator is given by

$$W = \frac{T_h - T_c}{T_c} Q_c$$

(b) Show that the coefficient of performance of the ideal refrigerator is given by

$$COP = \frac{T_c}{T_h - T_c}$$

Solution

(a)   For a complete cycle, $\Delta U = 0$ and

$$W = Q_h - Q_c = Q_c \left(\frac{Q_h}{Q_c} - 1 \right)$$

We have already shown for a Carnot cycle (and only for a Carnot cycle) that

$$\frac{Q_h}{Q_c} = \frac{T_h}{T_c}$$

Therefore,

$$W = Q_c \frac{(T_h - T_c)}{T_c}$$

(b)   We have from Eq. 22.7, $COP = \frac{Q_c}{W}$.   Using the result from part (a), this becomes

$$COP = \frac{T_c}{T_h - T_c}$$

---

33. One mol of an ideal monatomic gas is heated quasi-statically at constant volume from 300 K to 400 K. What is the change in entropy of the gas?

Solution

$$\Delta S = \int_{T_1}^{T_2} \frac{nC_v dT}{T} = nC_v Ln\left(\frac{T_2}{T_1}\right) \quad \text{and} \quad C_v = \frac{3R}{2}, \quad \text{so that}$$

$$\Delta S = n\left(\frac{3R}{2}\right)Ln\left(\frac{T_2}{T_1}\right) = (1 \text{ mol})\left(\frac{3}{2}\right)(8.31 \text{ J/mol·K})Ln\left(\frac{400 \text{ K}}{300 \text{ K}}\right) = \boxed{3.59 \text{ J/K}}$$

---

45. If 200 g of water at 20°C is mixed with 300 g of water at 75°C, find (a) the final equilibrium temperature of the mixture and (b) the change in entropy of the system.

Solution

(a)　$\Delta Q_1 = \Delta Q_2$　or　$(mc\Delta T)_1 = - (mc\Delta T)_2$,　and　$(200 \text{ g})(T - 20°C) = (300 \text{ g})(75°C - T)$

This gives　$T = \boxed{53°C}$　or　$\boxed{326 \text{ K}}$

(b)　$\Delta S = mcLn\left(\dfrac{T_2}{T_1}\right)$　for each material

$\Delta S = (200 \text{ g})(1 \text{ cal/g·C°})Ln\left(\dfrac{53 + 273}{20 + 273}\right) + (300 \text{ g})(1 \text{ cal/g·C°})Ln\left(\dfrac{53 + 273}{75 + 273}\right)$

$\Delta S = 1.75 \text{ cal/deg} = \boxed{7.33 \text{ J/K}}$

---

51. One mol of an ideal monatomic gas is taken through the cycle shown in Figure 22.18. The process AB is an isothermal expansion. Calculate (a) the net work done by the gas, (b) the heat added to the gas, (c) the heat expelled by the gas, and (d) the efficiency of the cycle.

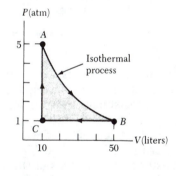

Figure 22.18

Solution

(a)　For the isothermal process AB,

$$W_{AB} = P_A V_A Ln\left(\frac{V_B}{V_A}\right)$$

$$W_{AB} = (5)(1.013 \times 10^5 \text{ N/m}^2)(10 \times 10^{-3} \text{ m}^3)Ln\left(\frac{50}{10}\right) = 8.15 \times 10^3 \text{ J}$$

where we used　$1 \text{ atm} = 1.013 \times 10^5 \text{ N/m}^2$　and　$1 \text{ L} = 10^{-3} \text{ m}^3$

$$W_{BC} = P_B \Delta V = (1.013 \times 10^5 \text{ N/m}^2)[(10 - 50) \times 10^{-3}] \text{ m}^3 = -4.05 \times 10^3 \text{ J}$$

$$W_{CA} = 0 \quad \text{and} \quad W = W_{AB} + W_{BC} = \boxed{4.11 \text{ kJ}}$$

(b) and (c) Since AB is an isothermal process, $\Delta U_{AB} = 0$,   and

$$Q_{AB} = W_{AB} = 8.15 \times 10^3 \text{ J}$$

For an ideal monatomic gas, $C_v = \dfrac{3R}{2}$ and $C_p = \dfrac{5R}{2}$

$$T_B = T_A = \frac{P_B V_B}{nR} = \frac{(1.013 \times 10^5)(50 \times 10^{-3})}{R} = \frac{5.05 \times 10^3}{R}$$

$$T_C = \frac{P_C V_C}{nR} = \frac{(1.01 \times 10^5)(10 \times 10^{-3})}{1(R)} = \frac{1.01 \times 10^3}{R}$$

The heat expelled by the gas < 0 and

$$Q_C = Q_{BC} = nC_p\Delta T = \left(\tfrac{5}{2}\right)(1.01 \times 10^3)(1 - 5) = \boxed{-10.1 \text{ kJ}}$$

The heat added to the gas (Q_h) is > 0, therefore

$$Q_h = Q_{CA} + Q_{AB} = \boxed{14.2 \text{ kJ}}$$

(d) $e = \dfrac{W}{Q_h} = \dfrac{4.11 \text{ kJ}}{14.2 \text{ kJ}} = 0.289$ or $\boxed{28.9\%}$

---

53. Figure 22.19 represents n mols of an ideal monatomic gas being taken through a reversible cycle consisting of two isothermal processes at temperatures $3T_0$ and T_0 and two constant-volume processes. For each cycle, determine in terms of n, R, and T_0 (a) the net heat transferred to the gas and (b) the efficiency of an engine operating in this cycle.

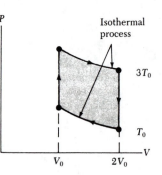

Figure 22.19

Solution

(a)   For an isothermal process, $Q = nRT Ln\left(\dfrac{V_2}{V_1}\right)$

Therefore, $Q_1 = nR(3T_0)Ln2$ and $Q_3 = nR(T_0)Ln\left(\tfrac{1}{2}\right)$

For the constant volume processes we have

$$Q_2 = nC_v(T_0 - 3T_0)   \text{ and }   Q_4 = nC_v(3T_0 - T_0)$$

The net heat transferred is then

$$Q = Q_1 + Q_2 + Q_3 + Q_4 \quad \text{or} \quad Q = \boxed{2nRT_oLn2}$$

(b) Heat > 0 is the heat added to the system. Therefore,

$$Q_h = Q_1 + Q_4 = 3nRT_0(1 + Ln2)$$

Since the change in temperature for the complete cycle is zero, $\quad \Delta U = 0 \quad$ and $\quad W = Q$.

Therefore, the efficiency is

$$e = \frac{W}{Q_h} = \frac{Q}{Q_h} = \frac{2Ln2}{3(1 + Ln2)} = \boxed{0.273} \quad \text{or} \quad \boxed{27.3\%}$$

---

55. One mol of a monatomic ideal gas is taken through the reversible cycle shown in Figure 22.20. At point A, the pressure, volume, and temperature are P_0, V_0, and T_0, respectively. In terms of R and T_0, find (a) the total heat entering the system per cycle, (b) the total heat leaving the system per cycle, (c) the efficiency of an engine operating in this reversible cycle, and (d) the efficiency of an engine operating in a Carnot cycle between the same temperature extremes for this process.

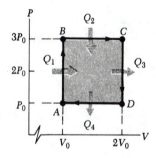

Figure 22.20

Solution

At point A, $\quad P_0V_0 = nRT_0, \quad n = 1$

At point B, $\quad 3P_0V_0 = RT_B \quad$ and $\quad T_B = 3T_0$

At point C, $\quad (3P_0)(2V_0) = RT_C \quad$ and $\quad T_C = 6T_0$

At point D, $\quad P_0(2V_0) = RT_0 \quad$ and $\quad T_D = 2T_0$

The heat transfer for each step in the cycle is found using

$$C_v = \frac{3R}{2} \quad \text{and} \quad C_p = \frac{5R}{2}$$

$$Q_{AB} = C_v(3T_0 - T_0) = 3RT_0 \qquad Q_{BC} = C_p(6T_0 - 3T_0) = 7.5RT_0$$

$$Q_{CD} = C_v(2T_0 - 6T_0) = -6RT_0 \qquad Q_{DA} = C_p(T_0 - 2T_0) = -2.5RT_0$$

Therefore,

(a) $Q_{in} = Q_h = Q_{AB} + Q_{DA} = \boxed{10.5RT_o}$

(b) $Q_{out} = Q_C = |Q_{CD} + Q_{DA}| = \boxed{8.5RT_o}$

(c) $e = \dfrac{Q_h - Q_C}{Q_h} = \boxed{0.190}$ or $\boxed{19\%}$

(d) Carnot Eff, $e_C = 1 - \dfrac{T_C}{T_h} = 1 - \dfrac{T_o}{6T_o} = \boxed{0.833}$ or $\boxed{83.3\%}$

---

57. A system consisting of n mols of an ideal gas ungoes a reversible, *isobaric* process from a volume V_0 to a volume $3V_0$. Calculate the change in entropy of the gas. (Hint: Imagine that the system goes from the initial state to the final state first along an isotherm and then along an adiabatic curve, for which there is no change in entropy).

Solution

The isobaric process (AB) is shown along with an isotherm (AC) and an adiabat (CB) in the PV diagram. Since the change in entropy is path independent, $\Delta S_{AB} = \Delta S_{AC} + \Delta S_{CB}$ and $\Delta S_{CB} = 0$ for an adiabatic process. For isotherm (AC), $P_A V_A = P_C V_C$ and for adiabat (CB), $P_C V_C{}^\gamma = P_B V_B{}^\gamma$. Combining these give

$$V_C = \left(\frac{P_B V_B{}^\gamma}{P_A V_A}\right)^{1/(\gamma-1)} = \left[\left(\frac{P_o}{P_o}\right)\frac{(3V_o)^\gamma}{V_o}\right]^{1/(\gamma-1)} = 3^{\gamma/\gamma-1}V_o$$

Therefore,

$$\Delta S_{AC} = \left(\frac{P_A V_A}{T}\right)Ln\left(\frac{V_C}{V_A}\right) = nRLn[3^{(\gamma/\gamma-1)}] = \frac{nR\gamma Ln3}{\gamma - 1}$$

---

61. An idealized Diesel engine operates in a cycle known as the *air-standard Diesel cycle*, shown in Figure 22.21. Fuel is sprayed into the cylinder at the point of maximum compression, B. Combustion occurs during the expansion B → C, which is approximated as an isobaric process. The rest of the cycle is the same as in the gasoline engine, described in Figure 22.11. Show that the efficiency of an engine operating this idealized Diesel cycle is given by

$$e = 1 - \frac{1}{\gamma}\left(\frac{T_D - T_A}{T_C - T_B}\right)$$

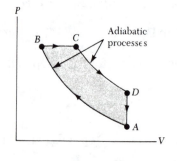

Figure 22.21

Solution

The heat transfer over the paths CD and BA are zero since they are adiabats.

Over path BC: $Q_{BC} = nC_p(T_C - T_B) > 0$

Over path DA: $Q_{DA} = nC_v(T_A - T_D) < 0$

Therefore,

$$Q_c = |Q_{DA}| \quad \text{and} \quad Q_h = Q_{BC}$$

Hence, the efficiency is

$$e = 1 - \frac{Q_c}{Q_h} = 1 - \left(\frac{T_D - T_A}{T_C - T_B}\right)\frac{C_v}{C_p}$$

$$e = 1 - \frac{1}{\gamma}\left(\frac{T_D - T_A}{T_C - T_B}\right)$$

---

63. Consider one mol of an ideal gas that undergoes a quasi-static, reversible process in which the heat capacity C remains constant. Show that the pressure and volume of the gas obey the relation

$$PV^{\gamma'} = \text{constant} \quad \text{where} \quad \gamma' = \frac{C - C_p}{C - C_v}$$

(Hint: Start with the first law of thermodynamics, $dQ = dU + PdV$, and use the fact that $dQ = CdT$ and $dU = C_v dT$. Furthermore, $PV = RT$ applies to the gas, so it follows that $PdV + VdP = RdT$.)

Solution

From the first law of thermodynamics, $dQ = dU + PdV$ or $CdT = C_v dT + PdV$. Also, $PdV + VdP = RdT$. Solving for dT in the first equation and substituting into the second gives

$$PdV + VdP = \frac{RdV}{C - C_v}$$

Rearranging terms and using the fact that $R = C_p - C_v$, we get

$$\left(\frac{C - C_p}{C - C_v}\right)\frac{dV}{V} + \frac{dP}{P} = 0$$

Since $\frac{dx}{x} = d(Ln\,x)$, we have

$$Ln\,V^{\gamma'} + Ln\,P = Ln(k) \quad \text{and} \quad PV^{\gamma'} = k \quad \text{where k is constant.}$$

23

Electric Fields

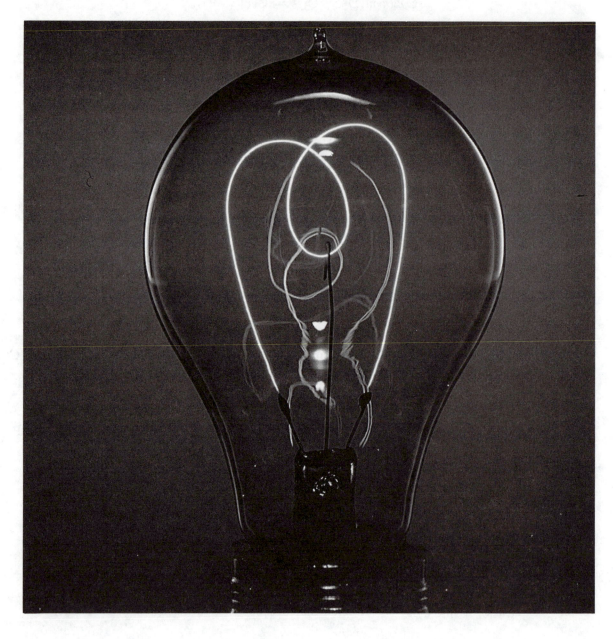

OBJECTIVES

1. Describe the fundamental properties of electric charge and the nature of electrostatic forces between charged bodies.

2. Describe the processes involved in charging a conductor by contact and by induction.

3. Use Coulomb's law to determine the net electrostatic force on a point electric charge due to a known distribution of a finite number of point charges.

4. Calculate the electric field **E** (magnitude and direction) at a specified location in the vicinity of a group of point charges.

5. Calculate the electric field due to a continuous charge distribution. The charge may be distributed uniformly or nonuniformly along a line, over a surface, or throughout a volume.

6. Visualize qualitatively the electric field throughout a region of space in terms of electric field lines.

7. Describe quantitatively the motion of a charged particle in a uniform electric field.

SKILLS

1. Remember that the electric field at some point in space and electric forces between charged particles are both *vector* quantities and must be treated accordingly. The principle of superposition can be applied to electrostatic forces and to electric fields. This means that when several charges are present, the resultant force on any one is the *vector sum* of the forces due to the several charges acting separately.

   To find the total electric field at a given point, you must first calculate the electric field at the point due to each individual charge. The resultant field at the point is the vector sum of the fields due to the individual charges.

2. To evaluate the electric field of a continuous charge distribution, it is convenient to employ the concept of charge density. Charge density can be written in different ways: charge per unit volume, ρ; charge per unit area, σ; or charge per unit length, λ. The total charge distribution is then subdivided into a small element of volume dV, area dA, or length dx. Each element contains an increment of charge dq (dq = ρdV or σdA or λdx). If the charge is *nonuniformly* distributed over the region, then the charge densities must be written as functions of position. For example, if the charge density along a line or long bar is proportional to the distance from one end of the bar, then the linear charge density could be written as λ = bx and the charge increment dq becomes dq = bx dx.

3. The motion of a charged particle in a uniform electric field can be described by using the equations of projectile motion as developed in Chapter 4 where the acceleration $\mathbf{a} = \dfrac{q}{m}\mathbf{E}$. (For an electron q = -e and the electron experiences an acceleration which is directed opposite the direction of the field.)

NOTES FROM SELECTED CHAPTER SECTIONS_____

23.1 PROPERTIES OF ELECTRIC CHARGES

Electric charge has the following important properties:

1. There are two kinds of charges in nature, with the property that unlike charges attract one another and like charges repel one another.

2. The force between charges varies as the inverse square of their separation.

3. Charge is conserved.

4. Charge is quantized.

23.2 INSULATORS AND CONDUCTORS

Conductors are materials in which electric charges move freely under the influence of an electric field; *insulators* are materials that do not readily transport charge.

23.4 THE ELECTRIC FIELD

The electric field vector **E** at some point in space is defined as the electric force **F** acting on a positive test charge placed at that point divided by the magnitude of the test charge q_o.

An electric field exists at some point if a test charge at rest placed at that point experiences an electrical force.

The total electric field due to a group of charges equals the vector sum of the electric fields of all the charges at some point.

23.6 ELECTRIC FIELD LINES

A convenient aid for visualizing electric field patterns is to draw lines pointing in the same direction as the electric field vector at any point. These lines, called electric field lines, are related to the electric field in any region of space in the following manner:

1. The electric field vector **E** is *tangent* to the electric field line at each point.

2. The number of lines per unit area through a surface perpendicular to the lines is proportional to the strength of the electric field in that region. Thus **E** is large when the field lines are close together and small when they are far apart.

The rules for drawing electric field lines for any charge distribution are as follows:

1. The lines must begin on positive charges and terminate on negative charges, or at infinity in the case of an excess of charge.

2. The number of lines drawn leaving a positive charge or approaching a negative charge is proportional to the magnitude of the charge.

3. No two field lines can cross.

EQUATIONS AND CONCEPTS

The *magnitude* of the *electrostatic force* between two stationary point charges, q_1 and q_2, separated by a distance r is given by Coulomb's law.

$$F = k \frac{q_1 q_2}{r^2}$$

(23.1)

In calculations, an approximate value for k may be used.

$$k = 9.0 \times 10^9 \ \frac{N \cdot m^2}{C^2}$$

(23.3)

The *direction* of the force on each charge is determined from the experimental observation that like charges repel each other and charges of unlike sign attract each other.

When more than one point charge is present, the *total electrostatic force* exerted on the ith charge is the vector sum of the forces exerted on that charge by the others individually.

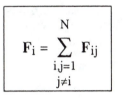

$$\mathbf{F}_i = \sum_{\substack{i,j=1 \\ j \neq i}}^{N} \mathbf{F}_{ij}$$

The electric force between two charges can be expressed in vector form. In Eq. 23.6, $\mathbf{F}_{1,2}$ is the force *on q_1 due to q_2* and $\hat{\mathbf{r}}_{1,2}$ is a unit vector directed from q_1 to q_2.

$$\mathbf{F}_{1,2} = k \frac{q_1 q_2}{r^2} \hat{\mathbf{r}}_{1,2}$$

(23.6)

The *electric field* at any point in space is defined as the ratio of electric force per unit charge exerted on a small positive test charge placed at the point where the field is to be determined.

$$\mathbf{E} = \frac{\mathbf{F}}{q_0}$$

(23.7)

This definition together with Coulomb's law leads to an expression for calculating the *electric field a distance r from a point charge, q*. In this case the unit vector $\hat{\mathbf{r}}$ is directed away from q and toward the point P where the field is to be calculated.

$$\mathbf{E} = k \frac{q}{r^2} \hat{\mathbf{r}}$$

(23.8)

The *electric field* at a point *due to a group of point charges* is the vector sum of the electric fields due to the individual charges.

$$\mathbf{E} = \sum_i k \frac{q_i}{r_i^2} \hat{\mathbf{r}}_i$$

(Vector Sum)

(23.9)

When the electric field is due to a *continuous charge distribution*, the contribution to the field by each element of charge must be integrated over the total line, surface, or volume which contains the charge.

$$E = k \int \frac{dq}{r^2} \hat{r}$$

(23.11)

In order to perform the integration described above, it is convenient to represent a charge increment dq as the product of an element of length, area, or volume and the *charge density over that region*. Note: For those cases in which the charge is not uniformly distributed, the densities λ, σ, and ρ must be stated as functions of position.

For an element of length
$$dq = \lambda \, dx$$
For an element of area
$$dq = \sigma \, dA$$
For an element of volume
$$dq = \rho \, dV$$

A particle of mass m and charge q experiences a *constant acceleration in a uniform electric field*. The motion of the particle can be described by using the equations for projectile motion (from Chapter 4).

$$a = \frac{qE}{m}$$

(23.19)

EXAMPLE PROBLEM SOLUTION

Example 23.1 A small diameter dielectric rod is shaped into a semicircle of radius R. The top quadrant has a uniform negative charge density λ while the bottom quadrant has an equal uniform positive charge density. Determine the electric field at point 0.

Solution

Let an element of charge dq be that charge on an element of arc length ds so that $dq = \lambda ds$. Since $ds = Rd\theta$, $dq = \lambda Rd\theta$.

Note that as a result of the symmetry of the charge distribution, the *horizontal* components of the field contributions due to a given positive charge element and the corresponding negative charge element will cancel. The net field will then be found by integrating the vertical components of the field contributions due to each charge element.

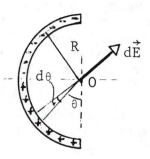

Figure 23.1

286

The charge element shown in the figure gives rise to a field contribution at point 0 of magnitude

$$dE = k \frac{dq}{R^2} = k \frac{\lambda R d\theta}{R^2}$$

and the vertical component of dE is

$$dE_y = dE \cos\theta = \frac{k\lambda}{R} \cos\theta d\theta$$

so the field at point 0 has a value

$$E = \int dE_y = \frac{k}{R} \int \lambda \cos\theta d\theta$$

Since the top and bottom segments of the semicircle make equal contributions to the net field, the limits of the integration can be taken from zero to $\pi/2$ and the result multiplied by 2. Therefore

$$E = \frac{2k\lambda}{R} \int_0^{\pi/2} \cos\theta d\theta$$

$$E = \frac{2k\lambda}{R}$$

Due to the symmetry of the charge distribution, the direction of E is *vertically upward*.

---

1. Sparks are often observed (or heard) on a dry day when clothes are removed in the dark. Explain

Answer: Static charge is built up on the clothes as they rub against your body. Electrical discharge occurs to ground when the charge reaches a certain limit.

3. A balloon is negatively charged by rubbing and then clings to a wall. Does this mean that the wall is positively charged? Why does the balloon eventually fall?

Answer: No. The balloon induces charge of opposite sign in the wall, causing it to be attracted. The balloon eventually falls since its charge slowly diminishes as it leaks to ground. Some of the charge could also be lost due to ions of opposite sign in the surrounding atmosphere which would tend to neutralize the charge.

7. A charged comb will often attract small bits of dry paper that fly away when they touch the comb. Explain?

Answer: When the comb is nearby, charges separate on the paper, resulting in the paper being attracted. After contact, charges from the comb are transferred to the paper so that it has the same type charge as the comb. It is thus repelled.

9. A large metal sphere insulated from ground is charged with an electrostatic generator while a person standing on an insulating stool holds the sphere while it is being charged. Why is it safe to do this? Why wouldn't it be safe for another person to touch the sphere after it has been charged?

Answer: The person holding the sphere would become charged as evidenced by the person's hair standing on end. The insulating stool does not allow the charge to flow to ground, so there is no current path. However, a person standing on the floor could receive an electrical shock if he is making good electrical contact with ground. In order to prevent electrical shock, the charged sphere should be grounded before the person holding the sphere leaves the stool.

12. Assume that someone proposes a theory that says people are bound to the earth by electric forces rather than by gravity. How could you prove this theory wrong?

Answer: This could only happen if the earth and the person were of opposite charge. Since only two charges exist, + and –, all people would then be of the same charge and would repel one another.

13. Would life be different if the electron were positively charged and the proton were negatively charged? Does the choice of signs have any bearing on physical and chemical interactions? Explain.

Answer: No. The assignment of positive and negative charge is completely arbitrary.

30. If a metal object receives a positive charge, does its mass increase, decrease, or stay the same? What happens to the mass if the object is given a negative charge?

Answer: An object's mass decreases very slightly (immeasurably) when it is given a positive charge, because it loses electrons. When charged negatively, it gains mass very slightly because it gains electrons.

33. Are the occupants of a steel-frame building safer than those in a wood-frame house during an electrical storm or vice-versa? Explain.

Answer: The inhabitants of the steel building are safer because any electrical charge on the building will be on the outer surface, and any electric current passing through the building will pass through the steel (which is a good conductor.)

SOLUTIONS TO SELECTED END-OF-CHAPTER PROBLEMS_____

15. What are the magnitude and direction of the electric field that will balance the weight of (a) an electron and (b) a proton? (Use the data in Table 23.1.)

Solution

Require $m\mathbf{g} + q\mathbf{E} = 0$　or　$\mathbf{E} = -\dfrac{m\mathbf{g}}{q}$

(a) For an electron:　$\mathbf{E} = \dfrac{-(9.1 \times 10^{-31} \text{ kg})(9.8 \text{ m/s}^2)(-\mathbf{j})}{-1.6 \times 10^{-19} \text{ C}} = (-5.57 \times 10^{-11} \mathbf{j}) \text{ N/C}$

(b) For a proton:　$\mathbf{E} = \dfrac{-(1.6 \times 10^{-27} \text{ kg})(9.8 \text{ m/s}^2)(-\mathbf{j})}{-1.6 \times 10^{-19} \text{ C}} = (1.02 \times 10^{-7} \mathbf{j}) \text{ N/C}$

---

25. Four charges are at the corners of a square as in Figure 23.27. (a) Find the magnitude and direction of the electric field at the position of the charge $-q$, the coordinates of which are $x = a$, $y = a$. (b) What is the electric force on this charge?

Solution

$q_2 = +q;$　$q_1 = q_3 = -q$

(a) Point (a,a); $\mathbf{E} = \mathbf{E}_1 + \mathbf{E}_2 + \mathbf{E}_3$

$E_x = E_{1x} + E_{2x} + E_{3x}$

$E_x = -\dfrac{kq}{a^2} + \dfrac{kq}{2a^2}\left(\dfrac{\sqrt{2}}{2}\right) = \dfrac{kq}{a^2}\left(-1 + \dfrac{\sqrt{2}}{4}\right)$

$E_y = E_{1y} + E_{2y} + E_{3y}$

$E_y = 0 + \dfrac{kq}{2a^2}\sin 45° - \dfrac{kq}{a^2} = \dfrac{kq}{a^2}\left(-1 + \dfrac{\sqrt{2}}{4}\right)$

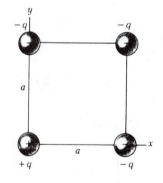

Figure 23.27

$E = \sqrt{E_x^2 + E_y^2} = \boxed{0.914\,\dfrac{kq}{a^2}}$

$0 = \tan^{-1}\left(\dfrac{E_y}{E_x}\right) = \tan^{-1}(1) = \boxed{225°}$　　or　　$\mathbf{E} = \boxed{-0.646\,\dfrac{kq}{a^2}(\mathbf{i} + \mathbf{j})}$

(b) $\mathbf{F} = -q\mathbf{E} = \boxed{+0.646\,\dfrac{kq^2}{a^2}(\mathbf{i} + \mathbf{j})}$

---

27. In Figure 23.29, determine the point (other than ∞) at which the total electric field is zero.

Solution

Label the charges $q_1 = -2.5\ \mu C$; $q_2 = 6\ \mu C$. Then $\mathbf{E}_1$ and $\mathbf{E}_2$ will cancel at some point P a distance d to the left of q_1.

At point P:
$$E_x = \frac{-kq_1}{d^2} - \frac{kq_2}{(d+1)^2} = 0$$

or
$$\frac{q_1}{d^2} + \frac{q_2}{(d+1)^2} = 0$$

Solve for d using the positive sign to find
$d = \boxed{7.82\ \text{m}}$ (to the left of q_1)

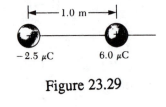

Figure 23.29

---

39. A uniformly charged rod of length 14 cm is bent into the shape of a semicircle as in Figure 23.32. If the rod has a total charge of $-7.5\ \mu C$, find the magnitude and direction of the electric field at 0, the center of the semicircle.

Solution

$$dq = \lambda ds = \lambda r d\theta$$

$$dE = \frac{kdq}{r^2}; \qquad dE_x = dE\cos\theta$$

$E_y = 0$ (from symmetry)

$$E_x = \int dE_x = \int \frac{k\lambda r\cos\theta\, d\theta}{r^2}$$

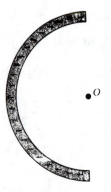

Figure 23.32

$$E_x = \frac{k\lambda}{r}\int_{-\pi/2}^{\pi/2}\cos\theta\, d\theta = \frac{2k\lambda}{r}$$

But $Q_{total} = \lambda\pi r$ and $r = \dfrac{\text{length}}{\pi}$

Thus
$$E_x = \frac{2kQ}{\pi r^2} = \frac{(2)\left(9\times10^9\ \dfrac{N\cdot m^2}{C^2}\right)(-7.5\times10^{-6}\ C)}{\pi\left(\dfrac{0.14\ m}{\pi}\right)^2}$$

$$\mathbf{E} = \boxed{(-2.16\times10^7\ \mathbf{i})\ \text{N/C}}$$

43. Figure 23.33 shows the electric field lines for two point charges separated by a small distance. (a) Determine the ratio q_1/q_2. (b) What are the signs of q_1 and q_2?

Solution

6 field lines terminate on q_1;

18 field lines originate from q_2

(a) $\dfrac{q_1}{q_2} = \dfrac{-N_1}{N_2} = -\dfrac{1}{3}$

(b) q_1 is negative; q_2 is positive

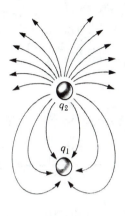

Figure 23.33

---

45. A proton accelerates from rest in a uniform electric field of 640 N/C. At some later time, its speed is 1.20×10^6 m/s (nonrelativistic since v is much less than the speed of light). (a) Find the acceleration of the proton. (b) How long does it take the proton to reach this velocity? (c) How far has it moved in this time? (d) What is its kinetic energy at this time?

Solution

(a) $a = \dfrac{F}{m} = \dfrac{qE}{m} = \dfrac{(1.6 \times 10^{-19}\ \text{C})(640\ \text{N/C})}{1.67 \times 10^{-27}\ \text{kg}}$

$a = \boxed{6.14 \times 10^{10}\ \text{m/s}^2}$

(b) $t = \dfrac{\Delta v}{a} = \dfrac{1.2 \times 10^6\ \text{m/s}}{6.14 \times 10^{10}\ \text{m/s}^2} = \boxed{19.5\ \mu\text{s}}$

(c) $x = v_o t + \dfrac{1}{2} at^2 = 0 + \left(\dfrac{1}{2}\right)(6.14 \times 10^{10}\ \text{m/s}^2)(1.95 \times 10^{-5}\ \text{s})^2$

$x = \boxed{11.7\ \text{m}}$

(d) $K = \dfrac{1}{2} mv^2 = \dfrac{1}{2}(1.67 \times 10^{-27}\ \text{kg})(1.2 \times 10^6\ \text{m/s})^2$

$K = \boxed{1.2 \times 10^{-15}\ \text{J}}$

51. A proton has an initial velocity of 4.50×10^5 m/s in the horizontal direction. It enters a uniform electric field of 9.60×10^3 N/C directed vertically. Ignore any gravitational effects and (a) find the time it takes the proton to travel 5.0 cm horizontally, (b) the vertical displacement of the proton after it has traveled 5.0 cm horizontally, and (c) the horizontal and vertical components of the proton's velocity after it has traveled 5.0 cm horizontally.

Solution

E is directed along the y direction, therefore $a_x = 0$ and $x = v_{ox}t$

(a) $t = \dfrac{x}{v_{ox}} = \dfrac{0.05 \text{ m}}{4.5 \times 10^5 \text{ m/s}} = \boxed{1.11 \times 10^{-7} \text{ s}}$

$a_y = \dfrac{qE_y}{m} = \dfrac{(1.6 \times 10^{-19} \text{ C})(9.6 \times 10^3 \text{ N/C})}{1.67 \times 10^{-27} \text{ kg}} = 9.2 \times 10^{11} \text{ m/s}^2$

(b) $y = v_{oy}t + \dfrac{1}{2} a_y t^2$

$y = \left(\dfrac{1}{2}\right)(9.2 \times 10^{11} \text{ m/s}^2)(1.11 \times 10^{-7} \text{ s})^2 = \boxed{5.67 \text{ mm}}$

(c) $v_x = v_{ox} = \boxed{4.5 \times 10^5 \text{ m/s}}$

$v_y = v_{oy} + a_y t = 0 + (9.2 \times 10^{11} \text{ m/s}^2)(1.11 \times 10^{-7} \text{ s}) = \boxed{1.02 \times 10^5 \text{ m/s}}$

---

55. A charged cork ball of mass 1 g is suspended on a light string in the presence of a uniform electric field as in Figure 23.36. When $E = (3\mathbf{i} + 5\mathbf{j}) \times 10^5$ N/C, the ball is in equilibrium at $\theta = 37°$. Find (a) the charge on the ball and (b) the tension in the string. Show forces in a free-body diagram.

Solution

(a) $E_x = 3 \times 10^5$ N/C

$E_y = 5 \times 10^5$ N/C

(1) $\Sigma F_x = qE_x - T \sin 37° = 0$

(2) $\Sigma F_y = qE_y + T \cos 37° - mg = 0$

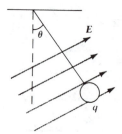

Figure 23.36

Substitute T from (1) into (2)

$$q = \frac{mg}{\left(E_y + \dfrac{E_x}{\tan 37°}\right)} = \frac{(10^{-3} \text{ kg})(9.8 \text{ m/s}^2)}{\left(5 + \dfrac{3}{\tan 37°}\right) \times 10^5 \text{ N/C}}$$

$$q = \boxed{1.09 \times 10^{-8} \text{ C}}$$

(b) From Eq. (1), $T = \dfrac{qE_x}{\sin 37°} = \boxed{5.43 \times 10^{-3} \text{ N}}$

---

63. Three charges of equal magnitude q reside at the corners of an equilateral triangle of side length a. Two of the charges are negative, and the other is positive, as shown in Figure 23.40. (a) Find the magnitude and direction of the electric field at point P, midway between the negative charges, in terms of k, q, and a. (b) Where must a $-4q$ charge be placed so that any charge located at point P will experience no net electrostatic force ($F_e = 0$)? In (b) let the distance between the +q charge and point P be *one meter*.

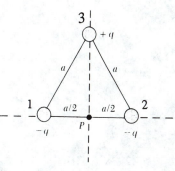

Figure 23.40

Solution

Label the charges and field contributions as shown in the figure. By symmetry $E_1 = E_2$.

(a) $E_x = -E_1 + E_2 + 0;$ $E_x = 0$

$$E_y = 0 - E_3 + 0 = -\frac{kq}{\left(\dfrac{3a^2}{4}\right)} = -\frac{4kq}{3a^2}$$

$$\mathbf{E} = \boxed{\dfrac{4kq}{3a^2} (-\mathbf{j})}$$

(b) Let $q_4 = -4q$ and require $E_4 = |\mathbf{E}|$.   Let y equal location of q_4 on axis above point P.

Then $\dfrac{kq_4}{y^2} = \dfrac{4kq}{3a^2}$ or $\dfrac{k(4q)}{y^2} = \dfrac{4kq}{3a^2}$ and $y = \sqrt{3}\, a$

but given height of triangle equal to 1 m $a = \dfrac{2}{\sqrt{3}}$ and $y = \boxed{2 \text{ m}}$

Gauss' Law

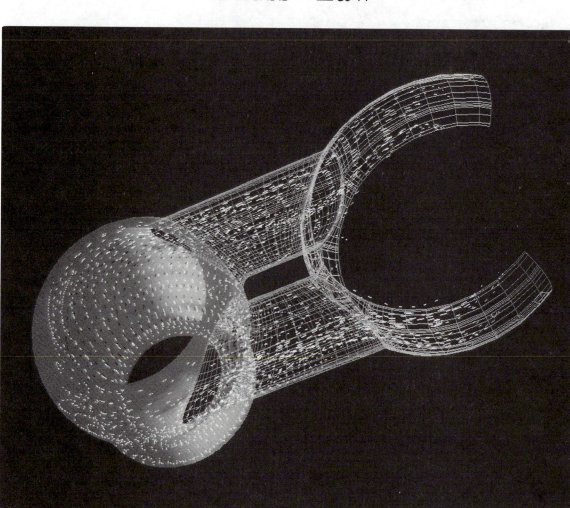

OBJECTIVES

1. Calculate the *electric* flux through a surface; in particular, find the net electric flux through a *closed* surface.

2. Understand that a gaussian surface must be a real or imaginary *closed* surface within a conductor, a dielectric, or in space. Also remember that the net electric flux through a closed gaussian surface is equal to the net charge enclosed by the surface divided by the constant ε_0.

3. Use Gauss' law as given by Eq. 24.6 to evaluate the electric field at points in the vicinity of charge distributions which exhibit spherical, cylindrical, or planar symmetry.

4. Describe the properties which characterize an electrical conductor in electrostatic equilibrium.

SKILLS

Gauss' law is a very powerful theorem which relates any charge distribution to the resulting electric field at any point in the vicinity of the charge. In this chapter you should learn how to apply Gauss' law to those cases in which the charge distribution has a sufficiently high degree of symmetry. As you review the examples presented in Section 24.3 of the text, observe how each of the following steps have been included in the application of the equation $\oint \mathbf{E} \cdot d\mathbf{A} = \dfrac{q}{\varepsilon_0}$ to that particular situation.

a. The gaussian surface should be chosen to have the *same symmetry as the charge distribution*.

b. The dimensions of the surface must be such that the surface includes the point where the electric field is to be calculated.

c. From the symmetry of the charge distribution, you should be able to correctly describe the direction of the electric field vector, **E**, relative to the direction of an element of surface area vector, d**A**, over each region of the gaussian surface.

d. From the symmetry of the charge distribution, you should also be able to identify one or more portions of the closed surface (and in some cases the entire surface) over which the magnitude of **E** remains constant.

e. Write **E·dA** as E dA cos θ and use the results of (c) and (d) to divide the surface into separate regions such that over each region:

E dA cos θ will equal 0 when **E** ⊥ d**A** (as is the case over each end of the cylindrical gaussian surface in Figure 24.1 or when **E** = 0 as is the case over a surface inside of a conductor).

E dA cos θ will equal E dA when **E** ‖ d**A** (as is the case over the curved portion of the cylindrical gaussian surface in Figure 24.1).

E dA cos θ = -E dA when **E** and d**A** are oppositely directed.

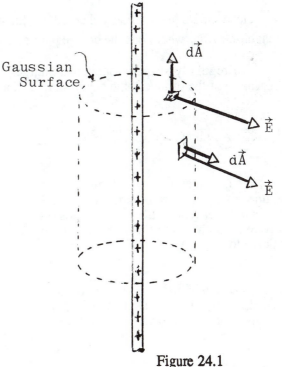

Figure 24.1

f. If the Gaussian surface has been chosen and subdivided so that the magnitude of **E** is *constant* over those regions where **E·dA** = E dA, then over each of those regions

$$\int \mathbf{E} \cdot d\mathbf{A} = E \int dA = E(\text{area of region})$$

g. The total charge enclosed by the Gaussian surface is $q = \int dq$. It is often convenient to represent the charge distribution in terms of the charge density ($dq = \lambda dx$ for a line of charge, $dq = \sigma dA$ for a surface of charge, or $dq = \rho dV$ for a volume of charge). The integral of dq is then evaluated only over that length, area, or volume which includes that portion of the charge *inside* the Gaussian surface.

NOTES FROM SELECTED CHAPTER SECTIONS

24.2 GAUSS' LAW

Gauss' law states that the net electric flux through a closed gaussian surface is equal to the net charge inside the surface divided by ε_0.

The gaussian surface should be chosen so that it has the same symmetry as the charge distribution.

24.4 CONDUCTORS IN ELECTROSTATIC EQUILIBRIUM

A conductor in electrostatic equilibrium has the following properties:

1. The electric field is zero everywhere inside the conductor.

2. Any excess charge on an isolated conductor must reside entirely on its surface.

3. The electric field just outside a charged conductor is perpendicular to the conductor's surface and has a magnitude σ/ε_0, where σ is the charge per unit area at that point.

4. On an irregularly shaped conductor, charge tends to accumulate at locations where the radius of curvature of the surface is the smallest, that is, at sharp points.

EQUATIONS AND CONCEPTS

The *electric flux* is a measure of the number of electric field lines that penetrate some surface. For a *plane surface* in a *uniform field*, the flux depends on the angle between the normal to the surface and the direction of the field.

$$\Phi = EA \cos \theta \tag{24.2}$$

In the case of a *general surface* in the region of a *nonuniform* field, the flux is calculated by integrating the normal component of the field over the surface in question.

$$\Phi = \int \mathbf{E} \cdot d\mathbf{A} \tag{24.4}$$

Gauss' law states that when the integral in Eq. 24.4 is evaluated over a *closed* surface (gaussian surface), the result equals *the net charge enclosed by the surface* divided by the constant ε_0. In this equation, the symbol $\oint$ indicates that the integral must be evaluated over a *closed* surface. Also **E** represents the electric field in a *region external* to a uniformly charged sphere of radius a. The field exterior to the charged sphere is equivalent to that of a point charge located at the center of the sphere.

$$\oint \mathbf{E} \cdot d\mathbf{A} = \frac{q_{in}}{\varepsilon_0}$$

(24.6)

$$E = k \frac{Q}{r^2}$$
$$r > a$$

(24.7)

At an *interior point* of a uniformly charged sphere, the electric field is proportional to the distance from the center of the sphere.

$$E = k \left(\frac{Q}{a^3}\right) r$$
$$r < a$$

(24.8)

The electric field due to a uniformly charged, *nonconducting, infinite plate* is uniform everywhere.

$$E = \frac{\sigma}{2\varepsilon_0}$$

(24.10)

EXAMPLE PROBLEM SOLUTION

Example 24.1 A dielectric in the shape of a long cylinder of radius R has a nonuniform positive charge density given by $\rho = br$, where b is a positive constant (see Figure 24.2).

Case I. Determine the electric field at a distance $r_1 > R$ from the axis of the cylinder.

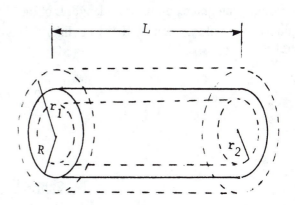

Figure 24.2

Solution

Since the charge distribution has cylindrical symmetry, the gaussian surface must also be cylindrical in shape and have a radius r_1 (so that the point where the field is to be evaluated will be contained in the gaussian surface) and be of arbitrary length. If it is assumed that the charged cylinder is very long (i.e. $L \gg r_1$), then the electric field lines will be directed radially outward from the axis. Therefore, the field lines will be *perpendicular to the curved walls* of the gaussian surface and will lie in planes *parallel to the ends* of the cylindrical gaussian surface. Furthermore, since all points on the curved wall of the cylindrical gaussian surface are at the same distance from the axis of the charge distribution, the magnitude of **E** will be constant over that surface. In applying Gauss' law to this situation, it is convenient to divide the closed gaussian surface into three regions: the left end of cylinder, the right end of cylinder, and the wall of the cylinder. The *left hand member* of the Gauss' law equation becomes

$$\oint_{\substack{\text{(closed} \\ \text{surface)}}} \mathbf{E} \cdot d\mathbf{A} = \int_{\substack{\text{(left} \\ \text{end)}}} E \, dA \cos(90°) + \int_{\substack{\text{(right} \\ \text{end)}}} E \, dA \cos(90°) + \int_{\substack{\text{(wall)}}} E \, dA \cos(0)$$

or

$$\oint_{\substack{\text{(closed} \\ \text{surface)}}} \mathbf{E} \cdot d\mathbf{A} = 0 + 0 + E \int_{\substack{\text{(wall)}}} dA$$

The integral of the element of area ($\int dA$) over the wall of the cylinder is just the area of a rectangle of length L and width $2\pi r_1$ (the circumference of the gaussian cylinder). That is,

$$\int \mathbf{E} \cdot d\mathbf{A} = E(2\pi r_1 L) \qquad (1)$$

The *right-hand member* of the Gauss' law equation becomes

$$\frac{q}{\varepsilon_o} = \frac{1}{\varepsilon_o} \int \rho \, dV = \frac{1}{\varepsilon_o} \int br \, dV$$

This volume integral extends over a length L of the charged cylinder of radius R (not r_1, the radius of the gaussian cylinder). The volume element dV is a cylindrical shell of radius r, thickness dr and length L so that $dV = 2\pi r L dr$. Therefore

$$\frac{q}{\varepsilon_o} = \frac{2\pi bL}{\varepsilon_o} \int_0^R r^2 dr$$

or

$$\frac{q}{\varepsilon_o} = \frac{2\pi bLR^3}{3\varepsilon_o} \qquad (2)$$

Combining equations (1) and (2) we have

$$E(2\pi r_1 L) = \frac{2\pi bLR^3}{3\varepsilon_o}$$

so

$$\boxed{E = \frac{bR}{3\varepsilon_o r_1}} \qquad \text{(for } r_1 > R)$$

This is the expression for the electric field at a point outside an infinitely long cylinder of charge.

Case II.   Find the electric field inside the cylinder at $r_2 < R$.

Solution

The gaussian surface is now a cylinder of radius r_2 and using the same procedure as for Case I we find

$$\oint \mathbf{E} \cdot d\mathbf{A} = E \oint dA = E2\pi r_2 L$$

also

$$\frac{q}{\varepsilon_o} = \frac{2\pi bL}{\varepsilon_o} \int_o^{r_2} r^2 dr$$

so

$$E2\pi r_2 L = \frac{2\pi bL}{\varepsilon_o} \left(\frac{r_2^3}{3} \right)$$

and

$$\boxed{E = \frac{br_2^2}{3\varepsilon_o}} \qquad\qquad (\text{for } r_2 < R)$$

ANSWERS TO SELECTED QUESTIONS

1. If the net flux through a gaussian surface is zero, which of the following statements are true? (a) There are no charges inside the surface. (b) The net charge inside the surface is zero. (c) The electric field is zero everywhere on the surface. (d) The number of electric field lines entering the surface equals the number leaving the surface.

Answer: Statements (b) and (d) are true. Statement (a) is not necessarily true since Gauss' Law says that the net flux through the closed surface equals the net charge inside the surface divided by ε_o. For example, you could have an electric dipole inside the surface. Statement (c) is not necessarily true. Although the net flux may be zero, we cannot conclude that the electric field is zero in that region.

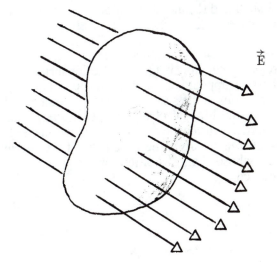

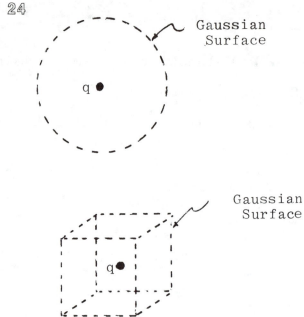

Gaussian Surface

3. A spherical gaussian surface surrounds a point charge q. Describe what happens to the flux through the surface if (a) the charge is tripled, (b) the volume of the sphere is doubled, (c) the shape of the surface is changed to that of a cube, and (d) the charge is moved to another position inside the surface.

Answer: (a) If the charge is tripled, the flux through the surface is tripled, since the net flux is proportional to the charge inside the surface. (b) The flux remains unchanged when the volume changes, since it still surrounds the same amount of charge. (c) The flux does not change when the shape of the closed surface changes. (d) The flux through the closed surface remains unchanged as the charge inside the surface is moved to another position. All of these conclusions are arrived at through an understanding of Gauss' Law.

Gaussian Surface

6. Explain why Gauss' Law cannot be used to calculate the electric field of (a) an electric dipole, (b) a charged disk, (c) a charged ring, and (d) three point charges at the corners of a triangle.

Answer: The electric field patterns of each of these configurations do not have enough symmetry to make the calculations practical. In order to apply Gauss' Law, you must be able to find a close surface surrounding the charge distribution, which can be subdivided so that the field over the separate regions is constant. Such a surface cannot be found for these cases.

7. If the total charge inside a closed surface is known but the distribution of the charge is unspecified, can you use Gauss' Law to find the electric field? Explain.

Answer: No. It is necessary to know the charge distribution before applying Gauss' Law. (See question 6.)

11. A point charge is placed at the center of an uncharged metallic spherical shell insulated from ground. As the point charge is moved off center, describe what happens to (a) the total induced charge on the shell and (b) the distribution of charge on the interior and exterior surfaces of the shell.

Answer: (a) The total induced charge on the shell will remain unchanged. (b) The interior surface will be charged opposite in sign to the point charge. Those portions of the surface nearest the point charge will have the highest charge per unit area, while those portions furthest from the charge will have the lowest charge density. The same is true for the outer surface, whose charge is of the same sign as that of the point charge.

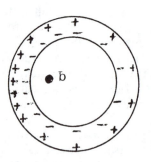

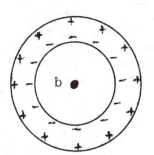

13. A person is placed in a large hollow metallic sphere that is insulated from ground. If a large charge is placed on the sphere, will the person be harmed upon touching the inside of the sphere? Explain what will happen if the person also has an initial charge whose sign is opposite to that of the charge on the sphere.

Answer: If the person is uncharged, the electric field inside the sphere would be zero. He or she would not be harmed by touching the inside of the sphere which is uncharged. On the other hand, if the person is charged, the electric field inside the sphere is no longer zero, and charge will be induced on the inner surface of the sphere. The person will get shocked when touching the sphere as charge is transferred between the person and the sphere.

SOLUTIONS TO SELECTED END-OF-CHAPTER PROBLEMS

13. A point charge of 12 μC is placed at the center of a *spherical* shell of radius 22 cm. What is the total electric flux through (a) the entire surface of the shell and (b) any hemispherical surface of the shell? (c) Do the results depend on the radius? Explain.

Solution

(a) $\Phi = 4\pi kq = 4\pi \left(9 \times 10^9 \ \frac{\text{N·m}^2}{\text{C}^2} \right) (12 \times 10^{-6} \ \text{C}) = \boxed{1.36 \times 10^6 \ \frac{\text{N·m}^2}{\text{C}}}$

(b) For half shell, $\qquad \Phi = \frac{1}{2} \left(1.36 \times 10^6 \ \frac{\text{N·m}^2}{\text{C}} \right) = \boxed{6.79 \times 10^5 \ \frac{\text{N·m}^2}{\text{C}}}$

(c) $\Phi \propto EA \propto \left(\frac{1}{r^2} \right) (r^2)$

Φ is *independent of r*

---

33. A uniformly charged, straight filament 7 m in length has a total positive charge of 2 μC. An uncharged cardboard cylinder 2 cm in length and 10 cm in radius surrounds the filament at its center, with the filament as the axis of the cylinder. Using any reasonable approximations, find (a) the electric field at the surface of the cylinder and (b) the total electric flux through the cylinder.

Solution

The approximation in this case is that the filament length is so large when compared to the cylinder length that the "infinite line" of charge can be assumed.

(a) $E = \frac{2k\lambda}{r} \qquad$ where $\qquad \lambda = \frac{2 \times 10^{-6} \ \text{C}}{7 \ \text{m}} = 2.86 \times 10^{-7} \ \text{C/m}$

so $\qquad E = \dfrac{(2)(9 \times 10^9 \ N \cdot m^2/C)(2.86 \times 10^{-7} \ C/m)}{0.10 \ m} = \boxed{5.15 \times 10^4 \ N/C}$

(b) $\Phi = 2\pi r L E = 2\pi r L \left(\dfrac{2k\lambda}{r} \right) = 4\pi k\lambda L$

so $\qquad \Phi = 4\pi \left(9 \times 10^9 \ \dfrac{N \cdot m^2}{C^2} \right) (2.86 \times 10^{-7} \ C/m)(0.02 \ m) = \boxed{6.47 \times 10^2 \ \dfrac{N \cdot m^2}{C}}$

---

37. A large plane sheet of charge has a charge per unit area of 9.0 $\mu C/m^2$. Find the electric field intensity *just above the surface* of the sheet, measured from the sheet's midpoint.

Solution

For a *large* insulating sheet, $E = 2\pi k\sigma$

so $\qquad E = (2\pi) \left(9 \times 10^9 \ \dfrac{N \cdot m^2}{C^2} \right) (9 \times 10^{-6} \ C/m^2) = \boxed{5.09 \times 10^5 \ N/C}$

E will be *perpendicular* to the sheet.

---

39. A long, straight metal rod has a radius of 5 cm and a charge per unit length of 30 nC/m. Find the electric field at the following distances from the axis of the rod: (a) 3 cm, (b) 10 cm, (c) 100 cm.

Solution

(a) Inside the conductor, $\boxed{E = 0}$; $\qquad$ outside, $\boxed{E = \dfrac{2k\lambda}{r}}$

(b) At $r = 0.10 \ m$

$$E = \dfrac{(2) \left(9 \times 10^9 \ \dfrac{N \cdot m^2}{C^2} \right) (30 \times 10^{-9} \ C/m)}{0.1 \ m} = \boxed{5.4 \times 10^3 \ N/C}$$

(c) At $r = 1.0 \ m$

$$E = \dfrac{(2) \left(9 \times 10^9 \ \dfrac{N \cdot m^2}{C^2} \right) (30 \times 10^{-9} \ C/m)}{1 \ m} = \boxed{540 \ N/C}$$

41. A conducting plate 50 cm on a side lies in the xy plane. If a total charge of 4×10^{-8} C is placed on the plate, find (a) the charge density on the plate, (b) the electric field just above the plate, and (c) the electric field just below the plate.

Solution

In this problem ignore "edge" effects and assume that the total charge distributes *uniformly* over each side of the plate (one-half of the total charge on each side).

(a) $\sigma = \dfrac{q}{A} = \left(\dfrac{1}{2}\right)\dfrac{(4 \times 10^{-8} \text{ C})}{(0.5 \text{ m})^2} = \boxed{8 \times 10^{-8} \text{ C/m}^2}$

(b) Just above the plate

$$E = \dfrac{\sigma}{2\varepsilon_o} = \dfrac{8 \times 10^{-8} \text{ C/m}^2}{(2)(8.85 \times 10^{-12} \text{ C}^2/\text{N·m}^2)} = \boxed{4.52 \times 10^3 \text{ N/C}}$$

(c) Just below the plate

$$E = \dfrac{\sigma}{2\varepsilon_o} = \boxed{4.52 \times 10^3 \text{ N/C}}$$

---

43. A solid copper sphere 15 cm in radius has a total charge of 40 nC. Find the electric field at the following distances measured from the center of the sphere: (a) 12 cm, (b) 17 cm, (c) 75 cm. (d) How would your answers change if the sphere were hollow?

Solution

(a) Inside the conductor, $E = 0$.   Outside the sphere, $E = \boxed{\dfrac{kQ}{r^2}}$

(b) At $r = 0.17$ m, $E = \boxed{1.25 \times 10^4 \text{ N/C}}$

(c) At $r = 0.75$ m, $E = \boxed{640 \text{ N/C}}$

(d) No change

---

47. A long, straight wire is surrounded by a hollow metallic cylinder whose axis coincides with that of the wire. The solid wire has a charge per unit length of $+\lambda$, and the hollow cylinder has a *net* charge per unit length of $+2\lambda$. From this information, use Gauss' law to find (a) the charge per unit length on the inner and outer surfaces of the hollow cylinder and (b) the electric field outside the hollow cylinder, a distance r from the axis.

Solution

(a) Use a cylindrical Gaussian surface S_1 within the conducting cylinder where $E = 0$

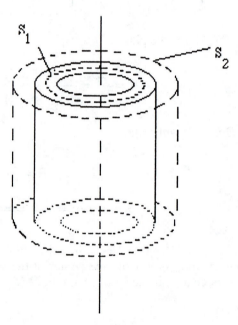

Thus $\oint E_n dA = \left(\dfrac{1}{\varepsilon_o}\right) q_{in} = 0$

and $\boxed{\lambda_{inner} = -\lambda}$

Also, $\lambda_{inner} + \lambda_{outer} = 2\lambda$

thus $\boxed{\lambda_{outer} = 3\lambda}$

(b) For a Gaussian surface S_2 outside the conducting cylinder

$\oint E_n dA = \left(\dfrac{1}{\varepsilon_o}\right) q_{in}$

$E(2\pi r L) = \dfrac{1}{\varepsilon_o} (\lambda - \lambda + 3\lambda)L$

$E = \boxed{\dfrac{3\lambda}{2\pi\varepsilon_o r}}$

59. Repeat the calculations for Problem 58 when both sheets have *positive* uniform charge densities of value σ.

Solution

For each sheet, the magnitude of the field at any point is

$$|\mathbf{E}| = \frac{\sigma}{2\varepsilon_0}$$

(a) At point to the left of the two parallel sheets

$$\mathbf{E} = E_1(-\mathbf{i}) + E_2(-\mathbf{i}) = 2E(-\mathbf{i})$$

$$\boxed{\mathbf{E} = -\frac{\sigma}{\varepsilon_0}\mathbf{i}}$$

(b) At point between the two sheets

$$\mathbf{E} = E_1\mathbf{i} + E_2(-\mathbf{i}) = 0$$

$$\boxed{\mathbf{E} = 0}$$

(c) At point to the right of the two parallel sheets

$$\mathbf{E} = E_1\mathbf{i} + E_2\mathbf{i} = 2E\mathbf{i}$$

$$\boxed{\mathbf{E} = \frac{\sigma}{\varepsilon_0}\mathbf{i}}$$

25

Electric Potential

OBJECTIVES

1. Understand that each point in the vicinity of a charge distribution can be characterized by a scalar quantity called the electric potential, V. The values of this potential function over the region (a scalar field) are related to the values of the electrostatic field over the region (a vector field).

2. Calculate the electric potential difference between any two points in a uniform *electric field*.

3. Calculate the electric potential difference between any two points in the vicinity of a *group of point charges*.

4. Calculate the electric *potential energy* associated with a group of point charges.

5. Calculate the electric potential due to *continuous charge distributions* of reasonable symmetry--such as a charged ring, sphere, line, or disk.

6. Obtain an expression for the electric field (a *vector* quantity) over a region of space if the scalar electric potential function for the region is known.

7. Calculate the work done by an external force in moving a charge q between any two points in an electric field when (a) an expression giving the field as a function of position is known, or when (b) the charge distribution (either point charges or a continuous distribution of charge) giving rise to the field is known.

SKILLS

The vector expressions giving the electric field **E** over a region can be obtained from the scalar function which describes the electric potential, V, over the region by using a vector differential operator called the gradient operator, ∇:

$$\mathbf{E} = -\nabla V$$

This is equivalent to

$$\mathbf{E} = -\mathbf{i}\,\frac{\partial V}{\partial x} - \mathbf{j}\,\frac{\partial V}{\partial y} - \mathbf{k}\,\frac{\partial V}{\partial z}$$

The derivatives in the above expression are called *partial derivatives*. This means that when the derivative is taken with respect to any one coordinate, any other coordinates which appear in the expression for the potential function are treated as constants.

Since the electrostatic force is a conservative force, the work done by the electrostatic force in moving a charge q from an initial point a to a final point b depends only on the location of the two points and is independent of the path taken between a and b. When calculating potential differences using the equation

$$V_b - V_a = -\int_a^b \mathbf{E} \cdot d\mathbf{s} \qquad (25.4)$$

any path between a and b may be chosen to evaluate the integral; therefore you should select a path for which the evaluation of the "line integral" in Eq. 25.4 will be as convenient as possible. For example; where a, b, and c are constants, if **E** is in the form

$$\mathbf{E} = ax\mathbf{i} + by\mathbf{j} + cz\mathbf{k}$$

then

$$\int_a^b \mathbf{E} \cdot d\mathbf{s} = \int_a^b (ax\,dx + by\,dy + cz\,dz)$$

and Eq. 25.4 becomes

$$V_b - V_c = -\left[a\int_{x_a}^{x_b} x\,dx + b\int_{y_a}^{y_b} y\,dy + c\int_{z_a}^{z_b} z\,dz \right]$$

$$= \frac{a}{2}(x_a{}^2 - x_b{}^2) + \frac{b}{2}(y_a{}^2 - y_b{}^2) + \frac{c}{2}(z_a{}^2 - z_b{}^2)$$

PROBLEM-SOLVING STRATEGY

1. When working problems involving electric potential, remember that potential is a *scalar quantity* (rather than a vector quantity like the electric field), so there are no components to worry about. Therefore, when using the superposition principle to evaluate the electric potential at a point due to a system of point charges, you simply take the algebraic sum of the potentials due to each charge. However, you must keep track of signs. The potential for each positive charge ($V = kq/r$) is positive, while the potential for each negative charge is negative.

2. Just as in mechanics, only *changes* in electric potential are significant, hence the point where you choose the potential to be zero is arbitrary. When dealing with point charges or a finite-sized charge distribution, we usually define $V = 0$ to be at a point infinitely far from the charges. However, if the charge distribution itself extends to infinity, some other nearby point must be selected as the reference point.

3. The electric potential at some point P due to a continuous distribution of charge can be evaluated by dividing the charge distribution into infinitesimal elements of charge dq located at a distance r from the point P. You then treat this element as a point charge, so that the potential at P due to the element is $dV = k\,dq/r$. The total potential at P is obtained by integrating dV over the entire charge distribution. In performing the integration for most problems, it is necessary to express dq and r in terms of a single variable. In order to simplify the integration, it is important to give careful consideration of the geometry involved in the problem.

4. Another method that can be used to obtain the potential due to a finite continuous charge distribution is to start with the definition of the potential difference given by Eq. 25.4. If $\mathbf{E}$ is known or can be obtained easily (say from Gauss' law), then the line integral of $\mathbf{E} \cdot d\mathbf{s}$ can be evaluated. An example of this method is given in Example 25.7.

5. Once you know the electric potential at a point, it is possible to obtain the electric field at that point by remembering that *the electric field is equal to the negative of the derivative of the potential with respect to some coordinate.*

NOTES FROM SELECTED CHAPTER SECTIONS_____

25.1 POTENTIAL DIFFERENCE AND ELECTRIC POTENTIAL

The potential difference between two points $V_B - V_A$ equals the work per unit charge that an *external agent* must perform in order to move a test charge, q_0, from point A to point B *without* a change in kinetic energy.

The potential at an arbitrary point is the work required per unit charge to bring a *positive test charge from infinity to that point*.

25.6 POTENTIAL OF A CHARGED CONDUCTOR

The surface of any charged conductor in equilibrium is an equipotential surface. Also (*since the electric field is zero inside the conductor*), the potential is constant everywhere inside the conductor and equal to its value at the surface.

EQUATIONS AND CONCEPTS_____

The *potential difference* between two points a and b in an electric field, $\Delta V = V_b - V_a$, can be found by integrating $\mathbf{E} \cdot d\mathbf{s}$ along *any path* from a to b.

$$V_b - V_a = -\int_a^b \mathbf{E} \cdot d\mathbf{s}$$

(25.4)

If the field is *uniform*, the potential difference depends only on the displacement d in the direction parallel to $\mathbf{E}$.

$$V_b - V_a = -Ed$$

The *change in potential energy*, ΔU, of a charge in moving from point a to point b in an electric field depends on the sign and magnitude of the charge as well as on the change in potential, ΔV.

$$\Delta U = q(V_b - V_a)$$

(25.8)

The *electric potential* at a point in the vicinity of several point charges is calculated in a manner which assumes that the potential is zero at infinity.

$$V = k \sum_{i=1}^{N} \frac{q_i}{r_i}$$

(25.13)

The *potential energy of a pair of charges* separated by a distance r represents the work required to assemble the charges from an infinite separation. Hence, the negative of the potential energy equals the minimum work required to separate them by an infinite distance. The electric potential energy associated with a system of two charged particles is positive if the two charges have the same sign, and negative if they are of opposite sign.

$$U = k \frac{q_1 q_2}{r_{12}}$$

(25.14)

If there are more than two charged particles in the system, the *total potential energy* is found by calculating U for each pair of charges and summing the terms algebraically.

$$U = \frac{k}{2} \sum_{i=1}^{N} \sum_{j=1}^{N} \frac{q_i q_j}{r_{ij}}$$

$$\text{when } i \neq j$$

If the scalar electric potential function throughout a region of space is known, then the vector electric field can be calculated from the potential function. The *components of the electric field* in rectangular coordinates are given in terms of partial derivatives of the potential.

$$E_x = -\frac{\partial V}{\partial x}$$

$$E_y = -\frac{\partial V}{\partial y}$$

$$E_z = -\frac{\partial V}{\partial z}$$

(25.25)

The vector expression for the electric field can be evaluated at any point P (x,y,z) within the region.

$$\mathbf{E} = -\mathbf{i}\frac{\partial V}{\partial x} - \mathbf{j}\frac{\partial V}{\partial y} - \mathbf{k}\frac{\partial V}{\partial z}$$

In other words, the electric field is equal to the negative gradient of the potential.

$$\mathbf{E} = -\nabla V$$

The *potential* (relative to zero at infinity) *for a continuous charge distribution* can be calculated by integrating the contribution due to a charge element dq over the line, surface, or volume which contains all the charge. Here, as in the case of calculating the field due to a continuous charge distribution, it is convenient to represent dq in terms of the appropriate charge density.

$$V = k \int \frac{dq}{r}$$

(25.17)

EXAMPLE PROBLEM SOLUTION_____

Example 25.1   Two charges of 2 μC and -6 μC are located at the positions (0, 0) m and (0, 3) m, respectively (Figure 25.1 on the following page).

(a) Find the total electric potential due to these charges at the point (4, 0) m.

Solution

For two charges, the sum in Equation 25.13 gives

$$V = k \left[\frac{q_1}{r_1} + \frac{q_2}{r_2} \right]$$

In this example, $q_1 = 2 \times 10^{-6}$ C, $r_1 = 4$ m, $q_2 = -6 \times 10^{-6}$ C, and $r_2 = 5$ m. Therefore, V at P reduces to

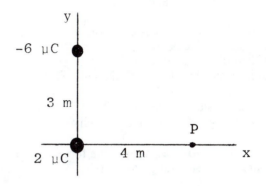

Figure 25.1

$$V_p = 9 \times 10^9 \text{ N}\frac{\text{m}^2}{\text{C}^2} \left[\frac{2 \times 10^{-6} \text{ C}}{4 \text{ m}} - \frac{6 \times 10^{-6} \text{ C}}{5 \text{ m}} \right]$$

$$V_p = \boxed{-6.3 \times 10^3 \text{ V}}$$

(b) How much work is required to bring a 3 μC charge from ∞ to the point P?

Solution

$$W = q_3 V_p = 3 \times 10^{-6} \text{ C } (-6.3 \times 10^3 \text{ V})$$

But 1 V = 1 J/C, so W becomes $W = \boxed{-18.9 \times 10\text{-}3 \text{ J}}$

The negative sign means that work is done by the charge for this displacement from ∞ to P. Therefore, positive work would have to be done to remove the charge from P back to ∞.

(c) What is the *potential energy* of the system of *three* charges?

Solution

For three charges, the total potential energy summed over all pairs is given by

$$U = U_{12} + U_{13} + U_{23}$$

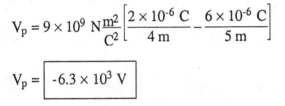

$$U = k \left[\frac{(2 \times 10^{-6})(-6 \times 10^{-6})}{3} + \frac{(2 \times 10^{-6})(3 \times 10^{-6})}{4} + \frac{(-6 \times 10^{-6})(3 \times 10^{-6})}{5} \right] = \boxed{-5.5 \times 10^{-2} \text{ J}}$$

ANSWERS TO SELECTED QUESTIONS

2. A negative charge moves in the direction of a uniform electric field. Does its potential energy increase or decrease? Does the electric potential increase or decrease?

Answer: Its potential energy increases, but the electric potential decreases.

3. If a proton is released from rest in a uniform electric field, does its electric potential increase or decrease? What about its potential energy?

Answer: The proton will be displaced in the direction of the electric field, so its potential and potential energy will *decrease*. The decrease in potential energy is accompanied by an equal increase in kinetic energy, as required by the law of conservation of energy. An electron would be displaced opposite to the electric field so its potential would increase, but its potential energy would decrease.

4. Give a physical explanation of the fact that the potential energy of a pair of like charges is positive whereas the potential energy of a pair of unlike charges is negative.

Answer: For two like charges, the force between them is repulsive, and it takes *positive* work to assemble them from an infinite separation to some separation r. The opposite is true for unlike charges.

5. A uniform electric field is parallel to the x axis. In what direction can a charge be displaced in this field without any external work being done on the charge?

Answer: No work is done when the charge is displaced in a direction perpendicular to the electric field.

7. Describe the equipotential surfaces for (a) an infinite line of charge and (b) a uniformly charged sphere.

Answer: The equipotential surfaces for an infinite line of charge are a family of cylindrical surfaces whose symmetry axes coincide with the line charge. (b) The equipotential surfaces for a uniformly charged sphere are a family of concentric spherical surfaces centered on the charged sphere.

9. If the electric potential at some point is zero, can you conclude that there are no charges in the vicinity of that point? Explain.

Answer: No. The potential at some point may be due to many charges whose individual potentials cancel at the point in question. For example, the potential at the midpoint of an electric dipole is zero.

10. If the potential is constant in a certain region, what is the electric field in that region?

Answer: If V is constant in a certain region, the electric field must be zero in that region, since the electric field is equal to the negative gradient of E. (In one dimension, E = –dV/dx, so if V = constant, E = 0.

11. The electric field inside a hollow, uniformly charged sphere is zero. Does this imply that the potential is zero inside the sphere? Explain.

Answer: No. When E is zero in some region, one can only conclude that the potential V is a constant in that region.

13. Two charged conducting spheres of different radii are connected by a conducting wire as in Figure 25.21. Which sphere has the greater charge density?

Answer: Both spheres are at the same potential, and the ratio of their charges is $q_1/q_2 = r_1/r_2$. However, the ratio of their charge densities is inversely proportional to the ratio of their radii. Therefore, the smaller sphere has the higher charge density.

17. Why is it important to avoid sharp edges, or points, on conductors used in high-voltage equipment?

Answer: The electric fields around such points is very intense and could lead to electrical discharge to ground.

19. Why is it relatively safe to stay in an automobile with a metal body during a severe thunderstorm?

Answer: The car serves as a shield since it is conducting. Even if a fallen power line were to hit the car, you would be advised to remain in the car until the power is turned off.

SOLUTIONS TO SELECTED END-OF-CHAPTER PROBLEMS

9. A positron, when accelerated from rest between two points at a fixed potential difference, acquires a speed of 30% of the speed of light. What speed will be achieved by a *proton* if accelerated from rest between the same two points?

Solution

$$\Delta K = qV \qquad \text{and} \qquad q_{e+} = q_H$$

$$\therefore \ \Delta K_{e+} = \Delta K_H \qquad \text{and} \qquad \left(\tfrac{1}{2} mv^2\right)_{e+} = \left(\tfrac{1}{2} mv^2\right)_H$$

Therefore,
$$v_H = v_{e+} \sqrt{\frac{m_{e+}}{m_H}} = (9 \times 10^7 \text{ m/s}) \sqrt{\frac{9.1 \times 10^{-31} \text{ kg}}{1.67 \times 10^{-27} \text{ kg}}}$$

and
$$v_H = \boxed{2.10 \times 10^6 \text{ m/s}}$$

17. A proton moves in a region of a uniform electric field. The proton experiences an increase in kinetic energy of 5×10^{-18} J after being displaced 2 cm in a direction parallel to the field. What is the magnitude of the electric field?

Solution

E is along the direction of Δs. Therefore $\Delta V = -E\Delta s$.

$$\Delta K = -\Delta U = -q\Delta V = -e(-E\Delta s)$$

$$E = \frac{\Delta K}{e\Delta s} = \frac{5 \times 10^{-18} \text{ J}}{(1.6 \times 10^{-19} \text{ C})(2 \times 10^{-2} \text{ m})} = \boxed{1.56 \times 10^3 \text{ N/C}}$$

19. For the situation described in Problem 18, calculate the change in electric potential while going from point A to point B along the direct red path AB. Which point is at the higher potential?

Solution

The field is uniform and along $-\mathbf{j}$.

$$V_{AB} = -\mathbf{E} \cdot (\mathbf{r}_A - \mathbf{r}_B)$$

$$\mathbf{r}_A - \mathbf{r}_B = (-0.2 - 0.5)\mathbf{i}\ m + (-0.3 - 0.5)\mathbf{j}\ m = (-0.6\mathbf{i} - 0.8\mathbf{j})\ m$$

so
$$V_A - V_B = -(-325\mathbf{j}\ V/m) \cdot (-0.6\mathbf{i} - 0.8\mathbf{j})\ m = \boxed{-260\ V}$$

Point B is at the higher potential.

---

27. The three charges shown in Figure 25.31 are at the vertices of an isosceles triangle. Calculate the electric potential at the *midpoint of the base*, taking $q = 7\ \mu C$.

Solution

Let $\qquad q_1 = q; \qquad q_2 = q_3 = -q$

where $q = 7 \times 10^{-6}\ C$

$$V_p = \frac{kq_1}{r_1} + \frac{kq_2}{r_2} + \frac{kq_3}{r_3}$$

$r_1 = r_2 = \sqrt{(0.04\ m)^2 - (0.01\ m)^2} = 3.87 \times 10^{-2}\ m$

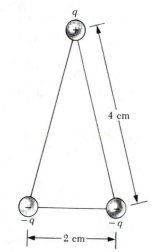

Figure 25.31

so

$$V_p = \left(9 \times 10^9\ \frac{N \cdot m^2}{C^2}\right)(7 \times 10^{-6}\ C)\left(\frac{1}{0.0387} - \frac{1}{0.01} - \frac{1}{0.01}\right)\ m^{-1}$$

$$V_p = \boxed{-10.9 \times 10^6\ V}$$

---

51. How many electrons should be removed from an initially uncharged spherical conductor of radius 0.3 m to produce a potential of 7.5 kV at the surface?

Solution

$$V = \frac{kQ}{R}; \qquad a = Ne \qquad \therefore N = \frac{VR}{ke} \quad \text{where } N = \text{number of electrons removed.}$$

$$N = \frac{(7.5 \times 10^3 \text{ V})(0.3 \text{ m})}{\left(9 \times 10^9 \frac{\text{N} \cdot \text{m}^2}{\text{C}^2}\right)(1.6 \times 10^{-19} \text{ C})} = \boxed{1.56 \times 10^{12} \text{ electrons}}$$

61. At a certain distance from a point charge, the field intensity is 500 V/m and the potential is −3000 V. (a) What is the distance to the charge? (b) What is the magnitude of the charge?

Solution

At a distance r from a point charge, $\quad V_r = \dfrac{kq}{r} \qquad$ and $\qquad E_r = \dfrac{kq}{r^2} \qquad \therefore E_r = \dfrac{rV_r}{r^2} = \dfrac{V_r}{r}$

(a) $\quad r = \dfrac{V}{E_r} = \dfrac{3000 \text{ V}}{500 \text{ N/C}} = 6 \dfrac{\text{N} \cdot \text{m/C}}{\text{N/C}} = \boxed{6 \text{ m}}$

(b) $\quad q = \dfrac{rV_r}{k} = \dfrac{(6 \text{ m})(-3000 \text{ V})}{9 \times 10^9 \text{ N} \cdot \text{m}^2/\text{C}^2}; \qquad q = \boxed{-2 \ \mu\text{C}}$

73. Calculate the work that must be done to charge a spherical shell of radius R to a total charge Q.

Solution

When the potential of the shell is V due to a charge q, the work required to add an additional increment of charge dq is

$$dW = Vdq \qquad \text{where} \qquad V = \frac{kq}{R}$$

$$dW = \left(\frac{kq}{R}\right)dq \qquad \text{and} \qquad W = \frac{k}{R}\int_o^Q qdq$$

Therefore $\qquad W = \boxed{\left(\frac{k}{R}\right)\left(\frac{Q^2}{2}\right)}$

79. It is shown in Example 25.6 that the potential at a point P a distance d above one end of a uniformly charged rod of length L lying along the x axis is given by

$$V = \frac{kQ}{L} \text{Ln}\left(\frac{L + \sqrt{L^2 + d^2}}{d}\right)$$

Use this result to derive an expression for the y component of the electric field at the point P. (Hint: Replace d with y.)

Solution

$$E_y = -\frac{\partial V}{\partial y} = -\frac{kQ}{L}\frac{d}{dy}\left[\text{Ln}\left(L + \sqrt{L^2 + y^2}\right) - \text{Ln } y\right]$$

$$E_y = \frac{-kQ}{L}\left[\frac{2y/(2\sqrt{L^2 + y^2})}{L + \sqrt{L^2 + y^2}} - \frac{1}{y}\right] = \frac{kQ}{Ly}\left[1 - \frac{y^2}{L^2 + y^2 + L\sqrt{L^2 + y^2}}\right]$$

---

81. A dipole is located along the y axis as in Figure 25.43. (a) At a point P, which is far from the dipole ($r \gg a$), the electric potential is given by

$$V = k\frac{p \cos \theta}{r^2}$$

where $p = 2qa$. Calculate the radial component of the associated electric field, E_r, and the azimuthal component, E_θ. Note that

$$E_\theta = \frac{1}{r}\left(\frac{\partial V}{\partial \theta}\right).$$

Do these results seem reasonable for $\theta = 90°$ and $0°$? For $r = 0$? (b) For the dipole arrangement shown, express V in terms of rectangular coordinates using $r = (x^2 + y^2)^{1/2}$ and

$$\cos \theta = \frac{y}{(x^2 + y^2)^{1/2}}$$

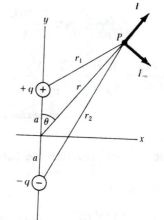

Figure 25.43

Using these results and taking $r \gg a$, calculate the field components E_x and E_y.

Solution

(a) $E_r = -\dfrac{\partial V}{\partial r} = -\dfrac{\partial}{\partial r}\left(\dfrac{kp \cos \theta}{r^2}\right) = \boxed{\dfrac{2kp \cos \theta}{r^3} = E_r}$

In spherical coordinates

$$E_\theta = \frac{1}{r}\left(\frac{\partial V}{\partial \theta}\right) = \frac{1}{r}\frac{\partial}{\partial \theta}\left(\frac{kp\cos\theta}{r^2}\right) = \boxed{\frac{kp\sin\theta}{r^3} = E_\theta}$$

(b) For $r = \sqrt{x^2 + y^2}$, $\qquad \cos\theta = \dfrac{y}{\sqrt{x^2 + y^2}} \qquad$ and $\qquad V = \dfrac{kpy}{(x^2 + y^2)^{3/2}}$

$$E_x = -\frac{\partial V}{\partial x} = -\frac{\partial}{\partial x}\left(\frac{kpy}{(x^2 + y^2)^{3/2}}\right) = \boxed{\frac{3kpxy}{(x^2 + y^2)^{5/2}} = E_x}$$

$$E_y = -\frac{\partial V}{\partial y} = -\frac{\partial}{\partial y}\left(\frac{kpy}{(x^2 + y^2)^{3/2}}\right) = \boxed{\frac{kp(2y^2 - x^2)}{(x^2 + y^2)^{5/2}}}$$

26

Capacitance and Dielectrics

OBJECTIVES

1. Use the basic definition of capacitance and the equation for finding the potential difference between two points in an electric field in order to calculate the capacitance of a capacitor for cases of relatively simple geometry-- parallel plates, cylindrical, spherical.

2. Determine the equivalent capacitance of a network of capacitors in series-parallel combination and calculate the final charge on each capacitor and the potential difference across each when a known potential is applied across the combination.

3. Make calculations involving the relationships among potential, charge, capacitance, stored energy, and energy density for capacitors, and apply these results to the particular case of a parallel plate capacitor.

4. Calculate the capacitance, potential difference, and stored energy of a capacitor which is partially or completely filled with a *dielectric*.

SKILLS

When analyzing a series-parallel combination of capacitors to determine the equivalent capacitance, you should make a sequence of circuit diagrams which show the successive steps in the simplification of the circuit; combine at each step those capacitors which are in simple parallel or simple series relationship to each other using Equation 26.8 or 26.11.

To calculate the charge on each capacitor when a known potential difference is applied across the series-parallel combination, use the sequence of circuit diagrams to find the equivalent capacitance. Remember that the charge on *each* capacitor in a group of capacitors connected in series is the same as the charge on the equivalent capacitor for that series combination. Also, the potential difference across *each* capacitor in a group of capacitors connected in parallel is the same as the potential difference across the equivalent capacitor for that parallel combination.

At each step, you know two of the three quantities Q, V, and C. You will be able to determine the remaining quantity using the relation $Q = CV$.

NOTES FROM SELECTED CHAPTER SECTIONS

26.1 DEFINITION OF CAPACITANCE

The capacitance of a capacitor depends on the physical characteristics of the device (size, shape, and separation of plates and the nature of the dielectric medium filling the space between the plates). Since the potential difference between the plates is proportional to the quantity of charge on each plate, the value of the capacitance is independent of the charge on the capacitor.

26.3 COMBINATIONS OF CAPACITORS
26.5 CAPACITORS WITH DIELECTRICS

When two or more unequal capacitors are connected in *series*, they carry the same charge, but the potential differences are not the same. Their capacitances add as reciprocals, and the equivalent capacitance of the combination is always *less* than the smallest individual capacitor.

When two or more capacitors are connected in *parallel*, the potential difference across each is the same. The charge on each capacitor is proportional to its capacitance, hence the capacitances add directly to give the equivalent capacitance of the parallel combination.

The effect of a dielectric on a capacitor is to *increase* its capacitance by a factor κ (the dielectric constant) over its empty capacitance. The reason for this is that induced surface charges on the dielectric reduces the electric field inside the material from E to E/κ.

EQUATIONS AND CONCEPTS

The *capacitance* of a capacitor is defined as the ratio of the charge on either conductor (or plate) to the magnitude of the potential difference between the conductors.

$$C = \frac{Q}{V}$$

(26.1)

The *capacitance of an air-filled parallel plate capacitor* is proportional to the area of the plates and inversely proportional to the separation of the plates.

$$C = \varepsilon_o \frac{A}{d}$$

(26.3)

When the region between the plates is completely filled by a material of dielectric constant κ, the capacitance increases by the factor κ.

$$C = \kappa \varepsilon_o \frac{A}{d}$$

(26.16)

The *equivalent capacitance* of a *parallel combination* of capacitors is larger than any individual capacitor in the group.

$$C_{eq} = C_1 + C_2 + C_3 + \ldots$$

(26.8)

The *equivalent capacitance* of a *series combination* of capacitors is smaller than the smallest capacitor in the group.

$$\frac{1}{C_{eq}} = \frac{1}{C_1} + \frac{1}{C_2} + \frac{1}{C_3} + \ldots$$

(26.11)

In the special case of only *two capacitors in series*, the equivalent capacitance is equal to the ratio of the product to the sum of their capacitance.

$$C_{eq} = \frac{C_1 C_2}{C_1 + C_2}$$

The *electrostatic energy* stored in the electrostatic field of a charged capacitor equals the work done (by a battery or other source) in charging the capacitor from q = 0 to q = Q.

$$U = \frac{Q^2}{2C} = \frac{1}{2} QV = \frac{1}{2} CV^2$$

(26.12)

The *energy density* at any point in the electrostatic field of a charged capacitor is proportional to the square of the electric field intensity at that point.

$$u = \frac{1}{2} \varepsilon_o E^2$$

(26.14)

Two equal and opposite charges of magnitude q separated by a distance 2a constitute an electric dipole. This configuration is characterized by an *electric dipole moment*, **p**. The direction of the vector **p** is from the negative to the positive charge.

$$|\mathbf{p}| = 2aq \qquad (26.17)$$

An external uniform electric field will exert a *net torque* on an electric dipole when the dipole moment makes an angle θ with the direction of the field.

$$\tau = \mathbf{p} \times \mathbf{E} \qquad (26.19)$$

There is *potential energy* associated with the dipole-electric field system.

$$U = -\mathbf{p} \cdot \mathbf{E} \qquad (26.21)$$

EXAMPLE PROBLEM SOLUTION

Example 26.1

(a) Find the equivalent capacitance between the points a and b for the group of capacitors shown in the figure below. $C_1 = 1\ \mu F$, $C_2 = 2\ \mu F$, $C_3 = 3\ \mu F$, $C_4 = 4\ \mu F$, $C_5 = 5\ \mu F$, and $C_6 = 6\ \mu F$.

(i)

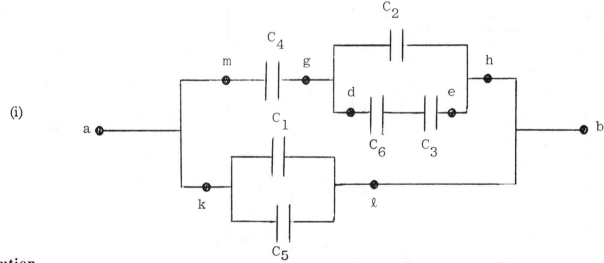

Solution

You should show sketches representing the successive steps in the simplification of the circuit. [In this example the original circuit will be labeled step (i).] Note that seven points in the circuit have been identified (d, e, g, h, k, L, m) so that they can be conveniently referred to at various steps in the solution. First, capacitors C_3 and C_6 are combined using the rule for capacitors in series so that

$$\frac{1}{C_{de}} = \frac{1}{6\ \mu F} + \frac{1}{3\ \mu F} \ ; \quad C_{de} = 2\ \mu F$$

Next, C_1 and C_5 can be combined using the rule for capacitors in parallel, so that

$$C_{kL} = 1\ \mu F + 5\ \mu F; \quad C_{kL} = 6\ \mu F$$

The reductions to this point are shown in step (ii). The values of C are given in μF.

(ii)

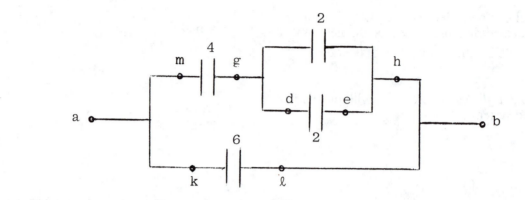

The circuit is further reduced to a single equivalent capacitor as shown in steps (iii), (iv), and (v).

(iii)

(iv)

(v)   Therefore $C_{eq} = 8\ \mu F$

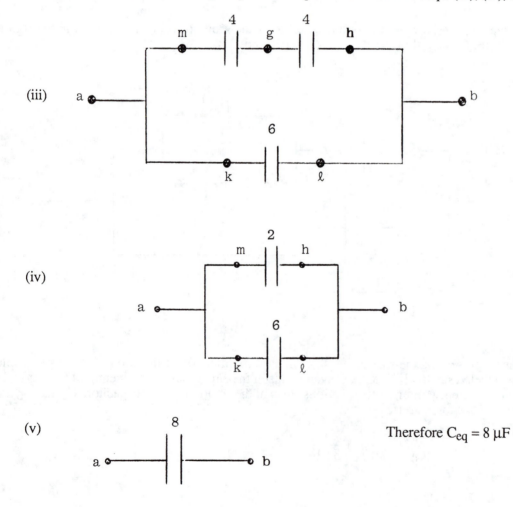

(b) Determine the potential difference across each capacitor and the charge accumulated on each capacitor if the total charge on the group of six capacitors is 384 μC.

Solution

First, consider the equivalent capacitor shown in step (v) to find the potential difference between points a and b.

$$V_{ab} = \frac{Q_{ab}}{C_{ab}} = \frac{384 \ \mu C}{8 \ \mu F} = 48 \ V$$

Next, notice that in step (i) the original circuit $V_{kL} = V_{ab}$ and since the two capacitors between k and L are in parallel, the potential difference across each is 48 V. Therefore for C_1,

$$V_1 = 48 \ V \quad \text{and} \quad Q_1 = C_1 V_1 = 48 \ \mu C$$

and for C_5

$$V_5 = 48 \ V \quad \text{and} \quad Q_5 = C_5 V_5 = 240 \ \mu C$$

Now return to step (iv) and notice that $V_{mh} = V_{ab} = 48$ V, and

$$Q_{mh} = C_{mh} \ V_{mh} = (2 \ \mu F) \ (48 \ V) = 96 \ \mu C$$

Since the two capacitors shown in step (iii) between points m and h are in series, each will have the same charge as that of the equivalent capacitor, or

$$Q_{mg} = Q_{gh} = Q_{mh} = 96 \ \mu C$$

$$V_{mg} = \frac{Q_{mg}}{C_{mg}} = \frac{96 \ \mu C}{4 \ \mu F} = 24 \ V$$

$$V_{gh} = \frac{Q_{ih}}{C_{ih}} = 24 \ V$$

Therefore for C_4, $V_4 = 24$ V and $Q_4 = 96 \ \mu C$.

In step (ii) the two capacitors shown between points g and h are in parallel so the potential difference across each is 24 V. This is also the potential difference across C_2, so that for C_2, $V_2 = 24$ V and $Q_2 = C_2 V_2 = 48 \ \mu C$. Also in step (ii), the potential difference

$$V_{de} = V_{gh} = 24 \ V$$

and

$$Q_{de} = C_{de} V_{de} = (2 \ \mu F) \ (24 \ V) = 48 \ \mu C$$

The two capacitors shown in step (i) between points d and a are in series, and therefore the charge on each is equal to Q_{de}. Therefore for C_6, $Q_6 = 48 \ \mu C$, and

$$V_6 = \frac{Q_6}{C_6} = 8 \ V$$

For C_3, $\qquad\qquad Q_3 = 48 \ \mu C \quad \text{and} \quad V_3 = \frac{Q_3}{C_3} = 16 \ V$

The results of the calculations can be summarized as follows:

Capacitor	Potential Difference	Charge
C_1	48 V	48 μC
C_2	24 V	48 μC
C_3	16 V	48 μC
C_4	24 V	96 μC
C_5	48 V	240 μC
C_6	8 V	48 μC
C_{eq} (8 μF)	48 V	384 μC

ANSWERS TO SELECTED QUESTIONS

1. What happens to the charge on a capacitor if the potential difference between the conductors is doubled?

Answer: Since $Q = CV$, doubling the potential difference will double the charge.

4. A pair of capacitors are connected in parallel while an identical pair are connected in series. Which pair would be more dangerous to handle after being connected to the same voltage source? Explain.

Answer: The pair connected in parallel would be more dangerous since it would have a larger equivalent capacitance, and hence would store more charge and energy.

11. If the potential difference across a capacitor is doubled, by what factor does the energy stored change?

Answer: Since $U = CV^2/2$, doubling V will quadruple the stored energy.

12. Why is it dangerous to touch the terminals of a high-voltage capacitor even after the applied voltage has been turned off? What can be done to make the capacitor safe to handle after the voltage source has been removed?

Answer: The capacitor very often remains charged long after the voltage is removed. This residual charge can be lethal. The capacitor can be safely handled after discharging the plates by short circuiting the device with a conductor, such as a screwdriver with an insulating handle.

22. If you were asked to design a capacitor where small size and large capacitance were required, what factors would be important in your design?

Answer: You should use a dielectric filled capacitor whose dielectric constant is very large. Furthermore, you should make the dielectric as thin as possible, keeping in mind that dielectric breakdown must also be considered.

SOLUTIONS TO SELECTED END-OF-CHAPTER PROBLEMS

7. An isolated charged conducting sphere of radius 12 cm creates an electric field of 4.9×10^4 N/C at a distance of 21 cm from its center. (a) What is its surface charge density? (b) What is its capacitance?

Solution (a) The electric field outside a spherical charge distribution of radius R is

$$E = \frac{kq}{r^2} \qquad \text{or} \qquad q = \frac{Er^2}{k}$$

The surface charge density $\sigma = \frac{q}{A}$, therefore

$$\sigma = \frac{Er^2}{k4\pi R^2} = \frac{\left(4.9 \times 10^4 \text{ N/C}\right)(0.21 \text{ m})^2}{\left(9 \times 10^9 \frac{\text{N} \cdot \text{m}^2}{\text{C}^2}\right)(4\pi)(0.12 \text{ m})^2} = \boxed{1.33 \ \mu\text{C/m}^2}$$

(b) For an isolated charged sphere of radius R,

$$C = 4\pi\varepsilon_o R = (4\pi)\left(8.85 \times 10^{-12} \frac{\text{C}^2}{\text{N} \cdot \text{m}^2}\right)(0.12 \text{ m}) = \boxed{13.3 \text{ pF}}$$

15. An air-filled capacitor consists of two parallel plates, each with an area of 7.6 cm^2, separated by a distance of 1.8 mm. If a 20-V potential difference is applied to these plates, calculate (a) the electric field between the plates, (b) the surface charge density, (c) the capacitance, and (d) the charge on each plate.

Solution (a) The potential difference between two points in a uniform electric field is V = Ed, so

$$E = \frac{V}{d} = \frac{20 \text{ V}}{1.8 \times 10^{-3} \text{ m}} = \boxed{1.11 \times 10^4 \text{ V/m}}$$

(b) The electric field near a charged conductor is $E = \frac{\sigma}{\varepsilon_o}$, therefore

$$\sigma = \varepsilon_o E = \left(8.85 \times 10^{-12} \frac{\text{C}^2}{\text{N} \cdot \text{m}^2}\right)\left(1.11 \times 10^4 \text{ V/m}\right) = \boxed{9.83 \times 10^{-8} \text{ C/m}^2 = 98.3 \text{ nC/m}^2}$$

(c) For a parallel plate capacitor,

$$C = \frac{\varepsilon_o A}{d} = \frac{\left(8.85 \times 10^{-12} \frac{\text{C}^2}{\text{N} \cdot \text{m}^2}\right)(7.6 \times 10^{-4} \text{ m}^2)}{1.8 \times 10^{-3} \text{ m}} = \boxed{3.74 \times 10^{-12} \text{ F} = 3.74 \text{ pF}}$$

(d) $Q = CV = (3.74 \times 10^{-12} \text{ F})(20 \text{ V}) = \boxed{7.48 \times 10^{-11} \text{ C} = 74.8 \text{ pC}}$

23. An air-filled spherical capacitor is constructed with inner and outer shell radii of 7 and 14 cm, respectively. (a) Calculate the capacitance of the device. (b) What potential difference between the spheres will result in a charge of 4 μC on each conductor?

Solution

(a) For a spherical capacitor $C = \dfrac{ab}{k(b-a)}$

$$C = \frac{(0.07 \text{ m})(0.14 \text{ m})}{\left(9 \times 10^9 \dfrac{\text{N·m}^2}{\text{C}^2}\right)(0.14 - 0.07) \text{ m}} = \boxed{1.56 \times 10^{-11} \text{ F} = 15.6 \text{ pF}}$$

(b) $V = \dfrac{Q}{C} = \dfrac{(4 \times 10^{-6} \text{ C})}{1.56 \times 10^{-11} \text{ F}} = \boxed{2.56 \times 10^5 \text{ V} = 256 \text{ kV}}$

---

33. Consider the circuit shown in Figure 26.28, where $C_1 = 6 \text{ μF}$, $C_2 = 3 \text{ μF}$, and $V = 20 \text{ V}$. C_1 is first charged by the closing of switch S_1. Switch S_1 is then opened, and the charged capacitor is connected to the uncharged capacitor by the closing of S_2. Calculate the initial charge acquired by C_1 and the final charge on each of the two capacitors.

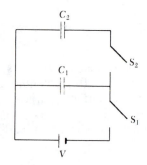

Solution

When S_1 is closed, the charge on C_1 will be

$$Q_1 = C_1 V_1 = (6 \text{ μF})(20 \text{ V}) = \boxed{120 \text{ μC}}$$

Figure 26.28

When S_1 is opened and S_2 is closed, the total charge will remain constant and be shared by the two capacitors:

$$Q_1' = 120 \text{ μC} - Q_2'$$

The potential across the two capacitors will be equal.

$$V' = \frac{Q_1'}{C_1} = \frac{Q_2'}{C_2}$$

or

$$\frac{120 \text{ μC} - Q_2'}{6 \text{ μF}} = \frac{Q_2'}{3 \text{ μF}}$$

and

$$Q_2' = \boxed{40 \ \mu C}$$

$$Q_1' = 120 \ \mu C - 40 \ \mu C = \boxed{80 \ \mu C}$$

45. Calculate the energy stored in an 18-μF capacitor when it is charged to a potential of 100 V.

Solution

The energy stored in a charged capacitor is

$$U = \frac{1}{2} CV^2 = \left(\frac{1}{2}\right)\!\left(18 \times 10^{-6} \ F\right)(100 \ V)^2 = \boxed{0.0900 \ J = 90.0 \ mJ}$$

57. A parallel-plate capacitor has a plate area of 0.64 cm². When the plates are in a vacuum, the capacitance of the device is 4.9 pF. (a) Calculate the value of the capacitance if the space between the plates is filled with nylon. (b) What is the maximum potential difference that can be applied to the plates without causing dielectric breakdown, or discharge?

Solution

For a capacitor filled with a dielectric, $\kappa = 3.4$ (nylon):

$$C = \kappa C_o = (3.4)(4.9 \ pF) = \boxed{16.7 \ pF}$$

$$C_o = \frac{\varepsilon_o A}{d}$$

so

$$d = \frac{\varepsilon_o A}{C_o} = \frac{\left(8.85 \times 10^{-12} \ \dfrac{C^2}{N \cdot m^2}\right)(6.4 \times 10^{-5} \ m^2)}{4.9 \times 10^{-12} \ F} = \boxed{1.16 \times 10^{-4} \ m}$$

and

$$V_{max} = E_{max} d$$

For nylon, $E_{max} = 14 \times 10^6$ V/m, therefore

$$V_{max} = (14 \times 10^6 \ V/m)(1.16 \times 10^{-4} \ m) = \boxed{1.62 \times 10^3 \ V = 1.62 \ kV}$$

73. Three capacitors of 8 μF, 10 μF, and 14 μF are connected to the terminals of a 12-volt battery. How much energy does the battery supply if the capacitors are connected (a) in series and (b) in parallel?

Solution

(a) For a series combination of capacitors,

$$\frac{1}{C} = \frac{1}{C_1} + \frac{1}{C_2} + \frac{1}{C_3} = \frac{1}{8 \ \mu F} + \frac{1}{10 \ \mu F} + \frac{1}{14 \ \mu F}$$

This gives

$$C = 3.37 \ \mu F$$

and

$$U = \frac{1}{2} CV^2 = \left(\frac{1}{2}\right)(3.37 \times 10^{-6} \ F)(12 \ V)^2 = \boxed{2.43 \times 10^{-4} \ J = 243 \ \mu J}$$

(b) For a parallel combination of capacitors,

$$C = C_1 + C_2 + C_3 = 8 \ \mu F + 10 \ \mu F + 14 \ \mu F = 32 \ \mu F$$

and

$$U = \frac{1}{2} CV^2 = \frac{1}{2} (32 \times 40^{-6} \ F)(12 \ V)^2 = \boxed{2.30 \times 10^{-3} \ J = 2.30 \ mJ}$$

Current and Resistance

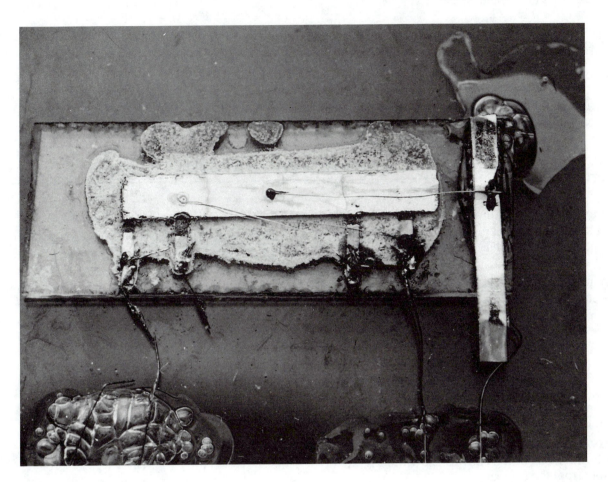

OBJECTIVES

1. Calculate the current density, electron drift velocity, and quantity of charge passing a point in a given time interval in a specified current-carrying conductor.

2. Determine the resistance of a conductor using Ohm's law. Also, calculate the resistance based on the physical characteristics of a conductor.

3. Make calculations of the variation of resistance with temperature which involves the concept of the temperature coefficient of resistivity.

4. Use Joule's law to calculate the power dissipated in a resistor.

5. Describe the classical model of electrical conduction in metals, and relate the resistivity to the mean time between collisions.

SKILLS

Equation 27.10, $R = \rho\left(\frac{L}{A}\right)$, can be used directly to calculate the resistance of a conductor of uniform cross-sectional area and constant resistivity. For those cases in which the area, resistivity, or both vary along the length of the conductor, the resistance must be determined as an integral of dR.

The conductor is subdivided into elements of length dx over which ρ and A may be considered constant in value and the total resistance is

$$R = \frac{\rho}{A} \int dx$$

Consider, for example, the case of a truncated cone of constant resistivity, radii a and b and height h. The conductor should be subdivided into disks of thickness dx, radius r, area $= \pi r^2$ and oriented parallel to the faces of the cone as shown in Figure 27.1. Note from the geometry that

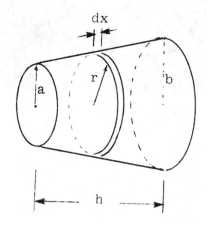

$$x = \left(\frac{r - a}{b - a}\right)h \quad \text{so that} \quad r = \frac{x}{h}(b - a) + a$$

and

$$R = \int_o^h \frac{\rho}{\pi r^2}\, dx$$

The remainder of this calculation is left as a problem for you to work out (see Problem 27.70 of the text).

Figure 27.1

CHAPTER 27

NOTES FROM SELECTED CHAPTER SECTIONS

27.2 ELECTRIC CURRENT

The direction of conventional current is designated as the direction of motion of positive charge. In an ordinary metal conductor, the direction of current will be *opposite* the *direction of flow of electrons* (which are the charge carriers in this case).

27.3 RESISTANCE AND OHM'S LAW

For ohmic materials, the ratio of the ratio of the current density and the electric field which gives rise to the current is equal to a constant, σ, which is the conductivity of the material. The reciprocal of the conductivity is called the resistivity, ρ. Each ohmic material has a characteristic resistivity which depends only on the properties of the specific material and is a function of temperature.

27.6 A MODEL FOR ELECTRICAL CONDUCTION

In the classical model of electronic conduction in a metal, electrons are treated like molecules in a gas and, in the absence of an electric field, have a *zero average velocity*.

Under the influence of an electric field, the electrons move along a direction opposite the direction of the applied field with a *drift velocity* which is proportional to the average time between collisions with atoms of the metal and inversely proportional to the number of free electrons per unit volume.

EQUATIONS AND CONCEPTS

Under the action of an electric field, electric charges will move through gases, liquids, and solid conductors. *Electric current*, I, is defined as the rate at which charge moves through a cross section of the conductor.

$$I = \frac{dQ}{dt} \tag{27.2}$$

The direction of the current is in the direction of the flow of positive charges. The SI unit of current is the *ampere* (A).

$$1 \text{ A} = 1 \text{ C/S} \tag{27.3}$$

The current in a conductor can be related to the number of mobile charge carriers per unit volume, n; the quantity of charge associated with each carrier, q; and the *drift velocity*, v_d, of the carriers.

$$I = nqv_dA \tag{27.4}$$

The *current density*, J, in a conductor is a vector quantity which is proportional to the electric field in the conductor. Equation 27.7 is a formal statement of Ohm's law. The constant, σ, called the *conductivity*, is a characteristic of a particular material. The magnitude of the *current density* is the ratio of current per unit cross-sectional area.

$$J = \frac{I}{A}$$
$$\text{or} \tag{27.5}$$
$$J = nqv_d$$

$$J = \sigma E \tag{27.7}$$

For many practical applications, a more useful form of Ohm's law relates the potential difference across a conductor and the current in the conductor to a composite of several physical characteristics of the conductor called the *resistance*, R.

$$R = \frac{V}{I}$$ (27.8)

The *resistance* of a given conductor of uniform cross section depends on the length, cross-sectional area, and a characteristic property of the material of which the conductor is made.   The parameter, ρ, is the *resistivity* of the material of which the conductor is made.   The resistivity is the inverse of the *conductivity* and has units of ohm-meters.   The unit of resistance is the ohm (Ω).

$$R = \rho \left(\frac{L}{A} \right)$$ (27.10)

The *resistivity* and therefore the resistance of a conductor varies with temperature in an approximately linear manner.   In these expressions, α is the *temperature coefficient of resistance* and T_0 is a stated reference temperature (usually 20°C).

$$\rho = \rho_o \left[1 + \alpha(T - T_o) \right]$$ (27.11)

$$R = R_o \left[1 + \alpha(T - T_o) \right]$$ (27.13)

Power will be supplied to a resistor or other current-carrying devices when a potential difference is maintained between the terminals of the circuit element.   Eq. 27.14 is often called Joule's law and the SI unit of power is the watt (W).   When the device obeys Ohm's law, the power dissipated can be expressed in alternative forms given by Eq. 27.15.

$$P = VI$$ (27.14)

$$P = I^2 R = \frac{V^2}{R}$$ (27.15)

The average time between collisions of atoms of a metal is an important parameter in the description of the classical model of electronic conduction in metals.   This characteristic time is denoted by τ and can be related to the drift velocity (see Eq. 27.4) or the resistivity (see Eq. 27.10) associated with the conductor.   In these equations, m and q represent the mass and charge of the electron, E is the magnitude of the applied electric field, and n is the number of free electrons per unit volume.

$$v_d = \frac{qE\tau}{m}$$ (27.16)

$$\rho = \frac{m}{nq^2\tau}$$ (27.19)

EXAMPLE PROBLEM SOLUTION

Example 27.1   In a certain accelerator, electrons emerge with energies of 40 MeV (1 MeV = 1.6×10^{-13} J). The electrons do not emerge in a "steady stream", but in pulses which are repeated 250 times per second.   Each pulse lasts for 200 ns (2×10^{-7} s) and the electrons in the pulse constitute a current of 250 mA.   The current is zero between pulses.

(a) How many electrons are delivered by the accelerator per pulse?

Solution

The manner in which the current varies with time can be represented in the figure below in which the time axis (horizontal axis) is not drawn to scale.

Figure 27.2

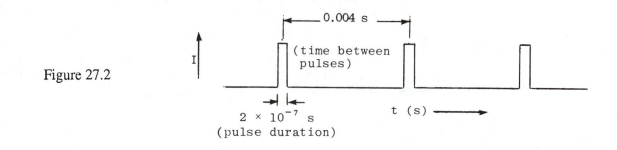

Eq. 27.2 can be written as dQ = I dt. During the time that the pulse is "on", the current is *constant*, therefore

$$Q = I \int dt \quad \text{or} \quad Q_{pulse} = It$$

$$Q_{pulse} = (250 \times 10^{-3} \text{ A}) (2 \times 10^{-7} \text{ s}) = 5 \times 10^{-8} \text{ C}$$

This quantity of charge per pulse divided by the electronic charge represents the number of electrons per pulse:

$$\frac{\text{\# electrons}}{\text{per pulse}} = \frac{5 \times 10^{-8} \text{ C/pulse}}{1.6 \times 10^{-19} \text{ C/electron}} = 3.13 \times 10^{11} \frac{\text{electrons}}{\text{pulse}}$$

(b) What is the *average* current delivered by the accelerator?

Solution

The average current is given by Eq. 27.1:

$$I = \frac{\Delta Q}{\Delta t}$$

where Δt extends over a full cycle $\left(\frac{2}{250}\right)$ s. Since there is only one current pulse each cycle, and using the value of Q found in part (a), we get

$$I_{av} = \frac{5 \times 10^{-8} \text{ C}}{0.004 \text{ s}} = 12.5 \text{ μA}$$

This is only 0.005% of the "peak" current!

(c) What maximum power is delivered in the electron beam?

Solution

To find the *maximum* power, it is necessary to calculate energy per unit time during the pulse period.

$$P = \frac{E}{t} = \frac{(3.13 \times 10^{11} \text{ electrons/pulse}) (40 \text{ MeV/electron}) (1.6 \times 10^{-13} \text{ J/MeV})}{2 \times 10^{-7} \text{ s/pulse}}$$

$$P = 10^7 \text{ W} = 10 \text{ MW}$$

ANSWERS TO SELECTED QUESTIONS

3. What factors affect the resistance of a conductor?

Answer: Temperature, sample dimensions, resistivity, and impurities are common factors.

4. What is the difference between resistance and resistivity?

Answer: Resistance of a conductor depends on the geometry of the conductor as well as its electronic properties. Resistivity of the conductor depends only on the electronic structure of the material.

5. We have seen that an electric field must exist inside a conductor that carries a current. How is this possible in view of the fact that in *electrostatics* we concluded that E must be zero inside a conductor?

Answer: In the electrostatic case, the internal field must be zero since a nonzero field would produce a current (by interacting with the "free" electrons in the conductor) which would violate the condition of static equilibrium. On the other hand, a conductor carries a current as the result of a potential difference between its ends, which produces an internal electric field.

6. Two wires A and B of circular cross section are made of the same metal and have equal lengths, but the resistance of wire A is three times greater than that of wire B. What is the ratio of their cross-sectional areas? How do their radii compare?

Answer: Since $R = \rho L / A$, the ratio of resistances is given by $R_1/R_2 = A_2/A_1$. Hence, the ratio of their areas is three. That is, the area of wire B is three times that of A. The radius of wire B is $\sqrt{3}$ times the radius of wire A.

9. When the voltage across a certain conductor is doubled, the current is observed to increase by a factor of 3. What can you conclude about the conductor?

Answer: The conductor does not obey Ohm's law since the current is not proportional to the applied voltage. This occurs, for example, in a device known as a diode.

12. Use the atomic theory of matter to explain why the resistance of a material should increase as its temperature increases.

Answer: As the temperature increases, the amplitude of atomic vibrations increase. This makes it more difficult for charges to move inside a conductor.

16. What would happen to the drift velocity of the electrons in a wire and to the current in the wire if the electrons could move freely without resistance through the wire?

Answer: Without resistance there would always be a net force on the electron, and it would continually accelerate. Thus, its drift velocity would continually increase, producing a continuously increasing current.

17. If charges flow very slowly through a metal, why does it not require several hours for a light to come on when you throw a switch?

Answer: Individual electrons move with a small average velocity through the conductor, but as soon as the voltage is applied, electrons all along the conductor start to move. Actually, the current does not flow "immediately" but is limited by the speed of light.

19. Two conductors of the same length and radius are connected across the same potential difference. One conductor has twice the resistance of the other. Which conductor will dissipate more power?

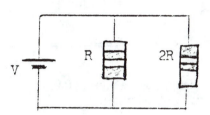

Answer: The voltage across each conductor is the same. Since the power dissipated is given by $P = V^2/R$, the conductor with the lower resistance will dissipate more power.

20. When incandescent lamps burn out, they usually do so just after they are switched on. Why?

Answer: The filaments are cold when the lamp is first turned on, hence they have a lower resistance and draw more current than when they are hot. The increased current can overheat the filament and destroy it.

22. Two light bulbs both operate from 110 V, but one has a power rating of 25 W and the other of 100 W. Which bulb has the higher resistance? Which bulb carries the greater current?

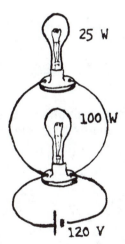

Answer: Since $P = V^2/R$, and V is the same for each bulb, the 25 W bulb would have the higher resistance. Since $P = IV$, we see that the 100 W bulb carries the greater current.

23. A typical monthly utility rate structure might go something like this: $1.60 for the first 16 kWh, 7.05 cents/kWh for the next 34 kWh used, 5.02 cents/kWh for the next 50 kWh, 3.25 cents/kWh for the next 200 kWh, 2.95 cents/kWh for the next 200 kWh, 2.35 cents/kWh for all in excess of 400 kWh. Based on these rates, what would be the charge for 227 kWh? From the standpoint of encouraging conservation of energy, what is wrong with this pricing method?

Answer:		
First 16 kWh	$ 1.60	
0.0705×34 kWh	2.40	
0.0502×50 kWh	2.51	
0.0325×127 kWh	10.64	Total cost = $17.15

The decreasing cost of power for increased usage discourages conservation.

SOLUTIONS TO SELECTED END-OF-CHAPTER PROBLEMS

5. The current I (in A) in a conductor depends on time as $I = 2t^2 - 3t + 7$, where t is in s. What quantity of charge moves across a section through the conductor during the interval $t = 2$ s to $t = 4$ s?

Solution

$$I = \frac{dq}{dt}; \qquad dq = I\,dt$$

$$q = \int I\,dt = \int_2^4 (2t^2 - 3t + 7)\,dt = \frac{2}{3}t^3 - \frac{3}{2}t^2 + 7t \Big|_2^4 = \boxed{33.3 \text{ C}}$$

17. A 2.4-m length of wire that is 0.031 cm^2 in cross section has a measured resistance of 0.24 Ω. Calculate the conductivity of the material.

Solution

$$R = \frac{\rho L}{A} \qquad \text{and} \qquad \rho = \frac{1}{\sigma}, \qquad \text{therefore}$$

$$\sigma = \frac{L}{RA} = \frac{2.4 \text{ m}}{(0.24 \text{ Ω})(3.1 \times 10^{-6} \text{ m}^2)} = \boxed{3.23 \times 10^6 /\text{Ω·m}}$$

25. A 0.9-V potential difference is maintained across a 1.5-m length of tungsten wire that has a cross-sectional area of 0.6 mm^2. What is the current in the wire?

Solution

From Ohm's law, $\qquad I = \frac{V}{R} \qquad$ where $\qquad R = \frac{\rho L}{A}, \qquad$ therefore

$$I = \frac{VA}{\rho L} = \frac{(0.9 \text{ V})(6 \times 10^{-7} \text{ m}^2)}{(5.6 \times 10^{-8} \text{ Ω·m})(1.5 \text{ m})} = \boxed{6.43 \text{ A}}$$

31. What is the fractional change in the resistance of an iron filament when its temperature changes from 25°C to 50°C?

Solution

$$R = R_o[1 + \alpha\Delta T] \quad \text{or} \quad R - R_o = R_o\alpha\Delta T$$

The fractional change in resistance $= f = \dfrac{R - R_o}{R_o}$, therefore

$$f = \frac{R_o\alpha\Delta T}{R_o} = \alpha\Delta T = (5 \times 10^{-3} \text{ C}^{\circ -1})(50 - 25) \text{ C}^{\circ} = \boxed{0.125}$$

---

41. Use data from Table 27.1 to calculate the collision mean free path of electrons in copper at a temperature corresponding to an average speed of 8.6×10^5 m/s.

Solution

The resistivity can be expressed as

$$\rho = \frac{m}{nq^2\tau}$$

where τ is the average time between collisions.

$$\tau = \frac{m}{n\rho q^2} = \frac{9.11 \times 10^{-31} \text{ kg}}{(8.48 \times 10^{28}/\text{m}^3)(1.7 \times 10^{-8} \text{ }\Omega\cdot\text{m})(1.60 \times 10^{-19} \text{ C})^2}$$

$$\tau = 2.46 \times 10^{-14} \text{ s}$$

The mean free path $L = \bar{v}\tau$ where $\bar{v}$ is the average thermal speed.

$$\therefore \quad L = (8.6 \times 10^5 \text{ m/s})(2.46 \times 10^{-14} \text{ s}) = \boxed{2.16 \times 10^{-8} \text{ m} = 21.6 \text{ nm}}$$

---

47. If a 55-Ω resistor is rated at 125 W (the maximum allowed power), what is the maximum allowed operating voltage?

Solution

Combining Joule's law and Ohm's law gives $\quad P = \dfrac{V^2}{R} \quad$ or

$$V = \sqrt{PR} = \sqrt{(125 \text{ W})(55 \text{ }\Omega)} = \boxed{82.9 \text{ V}}$$

---

51. What is the required resistance of an immersion heater that will increase the temperature of 1.5 kg of water from 10°C to 50°C in 10 min while operating at 110 V?

Solution

Assume $E_{(thermal)} = E_{(electrical)}$

$$E_{(thermal)} = mc\Delta T \quad \text{and} \quad E_{(electrical)} = \left(\frac{V^2}{R}\right)t, \quad \text{therefore}$$

$$R = \frac{V^2 t}{cm\Delta T} = \frac{(110 \text{ V})^2(600 \text{ s})}{\left(1\,\frac{cal}{g \cdot C°}\right)(1.5 \times 10^3 \text{ g})(40 \text{ C}°)(4.186 \text{ J/cal})} = \boxed{28.9 \text{ } \Omega}$$

Note the use of the conversion: $1 \text{ cal} = 4.186 \text{ J}$

63. A resistor is constructed by forming a material of resistivity ρ into the shape of a hollow cylinder of length L and inner and outer radii r_a and r_b, respectively (Figure 27.18). In use, a potential difference is applied between the ends of the cylinder, producing a current parallel to the axis. (a) Find a general expression for the resistance of such a device in terms of L, ρ, r_a, and r_b. (b) Obtain a numerical value for R when L = 4 cm, r_a = 0.5 cm, r_b = 1.2 cm, and the resistivity $\rho = 3.5 \times 10^5$ Ω·m.

Solution

(a) $$R = \frac{\rho L}{A} = \frac{\rho L}{\pi\left(r_b^2 - r_a^2\right)}$$

Figure 27.18

where A is the cross-sectional area of the conductor in the shape of a *hollow* cylinder.

(b) $$R = \frac{\rho L}{\pi\left(r_b^2 - r_a^2\right)} = \frac{(3.5 \times 10^5 \text{ }\Omega\text{·m})(0.04 \text{ m})}{\pi\left[(0.012 \text{ m})^2 - (0.005 \text{ m})^2\right]} = \boxed{3.74 \times 10^7 \text{ } \Omega = 37.4 \text{ M}\Omega}$$

28

Direct Current Circuits

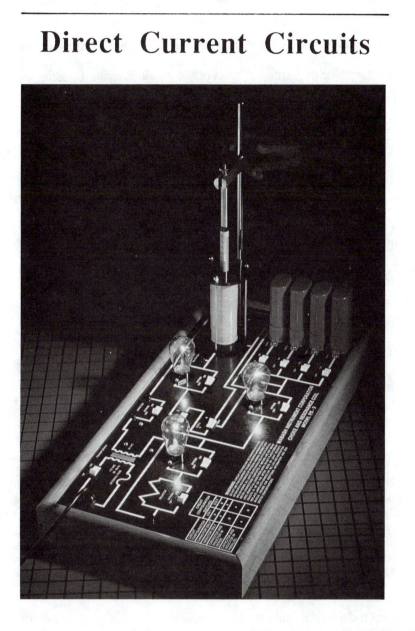

OBJECTIVES

1. Determine the terminal potential difference of a known source of emf (with internal resistance) when it is part of an open, closed, or short circuit.

2. Calculate the current in a single-loop circuit and the potential difference between any two points in the circuit.

3. Calculate the equivalent resistance of a group of resistors in parallel, series, or series-parallel combination.

4. Use Ohm's law to calculate the current in a circuit and the potential difference between any two points in a circuit which can be reduced to an equivalent single-loop circuit.

5. Use Joule's law to calculate the power dissipated by any resistor or group of resistors in a circuit.

6. Apply Kirchhoff's rules to solve multiloop circuits; that is, find the currents and the potential difference between any two points.

7. Calculate the charging (discharge) current i(t) and the accumulated (residual) charge q(t) during charging (and discharge) of a capacitor in an R-C circuit.

8. Calculate the energy expended by a source of emf while charging a capacitor.

9. Understand the circuitry and make calculations for an unknown resistance, R_x, using the ammeter-voltmeter method and the Wheatstone bridge method.

10. Determine the value of an unknown emf, $\mathcal{E}_x$, using a potentiometer circuit.

SKILLS

Many circuits which contain several resistors can be reduced to an equivalent single-loop circuit by successive step-by-step combinations of groups of resistors in series and parallel. However, in the most general case, such a reduction is not possible and you must solve a true multiloop circuit by use of Kirchhoff's rules.

In order to apply Kirchhoff's rules to solve a multiloop circuit, the currents in each part of the circuit must be assigned a direction. The assignment of directions of currents between junction points is *arbitrary*. However, you must adhere *rigorously* to the assigned current directions when writing out the equations corresponding to Kirchhoff's rules for the circuit.

Kirchhoff's first rule states that the algebraic sum of the currents entering any junction point in a circuit must equal zero. You should remember that the junction point rule should be applied a number of times which is *one fewer than the number of junction points in the circuit.*

Kirchhoff's second rule states that the algebraic sum of the potential differences across the circuit elements in any closed loop in the circuit must equal zero. In order to apply this rule, you must be able to correctly specify the change in potential as you cross each circuit element in traversing the closed loop.

Convenient "rules of thumb" which you may use to determine the increase or decrease in potential as you cross a resistor or seat of emf in traversing a circuit loop are illustrated in Figure 28.1. Notice that the potential *decreases* (changes by −IR) when the resistor is traversed *in the direction of the current*. There is an *increase* in potential of +IR if the direction of travel is *opposite* the direction of current. If a seat of emf is traversed *in* the direction of the emf (from − to + on the battery), the potential *increases* by $\mathcal{E}$. If the direction of travel is from + to −, the potential *decreases* by $\mathcal{E}$ (changes by − $\mathcal{E}$).

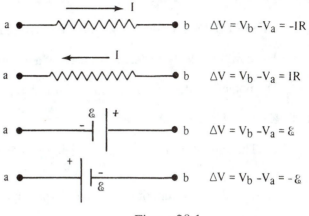

$\Delta V = V_b - V_a = -IR$

$\Delta V = V_b - V_a = IR$

$\Delta V = V_b - V_a = \mathcal{E}$

$\Delta V = V_b - V_a = -\mathcal{E}$

Figure 28.1

Finally, you must apply Kirchhoff's first and second rules to that number of loops and junction points in the circuit which will yield a set of independent equations equal in number to the number of unknown circuit parameters (I's, R's, or $\mathcal{E}$'s). Suppose, for example, that a particular circuit has the general form shown in Figure 28.2 (a) [where the actual circuit elements, R's and $\mathcal{E}$'s are not shown but assumed known]. There are six possible different values of I in the circuit; therefore you will need six independent equations to solve for the six values of I. There are four junction points in the circuit (at points a, d, f, and h). The first rule applied at *any three* of these points will yield three equations. The circuit can be thought of as a group of three "blocks" as shown in Figure 28.2 (b). Kirchhoff's second law, when applied to each of these loops (abcda, ahfga, and defhd), will yield three additional equations. You can then solve the total of six equations simultaneously for the six values of I_1, I_2, I_3, I_4, I_5, and I_6. You can, of course, expect that the sum of the changes in potential difference around *any other closed loop* in the circuit will be zero (for example, abcdefga or ahfedcba); however the equations found by applying Kirchhoff's second rule to these additional loops *will not be independent* of the six equations found previously.

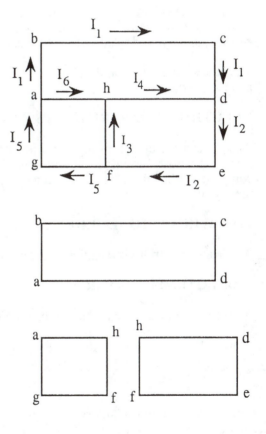

Figure 28.2

NOTES FROM SELECTED CHAPTER SECTIONS

28.2 RESISTORS IN SERIES AND PARALLEL

The current must be the same for each of a group of resistors connected in series.

The potential difference must be the same across each of a group of resistors in parallel.

28.3 KIRCHHOFF'S RULES

1. The sum of the currents entering any junction must equal the sum of the currents leaving that junction. (A junction is any point in the circuit where the current can split.)

2. The algebraic sum of the changes in potential around any closed circuit loop must be *zero*.

The first rule is a statement of *conservation of charge*; the second rule follows from the *conservation of energy*.

28.4 RC CIRCUITS

Consider an uncharged capacitor in series with a resistor, a battery, and a switch. In the charging process, charges are transferred from one plate of the capacitor to the other moving along a path *through the resistor, battery, and switch*. The charges *do not move across the gap between the plates of the capacitor*.

The battery does work on the charges to increase their electrostatic potential energy as they move from one plate to the other.

28.5 ELECTRICAL INSTRUMENTS

An ammeter is a current measuring instrument and must be placed in a circuit in series with the component (resistor) whose current is to be measured. A voltmeter measures potential difference across a circuit component when the meter is connected in parallel with the component whose potential difference is to be measured.

28.6 THE WHEATSTONE BRIDGE

A Wheatstone Bridge is a particular circuit that can be used to make accurate measurements of *resistance*.

28.7 THE POTENTIOMETER

A potentiometer is a circuit that can be used for accurate measurements of emf.

EQUATIONS AND CONCEPTS

The *current*, I, delivered by a battery in a simple dc circuit, depends on the value of the *emf* of the source, ε; the total *load resistance* in the circuit, R; and the *internal resistance* of the source, r.

$$I = \frac{\varepsilon}{R + r}$$ (28.3)

The *equivalent resistance* of *a group of resistors in series* (connected so that they have only one common point per pair) is equal to the sum of the values of individual resistors.

$$R_{eq} = R_1 + R_2 + R_3 + \ldots$$ (28.6)

The *equivalent resistance* of *two or more resistors in parallel* (connected so that the potential difference across each resistor is the same) is always less than the smallest resistor in the group.

$$\frac{1}{R_{eq}} = \frac{1}{R_1} + \frac{1}{R_2} + \frac{1}{R_3} + \ldots$$

(28.8)

When a potential difference is suddenly applied across an uncharged capacitor, the *current* in the circuit and the *charge* on the capacitor are functions of time. The *instantaneous values* of I and q depend on the capacitance and on the resistance in the circuit.

$$I(t) = \frac{\varepsilon}{R} e^{-t/(RC)}$$

(28.13)

$$q(t) = C\varepsilon\left(1 - e^{-t/(RC)}\right)$$

(28.14)

When a capacitor with an initial charge Q is discharged through a resistor, the *charge* and *current* decrease exponentially in time.

$$q(t) = Qe^{-t/(RC)}$$

(28.17)

$$I(t) = \frac{Q}{RC} e^{-t/(RC)}$$

(28.18)

The product RC is called the *time constant* of a circuit in which a capacitor is being charged (or discharged) through a resistor.

$$\tau = RC$$

During the charging process, the charge on an initially uncharged capacitor will reach 63% of its maximum value in a time $t = \tau$. Also, during one time constant, a capacitor being discharged through a resistor will lose 63% of its initial charge.

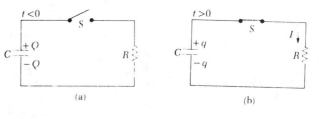

(a)   (b)

EXAMPLE PROBLEM SOLUTION

Example 28.1   Two-loop circuit: We wish to determine the currents I_1, I_2, and I_3 in the two-loop circuit shown in Figure 28.3. First, a direction is assigned to the currents. We must adhere to these directions when writing the equations corresponding to Kirchhoff's rules.

Solution

Applying Kirchhoff's first rule to the junction at b gives

$$I_1 - I_2 - I_3 = 0 \qquad (1)$$

where currents going into the junction are called positive, and currents leaving the junction are negative.

Now we apply Kirchhoff's second rule to the loop abda. Starting at d, going in the clockwise direction, gives

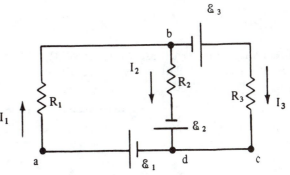

Figure 28.3

$$\varepsilon_1 - I_1 R_1 - I_2 R_2 + \varepsilon_2 = 0 \qquad (2)$$

Likewise, Kirchhoff's second rule applied to the loop dbcd, starting at d and going clockwise, gives

Likewise, Kirchhoff's second rule applied to the loop dbcd, starting at d and going clockwise, gives

$$-\mathcal{E}_2 + I_2R_2 + \mathcal{E}_3 - I_3R_3 = 0 \tag{3}$$

A third loop equation could be written for the loop abcda; however, it would provide no new information. In other words, the equation would not be independent of (2) and (3). Equations (1), (2), and (3) represent three linear equations with three unknowns, namely I_1, I_2, and I_3. Solving these equations simultaneously gives (after some cumbersome algebra)

$$I_1 = \frac{(R_2 + R_3)\mathcal{E}_1 + R_2\mathcal{E}_3 + R_3\mathcal{E}_2}{R_1R_2 + R_1R_3 + R_2R_3} \tag{4}$$

$$I_2 = \frac{(R_1 + R_3)\mathcal{E}_2 + R_3\mathcal{E}_1 - R_1\mathcal{E}_3}{R_1R_2 + R_1R_3 + R_2R_3} \tag{5}$$

$$I_3 = \frac{(\mathcal{E}_3 - \mathcal{E}_2)R_1 + (\mathcal{E}_1 + \mathcal{E}_3)R_2}{R_1R_2 + R_1R_3 + R_2R_3} \tag{6}$$

Note that in the *limit* $R_2 \rightarrow \infty$, the limiting values of the current are $I_2 = 0$ and

$$I_1 = I_3 = \frac{\mathcal{E}_1 + \mathcal{E}_3}{R_1 + R_3}$$

Likewise, as $R_1 \rightarrow \infty$, $I_1 = 0$ and

$$I_2 = -I_3 = \frac{\mathcal{E}_2 - \mathcal{E}_3}{R_2 + R_3}$$

Finally, as $R_3 \rightarrow \infty$, $I_3 = 0$ and

$$I_1 = I_2 = \frac{\mathcal{E}_1 + \mathcal{E}_2}{R_1 + R_2}$$

You should verify these limiting results.

(b) Find the currents if $R_1 = 5\ \Omega$, $R_2 = 8\ \Omega$, $R_3 = 6\ \Omega$, $\mathcal{E}_1 = 4$ V, $\mathcal{E}_2 = 6$ V, and $\mathcal{E}_3 = 8$ V.

Solution

Direct substitution of these values into (4), (5), and (6) gives $I_1 = 1.32$ A, $I_2 = 0.42$ A, and $I_3 = 0.90$ A. Since the values are all positive, the correct directions were chosen for the currents. Note also that the junction rule $I_1 = I_2 + I_3$ is verified.

ANSWERS TO SELECTED QUESTIONS

3. Is the direction of current through a battery always from negative to positive on the terminals? Explain.

Answer: No. This is only true if there is only one battery in the circuit. If other batteries are in the circuit, they could be charging the battery in question, which means a reversal in the current from positive to negative.

4. Two different sets of Christmas-tree lights are available. For set A, when one bulb is removed (or burns out), the remaining bulbs remain illuminated. For set B, when one bulb is removed, the remaining bulbs will not operate. Explain the difference in wiring for the two sets of lights.

Answer: The bulbs in set A are connected in parallel, so they operate independently. The bulbs in set B are connected in series, so removing one bulb causes an open circuit condition.

7. Given three light bulbs and a battery, sketch as many different electric circuits as you can.

Answer: See the sketches below.

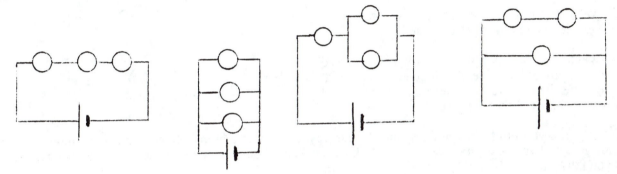

11. Are the two headlights on a car wired in series or in parallel? How can you tell?

Answer: They are in parallel. If they were in series and one burned out, the other would go out. (Bad news!)

12. An incandescent lamp connected to a 120-V source with a short extension cord will provide more illumination than if it were connected to the same source with a long extension cord. Explain.

Answer: The long cord has more resistance than the short cord, hence there is a larger voltage drop across the longer cord. Since the cord is in series with the lamp, the current in the circuit will be reduced when the longer cord is used, thereby reducing the illumination.

13. Embodied in Kirchhoff's rules are two conservation laws. What are they?

Answer: The junction rule is a statement of conservation of charge, while the loop rule is a statement of conservation of energy.

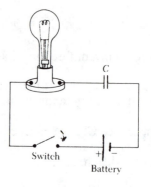

16. With reference to Figure 28.27, describe what happens to the light bulb after the switch is closed. Assume the capacitor is initially uncharged and assume that the light will illuminate when connected directly across the battery terminals.

Answer: The bulb will light up for an instant as the capacitor is being charged and there is a current in the circuit. As soon as the capacitor is fully charged, the current in the circuit is zero and the illumination disappears.

Figure 28.27

20. Why is it dangerous to turn on a light when you are in the bathtub?

Answer: You are well-grounded when standing (sitting, etc.) in the tub. If the switch is not well insulated, you will provide an excellent path for the current from the switch to ground.

21. Why is it possible for a bird to sit on a high-voltage wire without being electrocuted?

Answer: The bird is resting on a wire of a fixed potential. In order to be electrocuted, a potential difference is required. There is no potential difference between the bird's feet.

22. Suppose you fall from a building and on the way down grab a high-voltage wire. Assuming that the wire holds you, will you be electrocuted? If the wire then breaks, should you continue to hold onto the end of the wire as you fall?

Answer: No. Because you are not completing a circuit--you are not in contact with another conductor. You should not be holding the wire when you contact something else, for then you may complete the circuit (and be part of it).

23. Would a fuse work successfully if it were placed in parallel with the device it is supposed to protect?

Answer: Fuses are placed in series with the devices they are to protect so that when the fuse burns out, the devices are turned off. This would not happen if the fuse were in parallel.

25. When electricians work with potentially live wires, they often use the backs of their hands or fingers to move wires. Why do you suppose they use this technique?

Answer: If they touch a live wire with the back of the hand, the hand will be thrown away from the wire when the muscles contract. On the other hand, if they grab a live wire with a normal grip, contracting muscles cause the hand to be "frozen" to the live wire.

28. If it is the current flowing through the body that determines how serious a shock will be, why do we see warnings of high voltage rather than high current near electric equipment?

Answer: From the relationship V = IR, we see that if the voltage is great, the current will likewise be great.

29. Suppose you are flying a kite when it strikes a high-voltage wire. What factors determine how great a shock you receive?

Answer: Among the factors involved: how good a conductor the string is (is it wet?); how well insulated you are from ground (thick rubber soles?); how great the voltage is.

CHAPTER 28

SOLUTIONS TO SELECTED END-OF-CHAPTER PROBLEMS

7. A battery has an emf of 15 V.   The terminal voltage of the battery is 11.6 V when it is delivering 20 W of power to an external load resistor R.   (a) What is the value of R?   (b) What is the internal resistance of the battery?

Solution

(a)   Combining Joule's law, $P = VI$ and Ohm's law, $V = IR$ gives

$$R = \frac{V^2}{P} = \frac{(11.6 \text{ V})^2}{20 \text{ W}} = \boxed{6.73 \ \Omega}$$

(b) $\mathcal{E} = IR + Ir$

$$r = \frac{\mathcal{E} - IR}{I} \qquad \text{where } I = \frac{V}{R}, \qquad \text{therefore}$$

$$r = \frac{(\mathcal{E} - V)R}{V} = \frac{(15 \text{ V} - 11.6 \text{ V})(6.73 \ \Omega)}{11.6 \text{ V}} = \boxed{1.97 \ \Omega}$$

17. Evaluate the effective resistance of the network of identical resistors, each having resistance R, shown in Figure 28.30.

Solution

The network of resistors is equivalent to three resistors having values R, 2R, and 3R in *parallel*. Therefore,

$$\frac{1}{R_{\text{eff}}} = \frac{1}{R} + \frac{1}{2R} + \frac{1}{3R} = \frac{1}{R}\left(1 + \frac{1}{2} + \frac{1}{3}\right) = \frac{11}{6R}$$

$$R_{\text{eff}} = \boxed{\frac{6}{11} R = 0.545 \text{ R}}$$

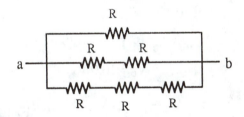

Figure 28.30

35. (a) Find the value of I_1 and I_3 in the circuit of Figure 28.44 if the 4-V battery is replaced by a 5-μF capacitor. (b) Determine the charge on the 5-μF capacitor.

Solution

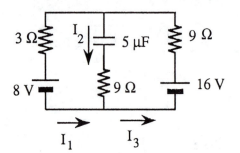

Figure 28.44

(a) In the *steady-state* condition (after the capacitor is charged), $I_2 = 0$ and $I_1 = I_2 = I$. Traversing the outer loop counterclockwise gives

$$16 \text{ V} - (9 \text{ }\Omega)I - (3 \text{ }\Omega)I - 8 \text{ V} = 0$$

$$I = \boxed{0.667 \text{ A}}$$

(b) Let V_c = the potential difference across the capacitor and traverse the right-hand loop counterclockwise.

$$16 \text{ V} - (9 \text{ }\Omega)(0.67 \text{ A}) - V_c - (9 \text{ }\Omega)(0 \text{ A}) = 0$$

$$V_c = 10.0 \text{ V}$$

$$Q = CV = (5 \times 10^{-6} \text{ F})(10.0 \text{ V}) = 5.00 \times 10^{-5} \text{ C} = \boxed{50.0 \text{ μC}}$$

37. For the circuit shown in Figure 28.46, calculate
(a) the current in the 2-Ω resistor and (b) the
potential difference between points a and b.

Solution

Use current directions as labeled in the figure
below.

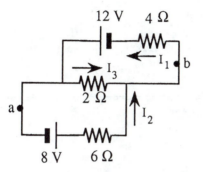

Figure 28.46

(a)   From the junction point rule, we have

$$I_1 = I_2 + I_3 \qquad (1)$$

Traversing the top loop counterclockwise gives

$$12 \text{ V} - (2 \text{ }\Omega)I_3 - (4 \text{ }\Omega)I_1 = 0 \qquad (2)$$

Traversing the bottom loop counterclockwise,

$$8 \text{ V} - (6 \text{ }\Omega)I_2 + (2 \text{ }\Omega)I_3 = 0 \qquad (3)$$

From Equation (2), $\qquad\qquad I_1 = 3 - \frac{1}{2}I_3$

From Equation (3), $\qquad\qquad I_2 = \frac{4}{3} - \frac{1}{3}I_3$

Substituting these values into Equation (1), we find $I_3 = 0.909$ A.

Therefore, the current in the 2-Ω resistor is $\boxed{0.909 \text{ A}}$.

(b) $V_a - (0.909 \text{ A})(2 \text{ }\Omega) = V_b,$ $\qquad$ therefore

$V_a - V_b = \boxed{1.82 \text{ V}}$; $\qquad V_a > V_b$

43. The switch in the RC circuit described in Problem 42 is closed at $t = 0$. Find the current in the resistor R at a time 10 s after the switch is closed.

Solution

$$I = I_o e^{-t/\tau} \quad \text{where} \quad I_o = \frac{\mathcal{E}}{R} \quad \text{and} \quad \tau = RC$$

$$I = \frac{\mathcal{E}}{R} e^{-t/(RC)} = \left(\frac{30 \text{ V}}{1 \times 10^6 \text{ }\Omega}\right) e^{-\left(\frac{10 \text{ s}}{(1 \times 10^6 \text{ }\Omega)(5 \times 10^{-6} \text{ F})}\right)}$$

$$I = 4.06 \times 10^{-6} \text{ A} = \boxed{4.06 \text{ }\mu\text{A}}$$

53. The same galvanometer movement as used in Problem 52 may be used to measure voltages. In this case a large resistor is wired in series with the meter movement similar to Figure 28.22b, which in effect limits the current that flows through the movement when large voltages are applied. Most of the potential drop occurs across the resistor placed in series. Calculate the value of the resistor that enables the movement to measure an applied voltage of 25 V at full-scale deflection.

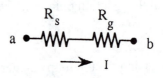

Figure 28.22b

Solution

Require $V_{ab} = 25$ V and $I = 1.5 \times 10^{-3}$ A

$R_g = 75 \text{ }\Omega$

$V_{ab} = I(R_s + R_g)$

$$R_g = \frac{V_{ab}}{I} - R_g = \frac{25 \text{ V}}{1.5 \times 10^{-3} \text{ A}} - 75 \text{ }\Omega = 16,590 \text{ }\Omega = \boxed{16.6 \text{ k}\Omega}$$

79. A battery has an emf $\mathcal{E}$ and internal resistance r. A variable resistor R is connected across the terminals of the battery. Find the value of R such that (a) the potential difference across the terminals is a maximum, (b) the current in the circuit is a maximum, (c) the power delivered to the resistor is a maximum.

Solution

(a) $V_{terminal} = \mathcal{E} - Ir = \dfrac{\mathcal{E}R}{R + r}$

Therefore,

$$V_{terminal} \rightarrow \mathcal{E} \qquad as \qquad R \rightarrow \infty$$

(b) $I = \dfrac{\mathcal{E}}{R + r}$

Therefore,

$$I \rightarrow \dfrac{\mathcal{E}}{r} \qquad as \qquad R \rightarrow 0$$

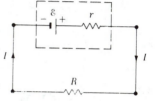

(c) $P = I^2 R = \dfrac{\mathcal{E}^2 R}{(R + r)^2}$

Therefore,

$$P \rightarrow \dfrac{\mathcal{E}^2}{4r} \qquad as \qquad R \rightarrow r$$

81. The values of the components in a simple RC circuit (Figure 28.14) are as follows:

$C = 1\ \mu F$, $R = 2 \times 10^6\ \Omega$, and $\mathcal{E} = 10$ V. At the instant 10 s after the switch in the circuit is closed, calculate (a) the charge on the capacitor, (b) the current in the resistor, (c) the rate at which energy is being stored in the capacitor, and (d) the rate at which energy is being delivered by the battery.

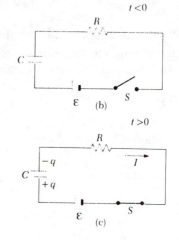

Figure 28.14

Solution

(a) $q = CV(1 - e^{-t/(RC)})$

$$q = (10^{-6}\ F)(10\ V)\left[1 - e^{-\left(\frac{10\ s}{(2 \times 10^6\ \Omega)(10^{-6}\ F)}\right)}\right]$$

$$q = 9.93 \times 10^{-6}\ C = \boxed{9.93\ \mu C}$$

(b) $I = \dfrac{dq}{dt} = \dfrac{d}{dt}\left[CV(1 - e^{-t/(RC)})\right]$

$$I = \left(\frac{V}{R}\right)e^{-t/(RC)} = \left(\frac{10\ V}{2 \times 10^{-6}\ \Omega}\right)e^{-(10/2)} = \boxed{3.37 \times 10^{-8}\ A}$$

(c) Since the energy stored in the capacitor $U = \dfrac{q^2}{2C}$,

the rate of storing energy $= \dfrac{dU}{dt} = \dfrac{q}{C}\dfrac{dq}{dt} = \left(\dfrac{q}{C}\right)I$

$$Rate = \left(\frac{9.9 \times 10^{-6}\ C}{1 \times 10^{-6}\ F}\right)(3.37 \times 10^{-8}\ A) = \boxed{3.34 \times 10^{-7}\ W}$$

(d) $P_{batt} = I\mathcal{E} = (3.37 \times 10^{-8}\ A)(10\ V) = \boxed{3.37 \times 10^{-7}\ W}$

Magnetic Fields

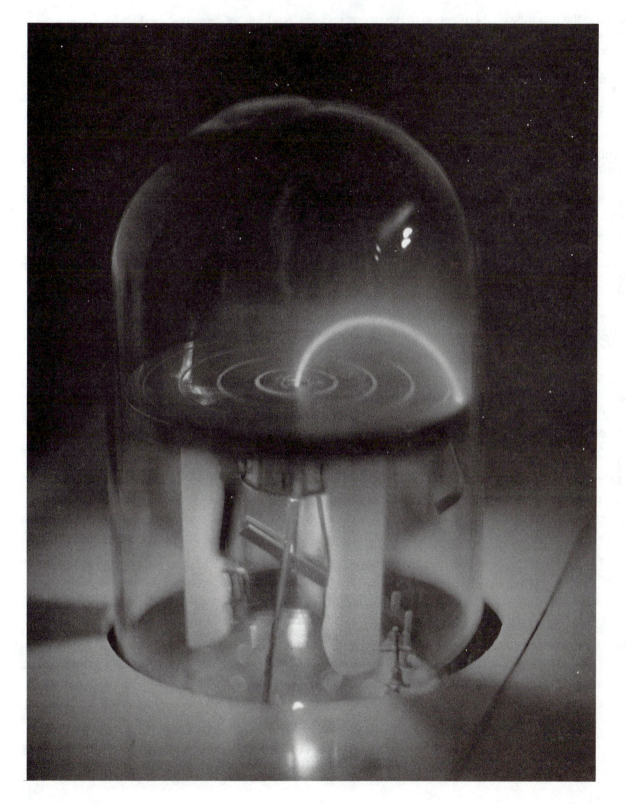

OBJECTIVES

1. Use the defining equation for a magnetic field **B** to determine the magnitude and direction of the magnetic force exerted on an electric charge moving in a region where there is a magnetic field. You should understand clearly the important differences between the forces exerted on electric charges by electric fields and those forces exerted on moving charges by magnetic fields.

2. Calculate the magnitude and direction of the magnetic force on a current-carrying conductor when placed in an external magnetic field. You should be able to perform such calculations for either a straight conductor or one of arbitrary shape.

3. Determine the magnitude and direction of the torque exerted on a closed current loop in an external magnetic field. You should understand how to correctly designate the direction of the area vector corresponding to a given current loop, and to incorporate the magnetic moment of the loop into the calculation of the torque on the loop.

4. Calculate the radius of the circular orbit of a charged particle moving in a uniform magnetic field, and also determine the period of the circulating charge.

5. Understand the essential features of the mass spectrometer and the cyclotron, and make appropriate quantitative calculations regarding the operation of these instruments. Note that these two devices are special applications of the motion of charged particles in a magnetic field.

6. Understand the principle of the Hall effect and use appropriate rearrangements of the Hall voltage equation to make calculations of magnetic field strengths and Hall coefficient values for various conductors.

SKILLS

Equation 29.1, $\mathbf{F} = q(\mathbf{v} \times \mathbf{B})$, serves as the definition of the magnetic field vector **B**. The direction of the magnetic force **F** is determined by the right-hand rule for the cross product which you have used before. This assumes that the charge q is a positive charge. If the vectors **v** and **B** are given in unit vector notation then $\mathbf{v} \times \mathbf{B}$ can be written as

$$\mathbf{v} \times \mathbf{B} = \mathbf{i}(v_y B_z - v_z B_y) + \mathbf{j}(v_z B_x - v_x B_z) + \mathbf{k}(v_x B_y - v_y B_x)$$

This means that the components of the magnetic force are given by

$$F_x = q(v_y B_z - v_z B_y)$$

$$F_y = q(v_z B_x - v_x B_z)$$

$$F_z = q(v_x B_y - v_y B_x)$$

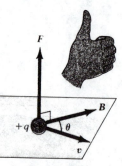

The right-hand rule and the vector cross product are also used to determine the direction of the torque and the direction of the resulting rotation for a closed current loop in a magnetic field. When the four fingers on your right hand circle the current loop in the direction of the current, the thumb will point in the direction of the area vector **A**.

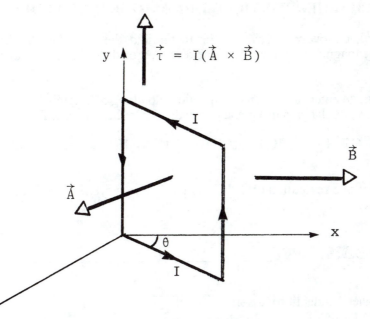

$$\vec{\tau} = I(\vec{A} \times \vec{B})$$

Figure 29.1

Applying this rule to the situation shown in Figure 29.1, the vector **A** is directed out of the plane of the rectangular loop facing you as shown. The resulting torque is parallel to the direction of **A** × **B** and is along the positive y-axis as shown in the figure. This is consistent with the general rule for determining the direction of the cross product of two vectors. If the loop is considered to be hinged along the edge joining the y-axis, then the rotation will be such that the angle θ decreases and the plane of the rectangular loop becomes parallel to the x-y plane. This is shown in Fig. 29.1 as a counterclockwise rotation about the y-axis as seen from above.

NOTES FROM SELECTED CHAPTER SECTIONS

29.2 DEFINITION AND PROPERTIES OF THE MAGNETIC FIELD

Particles with charge q, moving with speed v in a magnetic field **B**, experience a magnetic force **F**:

1. The magnetic force is proportional to the charge q and speed v of the particle.
2. The magnitude and direction of the magnetic force depend on the velocity of the particle and on the magnitude and direction of the magnetic field.
3. When a charged particle moves in a direction *parallel* to the magnetic field vector, the magnetic force **F** on the charge is *zero*.
4. The magnetic force acts in a direction perpendicular to both **v** and **B**; that is, **F** is perpendicular to the plane formed by **v** and **B**.
5. The magnetic force on a positive charge is in the direction opposite to the force on a negative charge moving in the same direction.
6. If the velocity vector makes an angle θ with the magnetic field, the magnitude of the magnetic force is proportional to sin θ.

29.3 MAGNETIC FORCE ON A CURRENT-CARRYING CONDUCTOR

The total magnetic force on a *closed* current loop in a *uniform* magnetic field is *zero*.

29.5 MOTION OF A CHARGED PARTICLE IN A MAGNETIC FIELD

When a charged particle moves in an external magnetic field, the *work done* by the magnetic force on the particle is *zero*. The magnetic force changes the direction of the velocity vector but does not change its magnitude.

When a charged particle enters an external magnetic field along a direction perpendicular to the field, the particle will move in a circular path in a plane perpendicular to the magnetic field.

29.6 APPLICATIONS OF THE MOTIONS OF CHARGED PARTICLES IN A MAGNETIC FIELD

The time required for one revolution of a charged particle in a *cyclotron* is independent of the speed (or radius) of the particle.

EQUATIONS AND CONCEPTS

The magnetic field (or magnetic induction) at some point in space is defined in terms of the *magnetic force* exerted on a moving positive electric charge at that point. The SI unit of the magnetic field is the tesla (T) or weber per square meter (Wb/m^2).

$$\boxed{F = q(v \times B)} \tag{29.1}$$

If a *straight* wire carrying a current is placed in an external magnetic field, a *magnetic force* will be exerted on the wire. The magnetic force on a wire of arbitrary shape is found by integrating over the length of the wire. In these equations the direction of *L* and ds is that of the current.

$$\boxed{F = I(L \times B)} \tag{29.5}$$

$$\boxed{F = I \int ds \times B} \tag{29.7}$$

When a closed conducting loop carrying a current is placed in an external magnetic field, there is a *net torque* exerted on the loop. In Eq. 29.12, the area vector **A** is directed perpendicular to the area of the loop with a sense given by the right-hand rule. The magnitude of **A** is numerically equal to the area of the loop.

$$\boxed{\tau = I(A \times B)} \tag{29.12}$$

When a charged particle enters the region of a uniform magnetic field with its velocity vector initially perpendicular to the direction of the field, the particle will move in a circular path in a plane perpendicular to the direction of the field. The *radius* of the circular path will be proportional to the linear momentum of the charged particle. The *angular frequency* (or cyclotron frequency) of the particle will be proportional to the ratio of charge to mass. Note that the frequency, and hence the period of rotation do *not* depend on the radius of the path.

$$\boxed{r = \frac{mv}{Bq}} \tag{29.15}$$

$$\boxed{\omega = \frac{qB}{m}} \tag{29.16}$$

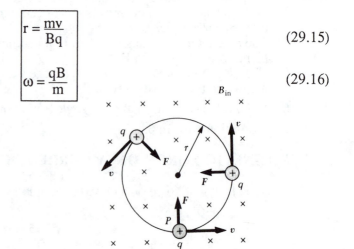

There are several important applications of the motion of charged particles in a magnetic field:

Velocity Selector --When a beam of charged particles is directed into a region where uniform electric and magnetic fields are perpendicular to each other and to the initial direction of the particle beam, those particles emerging along the initial beam direction will have a common velocity.

$$v = \frac{E}{B}$$

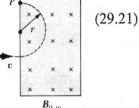

(29.19)

Mass Spectrometer--If an ion beam, after passing through a velocity selector, is directed perpendicularly into a second uniform magnetic field, the *ratio of charge to mass* for the isotopic species can be determined by measuring the radius of curvature of the beam in the field.

$$\frac{m}{q} = \frac{rB_0B}{E}$$

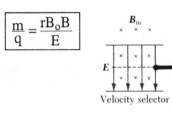

(29.21)

Cyclotron--The *maximum kinetic energy* acquired by an ion in a cyclotron depends on the radius of the dees and the intensity of the magnetic field. This relationship holds until the ion reaches relativistic energies ($\cong$ 20 MeV).

$$K = \frac{q^2B^2R^2}{2m}$$

(29.22)

When the *Hall Coefficient* (1/nq) is known for a calibrated sample, magnetic field strengths can be determined by accurate measurements of the *Hall voltage*, V_H.

$$V_H = \frac{IBd}{nqA}$$

(29.25)

EXAMPLE PROBLEM SOLUTION_____

Example 29.1   A cyclotron with a "dee" radius of 0.6 m is accelerating singly charged helium ions at an oscillator frequency of 10 MHz.

(a) What magnetic field is required?

Solution

The cyclotron frequency is given by Eq. 29.16 as

$$\omega = \frac{qB}{m} \quad \text{where} \quad \omega = 2\pi f$$

Therefore

$$B = 2\pi f\left(\frac{m}{q}\right)$$

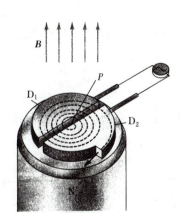

For helium ions, we have

$$m = 6.68 \times 10^{-27} \text{ kg}, \quad q = 1.6 \times 10^{-19} \text{ C}$$

Therefore

$$B = (2\pi)(10 \times 10^6 \text{ Hz}) \frac{(6.68 \times 10^{-27} \text{ kg})}{(1.6 \times 10^{-19} \text{ C})} = 2.62 \text{ T}$$

(b) What maximum kinetic energy is acquired by the ions when they emerge from the cyclotron?

Solution

From Eq. 29.22,

$$K = \frac{q^2 B^2 R^2}{2m}$$

or

$$K = \frac{(1.6 \times 10^{-19} \text{ C})^2 (2.62 \text{ T})^2 (0.6 \text{ m})^2}{(2)(6.68 \times 10^{-27} \text{ kg})} = 4.74 \times 10^{-12} \text{ J}$$

This is equivalent to $K = 29.6$ MeV and is approximately the energy at which the ion becomes relativistic.

(c) Suppose that the accelerating potential between the dees of the cyclotron is 60 kV. How many orbits are required in order that the helium ions reach the maximum kinetic energy value found in part (b)?

Solution

Each ion moves across the gap between the dees twice during each orbit and experiences an increase in energy per orbit equal to

$$\Delta K = 2qV$$

In this case,

$$\Delta K = (2)(1.6 \times 10^{-19} \text{ C})(6 \times 10^4 \text{ V}) = 1.92 \times 10^{-14} \text{ J}$$

Hence, the number of orbits required will be

$$N = \frac{K_{max}}{\Delta K} = \frac{4.74 \times 10^{-12} \text{ J}}{1.92 \times 10^{-14} \text{ J/orbit}} = 247 \text{ orbits}$$

It is interesting to note that when the polarity between the dees is being switched at a frequency of 10 MHz, the period of circulation of the ions is 10^{-7} s and therefore the 247 orbits will be completed in only 24.7 μs!

ANSWERS TO SELECTED QUESTIONS

2. Two charged particles are projected into a region where there is a magnetic field perpendicular to their velocities. If the charges are deflected in opposite directions, what can you say about them?

Answer: The charges must be of opposite sign.

3. If a charged particle moves in a straight line through some region of space, can you say that the magnetic field in that region is zero.

Answer: No. There may be another field such as an electric or gravitational field which produces a force which balances the magnetic force. Also, a charge moving parallel to a magnetic field would be undeflected.

4. Suppose an electron is chasing a proton up this page when suddenly a magnetic field is formed perpendicular to the page. What will happen to the particles?

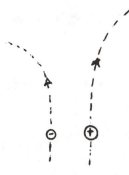

Answer: One will be deflected toward the left margin, the other toward the right margin.

5. Why does the picture on a TV screen become distorted when a magnet is brought near the screen?

Answer: The magnet produces a magnetic field which interacts with the moving electrons making up the beam in the tube. The magnetic force on the electrons causes them to be deflected to different directions.

6. How can the motion of a moving charged particle be used to distinguish between a magnetic field and an electric field? Give a specific example to justify your argument.

Answer: The magnetic force on a moving charged particle is always perpendicular to the direction of motion. There is no magnetic force on the charge when it moves in the direction of the magnetic field. On the other hand, the force on an electric charge moving through an electric field is never zero. Therefore, by projecting the particle in different directions, it is possible to determine the nature of the field.

7. List several similarities and differences in electric and magnetic forces.

Answer: *Similarities:* They can be attractive or repulsive. Both can cause electric/magnetic effects in uncharged/unmagnetized objects., Both decrease with increasing distances (in fact, according to the inverse square law). *Differences:* Electric charges can exist separately from other charges, whereas magnetic poles come in pairs; electric forces are exerted on moving charges only.

11. Is it possible to orient a current loop in a uniform magnetic field such that the loop will not tend to rotate? Explain.

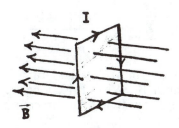

Answer: Yes. If the magnetic field is perpendicular to the plane of the loop, the forces on opposite sides will be equal and opposite, but produce no net torque.

12. How can a current loop be used to determine the presence of a magnetic field in a given region of space?

Answer: The loop can be mounted on a rotatable axis. The loop will rotate about this axis when placed in an external magnetic field for some arbitrary orientation. As the current through the loop is increased, the torque on it will also increase.

19. Can a magnetic field set a resting electron into motion? If so, how?

Answer: A magnetic force is exerted on a charge only when the charge is in motion, or if the source of the magnetic field moves.

SOLUTIONS TO SELECTED END-OF-CHAPTER PROBLEMS

5. A proton moving with a speed of 4×10^6 m/s through a magnetic field of 1.7 T experiences a magnetic force of magnitude 8.2×10^{-13} N. What is the angle between the proton's velocity and the field?

Solution

Since the magnitude of the force on a moving charge in a magnetic field is $F = qvB \sin \theta$,

$$\theta = \sin^{-1} \left[\frac{F}{qvB} \right]$$

$$\theta = \sin^{-1} \left[\frac{8.2 \times 10^{-13} \text{ N}}{(1.6 \times 10^{-19} \text{ C})(4 \times 10^6 \text{ m/s})(1.7 \text{ T})} \right] = \boxed{48.9°}$$

19. A current I = 15 A is directed along the positive x axis in a wire perpendicular to a magnetic field. The current experiences a magnetic force per unit length of 0.63 N/m in the negative y direction. Calculate the magnitude and direction of the magnetic field in the region through which the current passes.

Solution

The force on a current segment in a magnetic field is $\mathbf{F} = I(\mathbf{L} \times \mathbf{B})$, which has a magnitude of $F = ILB \sin \theta$. In this case, $\theta = 90°$ and $B = \frac{F/L}{I}$.

$$B = \frac{0.63 \text{ N/m}}{15 \text{ A}} = \boxed{0.042 \text{ T}}$$

We can use the right-hand rule to find the direction of $\mathbf{B}$. When $\mathbf{L}$ is along + x axis and $\mathbf{F}$ is along – y, $\mathbf{B}$ will be directed along + z. That is,

$$\boxed{\mathbf{B} = 0.042\mathbf{k} \text{ T}}$$

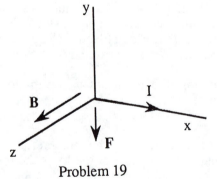

Problem 19

27. A circular coil of 100 turns has a radius of 0.025 m and carries a current of 0.1 A while in a uniform external magnetic field of 1.5 T. How much work must be done to rotate the coil from a position where the magnetic moment is parallel to the field to a position where the magnetic moment is opposite the field?

Solution

$$dW = \tau d\theta \qquad \text{and} \qquad W = \int_0^\pi \tau d\theta$$

where $\tau = NIAB \sin \theta$ (when the magnetic moment is parallel to the field $\theta = 0$)

so

$$W = NIAB \int_0^\pi \sin \theta d\theta = NIAB \left[- \cos \theta\right]_0^\pi$$

$$W = NIAB \left[- \cos \pi - (- \cos 0)\right] = 2NIAB$$

$$W = (2)(100)(0.1 \text{ A})(\pi)(0.025 \text{ m})^2(1.5 \text{ T}) = \boxed{0.059 \text{ J}}$$

31. What magnetic field would be required to constrain an electron whose energy is 725 eV to a circular path of radius 0.5 m?

Solution

The radius of the path $R = \dfrac{mv}{Bq}$ hence $B = \dfrac{mv}{Rq}$

Since $K = \dfrac{1}{2} mv^2$, $v = \sqrt{\dfrac{2K}{m}}$

Therefore,

$$B = \frac{m}{Rq} \sqrt{\frac{2K}{m}} = \sqrt{\frac{2mK}{R^2q^2}}$$

$$B = \sqrt{\frac{(2)(9.11 \times 10^{-31} \text{ kg})(725 \text{ eV})\left(1.60 \times 10^{-19} \frac{\text{J}}{\text{eV}}\right)}{(0.5 \text{ m})^2(1.6 \times 10^{-19} \text{ C})^2}}$$

$$B = 1.82 \times 10^{-4} \text{ T} \qquad \text{or} \qquad B = \boxed{182 \ \mu\text{T}}$$

47. A cyclotron designed to accelerate protons is provided with a magnetic field of 0.45 T and has a radius of 1.2 m. (a) What is the cyclotron frequency? (b) What is the maximum speed acquired by the protons?

Solution

(a)   The cyclotron frequency, $\omega = \dfrac{qB}{m}$

For protons,

$$\omega = \frac{(1.60 \times 10^{-19}\ \text{C})}{(1.67 \times 10^{-27}\ \text{kg})}(0.45\ \text{T}) = \boxed{4.31 \times 10^7\ \text{rad/s}}$$

(b) $R = \dfrac{mv}{Bq}$; $v = \dfrac{BqR}{m}$

$$v = \frac{(0.45\ \text{T})(1.6 \times 10^{-19}\ \text{C})(1.2\ \text{m})}{1.67 \times 10^{-27}\ \text{kg}} = \boxed{5.17 \times 10^7\ \text{m/s}}$$

53. In an experiment designed to measure the earth's magnetic field using the Hall effect, a copper bar 0.5 cm in thickness is positioned along an east-west direction. If a current of 8 A in the conductor results in a measured Hall voltage of 5.1×10^{-12} V, what is the calculated value of the earth's magnetic field? (Assume that $n = 8.48 \times 10^{28}$ electrons/m^3 and that the plane of the bar is rotated to be perpendicular to the direction of **B**.)

Solution

The Hall voltage is given by $V_H = \dfrac{IB}{nqt}$

$$B = \frac{nqtV_H}{I} = \frac{(8.48 \times 10^{28}/\text{m}^3)(1.6 \times 10^{-19}\ \text{C})(0.005\ \text{m})(5.1 \times 10^{-12}\ \text{V})}{8\ \text{A}}$$

$$B = 4.32 \times 10^{-5}\ \text{T} \quad \text{or} \quad B = \boxed{43.2\ \mu\text{T}}$$

59. A positive charge $q = 3.2 \times 10^{-19}$ C moves with a velocity $\mathbf{v} = (2\mathbf{i} + 3\mathbf{j} - \mathbf{k})$ m/s through a region where both a uniform magnetic field and a uniform electric field exist. (a) Calculate the total force on the moving charge (in unit-vector notation) if $\mathbf{B} = (2\mathbf{i} + 4\mathbf{j} + \mathbf{k})$ T and $\mathbf{E} = (4\mathbf{i} - \mathbf{j} - 2\mathbf{k})$ V/m. (b) What angle does the force vector make relative to the positive x axis?

Solution   The total force is the Lorentz force,

(a) $\mathbf{F} = q\mathbf{E} + q(\mathbf{v} \times \mathbf{B}) = q(\mathbf{E} + \mathbf{v} \times \mathbf{B})$

$\mathbf{F} = q[(4\mathbf{i} - \mathbf{j} - 2\mathbf{k})\ \text{V/m} + (2\mathbf{i} + 3\mathbf{j} - \mathbf{k})\ \text{m/s} \times (2\mathbf{i} + 4\mathbf{j} + \mathbf{k})\ \text{T}]$

$$F = q\left[(4\mathbf{i} - \mathbf{j} - 2\mathbf{k})\ \text{V/m} + (7\mathbf{i} - 4\mathbf{j} + 2\mathbf{k})\ \frac{\text{m·T}}{\text{s}}\right]$$

$$F = q[11\mathbf{i} - 5\mathbf{j}]\ \frac{\text{V·T}}{\text{s}}$$

$$F = (3.2 \times 10^{-19}\ \text{C})(11\mathbf{i} - 5\mathbf{j})\ \frac{\text{V·T}}{\text{s}}$$

$$F = \boxed{(3.52\mathbf{i} - 1.6\mathbf{j}) \times 10^{-18}\ \text{N}}$$

(b) $\mathbf{F·i} = F\cos\theta = F_x$

$$\theta = \cos^{-1}\left(\frac{F_x}{F}\right) = \cos^{-1}\left(\frac{3.52}{3.87}\right) = \boxed{24.4°}$$

---

69. Consider an electron orbiting a proton and maintained in a fixed circular path of radius equal to $R = 5.29 \times 10^{-11}$ m by the Coulomb force of mutual attraction. Treating the orbiting charge as a current loop, calculate the resulting torque when the system is in an external magnetic field of 0.4 T directed perpendicular to the magnetic moment of the orbiting electron.

Solution

$|\tau| = IAB$ where the effective current due to the orbiting electron is $I = \frac{q}{T}$. The period of the motion, $T = \frac{2\pi R}{v}$ where v is the speed in the circular path of radius, R. The Coulomb force on the electron must equal the centripetal force;

$$\frac{kq^2}{R^2} = \frac{mv^2}{R} \qquad \text{or} \qquad v = q\sqrt{\frac{k}{mR}}$$

Therefore,

$$T = 2\pi\sqrt{\frac{mR^3}{q^2k}} = 2\pi\sqrt{\frac{(9.1 \times 10^{-31}\ \text{kg})(5.29 \times 10^{-11}\ \text{m})^3}{(1.6 \times 10^{-19}\ \text{C})^2\left(9 \times 10^9\ \frac{\text{N·m}^2}{\text{C}^2}\right)}} = \boxed{1.52 \times 10^{-16}\ \text{s}}$$

and

$$|\tau| = \left(\frac{q}{T}\right)AB = \frac{(1.6 \times 10^{-19}\ \text{C})}{(1.52 \times 10^{-16}\ \text{s})}(\pi)(5.29 \times 10^{-11}\ \text{m})^2(0.4\ \text{T}) = \boxed{3.70 \times 10^{-24}\ \text{N·m}}$$

30

Sources of the Magnetic Field

OBJECTIVES

1. Use the Biot-Savart law to calculate the magnetic induction at a specified point in the vicinity of a current element, and by integration find the total magnetic field due to a number of important geometric arrangements. Your use of the Biot-Savart law must include a clear understanding of the *direction* of the magnetic field contribution relative to the direction of the current element which produces it and the direction of the vector which locates the point at which the field is to be calculated.

2. Understand the basis for defining the ampere and the coulomb in terms of the magnetic force between parallel current-carrying conductors.

3. Use Ampere's law to calculate the magnetic field due to steady current configurations which have a sufficiently high degree of symmetry such as a long straight conductor, a long solenoid, and a toroidal coil.

4. Calculate the magnetic field at interior points and at exterior axial points of a solenoid.

5. Calculate the magnetic flux through a surface area placed in either a uniform or nonuniform magnetic field.

6. Understand, via the generalized form of Ampere's law, that magnetic fields are produced both by *conduction currents* and by *changing electric fields*.

SKILLS

It is important to remember that the *Biot-Savart law*, given by Eq. 30.4

$$d\mathbf{B} = \frac{\mu_o I}{4\pi} \frac{d\mathbf{s} \times \hat{\mathbf{r}}}{r^2}$$

is a *vector* expression. The unit vector $\hat{\mathbf{r}}$ is directed from the element of conductor d**s** to the point P where the magnetic field is to be calculated, and r is the distance from d**s** to point P. For the arbitrary current element shown in Figure 30.1, the direction of **B** at point P, as determined by the right-hand rule for the cross product, is directed *out of the plane*; while the magnetic field at point P' due to the current in the element d**s** is directed *into the plane*. In order to find the *total* magnetic field at any point due to a conductor of finite sign, you must sum up the contributions from all current elements making up the conductor. This means that the total **B** field is expressed as an integral over the entire length of the conductor:

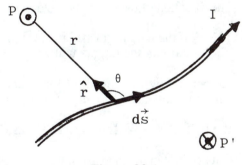

Figure 30.1

$$\mathbf{B} = \frac{\mu_o I}{4\pi} \int \frac{d\mathbf{s} \times \hat{\mathbf{r}}}{r^2}$$

365

NOTES FROM SELECTED CHAPTER SECTIONS

30.1 THE-BIOT-SAVART LAW

The *Biot-Savart law* says that if a wire carries a steady current I, the magnetic field d**B** at a point P associated with an element ds has the following properties:

1. The vector d**B** is perpendicular both to ds (which is in the direction of the current) and to the unit vector $\hat{r}$ directed from the element to the point P.

2. The magnitude of d**B** is inversely proportional to r^2, where r is the distance from the element to the point P.

3. The magnitude of d**B** is proportional to the current and to the length ds of the element.

4. The magnitude of d**B** is proportional to sin θ, where θ is the angle between the vectors ds and $\hat{r}$.

30.2 THE MAGNETIC FORCE BETWEEN TWO PARALLEL CONDUCTORS

Parallel conductors carrying currents in the *same direction attract* each other, whereas parallel conductors carrying currents in *opposite directions repel* each other.

The force between two parallel wires each carrying a current is used to define the ampere as follows:

If two long, parallel wires 1 m apart carry the same current and the force per unit length on each wire is 2×10^{-7} N/m, then the current is defined to be 1 A.

30.3 AMPERE'S LAW

The direction of the magnetic field due to a current in a conductor is given by the right-hand rule:

If the wire is grasped in the right hand with the thumb in the direction of the current, the fingers will wrap (or curl) in the direction of **B**.

Ampere's law is valid only for *steady* currents and is useful only in those cases where the current configuration has a *high degree* of *symmetry*.

30.7 GAUSS'S LAW IN MAGNETISM

Gauss's law in magnetism states that the net magnetic flux through any closed surface is always zero.

30.8 DISPLACEMENT CURRENT AND THE GENERALIZED AMPERE'S LAW

Magnetic fields are produced both by conduction currents and by changing electric fields.

30.9 MAGNETISM IN MATTER

In order to describe the magnetic properties of materials, it is convenient to classify the material into three categories: paramagnetic, ferromagnetic, and diamagnetic. *Paramagnetic and ferromagnetic materials* are those that have atoms with permanent magnetic dipole moments. *Diamagnetic materials* are those whose atoms have no permanent magnetic dipole moments. For materials whose atoms have permanent magnetic moments, the diamagnetic contribution to the magnetism is usually overshadowed by a paramagnetic or ferromagnetic effect.

Paramagnetic substances have a positive but small susceptibility which is due to the presence of atoms (or ions) with *permanent* magnetic dipole moments. These dipoles interact only weakly with each other and are randomly oriented in the absence of an external magnetic field. Experimentally, one finds that the magnetization of a paramagnetic substance is proportional to the applied field and inversely proportional to the absolute temperature under a wide range of conditions.

A diamagnetic substance is one whose atoms have no permanent magnetic dipole moment. When an external magnetic field is applied to a diamagnetic substance, a weak magnetic dipole moment is *induced* in the direction opposite the applied field.

The effect called magnetic hysteresis shows that the magnetization of a ferromagnetic substance depends on the history of the substance as well as the strength of the applied field.

The magnetization curve is useful for another reason. *The area enclosed by the magnetization curve represents the work required to take the material through the hysteresis cycle.* Ferromagnetic substances contain atomic magnetic moments that tend to align parallel to each other even in a weak external magnetic field. Once the moments are aligned, the substance will remain magnetized after the external field is removed.

EQUATIONS AND CONCEPTS

The *Biot-Savart law* gives the magnetic field at a point in space due to an element of conductor ds which carries a current I and is at a distance r away from the point.

$$d\mathbf{B} = \frac{\mu_o}{4\pi}\frac{I d\mathbf{s} \times \hat{\mathbf{r}}}{r^2}$$

(30.4)

The *total magnetic field* is found by integrating the Biot-Savart law expression over the entire length of the conductor.

$$\mathbf{B} = \frac{\mu_o I}{4\pi}\int \frac{d\mathbf{s} \times \hat{\mathbf{r}}}{r^2}$$

(30.5)

The magnetic field due to several important geometric arrangements of a current-carrying conductor can be calculated by use of the Biot-Savart law:

B at a distance a from a *long straight conductor*.

$$B = \frac{\mu_o I}{2\pi a}$$

(30.7)

B at the center of an *arc of radius R which subtends an angle θ (in radians) at the center*.

$$B = \frac{\mu_o I}{4\pi R}\theta$$

(30.8)

B on the axis of a *circular loop* of radius R and at a *distance x from the plane* of the loop.

$$B = \frac{\mu_0 I R^2}{2\left[x^2 + R^2\right]^{3/2}}$$ (30.9)

B at a distance a from a *straight wire* carrying a current I, where θ_1 and θ_2 are as shown in Figure 30.2.

$$B = \frac{\mu_0 I}{4\pi a}(\cos\theta_1 - \cos\theta_2)$$ (30.6)

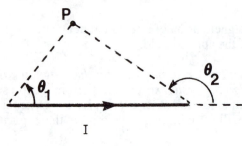

Figure 30.2

The magnitude of the *magnetic force per unit length* between parallel conductors depends on the distance a between the conductors and the magnitudes of the two currents.

$$\frac{F}{L} = 2k_m\frac{I_1 I_2}{a}$$

If the parallel currents I_1 and I_2 are in the same direction, the force between conductors will be one of attraction. Parallel conductors carrying currents in opposite directions will repel each other. In any case, the magnitude of the forces on the two conductors will be *equal*.

Ampere's law represents a relationship between the integral of the tangential component of the magnetic field around any closed path and the total current which threads the closed path.

$$\oint \mathbf{B}\cdot d\mathbf{s} = \mu_0 I$$ (30.15)

B *inside a toroidal coil* of N turns and at a distance r from the center.

$$B = \frac{\mu_0 N I}{2\pi r}$$ (30.18)

B near the center of a *solenoid* of n turns per unit length.

$$B = \mu_0 n I$$ (30.20)

The *magnetic flux* through a surface is the integral of the normal component of the field over the surface.

$$\Phi_m = \int \mathbf{B}\cdot d\mathbf{A}$$ (30.23)

The magnetic moment μ of an orbiting electron is proportional to its orbital angular momentum L.

$$\mu = \left(\frac{e}{2m}\right)L$$

$$L = 0, \hbar, 2\hbar, 3\hbar, \ldots$$

(30.30)

The intrinsic magnetic moment μ_s associated with the spin of the electron is called the Bohr magneton.

$$\mu_s = 9.27 \times 10^{-24} \text{ J/T}$$

(30.32)

The magnetic state of a substance is described by a quantity called the magnetization vector, **M**. For paramagnetic and diamagnetic substances, the magnetization is proportional to the magnetic field strength.

$$\mathbf{M} = \chi\mathbf{H}$$

(30.36)

In a region where the total magnetic field is due to that of a current-carrying conductor and the presence of a magnetic substance, the *total field* B can be expressed in terms of **M** and **H**.

$$\mathbf{B} = \mu_o(\mathbf{H} + \mathbf{M})$$

(30.34)

The total field can also be expressed in terms of the permeability κ_m of the substance.

$$\mathbf{B} = \kappa_m\mathbf{H}$$

(30.37)

The permeability of a magnetic substance is related to its magnetic susceptibility.

$$\kappa_m = \mu_o(1 + \chi)$$

(30.38)

The magnetization of a paramagnetic substance is proportional to the applied field and inversely proportional to the absolute temperature. This is known as Curie's law.

$$\mathbf{M} = C\left(\frac{\mathbf{B}}{T}\right)$$

(30.39)

EXAMPLE PROBLEM SOLUTION

Example 30.1 Find the magnetic field at the center of a rectangular conductor of dimensions 0.4 m × 0.25 m (Figure 30.3 on the following page) which carries a steady current of 5 A.

Solution

When the current has the direction shown in the figure, the magnetic field due to each segment of the rectangle at its center is directed *into* the page according to the right-hand rule. Therefore, the net magnetic field at the center of the rectangle is $B = 2B_L + 2B_w$ where B_L and B_w represent the field contributions due to a vertical segment and a horizontal segment, respectively.

Using Eq. 30.6, we find that

$$B_L = \frac{\mu_o I}{4\pi a} (\cos \theta_1 - \cos \theta_2)$$

where

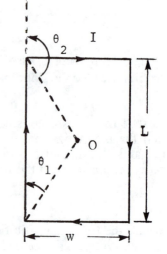

Figure 30.3

$$a = \frac{w}{2}, \quad \cos \theta_1 = \frac{L}{\sqrt{L^2 + w^2}}, \quad \text{and} \quad \cos \theta_2 = -\cos \theta_1$$

so

$$B_L = \frac{\mu_o I}{4\pi \left(\frac{w}{2}\right)} \left[\frac{L}{\sqrt{L^2 + w^2}} - \left(-\frac{L}{\sqrt{L^2 + w^2}} \right) \right] = \frac{\mu_o I}{\pi w} \left(\frac{L}{\sqrt{L^2 + w^2}} \right)$$

Again using Eq. 30.6, with $\quad a = \frac{L}{2} \quad$ and $\quad \cos \theta_1 = -\cos \theta_2 = \frac{w}{\sqrt{L^2 + w^2}}$

we find that

$$B_w = \frac{\mu_o I}{\pi L} \left(\frac{w}{\sqrt{L^2 + w^2}} \right)$$

so that

$$B_{total} = (2) \frac{\mu_o I}{\pi \sqrt{L^2 + w^2}} \left(\frac{L}{w} + \frac{w}{L} \right) = \frac{2\mu_o I}{\pi} \left(\frac{\sqrt{L^2 + w^2}}{Lw} \right)$$

For the numerical values given, we get

$$B_{total} = \frac{(2)(4\pi \times 10^{-7} \text{ Wb/A·m})(5 \text{ A})}{\pi} \left(\frac{\sqrt{(0.4)^2 + (0.25)^2} \text{ m}}{(0.4)(0.25) \text{ m}^2} \right) = 18.9 \ \mu\text{T}$$

That is, the total field at the center of the rectangle has a magnitude of 18.9 μT and is directed *into* the page.

ANSWERS TO SELECTED QUESTIONS

1. Is the magnetic field due to current loop uniform? Explain.

Answer: No. The sketch at the right gives an idea of the nonuniform magnetic field pattern of a current loop.

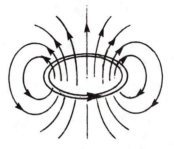

2. A current in a conductor produces a magnetic field which can be calculated using the Biot-Savart law. Since current is defined as the rate of flow of charge, what can you conclude about the magnetic field due to stationary charges? What about moving charges?

Answer: The Biot-Savart law says that the magnetic field of a current element is proportional to the current carried by that element. If charges are at rest, the current is zero, hence they produce no magnetic field. Magnetic fields are produced by charges in motion.

3. Two parallel wires carry currents in opposite directions. Describe the nature of the resultant magnetic field due to the two wires at points (a) between the wires and (b) outside the wires in a plane containing the wires.

Answer: The sketch at the right describes the directions of the fields in the various regions. (a) Between the wires, each wire produces a field into the paper, so the resultant field is into the paper in this region. (b) Outside the wires, the fields due to the two wires are in opposite directions. The *resultant* field is out of the paper in the left-hand region, and also out of the paper in the right-hand region.

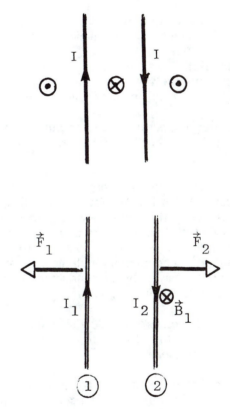

4. Explain why two parallel wires carrying currents in opposite directions repel each other.

Answer: The figure at the right will help you understand this result. The magnetic field due to wire 1 at the position of wire 2 is into the paper. Hence, the magnetic force on wire 2, given by $I_2 L_1 \times B_1$ must be to the right since $L_1 \times B_1$ is to the right. Likewise, you should show that the magnetic force on wire 1 due to the field of wire 2 is to the left.

5. Two wires carrying equal and opposite currents are twisted together in the construction of a circuit. Why does the technique reduce stray magnetic fields?

Answer: The magnetic fields due to the wires will tend to cancel each other since the currents are in opposite directions, and the wires follow the same paths.

6. Is Ampere's law valid for all closed paths surrounding a conductor? Why is it not useful for calculating **B** for all such paths?

Answer: Yes, it is valid for any closed path surrounding one or more conductors. However, it is only useful, in practice, for special situations in which a closed path can be found such that the magnetic field has a constant value everywhere on the path. This only occurs in highly symmetric cases, such as the long straight wire, the long solenoid, the toroid, and the infinite sheet of current.

7. Compare Ampere's law with the Biot-Savart law. Which is the more general method for calculating B for a current-carrying conductor?

Answer: The Biot-Savart law can be used to calculate the magnetic field of any current-carrying conductor, regardless of its geometry, so it represents a general procedure. Ampere's circuital law is valid only for steady currents, and is only useful for calculating the magnetic field of conductors with a high degree of symmetry.

8. Is the magnetic field inside a toroidal coil uniform? Explain.

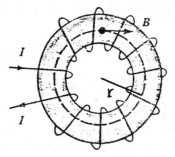

Answer: No. The magnetic field inside a toroidal coil varies 1/r, where r is the distance from the center of the configuration.

9. Describe the similarities between Ampere's law in magnetism and Gauss' law in electrostatics.

Answer: Both laws are useful for calculating the fields of highly symmetric current or charge distributions. Ampere's law involves a line integral over a closed path through which currents may pass, while Gauss' law involves a surface integral over a surface which may enclose some net charge.

10. A hollow copper tube carries a current. Why is **B** = 0 inside the tube? Is **B** nonzero outside the tube?

Answer: Let us apply Ampere's circuital law to the closed path labeled 1 in the figure. Since there is no current linking this path, and because of the symmetry of the configuration, we see that the magnetic field inside the tube must be zero. On the other hand, the net current through the path labeled 2 is I, the current carried by the conductor. Therefore, the field outside the tube is nonzero.

11. Why is **B** nonzero outside a solenoid? Why is **B** = 0 outside a toroid? (Note that the lines of **B** must form closed paths.)

Answer: Since the lines of **B** must form closed paths, we see that this can only occur if the lines of the solenoid return outside the solenoid. In the case of the toroid, all the closed paths forming the lines of **B** are contained within the ring.

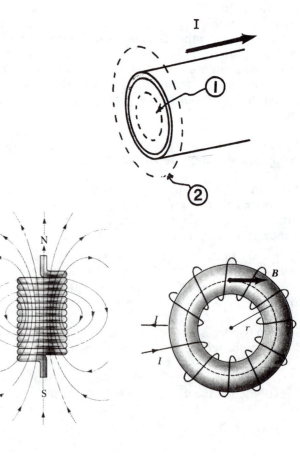

12. Describe the change in the magnetic field inside a solenoid carrying a steady current I if (a) the length of the solenoid is doubled, but the number of turns remains the same; and (b) the number of turns is doubled, but the length remains the same.

Answer: The magnetic field at the center of a long solenoid is given by $B = \mu_0 NI/L$. (a) If the length of L is doubled, but N remains the same, we see that the field is halved. (b) If the number of turns, N, is doubled, the field is doubled.

13. A plane conducting loop is located in a uniform magnetic field that is directed along the x axis. For what orientation of the loop is the flux through it a maximum? For what orientation is the flux a minimum?

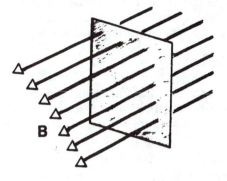

Answer: The flux is a maximum when the plane of the loop is perpendicular to the field as shown. The flux is zero when the plane of the loop is parallel with the field lines.

28. What is the difference between hard and soft ferromagnetic materials?

Answer: Magnetic flux is a measure of the number of magnetic field lines passing through a particular surface. The idea of a magnetic field refers to the intensity (or field strength) in some region due to a system of moving charges (or currents).

SOLUTIONS TO SELECTED END-OF-CHAPTER PROBLEMS

11. A closed current path shaped as shown in Figure 30.39 produces a magnetic field at P, the center of the arc. If the arc subtends an angle of 30° and the total length of wire in the closed path is 1.2 m, what are the magnitude and direction of the field produced at P if the current in the loop is 3 A?

Solution

For the radial segments, $ds \times \hat{r} = 0$; and for the arc, $|ds \times \hat{r}| = ds$. The magnetic field at point P is

$$B = \frac{\mu_0 I}{4\pi R}\theta \quad \text{where} \quad R = \frac{L}{2 + \theta} \quad (\theta \text{ in radians})$$

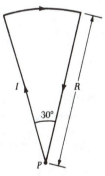

Figure 30.39

$$B = \frac{\mu_0 I(2 + \theta)\theta}{4\pi L} = \frac{\left(4\pi \times 10^{-7}\ \frac{N}{A^2}\right)(3\ A)\left(2 + \frac{\pi}{6}\right)\frac{\pi}{6}}{4\pi(1.2\ m)}$$

$$B = 3.31 \times 10^{-7}\ T \quad \text{or} \quad B = \boxed{0.331\ \mu T}$$

19. For the arrangement shown in Figure 30.42, the current in the long, straight conductor has the value $I_1 = 5$ A and lies in the plane of the rectangular loop, which carries a current $I_2 = 10$ A. The dimensions are $c = 0.1$ m, $a = 0.15$ m, and $L = 0.45$ m. Find the magnitude and direction of the *net force* exerted on the rectangle by the magnetic field of the straight current-carrying conductor.

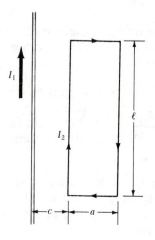

Solution

By symmetry the forces exerted on the segments of length, a, are equal and opposite and cancel. The magnetic field in the plane of I_2 to the right of I_1 is directed downward into the plane. By the right-hand rule, $\mathbf{F} = I(\mathbf{L} \times \mathbf{B})$ is directed toward the *left* for the near side of the loop and directed toward the *right* for the side at c + a.

Figure 30.42

$$\therefore \quad \mathbf{F} = \mathbf{F}_1 + \mathbf{F}_2 = \frac{\mu_0 I_1 I_2 L}{2\pi} \left(\frac{1}{c+a} - \frac{1}{c} \right) \mathbf{i}$$

$$\mathbf{F} = \frac{\mu_0 I_1 I_2 L}{2\pi} \left[\frac{a}{c(c+a)} \right] \mathbf{i}$$

$$\mathbf{F} = \frac{\left(4\pi \times 10^{-7} \frac{N}{A^2} \right)(5 \text{ A})(10 \text{ A})(0.45 \text{ m})}{2\pi} \left[\frac{0.15 \text{ m}}{(0.1 \text{ m})(0.25 \text{ m})} \right] \mathbf{i}$$

$$\mathbf{F} = (-2.70 \times 10^{-5} \mathbf{i}) \text{ N} \qquad \text{or} \qquad \boxed{\mathbf{F} = 2.70 \times 10^{-5} \text{ N}} \quad \text{toward the left}$$

27. The magnetic coils of a Tokamak fusion reactor are in the shape of a toroid having an inner radius of 0.7 m and outer radius of 1.3 m. Inside the toroid is the plasma. If the toroid has 900 turns of large diameter wire, each of which carries a current of 14000 A, find the magnetic field strength along (a) the inner radius of the toroid, and (b) the outer radius of the toroid.

Solution

From Ampere's law, the magnetic field at a distance r from the center of the toroid is found to be

$$B = \frac{\mu_0 NI}{2\pi r}$$

(a) Along the inner radius,

$$B_i = \frac{\left(4\pi \times 10^{-7} \frac{N}{A^2} \right)(900)(1.4 \times 10^4 \text{ A})}{2\pi (0.7 \text{ m})} = \boxed{3.60 \text{ T}}$$

(b) Along the outer radius, where r = 1.3 m, we find

$$B = \boxed{1.94 \text{ T}}$$

---

37. A cube of edge length $L = 2.5$ cm is positioned as shown in Figure 30.46. There is a uniform magnetic field throughout the region given by the expression $\mathbf{B} = (5\mathbf{i} + 4\mathbf{j} + 3\mathbf{k})$ T. (a) Calculate the flux through the shaded face of the cube. (b) What is the total flux through the six faces of the cube?

Solution

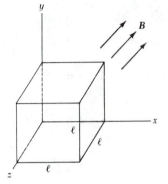

(a) $\Phi = \mathbf{B} \cdot \mathbf{A}$

$\Phi = B_x A_x + B_y A_y + B_z A_z$

$\Phi = (5 \text{ T})(0.025 \text{ m})^2 = 3.13 \times 10^{-3} \text{ Wb} = \boxed{3.13 \text{ mWb}}$

Figure 30.46

(b) For a closed surface, $\oint \mathbf{B} \cdot d\mathbf{A} = 0$ so $\Phi = 0$

---

41. The applied voltage across the plates of a 4-μF capacitor varies in time according to the expression

$$V_{app} = (8 \text{ V})(1 - e^{-t/4})$$

where t is in s. Calculate (a) the displacement current as a function of time and (b) the value of the current at t = 4 s.

Solution

(a) $I_d = C \dfrac{dV}{dt}$

$I_d = C \dfrac{d}{dt}\left[8\left(1 - e^{-t/4}\right)\right]$

$I_d = 8C\left(\frac{1}{4} e^{-t/4}\right) = 2(4 \times 10^{-6} \text{ F})e^{-t/4} \text{ V/s} = \boxed{(8 \times 10^{-6} \text{ A})e^{-t/4}}$

(b) At t = 4 s,

$$I_d = (8 \times 10^{-6} \text{ A})e^{-1} = 2.94 \times 10^{-6} \text{ A} = \boxed{2.94 \text{ }\mu\text{A}}$$

---

45. What is the relative permeability of a material that has a magnetic susceptibility of 10^{-4}?

Solution $\mu = \mu_o(1 + \chi)$

Relative permeability, $\dfrac{\mu}{\mu_o} = 1 + \chi = 1 + 10^{-4} = \boxed{1.0001}$

---

51. A coil of 500 turns is wound on an iron ring ($\kappa_m = 750\,\mu_o$) of 20 cm mean radius and 8 cm² cross-sectional area. Calculate the magnetic flux Φ in this Rowland ring when the current in the coil is 0.5 A.

Solution $\Phi = BA$ where $B = \mu nI$ so

$$\Phi = \mu nIA = (750)\left(4\pi \times 10^{-7}\,\frac{N}{A^2}\right)\left(\frac{500}{2\pi(0.2\text{ m})}\right)(0.5\text{ A})(8 \times 10^{-4}\text{ m}^2)$$

$$\Phi = 1.50 \times 10^{-4}\text{ Wb} = \boxed{150\ \mu\text{Wb}}$$

---

65. A very long, thin strip of metal of width w carries a current I along its length as in Figure 30.51. Find the magnetic field in the *plane* of the strip (at an external point P) a distance b from one edge.

Solution

Consider a long filament of the strip which has width dr and is a distance r from point b. The B field at a distance r from a long conductor is

$$B = \frac{\mu_o I}{2\pi r}$$

Thus, the field due to the thin filament is

$$dB = \frac{\mu_o\, dI}{2\pi r}$$

where $dI = I\left(\dfrac{dr}{w}\right)$ so

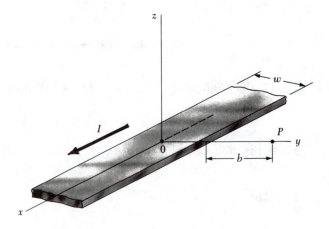

Figure 30.51

$$B = \int_{b}^{b+w} \frac{\mu_o}{2\pi r}\left(I\,\frac{dr}{w}\right) = \frac{\mu_o I}{2\pi w}\int_{b}^{b+w} \frac{dr}{r} = \boxed{\frac{\mu_o I}{2\pi w}\,Ln\left(1 + \frac{w}{b}\right)}$$

73. A long cylindrical conductor of radius R carries a current I as in Figure 30.54. The current density J, however, is *not* uniform over the cross section of the conductor but is a function of the radius according to J = br, where b is a constant. Find an expression for the magnetic field B (a) at a distance $r_1 < R$ and (b) at a distance $r_2 > R$, measured from the axis.

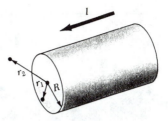

Solution

Figure 30.54

$$\oint \mathbf{B} \cdot d\mathbf{s} = \mu_o I \quad \text{where for nonuniform current density}$$

$$I = \int \mathbf{J} \cdot d\mathbf{A}$$

In this case **B** is parallel to ds and **J** is parallel to dA so Ampere's law gives

$$\oint B \, ds = \mu_o \int J \, dA$$

(a)   When $r = r_1 < R$,

$$2\pi r_1 B = \mu_o \int_o^{r_1} br(2\pi r \, dr)$$

$$B = \frac{\mu_o b r_1^2}{3} \quad \text{(inside)}$$

(b)   When $r = r_2 > R$,

$$2\pi r_2 B = \mu_o \int_o^R br(2\pi r \, dr)$$

$$B = \frac{\mu_o b R^3}{3 r_2} \quad \text{(outside)}$$

31

Faraday's Law

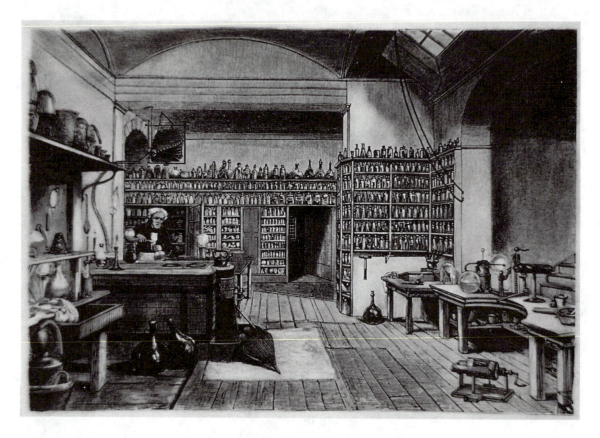

OBJECTIVES

1. Calculate the emf (or current) induced in a circuit when the magnetic flux through the circuit is changing in time. The variation in flux might be due to a change in (a) the area of the circuit, (b) the magnitude of the magnetic field, (c) the direction of the magnetic field, or (d) the orientation/location of the circuit in the magnetic field.

2. Calculate the emf induced between the ends of a conducting bar as it moves through a region where there is a constant magnetic field (motional emf).

3. Apply Lenz's law to determine the direction of an induced emf or current. You should also understand that Lenz's law is a consequence of the law of conservation of energy.

4. Calculate the maximum and instantaneous values of the sinusoidal emf generated in a conducting loop rotating in a constant magnetic field.

5. Calculate the electric field at various points in a charge free region when the time variation of the magnetic field over the region is specified.

SKILLS

It is important to distinguish clearly between the *instantaneous value* of emf induced in a circuit and the *average value* of the emf induced in the circuit over a finite time interval.

To calculate the average induced emf, it is often useful to write Eq. 31.3 as

$$\mathcal{E}_{av} = -N \left(\frac{d\Phi_m}{dt} \right)_{av} = -N \frac{\Delta\Phi_m}{\Delta t}$$

or

$$\left| \mathcal{E}_{av} \right| = N \frac{\left(\Phi_{mf} - \Phi_{mi} \right)}{\Delta t}$$

where the subscripts i and f refer to the magnetic flux through the circuit at the beginning and end of the time interval Δt. For a circuit (or loop) in a single plane, $\Phi_m = BA \cos \theta$, where θ is the angle between the direction of the normal to plane of the circuit (conducting loop) and the direction of the magnetic field.

Eq. 31.4 can be used to calculate the *instantaneous value of an induced emf*. For a multi-turn loop, the magnitude of the induced emf is

$$\left| \mathcal{E} \right| = N \frac{d}{dt} (BA \cos \theta)$$

where in a particular case B, A, θ, or any combination of those parameters can be time dependent while the others remain constant. The expression resulting from the differentiation is then evaluated using the values of B, A, and θ corresponding to the specified value.

NOTES FROM SELECTED CHAPTER SECTIONS

34.1 FARADAY'S LAW OF INDUCTION

The emf induced in a circuit is proportional to the time rate of change of magnetic flux through the circuit.

31.3 LENZ'S LAW

The polarity of the induced emf is such that it tends to produce a current that will create a magnetic flux to oppose the *change in flux* through the circuit.

EQUATIONS AND CONCEPTS

Faraday's law of induction states that the emf induced in a circuit is directly proportional to the time rate of change of magnetic flux through the circuit.

$$\mathcal{E} = -\frac{d\Phi_m}{dt} \tag{31.1}$$

Faraday's law can be written in a more *general form* in terms of the integral of the electric field around a closed path. In this form the electric field is a *non-conservative, time-varying field*.

$$\oint \mathbf{E} \cdot d\mathbf{s} = -\frac{d\Phi_m}{dt} \tag{31.9}$$

or

$$\int \mathbf{E} \cdot d\mathbf{s} = -\frac{d}{dt}\int \mathbf{B} \cdot d\mathbf{A}$$

A *"motional emf"* is induced in a conductor of length L moving with a speed v perpendicular to a uniform magnetic field **B**.

$$\mathcal{E} = -BLv \tag{31.5}$$

When a conducting coil of N turns and cross-sectional area A rotates at constant angular velocity ω in a magnetic field **B**, *the emf induced in the loop* will vary sinusoidally in time. For a given loop, the maximum value of the induced emf $\mathcal{E}_{max}$ will be proportional to the rate of rotation in the field.

$$\mathcal{E} = NBA\omega \sin \omega t \tag{31.10}$$

$$\mathcal{E}_{max} = NBA\omega \tag{31.11}$$

EXAMPLE PROBLEM SOLUTION

Example 31.1 Consider a 100-turn rectangular coil of area A located in a magnetic field B as shown in Figure 31.1.

(a) Let A = 0.08 m² and B = 0.5 T. If the loop is rotating about an axis in the plane of the loop and perpendicular to **B** with a frequency of 2 Hz, calculate the magnitude of the *average emf* induced in the loop as the coil rotates from a position where θ = 30° to a position where θ = 60°.

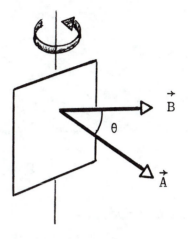

Figure 31.1

Solution

From Eq. 31.3,

$$\left| \mathcal{E}_{av} \right| = N\left(\frac{d\Phi_m}{dt}\right)_{av} = N\left(\frac{\Delta\Phi_m}{\Delta t}\right) = N\frac{\left|\Phi_{mf} - \Phi_{mi}\right|}{\Delta t}$$

since $\Phi_m = BA \cos \theta$

$$\left| \mathcal{E}_{av} \right| = N\frac{BA\left|\cos\theta_f - \cos\theta_i\right|}{\Delta t}$$

Since the frequency of rotation is 2 Hz, the angular frequency is $\omega = 2\pi f = 4\pi$ rad/s. Furthermore, $\Delta\theta = 30° = \pi/6$ rad. Using $\omega = \Delta\theta/\Delta t$ gives

$$\Delta t = \frac{\Delta\theta}{\omega} = \frac{\pi/6 \text{ rad}}{4\pi \text{ rad/s}} = 0.042 \text{ s}$$

Using this value of Δt and the given values for N, B, and A gives

$$\left| \mathcal{E}_{av} \right| = 100\frac{(0.5 \text{ T})(0.08 \text{ m}^2)\left|\cos 60° - \cos 30°\right|}{0.042 \text{ s}}$$

$$\left| \mathcal{E}_{av} \right| = 34.9 \text{ V}$$

(b) Using the values of A, B, and the frequency from part (a), calculate the magnitude of the *instantaneous emf* in the coil when θ = 30°.

Solution

From Eq. 31.4,

$$\mathcal{E} = -N \frac{d}{dt}(BA \cos \theta)$$

In this case, B and A have constant values, so

$$\mathcal{E} = -NBA \frac{d}{dt} \cos \theta = NBA \sin \theta \frac{d\theta}{dt}$$

Since f = 2 Hz, ω is given by

$$\omega = \frac{d\theta}{dt} = 2\pi f = 4\pi \text{ rad/s}$$

Therefore,

$$\mathcal{E} = 100(0.5\text{T})(0.08 \text{ m}^2)(\sin 30°)(4\pi \text{ rad/s})$$

$$\mathcal{E} = 25.1 \text{ V}$$

ANSWERS TO SELECTED QUESTIONS

2. A circular loop is located in a uniform and constant magnetic field. Describe how an emf can be induced in the loop in this situation.

Answer: An emf can be induced by either rotating the loop or by changing the dimensions of the loop.

4. As the conducting bar in Figure 31.21 moves to the right, an electric field is set up directed downward. If the bar were moving to the left, explain why the electric field would be upward.

Answer: If the bar were moving to the left, the magnetic force on the negative charge carriers in the bar would be upward, causing an accumulation of negative charge on the top and positive on the bottom. Hence, the electric field in the bar would be upwards.

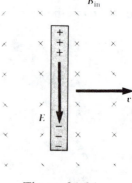

Figure 31.21

5. As the bar in Figure 31.21 moves perpendicular to the field, is an external force required to keep it moving with constant velocity?

Answer: No. Once the bar is in motion and the charges are separated, no external force is necessary to maintain the motion. In fact, an external force in the x direction will cause the bar to accelerate.

6. The bar in Figure 31.22 moves on rails to the right with a velocity **v**, and the uniform, constant magnetic field is *outward*. Why is the induced current clockwise? If the bar were moving to the left, what would be the direction of the induced current?

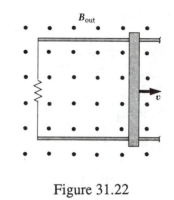

Answer: The external flux is out of the paper and is increasing as the bar moves to the right. Hence, the induced current in the loop must produce a flux into the paper *to oppose the increasing external flux out of the paper*. This corresponds to a clockwise current. If the bar were moving to the left, the induced current would be reversed, or counter-clockwise.

Figure 31.22

10. Will dropping a magnet down a long copper tube produce a current in the tube? Explain.

Answer: Yes. The current will be made to flow around the tube. As the south pole passes a particular point in the tube, the induced current will flow in one direction, and when the north pole passes that point, the current will flow in the opposite direction.

13. What happens when the coil of a generator is rotated at a faster rate?

Answer: A higher value of maximum voltage will be induced. In addition, if the generator is connected to produce alternating current, the direction of current will alternate at a higher frequency.

14. Could a current be induced in a coil by rotating a magnet inside the coil? If so, how?

Answer: Yes, if the magnet were rotated end-to-end. This would cause the magnetic field inside the coil to change, being first in one direction and then in the other.

15. When the switch in the circuit shown in Figure 31.23a is closed, a current is set up in the coil and the metal ring springs upward (see Figure 31.23b). Explain this behavior.

Answer: Eddy currents are induced in the conducting ring as a result of the *changing* field due to the solenoid. There is a force exerted on these eddy currents by the magnetic field of the solenoid.

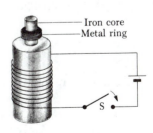

Figure 31.23

SOLUTIONS TO SELECTED END-OF-CHAPTER PROBLEMS_____

13. A coil, formed by wrapping 50 turns of wire in the shape of a square, is positioned in a magnetic field so that the normal to the plane of the coil makes an angle of 30° with the direction of the field. It is observed that if the magnitude of the magnetic field is increased uniformly from 200 μT to 600 μT in 0.4 s, an emf of 80 mV is induced in the coil. What is the total length of the wire?

Solution

$$\mathcal{E} = -N \frac{d\Phi}{dt} = -N \frac{d}{dt}(BA \cos \theta) = -NA \cos \theta \frac{dB}{dt}$$

$$|\overline{\mathcal{E}}| = NA \cos \theta \left(\frac{\Delta B}{\Delta t}\right) \qquad \text{and} \qquad A = \frac{|\overline{\mathcal{E}}|}{N \cos \theta \left(\frac{\Delta B}{\Delta t}\right)}$$

$$A = \frac{80 \times 10^{-3} \text{ V}}{(50)(\cos 30°)\left(\frac{600 - 200}{0.4}\right) \times 10^{-6} \text{ T/s}} = 1.85 \text{ m}^2$$

The edge length of the coil, $d = \sqrt{A}$, and the total length of the wire is $L = N(4d)$. Therefore,

$$L = (50)(4)\sqrt{1.85 \text{ m}^2} = \boxed{272 \text{ m}}$$

19. In the arrangement shown in Figure 31.28, a conducting bar moves to the right along parallel, frictionless conducting rails connected on one end by a 6-Ω resistor. A 2.5-T magnetic field is directed *into* the paper. Let $L = 1.2$ m and neglect the mass of the bar. (a) Calculate the applied force required to move the bar to the right at a *constant* speed of 2 m/s. (b) At what rate is energy dissipated in the resistor?

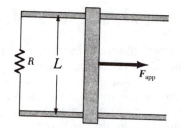

Figure 31.28

Solution

(a) At constant speed the net force on the moving bar equals zero, or

$$|\mathbf{F}_{app}| = I|\mathbf{L} \times \mathbf{B}|$$

where the current in the bar $I = \dfrac{\mathcal{E}}{R}$ and $\mathcal{E} = BLv$. Therefore,

$$F_{app} = \left(\frac{BLv}{R}\right)LB = \frac{B^2L^2v}{R} = \frac{(2.5 \text{ T})^2(1.2 \text{ m})^2(2 \text{ m/s})}{6\,\Omega} = \boxed{3.0 \text{ N}}$$

(b) $P = F_{app}v = (3 \text{ N})(2 \text{ m/s}) = \boxed{6.0 \text{ W}}$

29. A 0.15-kg wire in the shape of a closed rectangle 1 m wide and 1.5 m long has a total resistance of 0.75 Ω. The rectangle is allowed to fall through a magnetic field directed perpendicular to the direction of motion of the wire (Figure 31.30). The rectangle accelerates downward until it acquires a *constant* speed of 2 m/s with the top of the rectangle not yet in that region of the field. Calculate the value of B.

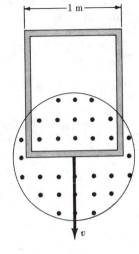

Solution   At constant speed,

$$F_g = F_B \quad \text{or} \quad mg = ILB.$$

Therefore, $B = \dfrac{mg}{IL}$

But $I = \dfrac{\mathcal{E}}{R} = \dfrac{BLv}{R}$

Figure 31.30

so

$$B = \sqrt{\frac{mgR}{vL^2}} = \sqrt{\frac{(0.15 \text{ kg})(9.8 \text{ m/s}^2)(0.75 \ \Omega)}{(2 \text{ m/s})(1 \text{ m})^2}} = \boxed{0.742 \text{ T}}$$

33. A magnetic field directed into the page changes with time according to $B = (0.03t^2 + 1.4)$ T, where t is in s. The field has a circular cross section of radius R = 2.5 cm (Figure 31.32). What are the magnitude and direction of the electric field at point P_1 when t = 3 s and $r_1 = 0.02$ m?

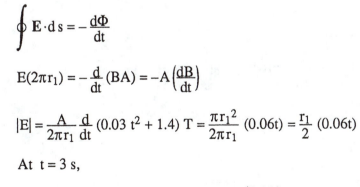

Solution

$$\oint \mathbf{E} \cdot d\mathbf{s} = -\frac{d\Phi}{dt}$$

Figure 31.32

$$E(2\pi r_1) = -\frac{d}{dt}(BA) = -A\left(\frac{dB}{dt}\right)$$

$$|E| = \frac{A}{2\pi r_1} \frac{d}{dt}(0.03 \ t^2 + 1.4) \text{ T} = \frac{\pi r_1^2}{2\pi r_1}(0.06t) = \frac{r_1}{2}(0.06t)$$

At t = 3 s,

$$E = \left(\frac{0.02}{2}\right)(0.06)(3) = 1.8 \times 10^{-3} \text{ N/C}$$

E will be perpendicular to R and counterclockwise.

41. A loop of area 0.1 m² is rotating at 60 rev/s with the axis of rotation perpendicular to a 0.2-T magnetic field. (a) If there are 1000 turns on the loop, what is the maximum voltage induced in the loop? (b) When the maximum induced voltage occurs, what is the orientation of the loop with respect to the magnetic field?

Solution   For a loop rotating in a magnetic field,

(a) $\varepsilon = NBA\omega \sin\theta$ and $\varepsilon_{max} = NBA\omega$

so

$$\varepsilon_{max} = (1000)(0.2 \text{ T})(0.1 \text{ m}^2)(60/\text{s})(2\pi \text{ rad}) = \boxed{7540 \text{ V}}$$

(b) $\varepsilon \rightarrow \varepsilon_{max}$ when $\sin\theta \rightarrow 1$ or $\theta = \pm \dfrac{\pi}{2}$

Therefore, at maximum emf, the plane of the loop is parallel to the field.

---

59. A solenoid wound with 2000 turns/m is supplied with current that varies in time according to $I = 4 \sin (120 \pi t)$, where I is in A and t is in s. A small coaxial circular coil of 40 turns and radius r = 5 cm is located inside the solenoid near its center. (a) Derive an expression that describes the manner in which the emf in the small coil varies in time. (b) At what average rate is energy dissipated in the small coil if the windings have a total resistance of 8 Ω?

Solution

(a)   The emf in the coil is given by

$$\varepsilon = -N \frac{d\Phi}{dt} = -NA\left(\frac{dB}{dt}\right)$$

where N = total number of turns in the coil and B is the magnetic field of the solenoid. Also, $B = \mu_o nI$ where n is the number of turns per unit length of the solenoid.

$$\varepsilon = -NA \frac{d}{dt}(\mu_o nI) = -NA\mu_o n \frac{dI}{dt}$$

$$\varepsilon = -NA\mu_o n \frac{d}{dt}[4 \sin (120 \pi t)]$$

$$\varepsilon = N\mu_o An(480 \pi) \cos (120 \pi t)$$

$$\varepsilon = (40)\left(4\pi \times 10^{-7} \frac{\text{N}}{\text{A}^2}\right)\left[\pi (0.05 \text{ m})^2\right](2 \times 10^3/\text{m})(480 \pi) \cos (120 \pi t)$$

$$\varepsilon = (1.19 \text{ V}) \cos (120 \pi t)$$

(b) $P_{ave} = \dfrac{\varepsilon_{ave}^2}{R} = \left(\dfrac{1}{R}\right)\left(\dfrac{1}{2} \varepsilon_{max}^2\right) = \left(\dfrac{1}{2}\right)\left(\dfrac{1}{8 \text{ }\Omega}\right)(1.19 \text{ V})^2$

$P_{ave} = 8.85 \times 10^{-2} \text{ W}$ or $P_{ave} = \boxed{88.5 \text{ mW}}$

---

65. A square loop of wire with edge length $a = 0.2$ m is perpendicular to the earth's magnetic field at a point where $B = 15$ μT, as in Figure 31.47. The total resistance of the loop and the wires connecting the loop to the galvanometer is 0.5 Ω. If the loop is suddenly collapsed by horizontal forces as shown, what total charge will pass through the galvanometer?

Solution

$$Q = \int Idt = \int \left(\frac{\mathcal{E}}{R}\right) dt$$

$$Q = \frac{1}{R} \int -\left(\frac{d\Phi}{dt}\right) dt = -\frac{1}{R} \int d\Phi$$

$$Q = -\frac{1}{R} \int d(BA) = -\frac{B}{R} \int_{A_1 = a^2}^{A_2 = 0} dA$$

$$Q = \frac{Ba^2}{R} = \frac{(15 \times 10^{-6}\ \text{T})(0.2\ \text{m})^2}{0.5\ \Omega} = 1.20 \times 10^{-6}\ \text{C} = \boxed{1.20\ \mu\text{C}}$$

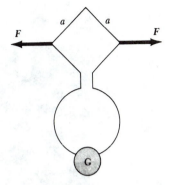

Figure 31.47

32

Inductance

OBJECTIVES

1. Calculate the inductance of a device of suitable geometry.

2. Calculate the magnitude and direction of the self-induced emf in a circuit containing one or more inductive elements when the current changes with time.

3. Determine instantaneous values of the current in an LR circuit while the current is either increasing or decreasing with time.

4. Calculate the total magnetic energy stored in a magnetic field. You should be able to perform this calculation if (1) you are given the values of the inductance of the device with which the field is associated and the current in the circuit, or (2) given the value of the magnetic field intensity throughout the region of space in which the magnetic field exists. In the latter case, you must integrate the expression for the energy density u_m over an appropriate volume.

5. Calculate the emf induced by mutual inductance in one winding due to a time-varying current in a nearby inductor.

6. Determine the expected angular frequency of oscillation of an LC circuit and write out expressions which show how the current in the inductor and the charge on the capacitor vary in time.

7. Understand the essential features of the damped harmonic behavior of an LRC circuit.

8. Determine the time constant of a circuit which contains two or more inductors in series parallel combination.

SKILLS

Equation 32.2, $L = \dfrac{N\Phi_m}{I}$, and Eq. 32.13, $U_m = \dfrac{LI^2}{2}$, provide two different approaches for the calculation of the inductance L of a particular device.

In order to use Eq. 32.2 to calculate L, take the following steps:

(a) Assume a current I to exist in the conductor for which you wish to calculate L (coil, solenoid, coaxial cable, or other device).

(b) Calculate the magnetic flux through the appropriate cross section using $\Phi_m = \displaystyle\int \mathbf{B} \cdot d\mathbf{A}$. Remember that in many cases, $\mathbf{B}$ will not be uniform over the area.

(c) Calculate L directly from the defining Eq. 32.2.

In order to use Eq. 32.13 to calculate L, take the following steps:

(a) Assume a current I in the conductor.

(b) Find an expression for $\mathbf{B}$ for the magnetic field produced by I.

(c) Use Eq. 32.15, $u_m = \dfrac{B^2}{2\mu_0}$ and integrate this value of u_m over the appropriate volume to find the total energy stored in the magnetic field of the inductor $U_m = \displaystyle\int u\, dV$.

(d) Substitute this value of U_m into Eq. 32.13 and solve for L. See Example 32.1 as an illustration of this method.

NOTES FROM SELECTED CHAPTER SECTIONS_____

32.1 SELF-INDUCTANCE

The self-induced emf is always proportional to the *time rate of change* of current in the circuit.

The *inductance* of a device (an inductor) depends on its *geometry*.

32.2 RL CIRCUITS

If a resistor and an inductor are connected in series to a battery, the current in the circuit will reach an *equilibrium* value ($\mathcal{E}/R$) after a time which is long compared to the *time constant* of the circuit, τ.

32.3 ENERGY IN A MAGNETIC FIELD

In an RL circuit, the rate at which energy is supplied by the battery equals the sum of the rate at which heat is dissipated in the resistor and the rate at which energy is stored in the inductor. The *energy density* is proportional to the *square of the magnetic field*.

32.4 MUTUAL INDUCTANCE

If two coils are near each other, a time-varying current in one coil will give rise to an induced emf in the other. The *mutual inductance* depends on the geometry of the two circuits and their orientation with respect to each other. The emf induced by mutual induction in one coil is always proportional to the rate of change of current in the other.

32.5 OSCILLATIONS IN AN LC CIRCUIT

The energy in an LC circuit continuously transfers between energy stored in the electric field of the capacitor and energy stored in the magnetic field of the inductor.

The angular frequency of the oscillations depends only on the values of inductance, L, and capacitance, C.

32.6 THE RLC CIRCUIT

The charge and current in an RLC circuit exhibit damped harmonic oscillations when the value of R in the circuit is small.

EQUATIONS AND CONCEPTS_____

A coil, solenoid, torroid, coaxial cable, or other conducting device is characterized by a parameter called its *inductance*, L. The inductance can be calculated knowing the current and magnetic flux.

$$L = \frac{N\Phi_m}{I} \tag{32.2}$$

The *inductance* of a particular circuit element can also be expressed as the ratio of the induced emf to the time rate of change of current in the circuit.

$$L = -\frac{\mathcal{E}}{dI/dt} \tag{32.3}$$

If the switch in the series circuit of Fig. 32.1 (which contains a battery, resistor, and inductor) is closed in position 1 at time t = 0, *current in the circuit* will increase in a characteristic fashion toward a *maximum* value of $\mathcal{E}/R$.

$$I(t) = \frac{\mathcal{E}}{R}\left(1 - e^{-t/\tau}\right) \tag{32.7}$$

Figure 32.1

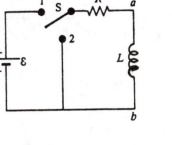

Let the switch in the circuit shown in Figure 32.1 be in position 1 with the current at its maximum value $I_0 = \mathcal{E}/R$. If the switch is thrown to position 2 at t = 0, the current will decay exponentially with time, according to Eq. 32.11.

$$I(t) = \frac{\mathcal{E}}{R} e^{-t/\tau} \tag{32.11}$$

In Eq. 32.7 and Eq. 32.11, the constant in the exponent is called the *time constant* of the circuit, τ.

$$\tau = \frac{L}{R} \tag{32.8}$$

The *energy stored* U_m in the magnetic field of an inductor is proportional to the square of the current in the inductor.

$$U_m = \frac{1}{2} L I^2 \tag{32.13}$$

It is often useful to express the energy in a magnetic field as *energy density* u_m; that is, energy per unit volume.

$$u_m = \frac{B^2}{2\mu_o} \tag{32.15}$$

When two coils are nearby, an emf is produced by mutual induction in one coil which is proportional to the rate of change of current in the other. The *mutual inductance* M depends on the coil geometries and the relative positions of the coils.

$$\mathcal{E}_1 = -M\frac{dI_2}{dt}$$

$$\mathcal{E}_2 = -M\frac{dI_1}{dt} \tag{32.19}$$

The situation in which two closely wound coils are adjacent to each other can be used to define the *mutual inductance*, M. In this expression Φ_{21} is the magnetic flux through coil 2 due to the current in coil 1. The unit of mutual inductance is the henry, H. Note that M is a *shared property of the pair of coils*.

$$M_{2,1} = \frac{N_2 \Phi_{21}}{I_1} \tag{32.16}$$

Figure 32.2 shows a charged capacitor which can be connected by switch S to an inductor in a circuit with zero resistance. When the switch S is closed, the circuit exhibits a continual transfer of energy (back and forth) between the electric field of the capacitor and the magnetic field of the inductor. The process occurs at an angular frequency ω, called the *frequency of oscillation of the circuit.*

$$\omega = \frac{1}{\sqrt{LC}}$$

(32.24)

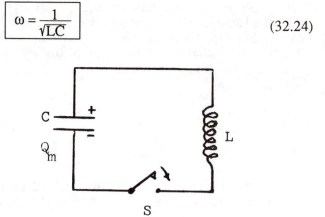

Figure 32.2

The *total energy* of this circuit can be expressed in terms of the charge on the capacitor and the current in the inductor at some arbitrary time (Eq. 32.20). Alternatively, the total energy can be written in terms of the maximum charge on the capacitor and the maximum current in the inductor at a specified time t after the switch is closed (Eq. 32.28).

$$U = \frac{Q^2}{2C} + \frac{1}{2}LI^2$$

(32.20)

$$U = \frac{Q^2_m}{2C}\cos^2 \omega t + \frac{LI^2_m}{2}\sin^2 \omega t$$

(32.28)

The *charge* on the capacitor and the *current* in the inductor vary sinusoidally in time and are 90° out of phase with each other. If the initial conditions are such that $I = 0$ and $Q = Q_m$ at $t = 0$, the charge and current vary with time according to Eqs. 32.26 and 32.27.

$$Q = Q_m \cos \omega t$$

(32.26)

$$I = -I_m \sin \omega t$$

(32.27)

EXAMPLE PROBLEM SOLUTION

Example 32.1 A cylindrical solid conductor of radius R carries a current I which is uniformly distributed over the cross section of the conductor. (See Fig. 32.3a.) Calculate the inductance of a length L of this conductor. Consider only the flux *inside* the cylinder.

Solution

In this example the calculation will be made by using the second method described in the SKILLS section and Eq. 32.13:

$$U_m = \frac{1}{2}LI^2$$

Figure 32.3a

The magnetic field *inside* the wire can be found using Ampere's law:

$$\oint \mathbf{B} \cdot d\mathbf{L} = \mu_o i$$

Integrating around a path of radius r gives $B(2\pi r) = \mu_o i$ where i is the fraction of the total current through that portion of the conductor bounded by the path of integration. In this case,

$$i = \left(\frac{\pi r^2}{\pi R^2}\right) I$$

Therefore, the B field inside the cylinder has a magnitude given by

$$B = \left(\frac{\mu_o I}{2\pi R^2}\right) r \qquad (r < R)$$

Next, the energy density is calculated using Eq. 32.15:

$$u_m = \frac{B^2}{2\mu_o} = \frac{\mu_o I^2 r^2}{8\pi^2 R^4} \qquad (r < R)$$

The total energy associated with the magnetic field inside the conductor is found by integrating this expression for the energy density over the volume of the cylindrical conductor of length L.

$$U_m = \int u_m dV$$

where the volume element dV is the volume of a cylindrical shell, $dV = 2\pi r L\, dr$. (See Fig. 32.3b.) The integration over the cylinder then gives

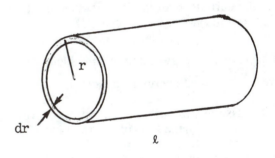

$$U_m = \frac{\mu_o I^2}{8\pi^2 R^4} (2\pi L) \int_0^R r^3\, dr = \frac{\mu_o I^2 L}{16\pi}$$

Substituting this result into Eq. 32.13, we find that

$$\boxed{L = \frac{2U_m}{I^2} = \frac{\mu_o L}{8\pi}}$$

Figure 32.3b

ANSWERS TO SELECTED QUESTIONS_____

2. A circuit containing a coil, resistor, and battery is in steady state; that is, the current has reached a constant value. Does the coil have an inductance? Does the coil affect the value of the current in the circuit?

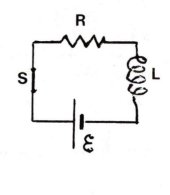

Answer: The coil has an inductance regardless of the nature of the current in the circuit. Inductance is only a function of the coil geometry and makeup. Since the current is constant, the self-induced emf of the coil is zero, so the coil does not affect the steady state current. (This assumes that the resistance of the coil is negligible.)

7. Suppose the switch in the RL circuit in Figure 32.20 has been closed for a long time and is suddenly opened. Does the current instantaneously drop to zero? Why does a spark tend to appear at the switch contacts when the switch is opened?

Answer: No. The current decays to zero exponentially with time due to the inductance in the circuit. A spark tends to appear at the switch when it is opened since the back emf is a maximum at this instant, which can cause breakdown in the air in the vicinity of the contacts.

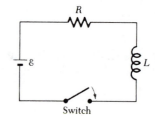

Switch

8. If the current in an inductor is doubled, by what factor does the stored energy change?

Figure 32.20

Answer: The energy stored in an inductor carrying a current I is given by $U = \frac{1}{2}LI^2$. Doubling the current will increase the stored energy by a factor of four.

9. Discuss the similarities between the energy stored in the electric field of a charged capacitor and the energy stored in the magnetic field of a current-carrying coil.

Answer: The energy stored in the electric field of a capacitor is proportional to the square of the electric field intensity, while the energy stored in the magnetic field of a coil is proportional to the square of the magnetic field intensity. Furthermore, the energy stored in a capacitor is proportional to its capacitance which depends on geometry and composition, while the energy stored in a coil is proportional to its inductance which also depends on geometry and composition.

12. The centers of two circular loops are separated by a fixed distance. For what relative orientation of the loops will their mutual inductance be a maximum? For what orientation will it be a minimum?

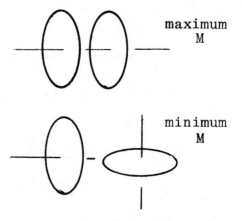

maximum M

minimum M

Answer: The mutual inductance will be a maximum when the coils have a common axis as shown, since the flux linking the two coils is a maximum in this case. The mutual inductance is a minimum when the coil axes are perpendicular to each other.

13. Two solenoids are connected in series such that each carries the same current at any instant. Is mutual induction present? Explain.

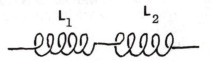

Answer: Yes. Mutual inductance is present as long as the flux of one coil links the other coil.

14. In the LC circuit shown in Figure 32.12, the charge on the capacitor is sometimes zero, even though there is current in the circuit. How is this possible?

Answer: When the capacitor is fully discharged, the current in the circuit is a maximum. At this time, all of the energy is stored in the magnetic field of the coil. This energy was originally stored in the charged capacitor.

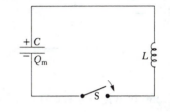

15. If the resistance of the wires in an LC circuit were not zero, would the oscillations persist? Explain.

Figure 32.12

Answer: No. some energy would be dissipated as joule heat, so the current and charge would decrease in time.

SOLUTIONS TO SELECTED END-OF-CHAPTER PROBLEMS

11. A current $I = I_0 \sin \omega t$, with $I_0 = 5$ A and $\frac{\omega}{2\pi} = 60$ Hz, flows through an inductor whose inductance is 10 mH. What is the back emf as a function of time?

Solution

$$\mathcal{E}_{back} = -\mathcal{E}_L = L\frac{dI}{dt} = L\frac{d}{dt}(I_0 \sin \omega t)$$

$$\mathcal{E}_{back} = L\omega I_0 \cos \omega t = (0.01 \text{ H})(120\pi \text{ s}^{-1})(5 \text{ A}) \cos (120\pi t)$$

$$\mathcal{E}_{back} = (18.85 \text{ V}) \cos (377t)$$

19. Calculate the inductance in an RL circuit in which $R = 0.5$ Ω and the current increases to one fourth its final value in 1.5 s.

Solution $\qquad I = I_0(1 - e^{-t/\tau})$

$$e^{-t/\tau} = 1 - \frac{I}{I_0}; \qquad \tau = \frac{-t}{Ln\left(1 - \frac{I}{I_0}\right)} = \frac{-1.5 \text{ s}}{Ln(1 - 0.25)} = 5.21 \text{ s}$$

Since $\tau = \frac{L}{R}$, $\quad L = \tau R = (5.21 \text{ s})(0.5 \text{ } \Omega) = \boxed{2.61 \text{ H}}$

33. An air-core solenoid with 68 turns is 8 cm long and has a diameter of 1.2 cm. How much energy is stored in its magnetic field when it carries a current of 0.77 A?

Solution

$U = \frac{1}{2}LI^2$ and for a solenoid of length d, $L = \frac{\mu_0 N^2 A}{d}$

$\therefore \quad U = \frac{\mu_0 N^2 A I^2}{2d} = \dfrac{\left(4\pi \times 10^{-7}\ \frac{N}{A^2}\right)(68)^2(\pi)(6 \times 10^{-3}\ m)^2(0.77\ A)^2}{2(0.08\ m)} = \boxed{2.44 \times 10^{-6}\ J}$

43. An RL circuit in which L = 4 H and R = 5 Ω is connected to a battery with $\mathcal{E}$ = 22 V at time t = 0. (a) What energy is stored in the inductor when the current in the circuit is 0.5 A? (b) At what rate is energy being stored in the inductor when I = 1 A? (c) What power is being delivered to the circuit by the battery when I = 0.5 A?

Solution

(a) $U_L = \frac{1}{2}LI^2 = \frac{1}{2}(4\ H)(0.5\ A)^2 = \boxed{0.50\ J}$

(b) $I = I_0(1 - e^{-t/\tau})$

Therefore, $t = -\tau\, Ln\left(1 - \frac{I}{I_0}\right)$

But $\tau = \frac{L}{R}$ and $I_0 = \frac{\mathcal{E}}{R}$.

When I = 1 A, we find

$$t = -\frac{L}{R}Ln\left(1 - \frac{IR}{\mathcal{E}}\right) = -\left(\frac{4\ H}{5\ \Omega}\right)Ln\left(1 - \frac{(1)(5)}{22}\right) = 0.206\ s$$

$$P_L = \mathcal{E}_L I = LI\frac{dI}{dt} \quad \text{and} \quad \frac{dI}{dt} = \left(\frac{\mathcal{E}}{R}\right)\left(\frac{R}{L}\right)e^{-t/\tau} = \frac{\mathcal{E}}{L}e^{-t/\tau}$$

so $P_L = \mathcal{E}\, I e^{-t/\tau} = (22\ V)(1\ A)e^{-\left(\frac{0.206}{0.8}\right)} = \boxed{17\ W}$

(c) $P_{batt} = \mathcal{E}I = (22\ V)(0.5\ A) = \boxed{11\ W}$

47. An emf of 96 mV is induced in the windings of a coil when the current in a nearby coil is increasing at the rate of 1.2 A/s. What is the mutual inductance of the two coils?

Solution

$$\mathcal{E}_2 = -M\left(\frac{dI_1}{dt}\right) \quad \text{and} \quad M = \frac{\mathcal{E}_2}{\left|\frac{dI_1}{dt}\right|}$$

$$M = \frac{96 \times 10^{-3} \text{ V}}{1.2 \text{ A/s}} = 80 \times 10^{-3} \text{ H} = \boxed{80 \text{ mH}}$$

57. A fixed inductance L = 1.05 μH is used in series with a variable capacitor in the tuning section of a radio. What capacitance will tune the circuit into the signal from a station broadcasting at a frequency of 96.3 MHz?

Solution

$$f_o = \frac{1}{2\pi\sqrt{LC}}; \quad \text{therefore} \quad C = \frac{1}{(2\pi f_o)^2 L}$$

$$C = \frac{1}{\left[(2\pi)(96.3 \times 10^6 \text{ Hz})\right]^2 (1.05 \times 10^{-6} \text{ H})} = \boxed{2.60 \text{ pF}}$$

65. Consider an LC circuit with L = 500 mH and C = 0.1 μF. (a) What is the resonant frequency (ω_0) of this circuit? (b) If a resistance of 1000 Ω is introduced into this circuit, what would be the frequency of the (damped) oscillations? (c) What is the percent difference between the two frequencies?

Solution

(a) $\quad \omega_o = \frac{1}{\sqrt{LC}} = \frac{1}{\sqrt{(0.50 \text{ H})(1 \times 10^{-7} \text{ F})}} = \boxed{4.47 \times 10^3 \text{ rad/s}}$

(b) $\quad \omega_d = \sqrt{\frac{1}{LC} - \left(\frac{R}{2L}\right)^2}$

$$\omega_d = \sqrt{\left(\frac{1}{(0.5 \text{ H})(1 \times 10^{-7} \text{ F})}\right) - \left(\frac{10^3 \Omega}{(2)(0.5 \text{ H})}\right)^2} = \boxed{4.36 \times 10^3 \text{ rad/s}}$$

(c) $\quad \frac{\Delta\omega}{\omega_o} = \frac{4.47 - 4.36}{4.47} = 0.0246$

Thus, the damped frequency is 2.46% lower than the undamped frequency.

33

Alternating Current Circuits

OBJECTIVES

1. Given an RLC series circuit in which values of resistance, inductance, capacitance, and the characteristics of the generator (source of emf) are known, calculate:
   - the maximum and instantaneous voltage drop across each component
   - the maximum and instantaneous current in the circuit
   - the phase angle by which the current leads or lags the voltage
   - the power expended in the circuit
   - resonance frequency and quality factor of the circuit

2. Understand the use of phasor diagrams for the description and analysis of ac circuits.

3. Sketch circuit diagrams for high-pass and low-pass filter circuits and make calculations of the ratio of output to input voltage in each case.

4. Understand the manner in which step-up and step-down transformers are used in the process of transmitting electrical power over large distances; and make calculations of primary to secondary voltage and current ratios for an ideal transformer.

SKILLS

The *phasor diagram* is a very useful technique to use in the analysis of ac RLC circuits. In such a diagram, each of the rotating quantities V_R, V_L, V_C, and I_m is represented by a separate phasor (rotating vector). A 'phasor' diagram which describes the ac circuit of Figure 33.1(a) is shown in Figure 33.1(b). Each phasor has a length which is proportional to the magnitude of the voltage or current which it represents and rotates counterclockwise about the common origin with an angular frequency which equals the angular frequency of the alternating source, ω. The direction of the phasor which represents the current in the circuit is used as the *reference* direction to establish the correct phase differences among the phasors, which represent the voltage drops across the resistor, inductor, and capacitor. The *instantaneous values* v_R, v_L, v_C, and i are given by the *projection onto the vertical axis of the corresponding phasor.*

Consider the phasor diagram in Figure 33.1, where the maximum voltage across the resistor, V_R, is greater than the maximum voltage across the inductor, V_L. At the instant shown, the *instantaneous* value of the voltage across the inductor is greater than that across the resistor. Also notice that as time increases and the phasors rotate counterclockwise, maintaining their constant relative phase, V_R, V_L, and V_C (the voltage amplitudes) will remain constant in magnitude but the instantaneous values v_R, v_L, and v_C will vary sinusoidally with time. For the case shown in Fig. 33.1(b), the phase angle ϕ is negative (this is because $X_C > X_L$ and

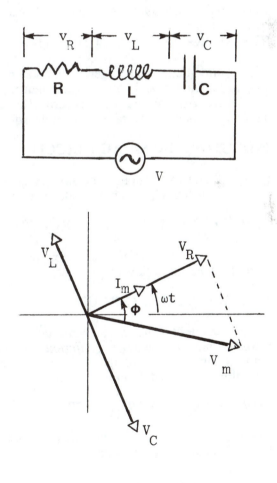

Figure 33.1

therefore $V_C > V_L$); hence, the current in the circuit *leads* the applied voltage in phase.

The maximum voltage across each element in the circuit is the product of I_m and the resistance or reactance of that component. It is possible, therefore, to construct an *impedance triangle* for any series circuit as shown in Figure 33.2.

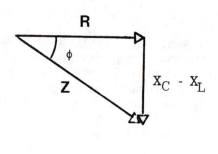

Figure 33.2

NOTES FROM SELECTED CHAPTER SECTIONS

33.1 AC SOURCES AND PHASORS

Phasor diagrams are graphical constructions in which alternating quantities such as current and voltage are represented by rotating vectors called *phasors*. The maximum value of a time-varying quantity is represented by the *length* of the phasor; and the instantaneous value is the *projection* of the phasor onto the vertical axis. Phasors rotate counterclockwise with an angular velocity, ω.

33.2 RESISTORS IN AN AC CIRCUIT

If an ac circuit consists of a generator and a resistor, the current in the circuit is in phase with the voltage. That is, the current and voltage reach their maximum values at the same time. The *average* value of the current over *one complete cycle* is zero. The rms current refers to *root mean square*, which simply means the square root of the average value of the square of the current.

33.3 INDUCTORS IN AN AC CIRCUIT

If an ac circuit consists of a generator and an inductor, the current *lags behind* the voltage by 90°. That is, the voltage reaches its maximum value one-quarter of a period before the current reaches its peak value.

33.4 CAPACITORS IN AN AC CIRCUIT

If an ac circuit consists of a generator and a capacitor, the current *leads* the voltage by 90°. That is, the current reaches its peak value one-quarter of a period before the voltage reaches its peak value.

33.5 THE RLC SERIES CIRCUIT

The ac current at all points in a series ac circuit has the same amplitude and phase. Therefore, the voltage across each element will have *different* amplitudes and phases: the voltage across the resistor is in phase with the current, the voltage across the inductor leads the current by 90°, and the voltage across the capacitor lags behind the current by 90°.

33.6 POWER IN AN AC CIRCUIT

The average power delivered by the generator is dissipated as heat in the resistor. There is no power loss in an ideal inductor or capacitor.

33.7 RESONANCE IN A SERIES RLC CIRCUIT

The current in a series RLC circuit reaches its peak value when the frequency of the generator equals ω_0; that is, when the "driving" frequency matches the resonance frequency.

33.9 THE TRANSFORMER AND POWER TRANSMISSION

A transformer is a device designed to raise or lower an ac voltage and current without causing an appreciable change in the product IV. In its simplest form, it consists of a primary coil of N_1 turns and a secondary coil of N_2 turns, both wound on a common soft iron core. In an ideal transformer, the power delivered by the generator must equal the power dissipated in the load.

EQUATIONS AND CONCEPTS

A simple series alternating current circuit with a sinusoidal source of emf is shown in Figure 33.3.

The rectangle _____ is used to represent the circuit element(s) which, in a particular case, may be (a) a resistor R, (b) a capacitor C, (c) an inductor L, or (d) some combination of two or more of these three components.

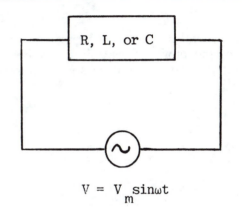

$$V = V_m \sin\omega t$$

Figure 33.3

The *applied sinusoidal voltage* of frequency ω has a maximum value of V_m.

$$v = V_m \sin \omega t \tag{33.1}$$

When the circuit element is a *resistor* of value R, the current and voltage across the resistor are *in phase*, and I_m is the maximum current.

$$v_R = V_m \sin \omega t \tag{33.2}$$

$$I_R = I_m \sin \omega t \tag{33.3}$$

When the circuit element is an *inductor* of value L, the *current lags the voltage* across the inductor by 90°.

$$v_L = V_m \sin \omega t \tag{33.13}$$

$$i_L = \frac{V_m}{\omega L} \sin\left(\omega t - \frac{\pi}{2}\right) \tag{33.10}$$

When the circuit element is a *capacitor* of value C, the *current leads the voltage* across the capacitor by 90°.

$$v_C = V_m \sin \omega t \qquad (33.14)$$

$$i_C = \omega C\, V_m \sin\left(\omega t + \frac{\pi}{2}\right) \qquad (33.17)$$

The maximum value of the current (or current amplitude) through each element is proportional to the instantaneous ac voltage across the element. In the case of an inductor and a capacitor, the maximum value of the current depends also on the angular frequency of the source of emf.

$$I_m = \begin{cases} \dfrac{V_m}{R}; & \text{Resistor} \qquad (33.4) \\[2mm] \dfrac{V_m}{\omega L}; & \text{Inductor} \qquad (33.11) \\[2mm] V_m\omega C; & \text{Capacitor} \qquad (33.18) \end{cases}$$

In the general case, the ac circuit will contain a resistor, inductor, and capacitor in series with a sinusoidally varying voltage source as shown in Figure 33.4.

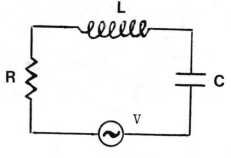

Figure 33.4

Since at any instant, the current has the same value (amplitude and phase) at every point in the circuit, it is convenient to express the phase relationship between the current and *instantaneous voltage* drops across R, L, and C relative to *the common current phase*.

$$v_R = V_R \sin \omega t \qquad (33.21)$$

$$v_L = V_L \sin\left(\omega t + \frac{\pi}{2}\right) \qquad (33.22)$$

$$v_C = V_C \sin\left(\omega t - \frac{\pi}{2}\right) \qquad (33.23)$$

Compare Eqs. 33.21, 33.22, and 33.23 to Eqs. 33.3, 33.10, and 33.17. You should convince yourself that *these two sets of equations express the same phase relationship between current and voltage.*

The *maximum voltage* across each circuit element can be written in the form of Ohm's law.

$$V_R = I_m R \qquad (33.24)$$

$$V_L = I_m X_L \qquad (33.25)$$

$$V_C = I_m X_C \qquad (33.26)$$

In the previous set of equations, R is of course the resistance while X_L and X_C represent the *inductive reactance* and the *capacitive reactance* respectively. The reactances are frequency dependent. The inductive reactance increases with increasing frequency, while the capacitive reactance decreases with increasing frequency.

$$X_L = \omega L \qquad (33.12)$$

$$X_C = \frac{1}{\omega C} \qquad (33.19)$$

where $\omega = 2\pi f$

The *maximum current* in the circuit depends on the angular frequency ω of the source of emf, as well as the values of V_m, R, L, and C.

$$I_m = \frac{V_m}{\sqrt{R^2 + \left(\omega L - \frac{1}{\omega C}\right)^2}}$$

It is useful to define an operating parameter of the circuit called the *impedance*, Z, defined by Eq. 33.29. V_m and I_m can be related in the form of Ohm's law, Eq. 33.30. This requires that Z have the SI unit of ohm (Ω).

$$Z \equiv \sqrt{R^2 + (X_L - X_C)^2} \qquad (33.29)$$

$$V_m = I_m Z \qquad (33.30)$$

The applied voltage (across the source) and the current in the circuit will differ in phase by some angle ϕ, the *phase angle* of the circuit, given by Eq. 33.31.

$$\tan \phi = \frac{X_L - X_C}{R} \qquad (33.31)$$

The *average power* delivered by a generator (source of emf) to an RLC series circuit is dissipated as heat in the resistor (there is zero power loss in ideal inductors and capacitors) and is directly proportional to $\cos \phi$, where ϕ is the phase angle. The quantity $\cos \phi$ is called the *power factor* of the circuit. Since ϕ is frequency dependent (see Eq. 33.31), the power factor also depends on frequency.

$$P_{av} = I_{rms} V_{rms} \cos \phi \qquad (33.35)$$

or

$$P_{av} = I^2_{rms} R \qquad (33.36)$$

When measuring values of current and voltage in ac circuits, it is customary to use instruments which respond to *root-mean-square* (rms) values of these quantities rather than their maximum or instantaneous values.

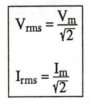

$$V_{rms} = \frac{V_m}{\sqrt{2}}$$

$$I_{rms} = \frac{I_m}{\sqrt{2}}$$

You should notice that V_{rms} and I_{rms} are in the same ratio as V_m and I_m. Compare this equation to Equation 33.30.

$$V_{rms} = I_{rms} Z$$

From Eq. 33.29 you can see that when the inductive reactance X_L equals the capacitive reactance X_C, the impedance of the circuit has its *minimum* value. Under these conditions, $Z = R$, and the current in the circuit will have its *maximum* value.

The condition that $Z = R$ occurs for a characteristic frequency of the circuit called the *resonance frequency*, ω_o, given by Eq. 33.39. This is obtained from the condition $X_L = X_C$, where $X_L = \omega L$ and $X_C = \dfrac{1}{\omega C}$.

$$\boxed{\omega_o = \frac{1}{\sqrt{LC}}} \qquad (33.39)$$

A simple transformer consists of a primary coil of N_1 turns and a secondary coil of N_2 turns wound on a common core. In Eq. 33.48, V_1 represents the *voltage across the primary* (input voltage), while V_2 represents the *voltage across the secondary* (output voltage). Likewise, I_1 and I_2 represent the currents in the primary and secondary circuits. In the *ideal* transformer, the ratio of voltages is equal to the ratio of turns, and the ratio of currents is equal to the inverse of the ratio of turns.

$$\boxed{\begin{aligned} V_2 &= \frac{N_2}{N_1} V_1 \\[2ex] I_2 &= \frac{N_1}{N_2} I_1 \end{aligned}} \qquad (33.48)$$

EXAMPLE PROBLEM SOLUTION

Example 33.1 In the RLC series circuit of Figure 33.1(a), the applied emf has a magnitude $V_m = 120$ V and oscillates at a frequency $f = 60$ Hz. Suppose that $R = 800\ \Omega$, $C = 4\ \mu F$, and assume the inductor can be varied in value. Determine the necessary value of L such that the voltage across the capacitor will be out of phase with the applied emf by 30° with V_m leading V_C.

Solution

The phase relationships for the voltage drops in the circuit are shown in Figure 33.5. From the figure, the phase angle is seen to be 60°. This is true since the phasors representing I_m and V_R are along the same direction (I_m and V_R are in phase). Using Eq. 33.31,

$$\tan \phi = \frac{X_L - X_C}{R}$$

we find that

$$X_L = X_C + R \tan \phi \qquad (1)$$

where

$$X_L = 2\pi f L \quad \text{and} \quad X_C = \frac{1}{2\pi f C}$$

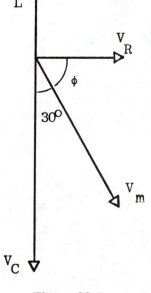

Figure 33.5

404

Substituting into Eq. (1) gives

$$2\pi fL = \frac{1}{2\pi fC} + R \tan \phi$$

or

$$L = \frac{1}{2\pi f}\left[\frac{1}{2\pi fC} + R \tan \phi\right] \tag{2}$$

Since $f = 60$ Hz, $C = 4$ μF, $R = 800$ Ω, and $\phi = 60°$, we get

$$L = \frac{1}{2\pi(60\ \text{Hz})}\left[\frac{1}{2\pi(60\ \text{Hz})(4 \times 10^{-6}\ \text{F})} + (800\ \Omega)(\tan 60°)\right]$$

$$L = 5.44\ \text{H}$$

ANSWERS TO SELECTED QUESTIONS

1. What is meant by the statement "the voltage across an inductor leads the current by 90°"?

Answer: This means that the voltage reaches its peak value one quarter of a cycle before the current reaches its peak value.

2. Explain why the reactance of a capacitor decreases with increasing frequency, whereas the reactance of an inductor increases with increasing frequency.

Answer: As the frequency of a capacitive circuit increases, the polarities of the charged plates must change more rapidly with time, corresponding to a larger current. The capacitive reactance varies as the inverse of the frequency, and hence approaches zero as f approaches infinity. The current is zero in a dc capacitive circuit which corresponds to zero frequency and an infinite reactance. On the other hand, the inductive reactance is proportional to the frequency, and therefore increases with increasing frequency. At higher frequencies, the current changes more rapidly, which according to Faraday's law produces an increase in the back emf associated with an inductive element, and a corresponding decrease in current.

6. Does the phase angle depend on frequency? What is the phase angle when the inductive reactance equals the capacitive reactance?

Answer: Yes. Since the phase angle is a function of the reactance, which depends on frequency, it must be frequency dependent. The phase angle is zero when the inductive reactance equals the capacitive reactance.

8. If the frequency is doubled in a series RLC circuit, what happens to the resistance, the inductive reactance, and the capacitive reactance?

Answer: There is no change in the resistance, the inductive reactance is doubled, and the capacitive reactance is halved.

12. What is the impedance of an RLC circuit at the resonance frequency?

Answer: At the resonance frequency, $X_L = X_C$, so the impedance equals the resistance in the circuit.

17. Will a transformer operate if a battery is used for the input voltage across the primary? Explain.

Answer: No. A voltage can only be induced in the secondary coil if the flux through the core changes in time.

22. Is the voltage applied to a circuit always in phase with the current through a resistor in the circuit?

Answer: No. The voltage across a resistor and the current through it are in phase when there are no capacitors or inductors in the circuit, or when the frequency corresponds to the resonance frequency of the circuit.

23. Would an inductor and a capacitor used together in an ac circuit dissipate any power?

Answer: No. The only element that dissipates any power in an ac circuit is a resistor.

24. Show that the peak current in an RLC circuit occurs when the circuit is in resonance.

Answer: At resonance the impedance is a minimum because $X_L = X_C$. Thus, I, which equals V/Z, is a maximum.

29. Why are the primary and secondary coils of a transformer wrapped on an iron core that passes through both coils?

Answer: The purpose of the iron core is to ensure that the flux passing through the secondary is the same as that passing through the primary.

SOLUTIONS TO SELECTED END-OF-CHAPTER PROBLEMS

15. What is the inductance of a coil that has an inductive reactance of 63 Ω at an angular frequency of 820 rad/s?

Solution

$X_L = \omega L$

$L = \dfrac{X_L}{\omega} = \dfrac{63\ \Omega}{820\ \text{rad/s}} = 7.68 \times 10^{-2}\ \text{H} = \boxed{76.8\ \text{mH}}$

21. What peak current will be delivered by an ac generator with $V_m = 48$ V and $f = 90$ Hz when connected across a 3.7-μF capacitor?

Solution

$I_m = \dfrac{V_m}{X_c} = V_m \omega C = V_m(2\pi f C)$

$I_m = (48\ \text{V})(2\pi)(90\ \text{Hz})(3.7 \times 10^{-6}\ \text{F}) = 0.100\ \text{A} = \boxed{100\ \text{mA}}$

29. A coil with an inductance of 18.1 mH and a resistance of 7 Ω is connected to a *variable*-frequency ac generator. At what frequency will the voltage across the coil lead the current by 45°?

Solution

$$\tan \phi = \frac{X_L - X_C}{R} = \frac{\omega L - 0}{R}$$

$$\omega = \frac{R \tan \phi}{L} = \frac{(7 \ \Omega)(\tan 45°)}{18.1 \times 10^{-3} \ H} = 387 \ rad/s$$

$$f = \frac{\omega}{2\pi} = \boxed{61.6 \ Hz}$$

33. An inductor (L = 400 mH), a capacitor (C = 4.43 μF), and a resistor (R = 500 Ω) are connected in series. A 50-Hz ac generator produces a peak current of 250 mA in the circuit. (a) Calculate the required peak voltage V_m. (b) Determine the angle by which the current in the circuit leads or lags behind the applied voltage.

Solution

(a) $\quad V_m = I_m Z = I_m \sqrt{R^2 + (X_L - X_C)^2}$

where $\qquad X_L = \omega L = 2\pi (50 \ Hz)(400 \times 10^{-3} \ H) = 126 \ \Omega$

$$X_C = \frac{1}{\omega C} = \frac{1}{(2\pi)(50 \ Hz)(4.43 \times 10^{-6} \ F)} = 719 \ \Omega$$

$$\therefore \quad Z = \sqrt{(500 \ \Omega)^2 + (126 \ \Omega - 719 \ \Omega)^2} = 776 \ \Omega$$

and $\qquad V_m = I_m Z = (0.25 \ A)(776 \ \Omega) = \boxed{194 \ V}$

(b) $\quad \tan \phi = \frac{X_L - X_C}{R}; \qquad \phi = \tan^{-1}\left(\frac{126 - 719}{500}\right) = \boxed{-49.9°}$

The current *leads* the voltage by 49.9°.

41. In a certain series RLC circuit, $I_{rms} = 9$ A, $V_{rms} = 180$ V, and the current leads the voltage by 37°. (a) What is the total resistance of the circuit? (b) Calculate the reactance of the circuit ($X_L - X_C$).

Solution

(a) $P_{av} = I_{rms}V_{rms} \cos \phi = I_{rms}^2 R$

$$R = \frac{V_{rms} \cos \phi}{I_{rms}} = \frac{(180 \text{ V}) \cos (-37°)}{9 \text{ A}} = \boxed{16 \ \Omega}$$

(b) $\tan \phi = \dfrac{X_L - X_C}{R}$

$$X_L - X_C = R \tan \phi = (16 \ \Omega) \tan (-37°) = -12 \ \Omega \qquad \text{or} \qquad |X_L - X_C| = \boxed{12 \ \Omega}$$

---

51. The RC high-pass filter shown in Figure 33.15 has a resistance $R = 0.50 \ \Omega$. (a) What capacitance will give an output signal with one-half the amplitude of a 300-Hz input signal? (b) What is the gain (V_{out}/V_{in}) for a 600-Hz signal?

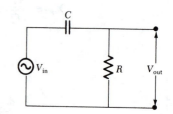

Solution

(a) $\dfrac{V_{out}}{V_{in}} = \dfrac{R}{\sqrt{R^2 + \left(\frac{1}{\omega C}\right)^2}}$

Solving for C gives

Figure 33.15

$$C = \frac{1}{\omega R \sqrt{\left(\frac{V_{in}}{V_{out}}\right)^2 - 1}} = \frac{1}{(2\pi)(300 \text{ Hz})(0.5 \ \Omega)\sqrt{(2)^2 - 1}} = \boxed{613 \ \mu\text{F}}$$

(b) Taking $\omega = 2\pi(600)$ rad/s, $R = 0.5 \ \Omega$, and $C = 613 \ \mu$F, we find

$$\frac{V_{out}}{V_{in}} = \boxed{0.756}$$

61. A particular transformer is 95% efficient and has twice as many secondary windings as primary windings. If the primary windings carry a 5 A current at an rms voltage of 120 V, what are the secondary current and rms voltage?

Solution

$$P_s = (0.95)P_p ; \qquad I_s V_s = (0.95)I_p V_p$$

$$\therefore \quad I_s = (0.95)I_p \left(\frac{V_p}{V_s}\right)$$

Since $\dfrac{V_p}{V_s} = \dfrac{N_p}{N_s} = \dfrac{1}{2}$, we find

$$I_s = (0.95)(5\text{ A})\left(\frac{1}{2}\right) = \boxed{2.38\text{ A}}$$

and

$$V_s = 2V_p = 2(120\text{ V}) = \boxed{240\text{ V}}$$

79. *Impedance matching:* A transformer may be used to provide maximum power transfer between two ac circuits that have different impedances. (a) Show that the ratio of turns N_1/N_2 needed to meet this condition is given by

$$\frac{N_1}{N_2} = \sqrt{\frac{Z_1}{Z_2}}$$

(b) Suppose you want to use a transformer as an impedance-matching device between an audio amplifier that has an output impedance of 8000 Ω and a speaker that has an input impedance of 8 Ω. What should be the ratio of primary to secondary turns on the transformer?

Solution

(a) $\dfrac{N_1}{N_2} = \dfrac{V_1}{V_2}; \qquad Z_1 = \dfrac{V_1}{I_1}; \qquad Z_2 = \dfrac{V_2}{I_2}$

$\therefore \dfrac{N_1}{N_2} = \dfrac{Z_1 I_1}{Z_2 I_2};$ since $\dfrac{I_1}{I_2} = \dfrac{N_2}{N_1}$, we find

$$\frac{N_1}{N_2} = \sqrt{\frac{Z_1}{Z_2}}$$

(b) $\dfrac{N_1}{N_2} = \sqrt{\dfrac{8000}{8}} = \boxed{31.6}$

34

Electromagnetic Waves

OBJECTIVES

1. Describe the essential features of the apparatus and procedure used by Hertz in his experiments leading to the discovery and understanding of the source and nature of electromagnetic waves.

2. Summarize the properties of electromagnetic waves.

3. Show by direct substitution that a sinusoidal plane wave solution (Eqs. 34.14 and 34.15) satisfies the linear differential wave equations for electromagnetic waves (Eqs. 34.10 and 34.11).

4. For a properly described plane electromagnetic wave, calculate the values for the Poynting vector (magnitude), wave intensity, and instantaneous and average energy densities.

5. Calculate the radiation pressure on a surface and the linear momentum delivered to a surface by an electromagnetic wave.

6. Using the geometry of the infinite current sheet as an example, describe the relative directions and the space and time dependences of the radiated electric and magnetic fields.

7. Understand the production of electromagnetic waves and radiation of energy by an oscillating dipole. Use a diagram to show the relative directions for **E**, **B**, and **S** and account for the intensity of the radiated wave at points near the dipole and at distant points.

8. Give a brief description (related to the source and typical use) of each of the "regions" of the electromagnetic spectrum.

NOTES FROM SELECTED CHAPTER SECTIONS

34.1 MAXWELL'S EQUATIONS AND HERTZ'S DISCOVERIES

Electromagnetic waves are generated by accelerating electric charges. The radiated waves consist of oscillating electric and magnetic fields, which are *at right angles to each other* and also *at right angles to the direction of wave propagation*.

The fundamental laws governing the behavior of electric and magnetic fields are Maxwell's equations. In this unified theory of electromagnetism, Maxwell showed that electromagnetic waves are a natural consequence of these fundamental laws.

34.2 PLANE ELECTROMAGNETIC WAVES

Following is a summary of the properties of electromagnetic waves:

1. The solutions of Maxwell's third and fourth equations are wavelike, where both **E** and **B** satisfy the same wave equation.

2. Electromagnetic waves travel through empty space with the speed of light, $c = 1/\sqrt{\varepsilon_o \mu_o}$.

3. The electric and magnetic field components of plane electromagnetic waves are perpendicular to each other and also perpendicular to the direction of wave propagation. The latter property can be summarized by saying that electromagnetic waves are transverse waves.

4. The relative magnitudes of **E** and **B** in empty space are related by $E/B = c$.

5. Electromagnetic waves obey the principle of superposition.

34.3 ENERGY CARRIED BY ELECTROMAGNETIC WAVES

The magnitude of the Poynting vector represents the rate at which energy flows through a unit surface area perpendicular to the flow.

For an electromagnetic wave, the instantaneous energy density associated with the magnetic field equals the instantaneous energy density associated with the electric field. Hence, in a given volume, the energy is equally shared by the two fields.

34.4 MOMENTUM AND RADIATION PRESSURE

Electromagnetic waves have momentum and exert pressure on surfaces on which they are incident. The pressure exerted by a normally incident wave on a *totally reflecting* surface is *double* that exerted on a surface which *completely absorbs* the incident wave.

34.5 RADIATION FROM AN INFINITE CURRENT SHEET

An *infinite conducting sheet* provides the most convenient geometry for the description of the radiated electric and magnetic fields of a time-varying current. Consider an infinite sheet in the y-z plane (see Figure 34.1) carrying a sinusoidal *surface current per unit length* J_s of magnitude $J_s = J_0 \cos \omega t$.

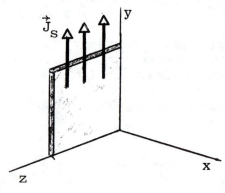

Figure 34.1

The direction of **E**, **B**, and **c** are always orthogonal and are shown in Figure 34.2 relative to the direction of the surface current at a particular instant.

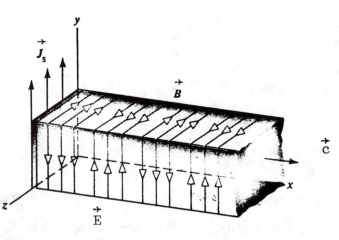

Figure 34.2   Linearly polarized electromagnetic wave due to an oscillating infinite sheet of current.

During the next half cycle, the directions of J_s, **B**, and **E** will reverse. The direction of **c** will remain along the positive x-axis. You should convince yourself that these directions are consistent with that required by the definition of the Poynting vector.

34.6 THE PRODUCTION OF ELECTROMAGNETIC WAVES BY AN ANTENNA

The fundamental mechanism responsible for radiation by an antenna is the acceleration of a charged particle. Whenever a charged particle undergoes an acceleration, it must radiate energy. An alternating voltage applied to the wires of an antenna forces electric charges in the antenna to oscillate.

EQUATIONS AND CONCEPTS

Maxwell's equations are the fundamental laws governing the behavior of electric and magnetic fields. Electromagnetic waves are a natural consequence of these laws.

$$\oint \mathbf{E} \cdot d\mathbf{A} = \frac{Q}{\varepsilon_o} \tag{34.1}$$

$$\oint \mathbf{B} \cdot d\mathbf{A} = 0 \tag{34.2}$$

You should notice that the integrals in *Eqs. 34.1 and 34.2 are surface integrals* in which the normal components of electric and magnetic fields are integrated over a *closed surface* while Eqs. 34.3 and 34.4 involve line integrals in which the tangential components of electric and magnetic fields are integrated around a *closed path*.

$$\oint \mathbf{E} \cdot d\mathbf{s} = -\frac{d\phi_m}{dt} \tag{34.3}$$

$$\oint \mathbf{B} \cdot d\mathbf{s} = \mu_o I + \mu_o \varepsilon_o \frac{d\phi_e}{dt} \tag{34.4}$$

Both **E** and **B** satisfy a differential equation which has the form of the general wave equation. These are the wave equations for electromagnetic waves in free space (where $Q = 0$ and $I = 0$). As stated here, they represent linearly polarized waves traveling with a speed c.

$$\frac{\partial^2 E}{\partial x^2} = \mu_o \varepsilon_o \frac{\partial^2 E}{\partial t^2} \tag{34.10}$$

$$\frac{\partial^2 B}{\partial x^2} = \mu_o \varepsilon_o \frac{\partial^2 E}{\partial t^2} \tag{34.11}$$

$$\mu_o \varepsilon_o = \frac{1}{c^2}$$

The electric and magnetic fields vary in position and time as *sinusoidal transverse waves*. Their planes of vibration are perpendicular to each other and perpendicular to the direction of propagation.

$$E = E_m \cos (kx - \omega t) \tag{34.14}$$

$$B = B_m \cos (kx - \omega t) \tag{34.15}$$

The ratio of the magnitude of the electric field to the magnitude of the magnetic field is constant and equal to the speed of light c.

$$\frac{E_m}{B_m} = \frac{E}{B} = c \tag{34.16}$$

The *Poynting vector* $\mathbf{S}$ describes the energy flow associated with an electromagnetic wave. The direction of $\mathbf{S}$ is along the direction of propagation and the magnitude of $\mathbf{S}$ is the rate at which electromagnetic energy crosses a unit surface area perpendicular to the direction of $\mathbf{S}$.

$$\mathbf{S} \equiv \frac{1}{\mu_o} \mathbf{E} \times \mathbf{B}$$

(34.23)

The *wave intensity* is the time average of the magnitude of the Poynting vector. E_m and B_m are the *maximum values* of the field magnitudes.

$$S_{av} = \frac{E_m{}^2}{2\mu_o c} = \frac{c}{2\mu_o} B_m{}^2$$

(34.26)

The electric and magnetic fields have *equal instantaneous energy densities*.

$$u_m = u_e = \frac{1}{2}\varepsilon_o E^2 = \frac{1}{2\mu_o} B^2$$

(34.27)

The total instantaneous energy density u is proportional to E^2 and B^2 while the *total average energy density* is proportional to $E_m{}^2$ and $B_m{}^2$. The average energy density is also proportional to the wave intensity.

$$u = \varepsilon_o E^2 = \frac{1}{\mu_o} B^2$$

(34.28)

$$u_{av} = \frac{1}{2}\varepsilon_o E_m{}^2 = \frac{1}{2\mu_o} B_m{}^2$$

(34.29)

$$u_{av} = \frac{1}{c} S_{av}$$

The *linear momentum p delivered to an absorbing surface* by an electromagnetic wave at normal incidence depends on the fraction of the total energy absorbed.

$$p = \frac{U}{c} \qquad \text{complete absorption}$$

(34.31)

$$p = \frac{2U}{c} \qquad \text{complete reflection}$$

(34.33)

An absorbing surface (at normal incidence) will experience a *radiation pressure* P which depends on the magnitude of the Poynting vector and the degree of absorption.

$$P = \frac{S}{c} \qquad \text{complete absorption}$$

(34.32)

$$P = \frac{2S}{c} \qquad \text{complete reflection}$$

(34.34)

The *magnetic field* due to an infinite current sheet in the y-z plane is in the x-z plane and varies in a sinusoidal fashion according to Eq. 34.35. Note that the electromagnetic wave associated with this sheet of current propagates along the x-axis as a linearly polarized wave. For small values of x, the magnetic field is independent of x.

$$B_z = -\frac{\mu_o J_o}{2} \cos(kx - \omega t)$$

(34.35)

$$B_z = -\frac{\mu_o J_o}{2} \cos \omega t \quad \text{(for } x \cong 0)$$

A *radiated electric field* vibrates in the y-z plane according to Eq. 34.36 and has the same space and time variations as the accompanying magnetic field.

$$E_y = -\frac{\mu_o J_o c}{2} \cos(kx - \omega t)$$

(34.36)

There is an outgoing electromagnetic wave on each side of the infinite sheet. In each half plane, the *rate of energy emission per unit area* (average intensity) is equal to the average of the Poynting vector.

$$S_{av} = \frac{\mu_o J_o^2 c}{8}$$

(34.38)

EXAMPLE PROBLEM SOLUTION

Example 34.1   The magnetic field of a plane, linearly polarized electromagnetic wave is described by the equation

$$B_z = (6 \times 10^{-6} \text{ T}) \sin\left[2\pi\left(\frac{x}{30} - 10^7 t\right)\right]$$

where x is in m and t in s.

(a) If the wave described here is due to an oscillating current in a plane sheet, find the expression for the surface current, J_s.

(b) Calculate the intensity of the electromagnetic wave (average value of the Poynting vector).

Solution

(a) Compare the wave expression given in this example with the general form for a magnetic field due to an infinite current sheet as stated by Eq. 34.35.

$$B_z = -\frac{\mu_o J_o}{2} \cos(kx - \omega t) \qquad (1)$$

This comparison shows that the amplitude of the magnetic field is

$$B_m = -\frac{\mu_o J_o}{2} = -6 \times 10^{-6} \text{ T} \qquad (2)$$

and

$$\omega = 2\pi \times 10^7 \text{ s}^{-1}$$

Substituting $\mu_o = 4\pi \times 10^{-7}$ Wb/A·m and solving for J_o in Eq. (2) gives

$$J_o = -\frac{2 B_m}{\mu_o} = -9.55 \frac{\text{T·A·m}}{\text{Wb}} = -9.55 \text{ A/m} \qquad (3)$$

Using this value of J_o and $\omega = 2\pi \times 10^7$ s^{-1}, the surface current per unit length is found to be

$$J_s = J_o \cos \omega t = (-9.55 \text{ A/m}) \cos(2\pi \times 10^7 t)$$

(b) From Eq. 34.26,

$$S_{av} = \frac{c}{2\mu_o} B_m^2$$

$$S_{av} = \frac{(3 \times 10^8 \text{ m/s})}{(2)(4\pi \times 10^{-7} \text{ Wb/A·m})} (6 \times 10^{-6} \text{ T})^2$$

$$S_{av} = 4.3 \times 10^5 \frac{A \cdot m^2 \cdot T^2}{Wb \cdot s} = 4.3 \times 10^3 \text{ W/m}^2$$

The electromagnetic wave intensity is 4.3×10^3 W/m². In assigning the correct SI units to this result, note that

$$1 \text{ T} = 1 \text{ Wb/m}^2 \quad \text{and} \quad 1 \text{ A·Wb} = 1 \text{ J}$$

ANSWERS TO SELECTED QUESTIONS

1. For a given incident energy of electromagnetic wave, why is the radiation pressure on a perfect reflecting surface twice as large as the pressure on a perfect absorbing surface?

Answer: Twice as much momentum is transferred to the surface when the wave is reflected.

4. What is the fundamental source of electromagnetic radiation?

Answer: The fundamental source of electromagnetic radiation is an accelerating charge.

7. Certain orientations of the receiving antenna on a TV give better reception than others. Furthermore, the best orientation varies from station to station. Explain these observations.

Answer: The best reception will occur when the conducting portion of the antenna is along the direction of electric field vibrations of the incident wave. This direction will vary from station to station.

8. Does a wire connected to a battery emit an electromagnetic wave? Explain.

Answer: No. The wire will only emit electromagnetic waves if the current varies in time. The radiation is the result of accelerating charges, which can only occur when the current has a nonsteady state value.

9. If you charge a comb by running it through your hair and then hold the comb next to a bar magnet, do the electric and magnetic fields produced constitute an electromagnetic wave?

Answer: There is an electric field set up by the charges on the comb, and there is a magnetic field set up by the bar magnet. However, these fields are constant in magnitude while the fields of an electromagnetic wave are continually changing in magnitude and direction. Hence, no electromagnetic wave is produced.

10. An empty plastic or glass dish removed from a microwave oven is cool to the touch. How can this be possible?

Answer: There are no water or fat molecules in the glass or plastic. Thus, the microwave radiation is not absorbed by these materials.

11. Often when you touch the indoor antenna on a television receiver, the reception instantly improves. Why?

Answer: You become part of the antenna and thus increase its receptive ability.

15. List as many similarities and differences as you can between sound waves and light waves.

Answer: This is an open-ended question, but you should consider such features as (1) their speed, (2) the medium in which they can propagate, (3) their origin, (4) whether they are transverse or longitudinal, and so forth.

16. What does a radio wave do to the charges in the receiving antenna to provide a signal for your car radio?

Answer: A car radio usually uses a metal rod for an antenna. Such a rod detects the electric field portion of the carrier wave. Variations in the amplitude of the carrier wave causes the electrons in the rod to vibrate with amplitudes emulating that of the carrier wave. Likewise, for frequency modulation, the variations of the frequency of the carrier wave cause constant amplitude vibrations of the electrons in the rod but at frequencies which emulate those of the carrier.

21. When light (or other electromagnetic radiation) travels across a given region, what is it that moves?

Answer: The vibrating electric and magnetic fields that make up the light wave.

22. Why should an infrared photograph of a person look different from a photograph taken with visible light?

Answer: The brightest portions of an infrared photograph correspond to those parts of the subject that are warmest.

SOLUTIONS TO SELECTED END-OF-CHAPTER PROBLEMS

9. Figure 34.3 shows a plane electromagnetic sinusoidal wave propagating in the x direction. The wavelength is 50 m, and the electric field vibrates in the xy plane with an amplitude of 22 V/m. Calculate (a) the sinusoidal frequency and (b) the magnitude and direction of B when the electric field has its maximum value in the negative y direction. (c) Write an expression for B in the form

$$B = B_m \cos (kx - \omega t)$$

with numerical values for B_m, k, and ω.

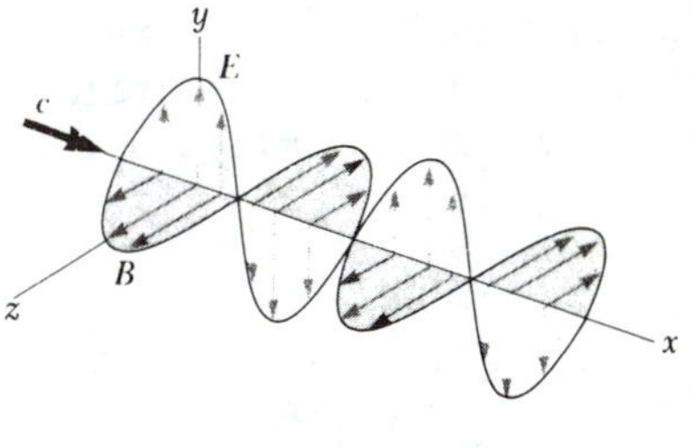

Solution

(a) $c = f\lambda$

$$f = \frac{c}{\lambda} = \frac{3 \times 10^8 \text{ m/s}}{50 \text{ m}} = \boxed{6 \times 10^6 \text{ Hz}}$$

Figure 34.3

(b) $c = \dfrac{E}{B}$

$B = \dfrac{E}{c} = \dfrac{22\ V/m}{3 \times 10^8\ m/s} = \boxed{7.33 \times 10^{-8}\ T = 73.3\ nT}$

B is directed along *negative z-direction*.

(c) $B = B_m \cos (kx - \omega t)$

$k = \dfrac{2\pi}{\lambda}$ and $\omega = 2\pi f$

$\therefore\ B = B_m \cos\left[\left(\dfrac{2\pi}{\lambda}\right)x - (2\pi f)t\right]$

$B = (73.3\ nT) \cos\left[\left(\dfrac{2\pi}{50}\right)x - (2\pi)(6 \times 10^6)t\right]$

$B = (73.3\ nT) \cos\left[2\pi\left(\dfrac{x}{50} - 6 \times 10^{-6}t\right)\right]$

---

15. What is the average magnitude of the Poynting vector at a distance of 5 miles from an isotropic radio transmitter, broadcasting with an average power of 250 kW?

Solution

$S_{av} = \dfrac{P}{A} = \dfrac{P}{4\pi r^2}$

$r = (5\ mi)(1609\ m/mi) = 8045\ m$

$S = \dfrac{250 \times 10^3\ W}{(4\pi)(8.045 \times 10^3\ m)^2} = \boxed{3.07 \times 10^{-4}\ W/m^2}$

---

21. A helium-neon laser intended for instructional use operates at a typical power of 5.0 mW. (a) Determine the maximum value of the electric field at a point where the cross section of the beam is 4 mm². (b) Calculate the electromagnetic energy in a 1-m length of the beam.

Solution

(a) $S_{av} = \dfrac{P}{A} = \dfrac{E_m^2}{2\mu_o c}$

$E_m = \sqrt{\dfrac{2P\mu_o c}{A}} = \sqrt{\dfrac{(2)(5 \times 10^{-3}\ W)(4\pi \times 10^{-7}\ N/A^2)(3 \times 10^8\ m/s)}{4 \times 10^{-6}\ m^2}} = \boxed{971\ V/m}$

(b) $U = (u)(\text{volume}) = \left(\frac{1}{2}\varepsilon_o E_m{}^2\right)(AL)$

$$U = \left(\frac{1}{2}\right)\left(8.85 \times 10^{-12} \frac{C^2}{N \cdot m^2}\right)(971 \text{ V/m})^2(4 \times 10^{-6} \text{ m}^2)(1 \text{ m}) = \boxed{1.67 \times 10^{-11} \text{ J}}$$

25. A radio wave transmits 25 W/m^2 of power per unit area. A plane surface of area A is perpendicular to the direction of propagation of the wave. Calculate the radiation pressure on the surface if the surface is a perfect absorber.

Solution

For complete absorption, $P = \dfrac{S}{c}$

$$P = \frac{25 \text{ W/m}^2}{3 \times 10^8 \text{ m/s}} = \boxed{8.33 \times 10^{-8} \text{ N/m}^2}$$

31. A rectangular surface of dimensions 120 cm $\times$ 40 cm is parallel to and 4.4 m from a very large conducting sheet in which there is a sinusoidally varying surface current which has a maximum value of 10 A/m^2. (a) Calculate the average power incident on the smaller sheet. (b) What power per unit area is radiated by the current-carrying sheet?

Solution

(a) $P = S_{av}A = \left(\dfrac{\mu_o J_o{}^2 c}{8}\right)A$

$$P = \frac{(4\pi \times 10^{-7} \text{ N/A}^2)(10 \text{ A/m}^2)^2(3 \times 10^8 \text{ m/s})}{8}(1.2 \text{ m})(0.4 \text{ m}) = \boxed{2.26 \times 10^3 \text{ W} = 2.26 \text{ kW}}$$

(b) $S_{av} = \dfrac{\mu_o J_o{}^2 c}{8}$

Using the same substitutions as in part (a) gives

$$S_{av} = \boxed{4.71 \text{ kW/m}^2}$$

45. A community plans to build a facility to convert solar radiation into electrical power. They require 1 MW of power (10^6 W), and the system to be installed has an efficiency of 30% (that is, 30% of the solar energy incident on the surface is converted to electrical energy). What must be the effective area of a perfectly absorbing surface used in such an installation, assuming a constant energy flux of 1000 W/m^2?

Solution

At 30% efficiency, $P = 0.3SA$

$$A = \frac{P}{0.3S} = \frac{1 \times 10^6 \text{ W}}{(0.3)(1000 \text{ W/m}^2)} = \boxed{3330 \text{ m}^2}$$

This area is approximately 0.75 acres.

---

57. A linearly polarized microwave of wavelength 1.5 cm is directed along the positive x axis. The electric field vector has a maximum value of 175 V/m and vibrates in the xy plane. (a) Assume that the magnetic field component of the wave can be written in the form $B = B_0 \sin (kx - \omega t)$ and give values for B_0, k, and ω. Also, determine in which plane the magnetic field vector vibrates. (b) Calculate the Poynting vector for this wave. (c) What radiation pressure would this wave exert if directed at normal incidence onto a perfectly reflecting sheet? (d) What acceleration would be imparted to a 500-g sheet (perfectly reflecting and at normal incidence) with dimensions 1 m $\times$ 0.75 m?

Solution

(a) $B = B_0 \sin (kx - \omega t)$

$$B_0 = \frac{E_0}{c} = \frac{175 \text{ V/m}}{3 \times 10^8 \text{ m/s}} = \boxed{5.83 \times 10^{-7} \text{ T}}$$

$$k = \frac{2\pi}{\lambda} = \frac{2\pi}{0.015 \text{ m}} = \boxed{419 \text{ m}^{-1}}$$

$$\omega = kc = (419 \text{ m}^{-1})(3 \times 10^8 \text{ m/s}) = \boxed{1.26 \times 10^{11} \text{ rad/s}}$$

(b) $$S = \frac{E_0 B_0}{\mu_0} = \frac{(175 \text{ V/m})(5.83 \times 10^{-7} \text{ T})}{4\pi \times 10^{-7} \text{ N/A}^2} = \boxed{81.2 \text{ W/m}^2}$$

(c)   For perfect reflection,

$$P_r = \frac{2S}{c} = \frac{(2)(81.2 \text{ W/m}^2)}{3 \times 10^8 \text{ m/s}} = \boxed{5.41 \times 10^{-7} \text{ N/m}^2}$$

(d) $$a = \frac{F}{m} = \frac{P_r A}{m} = \frac{(5.41 \times 10^{-7} \text{ N/m}^2)(0.75 \text{ m}^2)}{0.5 \text{ kg}} = \boxed{8.11 \times 10^{-7} \text{ m/s}^2}$$

35

The Nature of Light
and the Laws of Geometric Optics

OBJECTIVES

1. Understand Huygens' principle and the use of this technique to construct the subsequent position and shape of a given wavefront.

2. Describe the methods used by Roemer and Fizeau for the measurement of c and make calculations using sets of typical values for the quantities involved.

3. Determine the directions of the reflected and refracted rays when a light ray is incident obliquely on the interface between two optical media.

4. Understand the manner in which Fermat's principle of least time can be used as a basis of a derivation of the laws of reflection and refraction.

5. Calculate the fraction of the energy reflected and the fraction transmitted when a light ray is directed at near-normal incidence onto the interface of two media.

6. Calculate the intensity of a light beam as a function of length of travel in a homogeneous dielectric material.

7. Understand the conditions under which total internal reflection can occur in a medium and determine the critical angle for a given pair of adjacent media.

NOTES FROM SELECTED CHAPTER SECTIONS

35.3 THE RAY APPROXIMATION IN GEOMETRIC OPTICS

In the ray approximation, it is assumed that a wave travels through a medium along a straight line in the direction of its rays. A ray for a given wave is a straight line perpendicular to the wavefront. This approximation also neglects diffraction effects.

35.4 REFLECTION AND REFRACTION

A line drawn perpendicular to a surface at the point where an incident ray strikes the surface is called the *normal line*. Angles of reflection and refraction are measured relative to the normal.

When an incident ray undergoes partial reflection and partial refraction, the incident, reflected and refracted rays are all *in the same plane*.

The *path of a light ray through a refracting surface is reversible*.

As light travels from one medium into another, the *frequency does not change*.

35.5 DISPERSION AND PRISMS

Index of refraction is a function of wavelength.

35.6 HUYGENS' PRINCIPLE

Every point on a given wavefront can be considered as a point source for a *secondary wavelet*. At some later time, the new position of the wavefront is determined by the surface tangent to the set of secondary wavelets.

35.7 TOTAL INTERNAL REFLECTION

Total internal reflection is possible only when light rays traveling in one medium are incident on an interface bounding a second medium of *lesser* index of refraction than the first.

Total internal reflection is illustrated in Figure 35.1.

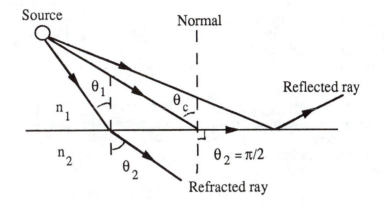

Figure 35.1   Total internal reflection of light occurs at angles of incidence $\theta_1 \geq \theta_c$, where $n_1 > n_2$.

35.8 FERMAT'S PRINCIPLE

A light ray traveling between any two points will follow a path which requires the *least time*.

EQUATIONS AND CONCEPTS

Consider the situation shown in Figure 35.2 in which a light ray is incident obliquely on a smooth, planar surface which forms the boundary between two transparent media of different optical densities. A portion of the ray will be reflected back into the original medium, while the remaining fraction will be transmitted into the second medium.

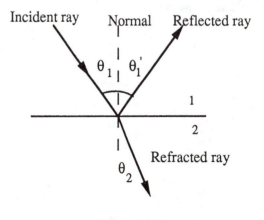

Figure 35.2

The reflection and refraction of light waves at an interface. The direction of the rays are perpendicular to the wave fronts.

423

The *law of reflection* states that the angle of incidence, θ_1, will be equal to the angle of reflection, θ_1'.

$$\theta_1' = \theta_1$$

(35.2)

The angle of refraction (angle between the refracted ray and the normal to the interface between the two media) depends on the angle of incidence and the speed of light in the two media.

$$\frac{\sin \theta_2}{\sin \theta_1} = \frac{v_2}{v_1}$$

(35.3)

The relationship between the angle of incidence and the angle of refraction is usually expressed as a form of *Snell's law*, Eq. 35.8. This expression involves a characteristic parameter of each medium known as the index of refraction, n.

$$n_1 \sin \theta_1 = n_2 \sin \theta_2$$

(35.8)

For a given medium, the *absolute index of refraction* is the ratio of the speed of light in vacuum to the speed of light in the medium.

$$n \equiv \frac{c}{v}$$

(35.4)

The frequency of a wave is characteristic of the source. Therefore, as a light ray travels from one medium to another of different index of refraction, the frequency remains constant but the wavelength changes. The *index of refraction* can be expressed as a ratio of the wavelength of light in vacuum, λ_o, to the wavelength in the medium, λ_n.

$$n = \frac{\lambda_o}{\lambda_n}$$

(35.7)

For angles of incidence equal to or greater than the critical angle, the incident ray will be totally reflected into the first medium.

$$\sin \theta_c = \frac{n_2}{n_1}$$

where $n_1 > n_2$

(35.10)

EXAMPLE PROBLEM SOLUTION

Example 35.1 Yellow light of wavelength $\lambda = 589$ nm travels from a vacuum, through a water layer, and finally through a thickness of crown glass. The path of the beam is shown in Figure 35.3. The angle of incidence at the vacuum-water interface is 30°. Calculate the angle θ_4 that the ray emerging from the glass surface makes with the normal.

Solution

Applying Snell's law to the first two media (vacuum and water) gives

$$\sin \theta_1 = n_2 \sin \theta_2 \qquad (1)$$

Likewise, for the next two interfaces (water-glass, glass-vacuum), we get

$$n_2 \sin \theta_2 = n_3 \sin \theta_3 \qquad (2)$$

$$n_3 \sin \theta_3 = \sin \theta_4 \qquad (3)$$

Comparing (1), (2), and (3), we see that $\sin \theta_1 = \sin \theta_4$, or $\theta_1 = \theta_4 = 30°$. That is, the *emerging light makes the same angle with the normal as the incident light*, since they both correspond to propagation in a vacuum. You should also verify that $\theta_2 = 22°$ and $\theta_3 = 19°$.

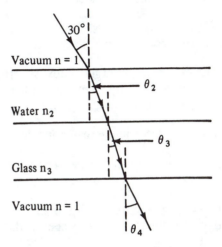

Figure 35.3

The transmission of light through a layer of water and a layer of glass.

ANSWERS TO SELECTED QUESTIONS

6. A laser beam ($\lambda = 632.8$ nm) is incident on a piece of Lucite as in Figure 35.28. Part of the beam is reflected and part is refracted. What information can you get from this photograph?

Answer: You could use a protractor to measure the angles of incidence and refraction, and then use Snell's Law to evaluate the index of refraction of Lucite, taking $n = 1$ for air.

8. The level of water in a clear, colorless glass is easily observed with the naked eye. The level of liquid helium in a clear glass vessel is extremely difficult to see with the naked eye. Explain.

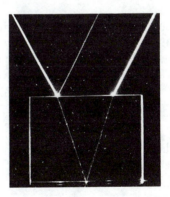

Figure 35.28

Answer: The index of refraction of water is 1.333, which is different from that of air (about 1). On the other hand, the index of refraction of liquid helium is closer to that of air. Consequently, light undergoes greater refraction in water than in liquid helium.

10. Why does a diamond show flashes of color when observed under ordinary white light?

Answer: The diamond acts like a prism in dispersing the light into its spectral components. Different colors are observed as a consequence of the manner in which the diamond is cut.

11. Explain why a diamond shows more "sparkle" than a glass crystal of the same shape and size.

Answer: Diamond has a larger index of refraction than glass and consequently a smaller critical angle. This results in more light being internally reflected.

12. Explain why an oar in the water appears bent.

Answer: The oar appears to be bent because of refraction.

13. Redesign the periscope of Figure 35.25c so that it can show you where you have been rather than where you are going.

Answer: The diagram at the right illustrates how this is done.

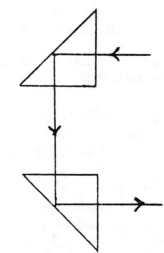

16. A solar eclipse occurs when the moon gets between the earth and the sun. Use a diagram to show why some areas of the earth see a total eclipse, other areas see a partial eclipse, and most areas see no eclipse.

Answer: Use the diagram at the right and imagine the appearance of the shadow that the moon casts on the sun when viewed from different locations on the earth.

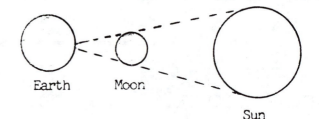

Earth   Moon   Sun

(Not to scale)

17. Some department stores have their windows slanted slightly inward at the bottom. This is to decrease the glare from streetlights or the sun, which would make it difficult for shoppers to see the display inside. Draw a sketch of a light ray reflecting off such a window to show how this technique works.

Answer: The diagram at the right illustrates how incident light will be reflected downward, and hence help decrease the glare.

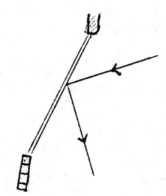

18. Suppose you are told only that two colors of light (X and Y) are sent through a glass prism and that X is bent more than Y. Which color travels more slowly in the prism?

Answer: The color traveling slowest is bent the most. Thus, X travels the slowest in the glass.

SOLUTIONS TO SELECTED END-OF-CHAPTER PROBLEMS

23. A crown glass prism has an apex angle of 15°. What is the angle of minimum deviation of this prism for light of wavelength 525 nm? See Figure 35.15 for the value of n.

Solution

From Equation 35.9 ,

$$n = \frac{\sin\left(\frac{\Phi + \delta_m}{2}\right)}{\sin\left(\frac{\Phi}{2}\right)}$$

Therefore,

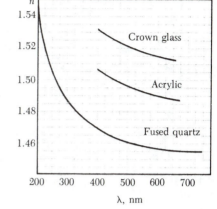

Figure 35.15

$$\delta_m = 2\left[\sin^{-1}\left(n \sin \frac{\Phi}{2}\right) - \frac{\Phi}{2}\right] = 2\left[\sin^{-1}\left(1.52 \sin \frac{\pi}{24}\right) - \frac{\pi}{24}\right]$$

$$\delta_m = 0.138 \text{ rad} = \boxed{7.89°}$$

---

29. A prism with apex angle 50° is made of cubic zirconia, with n = 2.20. What is the angle of minimum deviation δ_m for such a prism?

Solution

From Equation 35.9,
$$n = \frac{\sin\left(\frac{\Phi + \delta_m}{2}\right)}{\sin\left(\frac{\Phi}{2}\right)}$$

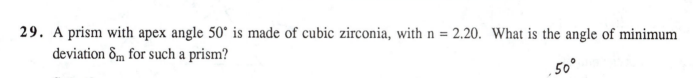

Solving for δ_m,
$$\delta_m = 2 \sin^{-1}\left(n \sin \frac{\Phi}{2}\right) - \Phi$$

$$\delta_m = 2 \sin^{-1}\left(2.2 \sin 25°\right) - 50° = \boxed{86.8°}$$

---

31. The index of refraction for violet light in silica flint glass is 1.66, and the refractive index for red light is 1.62. What is the angular dispersion of visible light passing through a prism of apex angle 60° if the angle of incidence is 50°? (See Figure 35.31.)

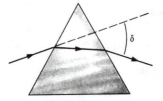

Figure 35.31

Solution

For the incoming ray, $\sin \theta_2 = \dfrac{\sin \theta_1}{n}$

Using Figure 36.19,

$$(\theta_2)_{\text{Violet}} = \sin^{-1}\left(\frac{\sin 50°}{1.66}\right) = 27.48°$$

$$(\theta_2)_{\text{Red}} = \sin^{-1}\left(\frac{\sin 50°}{1.62}\right) = 28.22°$$

For the outgoing ray,

$$\theta_2' = 60° - \theta_2 \quad \text{and} \quad \sin \theta_3 = n \sin \theta_2'$$

$$(\theta_3)_{\text{Violet}} = \sin^{-1}\left[1.66 \sin 32.52°\right] = 63.17°$$

$$(\theta_3)_{\text{Red}} = \sin^{-1}\left[1.62 \sin 31.78°\right] = 58.56°$$

The dispersion is

$$\Delta\theta_3 = 63.17° - 58.56° = \boxed{4.61°}$$

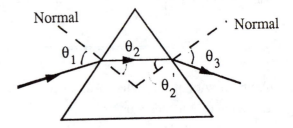

θ_1 = incident angle $\cong 50°$

θ_2 = angle of refraction

θ_2' = 2nd angle of refraction

θ_3 = angle of transmission

41. A specimen of glass has an index of refraction of 1.61 for the wavelength corresponding to the prominent bright line in the sodium spectrum. If an equiangular prism is made from this glass, what angle of incidence will result in minimum deviation of the sodium line?

Solution

Equation 35.9 gives the index of refraction in terms of the angle of minimum deviation and the apex angle of the prism:

$$n = \frac{\sin\left(\frac{\Phi + \delta_m}{2}\right)}{\sin\left(\frac{\Phi}{2}\right)}$$

This gives

$$\delta_m = 2\left[\sin^{-1}\left(n\sin\frac{\Phi}{2}\right) - \frac{\Phi}{2}\right] = 2\left[\sin^{-1}\left(1.61\sin\frac{\pi}{6}\right) - \frac{\pi}{6}\right] = 0.824 \text{ rad}$$

It can be shown from the geometry of the arrangement (see Example 35.7 of the text) that

$$\theta = \frac{1}{2}\left(\Phi + \delta_m\right) \qquad \text{or} \qquad \Phi = \frac{1}{2}\left(\frac{\pi}{3} + 0.824\right) \text{rad} = 0.674 \text{ rad} = \boxed{53.6°}$$

---

45. A drinking glass is 4 cm wide at the bottom, as shown in Figure 35.32. When an observer's eye is placed as shown, the observer sees the edge of the bottom of the glass. When this glass is filled with water, the observer sees the center of the bottom of the glass. Find the height of the glass.

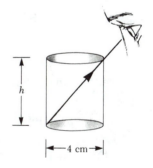

Figure 35.32

Solution

$$\tan\theta_1 = \frac{4}{h} \quad \Rightarrow \quad \tan\theta_1 = 2\tan\theta_2$$

$$\tan\theta_2 = \frac{2}{h} \quad \Rightarrow \quad \tan^2\theta_1 = 4\tan^2\theta_2$$

$$\frac{\sin^2\theta_1}{\left(1 - \sin^2\theta_1\right)} = \frac{4\sin^2\theta_2}{(1 - \sin^2\theta_2)} \qquad (1)$$

$n_1\sin\theta_1 = n_2\sin\theta_2$ or $\sin\theta_1 = 1.333\sin\theta_2$. Squaring both sides $\sin^2\theta_1 = 1.778\sin^2\theta_2$

Substitute this value into (1) which yields

$$\frac{1.778\sin^2\theta_2}{\left(1 - 1.778\sin^2\theta_2\right)} = \frac{4\sin^2\theta_2}{\left(1 - \sin^2\theta_2\right)}$$

Letting $\sin^2 \theta_2 = x$ gives

$$\frac{0.444}{(1 - 1.778x)} = \frac{1}{(1 - x)}$$

$0.444 - 0.444x = 1 - 1.778x \quad \Rightarrow \quad 1.333x = 1 - 0.444 = 0.556 \quad \text{so} \quad x = 0.417 = \sin^2 \theta_2$

$$\theta_2 = \sin^{-1} \sqrt{0.417} = 40.2°$$

$$h = \frac{(2 \text{ cm})}{\tan \theta_2} = \frac{(2 \text{ cm})}{\tan (40.2°)} = \boxed{2.37 \text{ cm}}$$

---

7. Derive the law of reflection (Equation 35.2) from Fermat's principle of least time. (See the procedure outlined in Section 35.9 for the derivation of the law of refraction from Fermat's principle.)

Solution

To derive law of *reflection*, locate point 0 so that the time of travel from point A to point B will be minimum. The *total* light path is

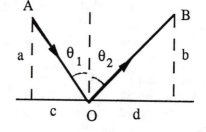

$$L = a \sec \theta_1 + b \sec \theta_2$$

The time of travel is

$$t = \left(\frac{1}{v}\right)\left(a \sec \theta_1 + b \sec \theta_2\right)$$

If point 0 is displaced by dx, then

$$dt = \left(\frac{1}{v}\right)\left(a \sec \theta_1 \tan \theta_1 \, d\theta_1 + b \sec \theta_2 \tan \theta_2 \, d\theta_2\right) = 0 \qquad (1)$$

(since for minimum time $dt = 0$)

Also, $\qquad\qquad c + d = a \tan \theta_1 + b \tan \theta_2 = \text{constant}$

so $\qquad\qquad a \sec^2\theta_1 \, d\theta_1 + b \sec^2 \theta_2 \, d\theta_2 = 0 \qquad (2)$

Combining (1) and (2) we find $\qquad \boxed{\theta_1 = \theta_2}$

---

49. A light ray of wavelength 589 nm is incident at an angle θ on the top surface of a block of polystyrene, as shown in Figure 35.33. (a) Find the maximum value of θ for which the refracted ray will undergo *total* internal reflection at the left vertical face of the block. (b) Repeat the calculation for the case in which the polystyrene block is immersed in water. (c) What happens if the block is immersed in carbon disulfide?

Solution

(a) For polystyrene *surrounded by air*, internal reflection requires

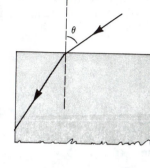

Figure 35.33

$$\theta_3 = \sin^{-1}\left(\frac{1}{1.49}\right) = 42.2°$$

and then from the geometry

$$\theta_2 = 90 - \theta_3 = 47.8°$$

From Snell's law, this would require that $\theta_1 > 90°$. Therefore, internal reflection in this case is *not possible*.

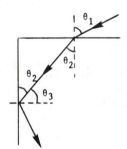

(b) For polystyrene surrounded *by water*, we have

$$\theta_3 = \sin^{-1}\left(\frac{1.33}{1.49}\right) = 63.2° \qquad \text{and} \qquad \theta_2 = 26.8°$$

and from Snell's law $\qquad\qquad \theta_1 = \boxed{30.3°}$

(c) *Not possible* since the beam is initially traveling in a medium of lower index of refraction.

53. A light ray is incident on a prism and refracted at the first surface as shown in Figure 35.35.

Let ϕ represent the apex angle of the prism and n its index of refraction. Find in terms of n and ϕ the smallest allowed value of the angle of incidence at the first surface for which the refracted ray will *not* undergo internal reflection at the second surface.

Solution

Refer to the figure at the right. We see that $\theta_2 + \alpha = 90°$ and $\theta_2 + \beta = 90°$, so $\theta_2 + \theta_3 + \alpha + \beta = 180°$. Also, from the figure we see $\alpha + \beta + \phi = 180°$; therefore $\phi = \theta_2 + \theta_3$. By applying Snell's law at the first and second refracting surfaces, we find

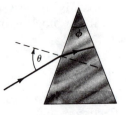

Figure 35.35

$$\theta_2 = \sin^{-1}\left(\frac{\sin \theta_1}{n}\right)$$

and

$$\theta_3 = \sin^{-1}\left(\frac{\sin \theta_4}{n}\right)$$

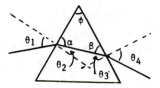

Substituting these values into the expression for ϕ

$$\phi = \sin^{-1}\left(\frac{\sin \theta_1}{n}\right) + \sin^{-1}\left(\frac{\sin \theta_4}{n}\right)$$

The limiting condition for internal reflection at the second surface is $\theta_4 \to 90°$. Under these conditions, we have

$$\sin \theta_1 = n \sin\left[\phi - \sin^{-1}\left(\frac{1}{n}\right)\right]$$

or

$$\theta_1 = \sin^{-1}\left[\sqrt{n^2 - 1} \, \sin \phi - \cos \phi\right]$$

Geometric Optics

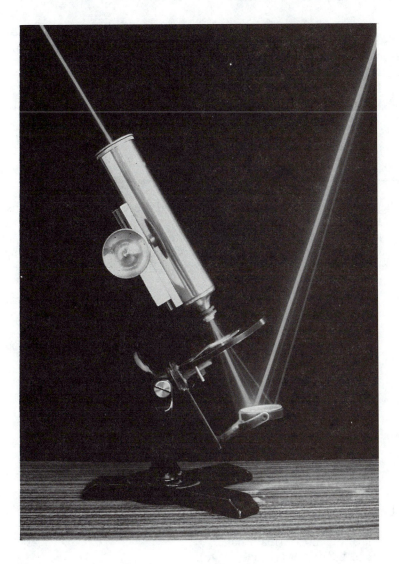

OBJECTIVES

1. Calculate the location of the image of a specified object as formed by a plane mirror, spherical mirror, plane refracting surface, spherical refracting surface, thin lens, or a combination of two or more of these devices.

2. Understand the relationship of the algebraic signs associated with calculated quantities to the nature of the image and object: real or virtual, erect or inverted.

3. Construct ray diagrams to determine the location and nature of the image of a given object when the geometrical characteristics of the optical device (lens or mirror) are known.

4. Describe the origin of each of the most frequently encountered lens aberrations.

5. Understand the geometry of the lens combination for each of several simple optical instruments: camera, compound microscope, and astronomical telescope.

SKILLS

A major portion of this chapter is devoted to the development and presentation of equations which can be used to determine the location and nature of images formed by various optical components acting either singly or in combination. It is essential that these equations be used with the correct algebraic sign associated with each quantity involved. You must understand clearly the sign conventions for mirrors, refracting surfaces, and lenses. The following discussion represents a review of these sign conventions.

SIGN CONVENTIONS FOR MIRRORS

Equation: $\frac{1}{s} + \frac{2}{s'} = \frac{1}{f}$

The front side of the mirror is the region on which light rays are incident and reflected.

s is + if the object is in front of the mirror (real object).
s is − if the object is in back of the mirror (virtual object).

s' is + if the image is in front of the mirror (real image).
s' is − if the image is in back of the mirror (virtual image).

Both f and R are + if the center of curvature is in front of the mirror (concave mirror). Both f and R are − if the center of curvature is in back of the mirror (convex mirror).

If M is positive, the image is erect.
If M is negative, the image is inverted.

You should check the sign conventions as stated against the situations described in Figure 36.1.

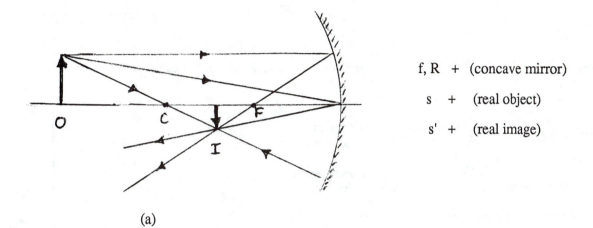

f, R + (concave mirror)

s   +   (real object)

s'   +   (real image)

(a)

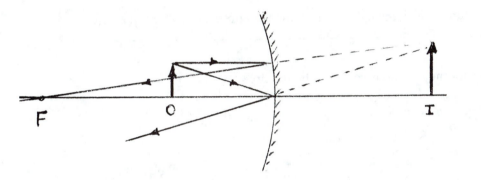

f, R   +   (concave mirror)

s   +   (real object)

s'   –   (virtual image)

(b)

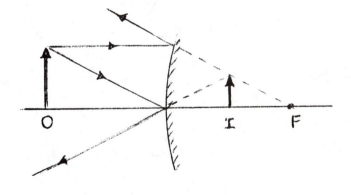

f, R   –   (convex mirror)

s   +   (real object)

s'   –   (virtual image)

(c)

Figure 36.1   Figures describing sign conventions for mirrors.

SIGN CONVENTIONS FOR REFRACTING SURFACES

Equation: $\dfrac{n_1}{s} + \dfrac{n_2}{s'} = \dfrac{n_2 - n_1}{R}$

In the following table, the *front* side of the surface is the side *from which the light is incident.*

---

s is + if the object is in front of the surface (real object).
s is − if the object is in back of the surface (virtual object).

s' is + if the image is in back of the surface (real image).
s' is − if the image is in front of the surface (virtual image).

R is + if the center of curvature is in back of the surface
R is − if the center of curvature is in front of the surface.

n_1 refers to the index of the medium on the side of the interface from which the light comes.

n_2 is the index of the medium into which the light is transmitted after refraction at the interface.

Review the above sign conventions for the situations shown in Figure 36.2.

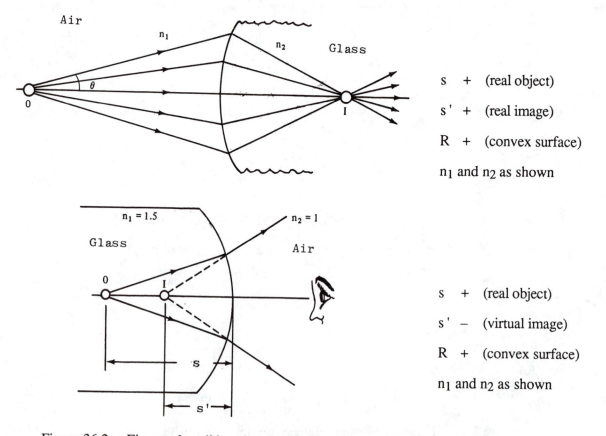

s + (real object)

s' + (real image)

R + (convex surface)

n_1 and n_2 as shown

s + (real object)

s' − (virtual image)

R + (convex surface)

n_1 and n_2 as shown

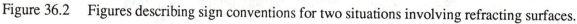

Figure 36.2   Figures describing sign conventions for two situations involving refracting surfaces.

SIGN CONVENTIONS FOR THIN LENSES

Equations: $\dfrac{1}{s} + \dfrac{1}{s'} = \dfrac{1}{f} = (n-1)\left(\dfrac{1}{R_1} - \dfrac{1}{R_2}\right)$

In the following table, the *front* of the lens is the *side from which the light is incident.*

---

s is + if the object is in front of the lens.
s is − if the object is in back of the lens.

s' is + if the image is in back of the lens.
s' is − if the image is in front of the lens.

R_1 and R_2 are + if the center of curvature is in back of the lens.
R_1 and R_2 are − if the center of curvature is in front of the lens.

The sign conventions for thin lenses are illustrated by the examples shown in Figure 36.3.

(a)

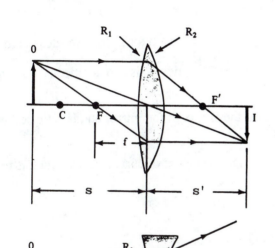

(a) Double-Convex Lens (Converging)
s, s', f, and R_1 positive
R_2 negative
Image real and inverted

(b)

(b) Double-Concave Lens (Diverging)
s, R_2 positive
s', R_1, f negative
Image virtual, erect, and diminished

(c)

(c) Double-Convex Lens
s, R_1 positive
s', R_2 negative
Image virtual, erect, and enlarged

Figure 36.3   Sign conventions for various thin lenses

NOTES FROM SELECTED CHAPTER SECTIONS

36.1 IMAGES FORMED BY PLANE MIRRORS

The image formed by a plane mirror has the following properties:

1. The image is as far behind the mirror as the object is in front.

2. The image is unmagnified, virtual, and erect. (By erect, we mean that, if the object arrow points upward as in Figure 36.2, so does the image arrow.)

3. The image has right-left reversal.

36.2 IMAGES FORMED BY SPHERICAL MIRRORS

A spherical mirror is a reflecting surface which has the shape of a segment of a sphere.

Real images are formed at a point when *reflected light actually passes through the point.*

Virtual images are formed at a point when light rays *appear to diverge from the point.*

The point of intersection of any two of the following rays in a ray diagram for mirrors locates the image:

1. The first ray is drawn from the top of the object parallel to the optical axis and is reflected back through the focal point, F.

2. The second ray is drawn from the top of the object to the vertex of the mirror and is reflected with the angle of incidence equal to the angle of reflection.

3. The third ray is drawn from the top of the object through the center of curvature, C, which is reflected back on itself.

36.4 THIN LENSES

The following three rays form the ray diagram for a thin lens:

1. The first ray is drawn parallel to the optic axis. After being refracted by the lens, this ray passes through (or appears to come from) one of the focal points.

2. The second ray is drawn through the center of the lens. This ray continues in a straight line.

3. The third ray is drawn through the focal point F, and emerges from the lens parallel to the optic axis.

36.5 LENS ABERRATIONS

Aberrations are responsible for the formation of imperfect images by lenses and mirrors. Spherical aberration is due to the variation in focal points for parallel incident rays that strike the lens at various distances from the optical axis. Chromatic aberration arises from the fact that light of different wavelengths focuses at different points when refracted by a lens.

36.6 THE CAMERA

The f-number of a lens is the ratio of the focal length to the diameter. The smaller the f-number, the faster the lens.

36.7 THE EYE

The near point represents the closest distance for which the lens will produce a sharp image on the retina. This distance usually increases with age and has an average value of around 25 cm.

The *power* of a lens in *diopters* equals the inverse of the focal length in meters.

EQUATIONS AND CONCEPTS

The *mirror equation* is used to locate the position of an image formed by reflection of paraxial rays. The focal point of a spherical mirror is located midway between the center of curvature and the vertex of the mirror.

$$\frac{1}{s} + \frac{1}{s'} = \frac{1}{f} \tag{36.6}$$

$$f = \frac{R}{2} \tag{36.5}$$

In using the equations related to the image forming properties of spherical mirrors, spherical refracting surfaces, and thin lenses, you must be very careful to use the correct algebraic sign for each physical quantity. The sign conventions appropriate for the form of the equations stated here are summarized in the SKILLS section.

The *lateral magnification* of a spherical mirror can be stated either as a ratio of image size to object size or in terms of the ratio of image distance to object distance.

$$M = \frac{h'}{h} = -\frac{s'}{s} \tag{36.2}$$

A magnified image of an object can be formed by a single spherical refracting surface of radius R which separates two media whose indices of refraction are n_1 and n_2.

$$\frac{n_1}{s} + \frac{n_2}{s'} = \frac{n_2 - n_1}{R} \tag{36.8}$$

$$M = \frac{h'}{h} = \frac{n_1 s'}{n_2 s}$$

A special case is that of the virtual image formed by a *planar refracting surface* ($R = \infty$).

$$\frac{n_1}{s} = -\frac{n_2}{s'}$$

$$s' = -\frac{n_2}{n_1} s \tag{36.9}$$

The *thin lens* is an important component in many optical instruments. The location of the image formed by a given object is determined by the characteristic properties of the lens (index of refraction n and radii of curvature R). If the lens is surrounded by a medium other than air, the index of refraction given in Eq. 36.10 must be *the index of the lens relative to the surrounding medium*. This means, for example, that a hollow convex lens ("air lens"), if immersed in water, would have a *negative* focal length. The lateral magnification of a thin lens has the same form as that of a spherical mirror (Eq. 36.2).

$$\frac{1}{s} + \frac{1}{s'} = (n - 1)\left(\frac{1}{R_1} - \frac{1}{R_2}\right) \qquad (36.10)$$

$$\frac{1}{s} + \frac{1}{s'} = \frac{1}{f} \qquad (36.12)$$

Thin lenses are often used in combination. A special case occurs when two thin lenses are in contact. The *focal length of the combination* given by Eq. 36.13 will be *less* than that of either lens individually.

$$\frac{1}{f} = \frac{1}{f_1} + \frac{1}{f_2} \qquad (36.13)$$

The use of lens and mirrors in the design and operation of optical instruments can be illustrated by several relatively simple examples.

Camera: The *light intensity* I incident on the film per unit area is inversely proportional to the square of a ratio of the diameter of the lens to its focal length. The *f-number* equals the ratio of the focal length to the lens diameter.

$$I \sim \frac{1}{(f/D)^2} \qquad (36.15)$$

$$f\text{-number} \equiv \frac{f}{D} \qquad (36.14)$$

Eye: The power of a lens in diopters is the reciprocal of the focal length measured in meters (including the correct algebraic sign).

$$P = \frac{1}{f}$$

Simple Magnifier: When an object is at the near point (25 cm), the angle subtended by the object at the eye is θ. When a convex lens of focal length f is placed between the eye and the object, an image which subtends an angle θ_0 can be formed at the near point. The *angular magnification* or magnifying power of the lens can be expressed in alternate forms.

$$m = \frac{\theta}{\theta_0} \qquad (36.16)$$

$$m = 1 + \frac{25\ \text{cm}}{f} \qquad (36.18)$$

Compound Microscope: This instrument contains an objective lens of short focal length f_0 and an eye piece of focal length f_e. The two lenses are separated by a distance L. When an object is located just beyond the focal point of the objective, the two lenses in combination form an enlarged, virtual and inverted image of lateral *magnification* M given by Eq. 36.20.

$$M = -\frac{L}{f_0}\left(\frac{25\ \text{cm}}{f_e}\right) \qquad (36.20)$$

Astronomical Telescope: Two convex lenses are separated by a distance equal to the sum of their focal lengths. The *angular magnification* is equal to the ratio of the two focal lengths.

$$m = \frac{f_o}{f_e}$$

(36.21)

EXAMPLE PROBLEM SOLUTIONS

Example 36.1 A long, thin-walled, hollow, air-filled glass cylinder has one end in the shape of a *concave* surface of radius 8 cm. The cylinder is immersed in water (n = 1.33) and forms an image of an object also in the water and located 20 cm from the vertex of the concave surface. Determine the location and nature of the image of the object.

Solution

We can make use of Eq. 36.8,

$$\frac{n_1}{s} + \frac{n_2}{s'} = \frac{n_2 - n_1}{R}$$

(36.8)

Referring to Fig. 36.4 in applying the following sign conventions for refracting surfaces, we find

$s = +20$ cm (object in front of the surface)
$R = -8$ cm (center of curvature in front of the surface)
$n_1 = 1.33$ (light is incident from the water side of the interface)
$n_2 = 1$ (light is transmitted after refraction into the air-filled cylinder)

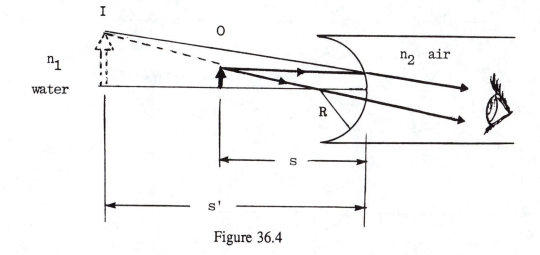

Figure 36.4

Substitution of these values into Eq. 36.8 yields:

$$\frac{1.33}{20\,\text{cm}} + \frac{1}{s'} = \frac{1 - 1.33}{-8\,\text{cm}}$$

so that

$$s' = -39.6\,\text{cm}$$

The negative value of s' indicates that the image is *virtual*, and is located 39.6 cm in front of (to the left of) the concave surface. The magnification produced by the refracting surface is given by Eq. 36.7:

$$M = -\frac{n_1 s'}{n_2 s}$$

so

$$M = -\frac{(1.33)(-39.6 \text{ cm})}{(1)(20 \text{ cm})} = +2.63$$

The positive sign for M signifies that the image is *erect* relative to the object.

Example 36.2 Two convex lenses are separated by a distance of 32 cm. The lens on the left has a focal length of 6 cm and the one on the right has a focal length of 24 cm. An object in the form of an arrow is located 10 cm to the left of the first lens. Determine the location and magnification of the image as formed by the combination of lenses.

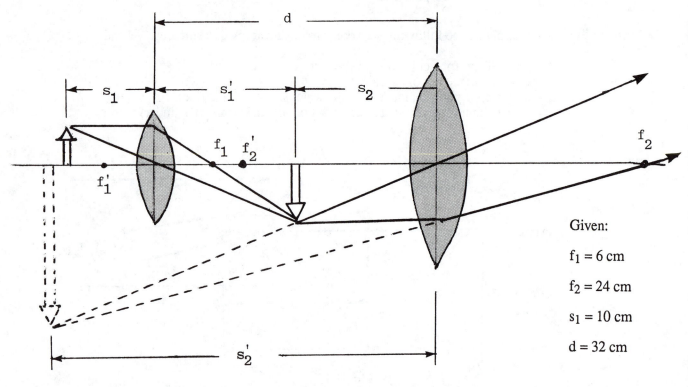

Given:

$f_1 = 6$ cm

$f_2 = 24$ cm

$s_1 = 10$ cm

$d = 32$ cm

Figure 36.5

Solution

The object and two lenses are positioned as shown in Figure 36.5. Applying Eq. 36.6 first to the lens on the left using subscript 1 gives

$$\frac{1}{s_1} + \frac{1}{s_1'} = \frac{1}{f_1} \quad \text{or} \quad s_1' = \frac{f_1 s_1}{s_1 - f_1}$$

From the sign conventions for thin lenses, $s_1 = +10$ cm and $f_1 = +6$ cm. Therefore,

$$s_1' = \frac{(6 \text{ cm})(10 \text{ cm})}{10 \text{ cm} - 6 \text{ cm}} = 15 \text{ cm}$$

The image of the real object formed by lens 1 is 15 cm to the right of lens 1. This image is located in Figure 36.5 using two rays.

The real image formed by lens 1 serves as the real object for lens 2. From the figure, we see that

$$s_2 = d - s_1' = 32 \text{ cm} - 15 \text{ cm} = 17 \text{ cm}$$

Substituting this value together with $f_2 = 24$ cm into Eq. 36.6 gives

$$s_2' = \frac{f_2 s_2}{s_2 - f_2} + \frac{(24 \text{ cm})(17 \text{ cm})}{17 \text{ cm} - 24 \text{ cm}} = -58.3 \text{ cm}$$

The negative sign for s_2' signifies that the image formed by the second lens is a *virtual image*. The final image is 58.3 cm to the left of the second lens and is located by two rays in Figure 36.5.

The lateral magnification for each lens is given by Eq. 36.2. The overall magnification due to the combination of lenses is

$$M = M_1 M_2 = \left(-\frac{s_1'}{s_1}\right)\left(-\frac{s_2'}{s_2}\right)$$

Substituting known values into this equation yields

$$M = -\left(\frac{15 \text{ cm}}{10 \text{ cm}}\right)\left(-\frac{-58.3 \text{ cm}}{17 \text{ cm}}\right) = -5.14$$

The final image is *inverted relative to the original object*.

ANSWERS TO SELECTED QUESTIONS

5. It is well known that distant objects viewed under water with the naked eye appear blurred and out of focus. On the other hand, the use of goggles provides the swimmer with a clear view of objects. Explain this, using the fact that the indices of refraction of the cornea, water, and air are 1.376, 1.333, and 1.00029, respectively.

Answer: Since water and the cornea have almost the same index of refraction, very little refraction of light occurs when viewed with the naked eye, hence the light from the object focuses behind the retina. When goggles are used, the air space in front of the eye acts as a lens, giving the usual amount of refraction at the eye-air interface, and the correct focusing condition.

6. Why does a clear stream always appear to be shallower than it actually is?

Answer: The apparent depth is less than the true depth due to refraction of light originating from objects such as stones at the bottom of the stream.

7. A person spearfishing in a boat sees a fish apparently located 3 m from the boat at a depth of 1 m. In order to hit the fish with his spear, should the person aim at the fish, above the fish, or below the fish? Explain.

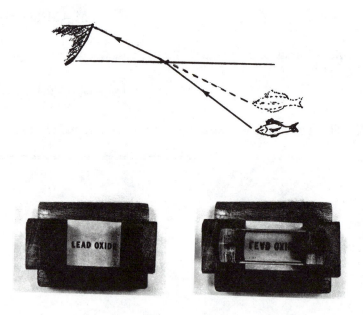

Answer: As the diagram shows, the fish appears to be at a depth which is less than its actual depth. Therefore, the person should aim below the fish in order to hit it.

10. If a cylinder of solid glass or clear plastic is placed above the words LEAD OXIDE and viewed from the side as shown in Figure 36.32, the word "LEAD" appears inverted but the word "OXIDE" does not. Explain.

Answer: Both words are actually inverted. However, the word "OXIDE" does not appear to be inverted because of the symmetry of the characters. That is, when the word "OXIDE" is inverted, it still looks the same.

Figure 36.32

12. Explain why a mirror cannot give rise to chromatic aberration.

Answer: The mechanism giving rise to chromatic aberration involves the refraction of light. Refraction does not take place in producing an image with a mirror. Images formed by mirrors are produced by reflection.

13. What is the magnification of a plane mirror? What is its focal length?

Answer: The magnification is unity, and the focal length is infinite.

14. Why do some emergency vehicles have the symbol "ƎƆИAⒾU8MA" written on the front?

Answer: When this word is viewed through the rear-view mirror of a car, the word is reversed so that it can be read as AMBULANCE.

15. Explain why a fish in a spherical goldfish bowl appears larger than it really is.

Answer: See Example 36.4. Note that the image is enlarged.

16. Lenses used in eyeglasses, whether converging or diverging, are always designed such that the middle of the lens curves away from the eye, like the center lenses of Figure 36.18a and 36.18b. Why?

Answer: This shape is used so the eyelashes of the wearer will not brush against the glasses when he or she blinks.

17. A mirage is formed when the air gets gradually cooler as the height above the ground increases. What might happen if the air grows gradually warmer as the height is increased? This often happens over bodies of water or snow-covered ground -- the effect is called looming.

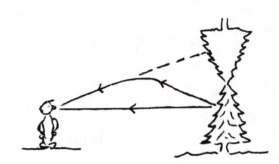

Answer: Light leaving the object and headed toward the sky would be refracted downward, and an image would be seen hovering in the air above the real object.

SOLUTIONS TO SELECTED END-OF-CHAPTER PROBLEMS_____

7. A concave mirror has a radius of curvature of 60 cm. Calculate the image position and magnification of an object placed in front of the mirror at distances of (a) 90 cm and (b) 20 cm. (c) Draw ray diagrams to obtain the image in each case.

Solution

(a) $\dfrac{1}{s} + \dfrac{1}{s'} = \dfrac{2}{R}$ or $\dfrac{1}{90\ cm} + \dfrac{1}{s'} = \dfrac{2}{60\ cm}$

$\dfrac{1}{s'} = \dfrac{2}{60\ cm} - \dfrac{1}{90\ cm} = 0.022\ cm^{-1}$

$s' = 45\ cm$

$M = -\dfrac{s'}{s} = -\dfrac{45\ cm}{90\ cm} = \boxed{-\dfrac{1}{2}}$

(b) $\dfrac{1}{s} + \dfrac{1}{s'} = \dfrac{2}{R}$ or $\dfrac{1}{20\ cm} + \dfrac{1}{s'} = \dfrac{2}{60\ cm}$

$\dfrac{1}{s'} = \dfrac{2}{60\ cm} - \dfrac{1}{20\ cm} = -0.0167\ cm^{-1}$

$s' = -60\ cm$

$M = -\dfrac{s'}{s} = -\left(\dfrac{-60\ cm}{20\ cm}\right) = \boxed{3}$

(c)

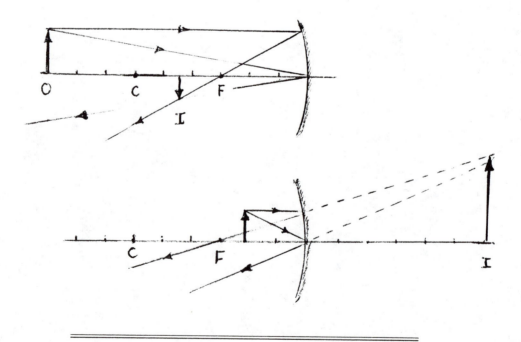

9. Calculate the image position and magnification for an object placed (a) 20 cm and (b) 60 cm in front of a convex mirror of focal length -40 cm. (c) Use ray diagrams to locate image positions corresponding to the object positions in (a) and (b).

Solution

(a) $\dfrac{1}{s}+\dfrac{1}{s'}=\dfrac{1}{f}$ or $\dfrac{1}{20\text{ cm}}+\dfrac{1}{s'}=-\dfrac{1}{40\text{ cm}}$

$\dfrac{1}{s'}=-0.075\text{ cm}^{-1}$ or $s'=-13.3\text{ cm}$

$M=-\dfrac{s'}{s}=-\left(\dfrac{-13.3\text{ cm}}{20\text{ cm}}\right)=\boxed{0.667}$

(b) $\dfrac{1}{s}+\dfrac{1}{s'}=\dfrac{1}{f}$ or $\dfrac{1}{60\text{ cm}}+\dfrac{1}{s'}=-\dfrac{1}{40\text{ cm}}$

$\dfrac{1}{s'}=-0.0417\text{ cm}^{-1}$ or $s'=-24\text{ cm}$

$M=-\dfrac{s'}{s}=-\dfrac{(-24\text{ cm})}{60\text{ cm}}=\boxed{0.400}$

(c)

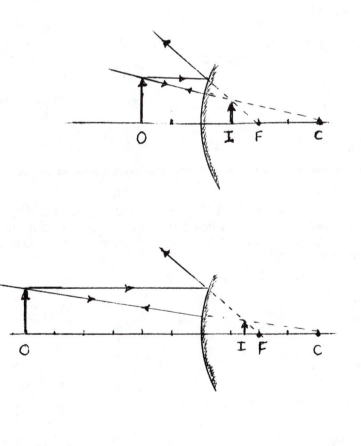

21. A glass sphere (n = 1.50) of radius 15 cm has a tiny air bubble located 5 cm from the center. The sphere is viewed along a direction parallel to the radius containing the bubble. What is the apparent depth of the bubble below the surface of the sphere?

Solution

From Equation 36.8,
$$\frac{n_1}{s} + \frac{n_2}{s'} = \frac{(n_2 - n_1)}{R}$$

Solve for s' to find
$$s' = \frac{n_2 R s}{s(n_2 - n_1) - n_1 R}$$

In this case $n_1 = 1.5$, $n_2 = 1$, $R = -15$ cm, and $s = 10$ cm; so

$$s' = \frac{(1)(-15 \text{ cm})(10 \text{ cm})}{(10 \text{ cm})(1 - 1.5) - (1.5)(-15 \text{ cm})} = \boxed{-8.57 \text{ cm}}$$

Therefore, the apparent depth is $\boxed{8.57 \text{ cm}}$

---

33. A real object is located 20 cm to the left of a diverging lens of focal length f = –32 cm. Determine (a) the location and (b) the magnification of the image.

Solution

(a) $\frac{1}{s} + \frac{1}{s'} = \frac{1}{f}$ or $\frac{1}{20 \text{ cm}} + \frac{1}{s'} = \frac{1}{-32 \text{ cm}}$

so $s' = -\left(\frac{1}{20 \text{ cm}} + \frac{1}{32 \text{ cm}}\right)^{-1} = \boxed{-12.3 \text{ cm}}$ (to the left of the lens)

(b) $M = -\frac{s'}{s} = \frac{12.3 \text{ cm}}{20 \text{ cm}} = \boxed{0.615}$

---

55. An object is located in a fixed position in front of a screen. A thin lens, placed between the object and the screen, produces a sharp image on the screen when it is in either of two positions that are 10 cm apart. The sizes of images in the two situations are in the ratio 3:2. (a) What is the focal length of the lens? (b) What is the distance from the screen to the object?

Solution

(a) $s_1 = s_2'$, $s_2 = s_1'$

$\frac{m_1}{m_2} = \frac{3}{2} = \frac{s_1'/s_1}{s_2'/s_2}$; therefore, $s_2 = \sqrt{\frac{3}{2}}\, s_1$

$s_1 = s_2 - 10 \text{ cm} = \sqrt{\frac{3}{2}}\, s_1 - 10 \text{ cm}$

$s_1 = \frac{10 \text{ cm}}{\sqrt{\frac{3}{2}} - 1} = 44.5 \text{ cm}$

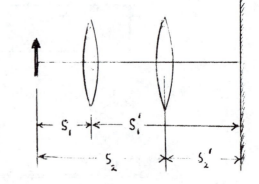

$s_1' = s_2 = s_1 + 10$ cm $= 54.5$ cm

$$\frac{1}{f} = \frac{1}{s} + \frac{1}{s'} = \frac{1}{44.5 \text{ cm}} + \frac{1}{54.5 \text{ cm}}$$

$f = \boxed{24.5 \text{ cm}}$

(b) $s + s' = 44.5$ cm $+ 54.5$ cm $= \boxed{99.0 \text{ cm}}$

---

60. A thin lens with refractive index n is immersed in a liquid with index n'. Show that the focal length f of the lens is given by

$$\frac{1}{f} = \left(\frac{n}{n'} - 1\right)\left(\frac{1}{R_1} - \frac{1}{R_2}\right)$$

Solution

$$\frac{n'}{s_1} + \frac{n}{s_1'} = \frac{n - n'}{R_1} \tag{1}$$

At the second interface,

$$\frac{n}{s_2} + \frac{n'}{s_2'} = \frac{n' - n}{R_2} \tag{2}$$

Since, for a thin lens, $s_2 = s_1'$,

$$\frac{-n}{s_1'} + \frac{n'}{s_2'} = \frac{n' - n}{R_2} \tag{3}$$

Adding (3) to (1) and dropping the subscripts on s and s',

$$\frac{n}{s} + \frac{n'}{s'} = (n - n')\left(\frac{1}{R_1} - \frac{1}{R_2}\right)$$

To find the focal length, let $s = \infty$.

$$\frac{n'}{s'} = (n - n')\left(\frac{1}{R_1} - \frac{1}{R_2}\right)$$

$$\frac{1}{s'} = \frac{1}{f} = \left(\frac{n}{n'} - 1\right)\left(\frac{1}{R_1} - \frac{1}{R_2}\right)$$

---

65. An object is placed 12 cm to the left of a diverging lens of focal length – 6 cm. A converging lens of focal length 12 cm is placed a distance d to the right of the diverging lens. Find the distance d such that the final image is at infinity. Draw a ray diagram for this case.

Solution

From Equation 36.12,

$$s_1' = \frac{f_1 s_1}{s_1 - f_1} = \frac{(-6 \text{ cm})(12 \text{ cm})}{12 \text{ cm} - (-6 \text{ cm})} = -4 \text{ cm}$$

When we require that $s_2' \to \infty$, Equation 36.12 becomes $s_2 = f_2$ and in this case $s_2 = d - (-4 \text{ cm})$. Therefore,

$$d + 4 \text{ cm} = f_2 = 12 \text{ cm} \quad \text{and} \quad d = \boxed{8 \text{ cm}}$$

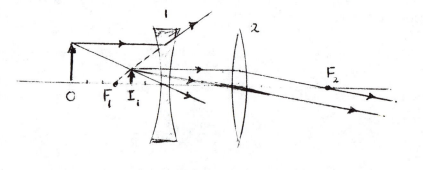

67. Repeat Problem 66 if the lens is diverging and has a focal length of 15 cm.

Solution

Use $s' = \dfrac{fs}{s-f}$ and $M = \dfrac{s'}{s}$ to calculate the image position and magnification in each case ($f = -15$ cm).

For $s = 50$ cm,

$$s' = \frac{(-15 \text{ cm})(50 \text{ cm})}{50 \text{ cm} - (-15 \text{ cm})} = \boxed{-11.54 \text{ cm}}$$

$$M = -\frac{(-11.54)}{50} = \boxed{0.231} \; ; \quad erect$$

For $s = 30$ cm,

$$s' = \frac{(-15 \text{ cm})(30 \text{ cm})}{30 \text{ cm} - (-15 \text{ cm})} = \boxed{-10 \text{ cm}}$$

$$M = -\frac{(-10)}{30} = \boxed{0.333} \; ; \quad erect$$

For $s = 10$ cm,

$$s' = \frac{(-15 \text{ cm})(10 \text{ cm})}{10 \text{ cm} - (-15 \text{ cm})} = \boxed{-6 \text{ cm}}$$

$$M = -\frac{(-6)}{10} = \boxed{0.6} \; ; \quad erect$$

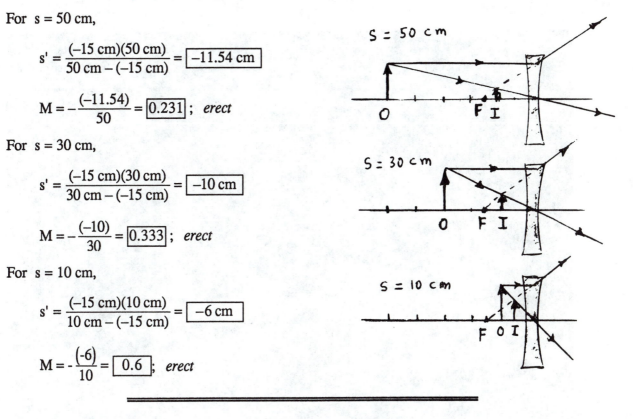

69. A colored marble is dropped in a large tank filled with benzene ($n = 1.50$). (a) What is the depth of the tank if the apparent depth of the marble when viewed from directly above the tank is 35 cm? (b) If the marble has a diameter of 1.5 cm, what is its apparent diameter when viewed from directly above, outside the tank?

Solution

(a) $\dfrac{n_1}{s} + \dfrac{n_2}{s'} = \dfrac{(n_2 - n_1)}{R}$ and when $R = \infty$, $s = -s'\left(\dfrac{n_1}{n_2}\right)$

Therefore, $s = -(-35 \text{ cm})\left(\dfrac{1.5}{1}\right) = \boxed{52.5 \text{ cm}}$

(Refer to the sign conventions in Table 36.2.)

(b) $M = -\dfrac{n_1 s'}{n_2 s} = \dfrac{h'}{h}$ or $h' = -h\left(\dfrac{n_1 s'}{n_2 s}\right)$

$$M = \frac{(-1.5 \text{ cm})(1.5)(-35 \text{ cm})}{(1)(52.5 \text{ cm})} = \boxed{1.5 \text{ cm}}$$

Interference of Light Waves

OBJECTIVES

1. Describe Young's double-slit experiment to demonstrate the wave nature of light. Account for the phase difference between light waves from the two sources as they arrive at a given point on the screen. State the conditions for constructive and destructive interference in terms of each of the following: path difference, phase difference, distance from center of screen, and angle subtended by the observation point at the source mid-point.

2. Outline the manner in which the superposition principle leads to the correct expression for the intensity distribution on a distant screen due to two coherent sources of equal intensity.

3. Describe the use of the phasor diagram method to determine the amplitude and phase of the wave which is the resultant of two or three coherent sources.

4. Calculate the intensity distribution due to N equally spaced coherent sources. Sketch the essential features of the intensity distribution due to N sources.

5. Account for the conditions of constructive and destructive interference in thin films considering both path difference and any expected phase changes due to reflection.

6. Describe the technique employed in the Michelson interferometer for precise measurement of length based on known values for the wavelength of light.

SKILLS

The technique of phasor addition offers a convenient alternative to the algebraic method for finding the resultant wave amplitude at some point on a screen. This is especially true when a large number of waves are to be combined. The method of phasor addition is outlined in the following steps and is illustrated in Fig. 37.1 for the case of two equal amplitude waves differing in phase by ϕ.

1. Draw the phasors representing the waves end to end. The angle between successive phasors is equal to the phase angle between the waves from successive source slits. The length of each phasor is proportional to the magnitude of the wave it represents.

2. The resultant wave is the vector sum (vector from the tail of the first phasor to the head of the last one) of the individual phasors.

3. The phase angle (α) of the resultant wave is the angle between the direction of the resultant phasor and the direction of the first phasor.

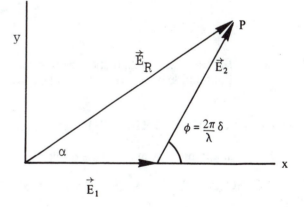

Figure 37.1   Phasor diagram for adding two electric field vectors of amplitudes E_1 and E_2 whose phase difference is ϕ.

NOTES FROM SELECTED CHAPTER SECTIONS_____

37.1 CONDITIONS FOR INTERFERENCE

In order to observe *sustained* interference in light waves, the following conditions must be met:

1. The sources must be coherent; they must maintain a *constant phase* with respect to each other.
2. The sources must be *monochromatic*--be of a *single wavelength*.
3. The *superposition principle* must apply.

37.2 YOUNG'S DOUBLE-SLIT EXPERIMENT

A schematic diagram illustrating the geometry used in Young's double-slit experiment is shown in Figure 37.2. The two slits S_1 and S_2 serve as coherent monochromatic sources. The *path difference* $\delta = r_2 - r_1 = d \sin \theta$.

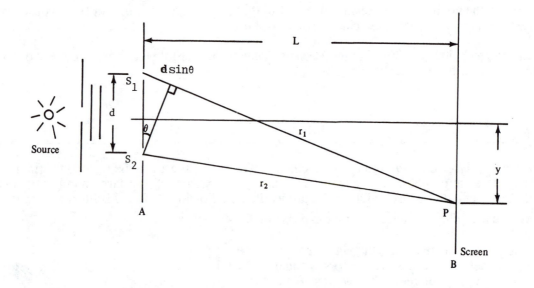

Figure 37.2 Schematic representation of Young's double-slit experiment. The light incident on screen A is a plane wave.

37.3 INTENSITY DISTRIBUTION OF THE DOUBLE-SLIT INTERFERENCE PATTERN

The resultant intensity at a point is proportional to the *square of the resultant amplitude*.

37.4 PHASOR ADDITION OF WAVES

The technique of *phasor addition* is described in the SKILLS section of this chapter.

The relative intensity as a function of ϕ is shown in Figure 37.3 on the following page. Note that the values of I/I_0 are not shown to the same relative size. *In fact, the intensity of each maxima is $N^2 I_0$.* The number of secondary maxima between principle maxima is $N - 2$ as shown in the figure.

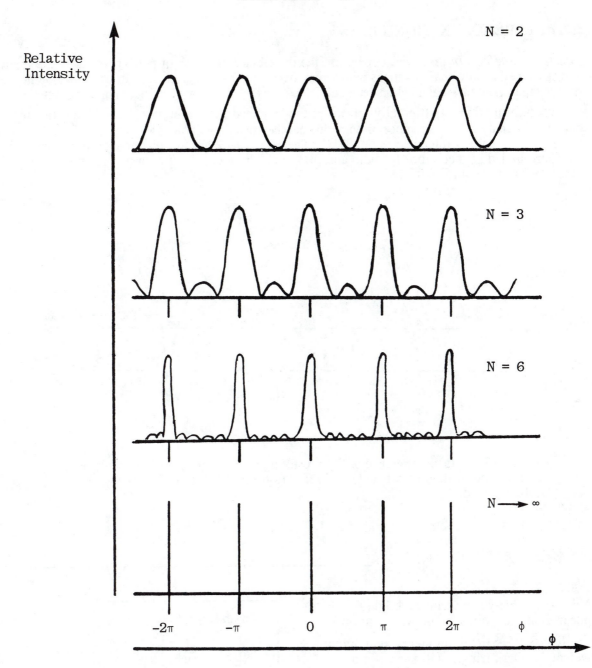

Figure 37.3   Effect of increasing the number of slits on the appearance of the diffraction pattern. The vertical scale is relative intensity for a particular value of N.

37.5 CHANGE OF PHASE DUE TO REFLECTION

An electromagnetic wave undergoes a *phase change of 180°* upon reflection from a medium that is *optically more dense* than the one in which it was traveling. There is also a 180° phase change upon reflection from a *conducting surface*.

37.6 INTERFERENCE IN THIN FILMS

Interference effects in thin films depend on the difference in length of path traveled by the interfering waves as well as any phase changes which may occur due to reflection. It can be shown by the Lloyd's mirror experiment that when light is reflected from a surface of index of refraction greater than the index of the medium in which it is initially traveling, the reflected ray undergoes a 180° (or π radians) phase change. In analyzing interference effects, this can be considered equivalent to the gain or loss of a half wavelength in path difference (see Eq. 37.9). Therefore, there are two different cases to consider: (a) a film surrounded by a common medium and (b) a thin film located between two different media. These cases are illustrated in Fig. 37.4.

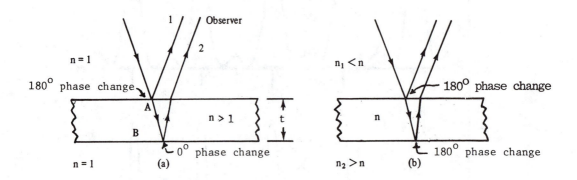

Figure 37.4   (a) Interference of light resulting from reflections at two surfaces of a thin film of thickness t, index of refraction n. (b) A thin film between two different media.

EQUATIONS AND CONCEPTS

In the Young's *double-slit* experiment, bright fringes will appear at those points P for which the path difference ($\delta = \overline{S_2P} - \overline{S_1P}$) is an integral multiple of the wavelength. Dark fringes will appear at points on the screen which correspond to path differences of an odd multiple of half wavelengths.

$$\delta = d\sin\theta = m\lambda \qquad (37.2)$$

$$m = 0, \pm 1, \pm 2, \pm 3, \ldots$$

constructive interference

$$\delta = d\sin\theta = \left(m + \frac{1}{2}\right)\lambda \qquad (37.3)$$

$$m = 0, \pm 1, \pm 2, \pm 3, \ldots$$

destructive interference

The positions of the bright and dark fringes can also be located by calculating their vertical distance (y) from the center of the screen.

$$y = \frac{\lambda L}{d} m \quad \text{bright} \tag{37.5}$$

$$y = \frac{\lambda L}{d}\left(m + \frac{1}{2}\right) \quad \text{dark} \tag{37.6}$$

Two waves which leave the slits initially in phase arrive at the screen *out of phase* by an amount which depends on the path difference.

$$\phi = \frac{2\pi}{\lambda} d \sin \theta \tag{37.8}$$

$$\phi = \frac{2\pi}{\lambda} \delta$$

Two waves initially in phase and of equal amplitude (E_o) will produce a resultant amplitude at some point on the screen which depends on the phase difference (and therefore the path difference).

$$E_p = 2E_o \cos\left(\frac{\phi}{2}\right) \sin\left(\omega t + \frac{\phi}{2}\right) \tag{37.10}$$

The average light intensity (I_{av}) at any point P on the screen is proportional to the square of the amplitude of the resultant wave. *The average intensity can be written:*

(1) as a function of phase difference ϕ;

$$I_{av} = I_o \cos^2\left(\frac{\phi}{2}\right) \tag{37.11}$$

(2) as a function of the angle (θ) subtended by the screen point at the source mid-point; or

$$I_{av} = I_o \cos^2\left(\frac{\pi d \sin \theta}{\lambda}\right) \tag{37.12}$$

(3) as a function of the vertical distance (y) from the center of the screen.

$$I_{av} = I_o \cos^2\left(\frac{\pi d}{\lambda L} y\right) \tag{37.13}$$

$$\text{where} \quad I_o = 4E_o^2$$

The intensity pattern observed on the screen will vary as the number of equally spaced sources is increased; however, the *positions of the principal maxima remain the same*.

The intensity pattern for N slits can be expressed in terms of I_0 (the intensity of a single source) and ϕ (the phase difference between adjacent sources).

$$I = I_0 \frac{\sin^2 (N\phi/2)}{\sin^2 (\phi/2)}$$

$$\text{where} \quad \phi = \frac{2\pi}{\lambda} d \sin \theta$$

In thin film interference, the wavelength of light in the film, λ_n, is not the same as the wavelength in the surrounding medium.

$$\lambda_n = \frac{\lambda}{n}$$

(37.14)

The conditions for interference in thin films can be stated in terms of the thickness (t) and index of refraction of the film. The conditions expressed by Eqs. 37.16 and 37.17 are valid when the film is surrounded by a common medium.

$$2nt = \left(m + \frac{1}{2}\right)\lambda$$

(37.16)

$$m = 0, 1, 2, 3, \ldots$$

constructive interference

$$2nt = m\lambda$$

(37.17)

$$m = 0, 1, 2, 3, \ldots$$

destructive interference

EXAMPLE PROBLEM SOLUTION

Example 37.1 It is often desirable that certain optical components be made "nonreflecting". This can be done by coating them with a thin film of transparent material whose index of refraction is intermediate between that of air and the material of which the component is manufactured. Determine the minimum thickness of material of index of refraction 1.32 which will provide a nonreflecting coating for glass of index of refraction 1.54. Assume a wavelength of 500 nm.

Solution

Consider the interference of rays reflected from the top and bottom surfaces of the film as shown in Figure 37.4(b). In order that the film be a nonreflecting coating, it is necessary that the two reflected rays interfere destructively. At both the top and bottom surfaces, the ray is reflected after incidence onto a material of greater index of refraction. This means that there will be a phase change of π radians (or 180°) in each case. *Hence, there is no relative phase difference introduced due to reflection only*. Therefore, in order to produce destructive interference, the geometric path difference (twice the film thickness) must be an odd multiple of half wave lengths in the film, or

$$2t = \left(m + \frac{1}{2}\right)\frac{\lambda}{n}; \qquad m = 0, 1, 2, 3, \ldots$$

Note that (λ/n) is the wavelength of the light in the film if λ is the wavelength in air and for *minimum* film thickness m = 0. This leads to

$$t = \frac{1}{2}\left(\frac{1}{2}\right)\left(\frac{\lambda}{n}\right) = \frac{1}{4}\left(\frac{\lambda}{n}\right)$$

For this reason such a coating is often referred to as a *"quarter wave" film*.

Using the values given for λ and n gives

$$d = \left(\frac{1}{4}\right)\frac{5 \times 10^{-7}\ m}{1.32} = 9.47 \times 10^{-8}\ m = 947\ \text{Å} = 94.7\ \text{nm}$$

ANSWERS TO SELECTED QUESTIONS

1. What is the necessary condition on the path length difference between two waves that interfere (a) constructively and (b) destructively?

Answer: (a) Two waves interfere constructively if their path difference is either zero or some integral multiple of the wavelength; that is, if the path difference equals $m\lambda$. (b) Two waves interfere destructively if their path difference is an odd multiple of one half of a wavelength; that is, if the path difference equals $\left(m = \frac{1}{2}\right)\lambda$.

2. Explain why two flashlights held close together will not produce an interference pattern on a distant screen.

Answer: The two sources are not coherent. That is, the phase difference between the two sources continually changes with time.

3. If Young's double-slit experiment were performed under water, how would the observed interference pattern be affected?

Answer: The wavelength of light under water would decrease, since the wavelength of light in a medium is given by $\lambda_n = \lambda_0/n$, where λ_0 is the wavelength in a vacuum and n is the index of refraction of the medium. Since the positions of the bright and dark fringes are proportional to the wavelength, the fringe separation would decrease.

8. If a soap film on a wire loop is held in air, it appears black in the thinnest regions when observed by reflected light and shows a variety of colors in thicker regions as in the photograph at the right. Explain.

Answer: The soap film becomes so thin that as the soap film evaporates and approaches zero thickness, there is destructive interference between light reflected from its outer surface and light reflected from its inner surface. The phase difference between the two waves is 180° phase change experienced by the wave reflected from the outer surface.

9. A simple way of observing an interference pattern is to look at a distant light source through a stretched handkerchief or an opened umbrella. Explain how this works.

Answer: Light passes between adjacent threads of the handkerchief. These openings act like slits.

SOLUTIONS TO SELECTED END-OF-CHAPTER PROBLEMS

7. A narrow slit is cut into each of two overlapping opaque squares. The slits are parallel, and the distance between them is adjustable. Monochromatic light of wavelength 600 nm illuminates the slits, and an interference pattern is formed on a screen 80 cm away. The third dark band is located 1.2 cm from the central bright band. What is the distance between the center of the central bright band and the center of the first bright band on either side of the central band?

Solution

For the dark bands, $\qquad y_{dark} = \dfrac{\lambda L \left(m + \frac{1}{2}\right)}{d}$ $\qquad$ or $\qquad$ $d = \dfrac{\lambda L \left(m + \frac{1}{2}\right)}{y_{dark}}$

For the third dark band, $\qquad m = 2$

$$d = \frac{(6 \times 10^{-7}\ m)(0.8\ m)(2.5)}{(1.2 \times 10^{-2}\ m)} = 1 \times 10^{-4}\ m$$

For the bright bands, $\qquad y_{bright} = \dfrac{\lambda L m}{d}$

For the first bright band, $\qquad m = 1$

$$y_{bright} = \frac{(6 \times 10^{-7}\ m)(0.8\ m)(1)}{(1 \times 10^{-4}\ m)} = \boxed{4.80 \times 10^{-3}\ m}$$

---

19. For the situation described in Problem 16, what is the value of θ for which (a) the phase difference will be equal to 0.333 rad and (b) the path difference will be $\dfrac{\lambda}{4}$?

Solution

(a) $\quad \phi = 0.333 = \dfrac{2\pi d \sin \theta}{\lambda}$

$$\theta = \sin^{-1}\left(\frac{\lambda \phi}{2\pi d}\right) = \sin^{-1}\left[\frac{(5 \times 10^{-7} m)(0.333)}{2\pi(1.2 \times 10^{-4}\ m)}\right] = \boxed{1.27 \times 10^{-2}\ deg}$$

(b) $\quad d \sin \theta = \dfrac{\lambda}{4}$

$$\theta = \sin^{-1}\left(\frac{\lambda}{4d}\right) = \sin^{-1}\left[\frac{5 \times 10^{-7} m}{4(1.2 \times 10^{-4}\ m)}\right] = \boxed{5.97 \times 10^{-2}\ deg}$$

---

23. Two narrow parallel slits are separated by 0.85 mm and are illuminated by light with $\lambda = 6000$ Å. (a) What is the phase difference between the two interfering waves on a screen (2.8 m away) at a point 2.50 mm from the central bright fringe? (b) What is the ratio of the intensity at this point to the intensity at the center of a bright fringe?

Solution

From Equation 37.8,

(a) $\phi = \dfrac{2\pi d}{\lambda} \sin\theta = \dfrac{2\pi d}{\lambda}\left(\dfrac{y}{\sqrt{y^2 + D^2}}\right) = \dfrac{2\pi dy}{\lambda D}$

$$= \dfrac{2\pi(0.85 \times 10^{-3}\ \text{m})(2.50 \times 10^{-3}\ \text{m})}{(6000 \times 10^{-10}\ \text{m})(2.8\ \text{m})} = 7.95\ \text{rad} = \boxed{95.5°}$$

(b) $\dfrac{I}{I_{max}} = \dfrac{\cos^2\left(\dfrac{\pi d}{\lambda}\sin\theta\right)}{\cos^2\left(\dfrac{\pi d}{\lambda}\sin\theta_{max}\right)} = \dfrac{\cos^2\left(\dfrac{\delta}{2}\right)}{\cos^2(m\pi)} = \cos^2\left(\dfrac{\delta}{2}\right) = \cos^2\left(\dfrac{95.5°}{2}\right) = \boxed{0.452}$

---

31. Consider N coherent sources described by

$$E_1 = E_o \sin(\omega t + \phi), \qquad\qquad E_2 = E_o \sin(\omega t + 2\phi),$$
$$E_3 = E_o \sin(\omega t + 3\phi),\ \ldots, \qquad\qquad E_N = E_o \sin(\omega t + N\phi).$$

Find the minimum value of ϕ for which $E_R = E_1 + E_2 + E_3 + \ldots\ E_N$ will be zero.

Solution

If N/M is an integer and if M is an integer $\neq 1$, then take

$$\phi = \dfrac{360°}{N}$$

and

$$E_R = \Sigma\ E_o \sin(\omega t + N\phi) = 0$$

where N defines the number of coherent sources. In essence, each set of M electric field components completes a full 360° circle and returns to zero.

---

45. The mirror on one arm of a Michelson interferometer is displaced a distance ΔL. During this displacement, 250 fringe shifts (formation of *successive dark or bright line fringes*) are counted. The light being used has a wavelength of 632.8 nm. Calculate the displacement ΔL.

Solution

When the mirror on one arm is displaced by ΔL, the path difference increases by $2\Delta L$. A shift resulting in the formation of successive dark (or bright) fringes requires a path length change of one-half wavelength.

Therefore, $2\Delta L = \dfrac{m\lambda}{2}$, where in this case $m = 250$.

$$\Delta L = \frac{m\lambda}{4} = \frac{(250)(6.328 \times 10^{-7}\ \text{m})}{4} = \boxed{3.96 \times 10^{-5}\ \text{m}}$$

55. In a Young's interference experiment, the two slits are separated by 0.15 mm and the incident light includes light of wavelengths $\lambda_1 = 540$ nm and $\lambda_2 = 450$ nm. The overlapping interference patterns are formed on a screen 1.4 m from the slits. Calculate the minimum distance from the center of the screen to the point where a bright line of the λ_1 light coincides with a bright line of the λ_2 light.

Solution

For Young's experiment, use Equation 37.2:

$$d \sin \theta = m_1 \lambda_1 = m_2 \lambda_2$$

$$\frac{\lambda_1}{\lambda_2} = \frac{540\ \text{nm}}{450\ \text{nm}} = \frac{m_2}{m_1} = \frac{6}{5}$$

$$\therefore \quad \sin \theta = \frac{6\lambda_2}{d} = \frac{6(450\ \text{nm})}{0.15\ \text{mm}} = 0.018$$

Since $\sin \theta \cong \theta$ and $L = 1.4$ m,

$$x = \theta L = (0.018)(1.4\ \text{m}) = \boxed{2.52\ \text{cm}}$$

56. Two sinusoidal vectors of the same amplitude A and frequency ω have a phase difference φ. Calculate the resultant amplitude of the two vectors both graphically and analytically if φ equals (a) 0, (b) 60°, (c) 90°.

Solution

(a) φ = 0

$A_R = A + A = 2A$

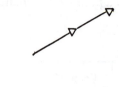

(b) φ = 60°

$A_R^2 = A^2 + A^2 - 2A^2\cos(120°)$

$$= A^2[2 - 2(-0.5)] = \sqrt{3}A$$

(c) φ = 90°

$A_R{}^2 = A^2 + A^2 = 2A^2 = \sqrt{2}A$

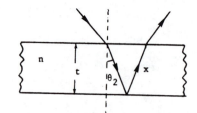

---

63. The condition for constructive interference by reflection from a thin film in air as developed in Section 37.6 assumes nearly normal incidence. (a) Show that if the light is incident on the film at an angle $\phi_1 \gg \theta$ (relative to the normal), then the condition for constructive interference is given by $2nt \sec \theta_2 = \left(m + \frac{1}{2}\right)\lambda$, where θ_2 is the angle of refraction. (b) Calculate the minimum thickness for constructive interference if sodium light ($\lambda = 5.9 \times 10^{-5}$ cm) is incident at an angle of 30° on a film with index of refraction 1.38.

Solution

(a) In this case constructive interference requires

$$2x = \frac{\lambda}{n}\left(m + \frac{1}{2}\right)$$

where $2x \equiv$ distance traveled in the film. From the sketch we have

$$\cos \theta_2 = \frac{t}{x}$$

Therefore,

$$\frac{2nt}{\cos \theta_2} = 2nt \sec \theta_2 = \lambda\left(m + \frac{1}{2}\right)$$

(b) $t = \cos \theta_2 \left(\dfrac{\lambda}{2n}\right)\left(m + \dfrac{1}{2}\right)$ and for minimum thickness choose $m = 0$.

Also from Snell's law, $\cos \theta_2 = \sqrt{1 - \dfrac{\sin^2 \theta_1}{n}}$

Therefore, $t = \dfrac{\lambda}{4n}\sqrt{1 - \dfrac{\sin^2 \theta_1}{n}}$

$$t = \frac{5.9 \times 10^{-7}\ m}{(4 \times 1.38)}\sqrt{1 - \frac{\sin^2(30°)}{1.38}} = \boxed{9.96 \times 10^{-8}\ m}$$

64. Use the method of phasor addition to find the resultant amplitude and phase constant when the following three harmonic functions are combined:

$$E_1 = \sin\left(\omega t + \frac{\pi}{6}\right), \quad E_2 = 3 \sin\left(\omega t + \frac{7\pi}{2}\right), \quad E_3 = 6 \sin\left(\omega t + \frac{4\pi}{3}\right)$$

Solution

The phasor diagram is shown at the right.
From the diagram we find:

resultant amplitude = $\boxed{8}$

and

phase constant = $\boxed{254°}$

$E_R = 8 \sin (\omega t + 254°)$

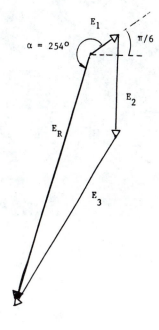

38

Diffraction and Polarization

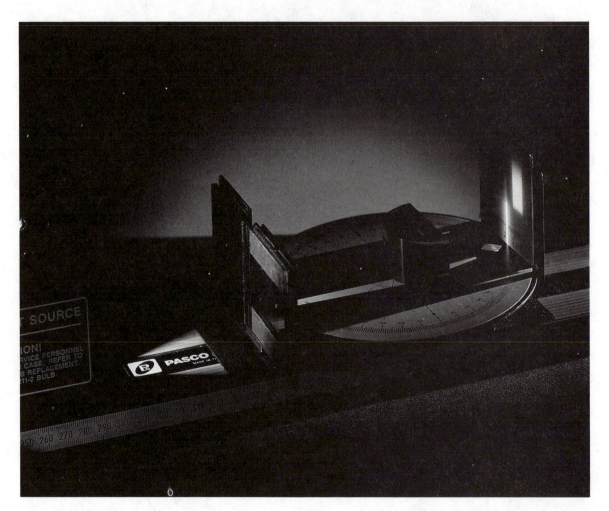

OBJECTIVES

1. Determine the positions of the maxima and minima in a single-slit diffraction pattern and calculate the intensities of the secondary maxima relative to the intensity of the central maximum.

2. Determine whether or not two sources under a given set of conditions are resolvable as defined by Rayleigh's criterion.

3. Determine the positions of the principal maxima in the interference pattern of a diffraction grating.

4. Understand what is meant by the resolving power and the dispersion of a grating, and calculate the resolving power of a grating under specified conditions.

5. Describe the technique of X-ray diffraction and make calculations of the lattice spacing using Bragg's law.

6. Understand how the state of polarization of a light beam can be determined by use of a polarizer-analyzer combination.

7. Describe qualitatively the polarization of light by selective absorption, reflection, scattering, and double refraction. Also, make appropriate calculations using Malus's law and Brewster's law.

NOTES FROM SELECTED CHAPTER SECTIONS

38.1 INTRODUCTION TO DIFFRACTION

Fraunhofer diffraction occurs when light rays reaching a point (on a screen) from a diffracting source (edge, slit, etc.) are approximately *parallel*.

Fresnel diffraction occurs when the observing screen is a *finite* distance from the slit (or edge) and the light rays are not rendered parallel by a lens.

38.2 SINGLE-SLIT DIFFRACTION

According to *Huygen's principle,* each portion of a slit acts as a *source* of waves; therefore, light from one portion of a slit can interfere with light from another portion.

In Figure 38.1, parallel light rays are shown incident on a slit of width a. In order to analyze the resulting Fraunhofer diffraction pattern, it is convenient to subdivide the width of the slit into many strips of equal width. Each substrip can be considered as a source represented by a phasor.

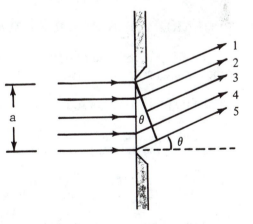

Figure 38.1   Diffraction of light by a narrow slit of width a.

A typical intensity distribution is shown in Figure 38.2.

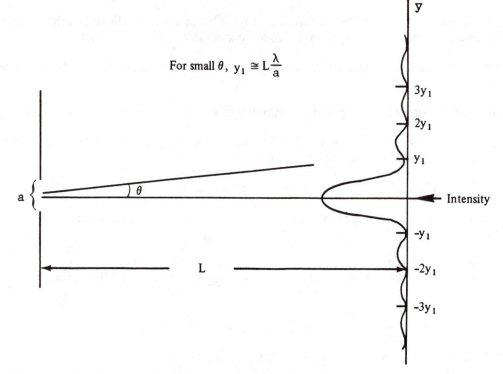

For small θ, $y_1 \cong L\dfrac{\lambda}{a}$

Figure 38.2   Intensity distribution of a single-slit diffraction pattern.

The positions of the minima in a single-slit diffraction pattern are determined by the conditions stated in Eq. 38.1. *The secondary maxima* (of progressively diminished intensity) *lie approximately halfway between the minima.*

38.3 RESOLUTION OF SINGLE-SLIT AND CIRCULAR APERTURES

When the central maximum of one image falls on the first minimum of another image, the images are said to be just resolved. This limiting condition of resolution is known as *Rayleigh's criterion.*

38.6 POLARIZATION OF LIGHT WAVES

If the electric field vector of the light wave emitted by the atoms or molecules of a light source vibrate in the *same direction at all times*, the wave is *plane polarized* or *linearly polarized.*

If the tip of the E vector rotates around a circle over time, the wave is *circularly polarized.*

If E_x and E_y are not equal in magnitude and differ in phase by 90°, the resulting wave is *elliptically polarized*--the tip of the E vector moves around an ellipse.

Polarization by selective absorption is illustrated in Figure 38.3 on the following page. A beam of unpolarized light is incident on a sheet of polarizing material. The electric field vectors of the transmitted light will have nonzero components only along a direction parallel to the axis of polarization of the polarizer (indicated by straight lines drawn on the disk). The light emerging from the polarizer is plane polarized and is allowed to fall on a second sheet of polarizing material (called the analyzer) such that there is an angle θ between the plane of polarization of the incident beam and the transmission axis of the analyzer.

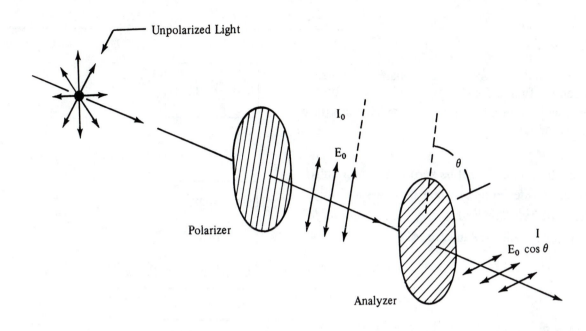

Figure 38.3   Only a fraction of the light is transmitted through the analyzer when θ lies between 0 and 180°.

Polarization by reflection when a light beam is reflected from a surface such that the reflected beam is perpendicular to the refracted beam, the *reflected component will be completely polarized* with its electric field vector parallel to the reflecting surface. Under these circumstances, the *refracted (or transmitted) beam will be partially polarized.* This situation is illustrated in Figure 38.4.

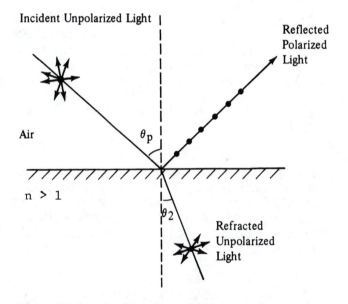

Figure 38.4   Light reflected from a surface at the angle θ_p given by $\tan \theta_p = n$ is completely polarized.

469

EQUATIONS AND CONCEPTS_____

In single-slit diffraction, the total phase difference β between the first and last phasors will depend on the angle θ which determines the direction to an arbitrary point on the screen.

$$\boxed{\beta = \frac{2\pi}{\lambda} a \sin\theta}$$ (38.3)

The resultant phasor will be zero (the condition for destructive interference) when β equals an integer multiple of 2π. This leads to a general *condition for destructive interference* in terms of θ.

$$\boxed{\begin{array}{l} \sin\theta = m\dfrac{\lambda}{a} \\[6pt] m = \pm 1, \pm 2, \pm 3, \ldots \\[6pt] \text{Destructive interference} \end{array}}$$ (38.1)

The *intensity* at any point on the screen is given in terms of the intensity I_0 at $\theta = 0$.

$$\boxed{I = I_0 \left| \frac{\sin(\beta/2)}{\beta/2} \right|^2}$$ (38.4)

Rayleigh's criterion states the condition for the *resolution* of two images due to nearby sources. For a slit, the angular separation between the sources must be greater than the ratio of the wavelength to slit width (Eq. 38.6). In the case of a *circular aperture*, the minimum angular separation depends on D, the diameter of the aperture (or lens), as given by Eq. 38.7.

$$\boxed{\begin{array}{l} \theta_m = \dfrac{\lambda}{a} \\[10pt] \theta_m = 1.22 \dfrac{\lambda}{D} \end{array}}$$

(38.6)

(38.7)

A grating of equally spaced parallel slits (separated by a distance d) will produce an interference pattern in which there is a series of maxima for each wavelength. Maxima due to wavelengths of different value comprise a spectral order denoted by order number m.

$$\boxed{\begin{array}{l} d \sin\theta = m\lambda \\[6pt] m = 0, 1, 2, 3, \ldots \end{array}}$$ (38.8)

The *resolving power* of a grating increases as the *number of lines illuminated* is increased and is proportional to the order in which the spectrum is observed.

$$\boxed{\begin{array}{l} R \equiv \dfrac{\lambda}{\Delta\lambda} \\[10pt] R = Nm \end{array}}$$

(38.9)

(38.10)

When the order number m is substituted from Eq. 38.8, $d \sin\theta = m\lambda$, into Eq. 38.10, it becomes clear that *the resolution depends on the width of the grating*, (Nd). It should also be noted that the angular width $\Delta\theta$ of a spectral line formed by a diffraction grating is inversely proportional to the width of the grating. This statement can be written as $\Delta\theta \sim 1/Nd$.

$$R = \frac{Nd}{\lambda} \sin\theta$$

In the zeroth order (central maximum) all wavelengths are indistinguishable. If in a particular order (m > 0) R = 10,000, the grating will produce a spectrum in which wavelengths differing in value by 1 part in 10,000 can be resolved.

Bragg's law gives the conditions for *constructive interference* of X-rays reflected from the parallel planes of a crystalline solid separated by a distance d. *The angle θ is the angle between the incident beam and the surface.*

$$2d \sin\theta = m\lambda$$
$$m = 1, 2, 3, \ldots$$

(38.11)

The fraction of the wave intensity transmitted by the analyzer is given by *Malus's law*.

$$I = I_o \cos^2\theta$$

(38.12)

Brewster's law gives the polarizing angle for a particular surface. The *polarizing angle* θ_p is the angle of incidence for which the reflected beam is *completely polarized*.

$$n = \tan\theta_p$$

(38.13)

EXAMPLE PROBLEM SOLUTION

Example 38.1 A diffraction grating has 2500 lines per cm and each slit has a width of 1 micron (10^{-6} m). The grating is illuminated with monochromatic light of wavelength 5000 Å. Determine the maximum number of spectral orders formed by the grating and make calculations to show that some of the expected maxima are "missing" because they fall at locations on the screen where diffraction by a *single slit* results in a minima.

Solution

According to Eq. 38.8, $d\sin\theta = m\lambda$, the grating will produce *maxima* in the interference pattern when

$$\sin\theta = \frac{m\lambda}{d}; \quad m = 0, 1, 2, 3, \ldots$$

The diffraction pattern for each individual slit will be governed by Eq. 38.1 and will result in *minima* when

$$\sin\theta = n\frac{\lambda}{a}; \quad n = 0, 1, 2, 3, \ldots$$

NOTE: In the equation above, the orders are denoted by n instead of m to avoid confusion between the two equations stating conditions for maxima and minima.

In the situation described here,

$$d = 4 \times 10^{-6} \text{ m} \quad \text{and} \quad a = 1 \times 10^{-6} \text{ m}$$

Since

$$\sin \theta = \frac{m\lambda}{d} = m \left(\frac{5 \times 10^{-7} \text{ m}}{4 \times 10^{-6} \text{ m}} \right) = (0.125)(m)$$

and $\sin \theta \leq 1$, we find that

$$m \leq \frac{1}{0.125} = 8$$

Therefore, the maximum number of spectral orders (largest allowed value of m) is 8.

The interference pattern formed on the screen will be the overlay or product of that formed by the grating (orders designated by m) and that formed by an individual slit (orders designated by n). Therefore, at those values of θ for which the *grating would produce a maximum* and (for the same angle) an individual *slit would produce a minimum* (zero), the anticipated principal maximum will be missing. This will occur when

$$m\frac{\lambda}{d} = n\frac{\lambda}{a} \quad \text{or} \quad m = n\left(\frac{d}{a}\right)$$

Again, using the given values for d and a, the "missing orders" occur for m = 4n.

Recalling that the largest allowed value of m is 8, it is necessary to find pairs of values for m and n such that

$$m = 4n \quad \begin{cases} \text{where} & m = 0, 1, 2, \ldots 8 \\ \text{and} & n = 1, 2, 3, \ldots \end{cases}$$

The acceptable sets of values are

$$n = 1, \quad m = 4 \quad \text{and} \quad n = 2, \quad m = 8$$

This means that in the diffraction pattern formed by this grating, the principal maxima corresponding to $\lambda = 5000$ Å are missing in the fourth (m = 4) and eighth (m = 8) orders. Sketches, shown on the following page in Figure 38.5, of the intensity vs. $\sin \theta$ verify the result found above, where

$$\sin \theta = m\frac{\lambda}{d}; \quad m = 0, 1, 2, \ldots 8$$

and

$$\sin \theta = n\frac{\lambda}{a}; \quad n = 1, 2, 3, \ldots$$

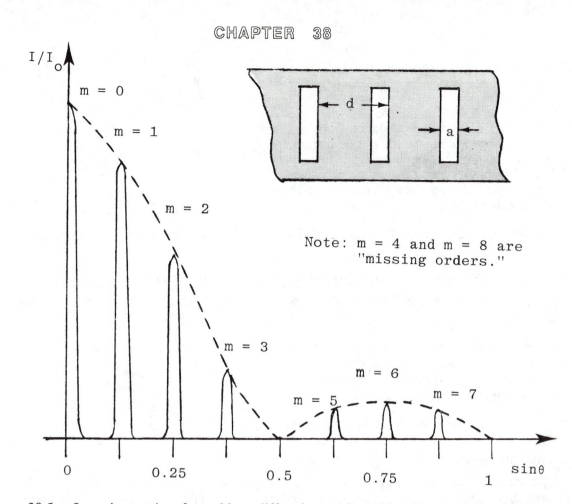

Note: m = 4 and m = 8 are "missing orders."

Figure 38.5   Intensity maxima formed by a diffraction grating with a large number of slits, N. The pattern is modulated by the diffraction due to an individual slit. The insert shows a section of the grating with d = 4a.

ANSWERS TO SELECTED QUESTIONS

1. If you place your thumb and index finger very close together and view light passing between them when they are a few cm in front of your eye, dark lines parallel to your thumb and finger will appear. Explain.

Answer: The space between your thumb and finger act as a slit. The dark lines you observe are the result of the diffraction of light passing through the slit.

2. Observe the shadow of your book or some other straight edge when it is held a few inches above a table with a lamp several feet above the book. Why is the shadow of the book somewhat fuzzy at the edges?

Answer: The fuzziness is due to the diffraction of light around the straight edge. A close inspection of the light intensity in the vicinity of the "shadow" would reveal a series of bright and dark bands as illustrated in the figure.

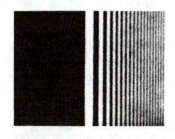

3. What is the difference between Fraunhofer and Fresnel diffraction?

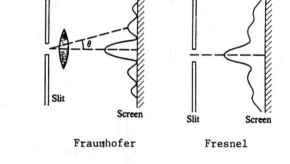

Fraunhofer   Fresnel

Answer: Fraunhofer diffraction occurs when the observing screen is at a large distance from the aperture or object producing the diffraction pattern, so that the rays reaching the screen are approximately parallel. A Fresnel diffraction pattern occurs when the screen is close to the aperture, corresponding to nonparallel rays reaching the screen.

5. Describe the change in width of the central maximum of the single-slit diffraction pattern as the width of the slit is made smaller.

Answer: The width of the central maximum increases as the width of the slit decreases. This can be seen by inspecting the condition for destructive interference, $\sin\theta = \lambda/a$, from which the width of the central maximum can be obtained.

6. Assuming that the headlights of a car are point sources, estimate the maximum distance from an observer to the car at which the headlights are distinguishable from each other.

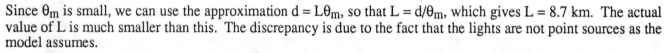

Answer: The distance between the headlights is estimated to be d = 2 m. The limiting angle of resolution of the human eye is $\theta_m = 2.3 \times 10^{-4}$ rad.

Since θ_m is small, we can use the approximation $d = L\theta_m$, so that $L = d/\theta_m$, which gives L = 8.7 km. The actual value of L is much smaller than this. The discrepancy is due to the fact that the lights are not point sources as the model assumes.

7. A laser beam is incident at a shallow angle on a machinist's ruler that has a finely calibrated scale. The rulings on the scale give rise to a diffraction pattern on a screen. Discuss how you can use this technique to obtain a measure of the wavelength of the laser light.

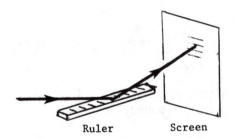

Ruler   Screen

Answer: The spacings between the dark and bright fringes of the diffraction pattern will be a function of the spacing between the rulings on the scale and the angle at which the beam is incident on the scale. If the spacings between the rulings are known to a high degree of accuracy, the wavelength can be determined with about the same accuracy.

8. Certain sunglasses use a polarizing material to reduce the intensity of light reflected from shiny surfaces, such as water or the hood of a car. What orientation of polarization should the material have in order to be most effective?

Answer: The reflected light is partially polarized, with the component parallel to the reflecting surface being the most intense. Therefore, the polarizing material should have its transmission axis oriented in the vertical direction in order to reduce the intensity of the reflected light.

9. Why is the sky black when viewed from the moon?

Answer: The moon has almost no atmosphere to cause scattering of light from the sun.

SOLUTIONS TO SELECTED END-OF-CHAPTER PROBLEMS

11. A diffraction pattern is formed on a screen 120 cm away from a 0.4-mm-wide slit. Monochromatic light of 546.1 nm is used. Calculate the fractional intensity $\dfrac{I}{I_o}$ at a point on the screen 4.1 mm from the center of the principal maximum.

Solution

$$\sin\theta \approx \frac{y}{L} = \frac{4.1 \times 10^{-3}\ m}{1.2\ m} = 3.417 \times 10^{-3}$$

$$\frac{\beta}{2} = \frac{\pi a \sin\theta}{\lambda} = \frac{\pi(4 \times 10^{-4}\ m)(3.417 \times 10^{-3})}{546.1 \times 10^{-9}\ m} = 7.863\ rad$$

$$\frac{I}{I_o} = \left[\frac{\sin\left(\frac{\beta}{2}\right)}{\left(\frac{\beta}{2}\right)}\right]^2 = \left[\frac{\sin(7.863\ rad)}{7.863}\right]^2 = \boxed{1.62 \times 10^{-2}}$$

29. The full width of a 3-cm-wide grating is illuminated by a sodium discharge tube. The lines in the grating are uniformly spaced at 775 nm. Calculate the angular separation in the first-order spectrum between the two wavelengths forming the sodium doublet ($\lambda_1 = 589.0$ nm and $\lambda_2 = 589.6$ nm).

Solution

$$m\lambda = d\sin\theta$$

$$\theta_2 = \sin^{-1}\left(\frac{m\lambda_2}{d}\right) = \sin^{-1}\left[\frac{(1)(5.896 \times 10^{-7}\ m)}{(7.75 \times 10^{-7}\ m)}\right] = 49.532°$$

$$\theta_1 = \sin^{-1}\left(\frac{m\lambda_1}{d}\right) = \sin^{-1}\left[\frac{(1)(5.890 \times 10^{-7}\ m)}{(7.75 \times 10^{-7}\ m)}\right] = 49.464°$$

$$\Delta\theta = \theta_2 - \theta_1 = \boxed{0.068°}$$

55. The critical angle for sapphire surrounded by air is 34.4°. Calculate the polarizing angle for sapphire.

Solution

$$n = \tan\theta_p \quad\text{and}\quad \theta_p = \tan^{-1} n; \quad \sin\theta_c = \frac{1}{n} \quad\text{or}\quad n = \frac{1}{\sin\theta_c}$$

Therefore,

$$\theta_p = \tan^{-1}\left(\frac{1}{\sin\theta_c}\right) = \tan^{-1}\left(\frac{1}{\sin 34.4°}\right) = \boxed{60.5°}$$

56. For a particular transparent medium surrounded by air, show that the critical angle for internal reflection and the polarizing angle are related by $\cot\theta_p = \sin\theta_c$.

Solution

$$\sin\theta_2 = \frac{1}{n} \quad\text{and}\quad \tan\theta_p = n. \quad\text{Thus,}\quad \sin\theta_2 = \frac{1}{\tan\theta_p} \quad\text{or}\quad \cot\theta_p = \sin\theta_c$$

67. Light of wavelength 500 nm is incident normally on a diffraction grating. If the third-order maximum of the diffraction pattern is observed at an angle of 32°, (a) what is the number of rulings per cm for the grating? (b) Determine the total number of primary maxima that can be observed in this situation.

Solution

(a) Use Equation 38.8, $\quad d\sin\theta = m\lambda$

$$d = \frac{m\lambda}{\sin\theta} = \frac{(3)(5\times10^{-7}\text{ m})}{\sin(32°)} = 2.83\times10^{-6}\text{ m}$$

Therefore, the number of lines per unit length $= \frac{1}{d} = 3.534\times10^5$ lines/m or $\boxed{3.53\times10^3 \text{ lines/cm}}$

(b) $\quad \sin\theta = m\left(\frac{\lambda}{d}\right) = \frac{m(5\times10^{-7}\text{ m})}{2.83\times10^{-6}\text{ m}} = m(0.177)$

For $\sin\theta \leq 1$, we require that $\quad m(0.177) \leq 1 \quad$ or $\quad m \leq 5.65$

Therefore, the total number of lines $= 2m + 1 = \boxed{11}$

68. Consider the case of a light beam containing two discrete wavelength components whose difference in wavelength $\Delta\lambda$ is small relative to the mean wavelength λ incident on a diffraction grating. A useful measure of the angular separation of the maxima corresponding to the two wavelengths is the *dispersion* D, given by $D = \dfrac{d\theta}{d\lambda}$. (The dispersion of a grating should not be confused with its resolving power R, given by Equations 38.9 and 38.10.) (a) Starting with Equation 38.8 show that the dispersion can be written

$$D = \frac{\tan\theta}{\lambda}$$

(b) Calculate the dispersion in the third order for the grating described in Problem 45. Give the answer in units of deg/nm.

Solution

(a) The dispersion of a grating is defined to be $D = \dfrac{d\theta}{d\lambda}$

From Equation 38.8, $d\sin\theta = m\lambda$ and $\sin\theta = \dfrac{m\lambda}{d}$

Differentiate to find

$$\cos\theta\, d\theta = \left(\frac{m}{d}\right)d\lambda \quad \text{or} \quad \frac{d\theta}{d\lambda} = \frac{m}{d\cos\theta}. \qquad \text{But} \quad \frac{m}{d} = \frac{\sin\theta}{\lambda}.$$

Therefore,

$$\frac{d\theta}{d\lambda} = \left(\frac{\sin\theta}{\lambda}\right)\left(\frac{1}{\cos\theta}\right) = \frac{\tan\theta}{\lambda} \quad \text{or} \quad D = \frac{\tan\theta}{\lambda}$$

(b) From Problem 45, $m = 3$, $\lambda = 4\times10^{-7}$ m and $d = 3.33\times10^{-6}$ m

$$\theta = \sin^{-1}\left(\frac{m\lambda}{d}\right) = 21.1° \quad \text{and}$$

$$D = \frac{\tan\theta}{\lambda} = \frac{\tan(21.1°)}{4\times10^{-7}\text{ m}} = 9.65\times10^5\text{ rad/m} = \boxed{0.0553\text{ deg/nm}}$$

71. Derive Equation 38.10 for the resolving power of a grating, $R = Nm$, where N is the number of lines illuminated and m is the order in the diffraction pattern. Remember that Rayleigh's criterion (Section 38.3) states that two wavelengths will be resolved when the principal maximum for one falls on the first minimum for the other.

Solution

For a diffraction grating, $\quad \sin \theta \approx \dfrac{y}{L} = \dfrac{m\lambda}{d}, \qquad$ and $\qquad d = \dfrac{a}{N}$

$\dfrac{\Delta y}{L} = \dfrac{Nm\Delta\lambda}{a}\quad$ where $\Delta\lambda$ is the difference in m order peak for two different wavelengths.

For a single slit, $\quad a \sin \theta = \lambda \quad$ or $\quad \sin \theta \approx \dfrac{y}{L} \quad$ where $\quad y = \Delta y$ from above.

Therefore,

$$\frac{\lambda}{a} = \frac{Nm\Delta\lambda}{a} \qquad \text{and} \qquad R = \frac{\lambda}{\Delta\lambda} = Nm$$

---

73. Suppose that the single-slit opening in Figure 38.7 is 6 cm wide and is placed in front of a microwave source operating at a frequency of 7.5 GHz. (a) Calculate the angle subtended by the first minimum in the diffraction pattern. (b) What is the relative intensity $\dfrac{I}{I_o}$ at $\theta = 15°$? (c) Consider the case when there are *two* such sources, separated laterally by 20 cm, behind the single slit. What is the maximum distance between the plane of the sources and the slit if the diffraction patterns are to be resolved? (In this case, the approximation that $\sin \theta \approx \tan \theta$ may not be valid because of the relatively small value of the ratio $\dfrac{a}{\lambda}$.)

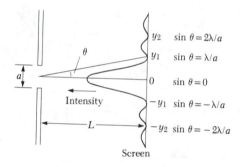

Figure 38.7

Solution

(a)   From Equation 38.1, $\qquad \theta = \sin^{-1}\left(\dfrac{m\lambda}{a}\right)$

In this case, $m = 1$;

$$\lambda = \frac{c}{f} = \frac{(3 \times 10^8 \text{ m/s})}{(7.5 \times 10^9 \text{ s})} = 4 \times 10^{-2} \text{ m} = 0.06 \text{ m}$$

so $\qquad \theta = \sin^{-1}(0.666) = \boxed{41.8°}$

(b) From Equation 38.4, $\dfrac{I}{I_o} = \left(\dfrac{\sin \beta}{\beta}\right)^2$

where $\beta = \dfrac{\pi a \sin \theta}{\lambda}$;

When $\theta = 15°$, $\beta = 1.22$ rad and $\dfrac{I}{I_o} = \boxed{0.593}$

(c) $\sin \theta = \dfrac{\lambda}{a}$ so $\theta = 41.81°$

This is the *minimum* angle subtended by the two sources at the slit. Let α = the half angle between the sources, each a distance L' from the center line and a distance L from the slip plane. Then

$$\sin \alpha = \dfrac{L'}{\sqrt{L'^2 + L^2}} = \sin\left(\dfrac{41.81}{2}\right) = 0.357$$

so

$$L^2 = \dfrac{\left[1 - (0.357)^2\right](0.1 \text{ m})^2}{(0.357)^2}$$

and $L = \boxed{0.262 \text{ m}}$

Special Theory of Relativity

OBJECTIVES

1. State the principle of newtonian relativity, and describe the galilean coordinate and velocity transformations, and their limitations.

2. Discuss Einstein's two postulates of the special theory of relativity.

3. Discuss the differences between the galilean transformation and the Lorentz transformation, and the significance of the Lorentz transformation equations within the framework of special relativity.

4. Describe some consequences of the Lorentz transformation equations; specifically, the effects of time dilation and length contraction.

5. Understand the idea of simultaneity, and the fact that simultaneity is not an absolute concept. That is, two events which are simultaneous in one reference frame are not simultaneous when viewed from a second frame moving with respect to the first.

6. State the correct relativistic expressions for the momentum, kinetic energy, and total energy of a particle.

7. Discuss the principle of the energy-mass equivalence, and its impact in the field of nuclear physics.

8. Discuss the Michelson-Morley experiment, its objectives, and the significance of its outcome.

NOTES FROM SELECTED CHAPTER SECTIONS

39.2 THE PRINCIPLE OF RELATIVITY

An *inertial system* is a system in which a free body exhibits no acceleration. A system moving with constant velocity with respect to an inertial system is also an inertial system.

39.3 THE MICHELSON-MORLEY EXPERIMENT

This experiment was designed to detect the velocity of the earth with respect to the *hypothetical ether*. The outcome of the experiment was *negative*, contracting the ether hypothesis.

39.4 EINSTEIN'S PRINCIPLE OF RELATIVITY

Einstein's special theory of relativity is based upon two postulates:

1. The laws of physics are the same in every inertial frame of reference.

2. The speed of light has the same value as measured by all observers, independent of the motion of the light source or observer.

The second postulate is consistent with the negative results of the Michelson-Morley experiment which failed to detect the presence of an ether and suggested that the speed of light is the same in all inertial frames.

39.6 SIMULTANEITY

Two events that are simultaneous in one reference frame are in general not simultaneous in another frame which is moving with respect to the first.

39.7 THE RELATIVITY OF TIME

The *proper time* is always the time measured by a clock at rest in the frame of reference of the measurement. According to a stationary observer, a moving clock runs slower than an identical clock at rest. This effect is known as *time dilation*.

39.8 THE RELATIVITY OF LENGTH

The *proper length* of an object is defined as the length of the object measured in the *reference frame in which the object is at rest*. The length of an object measured in a reference frame in which it is moving is always less than the proper length. This effect is known as *length contraction*. The contraction occurs only *along the direction of motion*.

EQUATIONS AND CONCEPTS

In order to explain the motion of particles moving at speeds approaching the speed of light, one must use the *Lorentz coordinate transformation equations*, given by Eq. 39.9. These equations represent the transformation between any two inertial frames in *relative* motion with velocity v in the x direction.

$$x' = \gamma(x - vt)$$
$$y' = y$$
$$z' = z$$
$$t' = \gamma\left[t - \left(\frac{v}{c^2}\right)x\right]$$

(39.9)

where

$$\gamma \equiv \frac{1}{\sqrt{1 - \frac{v^2}{c^2}}}$$

(39.10)

The *Lorentz velocity transformation equations* are given by Eq. 39.15, where an object has a velocity u in the S frame, and is observed in the S' frame. The S' frame has a velocity v along the x-axis, relative to the S frame.

$$u_x' = \frac{u_x - v}{1 - u_x \frac{v}{c^2}}$$

(39.15)

$$u_y' = \frac{u_y}{\gamma\left(1 - u_x \frac{v}{c^2}\right)}$$

$$u_z' = \frac{u_z}{\gamma\left(1 - u_x \frac{v}{c^2}\right)}$$

(39.16)

Consider a single clock in the S' frame in which an observer in this frame measures the time interval of an event Δt' (say the time it takes a light pulse to travel from a source to a mirror and back to the source). The time interval Δt for this event as measured by an observer in S is greater than Δt', since the observer in S must make clock readings at two different positions. That is, a *moving clock appears to run slower than an identical clock at rest with respect to the observer*. This is known as time dilation.

$$\Delta t = \gamma \Delta t' = \frac{\Delta t'}{\sqrt{1 - \frac{v^2}{c^2}}}$$ (39.7)

If an object moves along the x-axis with a speed v, and has a length L' as measured in the moving frame, the length L as measured by a stationary observer is shorter than L' by the factor $\frac{1}{\gamma}$, according to Eq. 39.8. This is called *length contraction*, and the length L' measured in the reference frame in which the object is at rest is called the *proper length*.

$$L = \frac{1}{\gamma} L' = \sqrt{1 - \frac{v^2}{c^2}} L'$$ (39.8)

The fact that observers in the S and S' frames do not reach the same conclusions regarding such measurements can be understood by recognizing that simultaneity is not an absolute concept. That is, *two events that are simultaneous in one reference frame are not simultaneous in another reference frame that is in motion relative to the first.*

The relativistic equation for the *momentum* of a particle of mass m moving with a speed u is given by Eq. 39.18. This expression satisfies the conditions that (1) momentum is conserved in all collisions and (2) p approaches the classical expression as u → 0. That is, as u → 0, p → mu.

$$p \equiv \frac{mu}{\sqrt{1 - \frac{u^2}{c^2}}}$$ (39.18)

The relativistic equation for the *kinetic energy* of a particle of mass m moving with a speed u is given by Eq. 39.22. In this expression, the term mc² is called the *rest energy*.

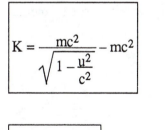

$$K = \frac{mc^2}{\sqrt{1 - \frac{u^2}{c^2}}} - mc^2$$ (39.22)

The *total energy* E of a particle is the sum of the kinetic energy and rest energy, and is given by Eq. 39.25. (This follows from Eq. 39.22.) This expression shows that *mass and energy are equivalent concepts*; that is, mass is a form of energy.

$$E = \frac{mc^2}{\sqrt{1 - \frac{u^2}{c^2}}}$$ (39.25)

The total energy and relativistic momentum of a particle are related according to Eq. 39.26.

$$E^2 = p^2c^2 + (mc^2)^2$$ (39.26)

EXAMPLE PROBLEM SOLUTION

Example 39.1 The kinetic energy of a proton is equal to *one-half* its rest energy. (a) What is the speed of the proton? (b) What is the momentum of the proton?

Solution

(a) The relativistic kinetic energy is given by Eq. 39.22:

$$K = \frac{mc^2}{\sqrt{1 - \frac{u^2}{c^2}}} - mc^2$$

In this case, $K = \frac{1}{2} mc^2$, so

$$\frac{1}{2} mc^2 = \frac{mc^2}{\sqrt{1 - \frac{u^2}{c^2}}} - mc^2$$

$$\frac{1}{\sqrt{1 - \frac{u^2}{c^2}}} = \frac{3}{2}$$

$$1 - \frac{u^2}{c^2} = \frac{4}{9}$$

$$u = \sqrt{\frac{5}{9}}c = 0.745c$$

(b) According to Eq. 39.18, the relativistic momentum is

$$p = \frac{mu}{\sqrt{1 - \frac{u^2}{c^2}}}$$

In this case, m = 1.67×10^{-27} kg, $\frac{u^2}{c^2} = \frac{5}{9}$, and u = 0.745c. Therefore

$$p = \frac{mc\sqrt{\frac{5}{9}}}{\sqrt{1 - \left(\frac{5}{9}\right)}} = 1.12mc = 1.12\,(1.67 \times 10^{-27} \text{ kg})\,(3 \times 10^8 \text{ m/s}) = 5.61 \times 10^{-19} \text{ kg·m/s}$$

ANSWERS TO SELECTED QUESTIONS

1. What one measurement will two observers in relative motion *always* agree upon?

Answer: The speed of light.

2. A spaceship in the shape of a sphere moves past an observer on earth with a speed 0.5c. What shape will the observer see as the spaceship moves past?

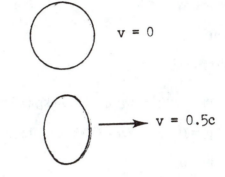

Answer: The spaceship will appear egg-shaped, with its shorter dimension in the direction of the motion of the spaceship.

3. An astronaut moves away from the earth at a speed close to the speed of light. If an observer on earth could make measurements of the astronaut's size and pulse rate, what changes (if any) would he or she measure? Would the astronaut measure any changes?

Answer: The earth observer would see a reduction in both pulse rate and size (for the dimension along the direction of motion). The astronaut would not measure any changes.

8. List some ways our day-to-day lives would change if the speed of light were only 50 m/s.

Answer: All of the relativity effects would be obvious in our lives. Time dilation, length contraction, and mass increase would all occur.

9. Give a physical argument which shows that it is impossible to accelerate an object of mass m to the speed of light, even with a continuous force acting on it.

Answer: As an object with finite rest mass approaches the speed of light, its energy approaches infinity. Hence, it would take an infinite amount of work to accelerate the object to the speed of light under the action of a continuous force or it would take an infinitely large force.

11. Since mass is a form of energy, can we conclude that a compressed spring has more mass than the same spring when it is not compressed? On the basis of your answer, which has more mass, a spinning planet or an otherwise identical but nonspinning planet?

Answer: Yes. The increase in mass equals the energy stored in the compressed spring $\left(\frac{1}{2}kx^2\right)$ divided by c^2, which is a *very* small number. A spinning planet has more mass.

12. Suppose astronauts were paid according to the time spent traveling in space. After a long voyage at a speed near that of light, a crew of astronauts returns and opens their pay envelopes. What will their reaction be?

Answer: Assuming that their on-duty time was kept on earth, they will be pleasantly surprised (unless they understand relativity). Less time will have passed for the astronauts than for their employer back on earth.

13. What happens to the density of an object as its speed increases?

Answer: Its mass increases, but its volume decreases (since length decreases). Therefore, its density (mass/volume) must increase because of the change in each of the factors.

SOLUTIONS TO SELECTED END-OF-CHAPTER PROBLEMS

3. A 2000-kg car moving with a speed of 20 m/s collides with and sticks to a 1500-kg car at rest at a stop sign. Show that momentum is conserved in a reference frame moving with a speed of 10 m/s in the direction of the moving car.

Solution

In the rest frame,

$$p_i = m_1v_{1i} + m_2v_{2i} = (2000 \text{ kg})(20 \text{ m/s}) + (1500 \text{ kg})(0 \text{ m/s}) = 4.0 \times 10^4 \text{ kg·m/s}$$

$$p_f = (m_1 + m_2)v_f = (2000 \text{ kg} + 1500 \text{ kg})v_f$$

$$p_i = p_f$$

$$v_f = \frac{4.0 \times 10^4 \text{ kg·m/s}}{(2000 \text{ kg} + 1500 \text{ kg})} = 11.4 \text{ m/s}$$

In the moving frame,

$$v_{1i}' = v_{1i} - v' = 20 \text{ m/s} - (-10 \text{ m/s}) = 30 \text{ m/s}$$

$$v_{2i}' = v_{2i} - v' = 0 \text{ m/s} - (-10 \text{ m/s}) = 10 \text{ m/s}$$

$$p_i' = m_1v_{1i}' + m_2v_{2i}' = (2000 \text{ kg})(30 \text{ m/s}) + (1500 \text{ kg})(10 \text{ m/s}) = 7.5 \times 10^4 \text{ kg·m/s}$$

$$p_f' = (2000 \text{ kg} + 1500 \text{ kg})v_f'$$

$$p_i = p_f$$

$$v_f' = \frac{7.5 \times 10^4 \text{ kg·m/s}}{(2000 \text{ kg} + 1500 \text{ kg})} = 21.4 \text{ m/s}$$

$$v_{1f} = v_{1f}' + v' = 21.4 \text{ m/s} - 10 \text{ m/s} = 11.4 \text{ m/s}$$

or $\quad v_{1f}' = v_{1f} - v' = 11.4 \text{ m/s} - (-10 \text{ m/s}) = 21.4 \text{ m/s}$

---

19. A cube of steel has a volume of 1 cm³ and a mass of 8 g when at rest on the earth. If this cube is now given a speed v = 0.9c, what is its density as measured by a stationary observer?

Solution

The measured volume will now be

$$V = (1 \text{ cm}) \times (1 \text{ cm}) \times \left(\sqrt{1 - \frac{v^2}{c^2}} \text{ cm}\right) = 0.436 \text{ cm}^3$$

The measured mass will be increased to

$$M = \gamma M_0 = \frac{8 \text{ g}}{\sqrt{1 - \frac{v^2}{c^2}}} = 18.34 \text{ g}$$

The density measured by the stationary observer is

$$\rho = \frac{M}{V} = \frac{18.34 \text{ g}}{0.436 \text{ cm}^3} = \boxed{42.1 \text{ g/cm}^3}$$

---

23. Show that the energy-momentum relationship given by $E^2 = p^2c^2 + (mc^2)^2$ follows from the expressions $E = \gamma mc^2$ and $p = \gamma mu$.

Solution

$$E = \gamma mc^2, \quad p = \gamma mu$$

$$E^2 = (\gamma mc^2)^2 \qquad p^2 = (\gamma mu)^2$$

$$E^2 - p^2c^2 = (\gamma mc^2)^2 - (\gamma mu)^2 c^2 = \gamma^2 \left\{ (mc^2)^2 - (mc)^2 u^2 \right\} = (mc^2)^2 \left\{ 1 - \frac{u^2}{c^2} \right\} \left(1 - \frac{u^2}{c^2} \right)^{-1} = (mc^2)^2$$

---

33. A pion at rest ($m_\pi = 270\, m_e$) decays into a muon ($m_\mu = 206 m_e$) and an antineutrino ($m_v = 0$) according to the following: $\pi^- \rightarrow \mu^- + \overline{v}$. Find the kinetic energy of the muon and the antineutrino in MeV. (Hint: Relativistic momentum is conserved.)

Solution

$$m_\pi c^2 = \lambda m_\mu c^2 + p_v c \quad \text{[conservation of energy]}$$

$$p_v = -p_\mu = -\lambda m_\mu v \quad \text{[conservation of momentum]}$$

$$138 = \gamma(105) + 105 \frac{\gamma v}{c} \quad \text{[E in MeV]}$$

$$\frac{v}{c} = 0.266$$

$$\gamma = \frac{1}{\sqrt{1 - \frac{v^2}{c^2}}} = 1.0374$$

$$KE_{muon} = (0.0374)(105) = \boxed{4 \text{ MeV}}$$

$$KE_{\overline{v}} = 138 - (105 + 4) = \boxed{29 \text{ MeV}}$$

---

39. An astronaut is traveling in a space vehicle that has a speed of 0.50c relative to the earth. The astronaut measures his pulse rate to be 75 per minute. Signals generated by the astronaut's pulse are radioed to earth when the vehicle is moving perpendicular to a line that connects the vehicle with an earth observer. (a) What pulse rate does the earth observer measure? (b) What would be the pulse rate if the speed of the space vehicle were increased to 0.99c?

Solution

(a) $\gamma = -\sqrt{\left(1 - \dfrac{v^2}{c^2}\right)} = -\sqrt{[1 - (0.50)^2]} = 1.155$

According to Equation 39.7, the time interval between pulses as measured by the earth observer is

$$\Delta t = \gamma \Delta t' = (1.155)(0.80 \text{ s}) = 0.924 \text{ s}$$

Thus, the earth observer records a pulse rate of

$$\frac{(60 \text{ s/min})}{0.924 \text{ s}} = \boxed{64.9 \text{ min}^{-1}}$$

(b)   At a relative speed u = 0.99c, the relativistic factor γ increases to 7.09 and the pulse rate recorded by the earth observer decreases to $\boxed{10.6 \text{ min}^{-1}}$. That is, the lifespan of the astronaut (reckoned by the total number of his heartbeats) is much longer as measured by an earth clock than by a clock aboard the space vehicle.

---

43. An electron has a speed of 0.75c. Find the speed of a proton which has (a) the same kinetic energy as the electron and (b) the same momentum as the electron.

Solution

(a)   When $K_e = K_p$, $m_e c^2 (\gamma_e - 1) = m_p c^2 (\gamma_p - 1)$

In this case, $m_e c^2 = 0.5117 \text{ MeV}$ and $m_p c^2 = 939 \text{ MeV}$

$$\gamma_e = \frac{1}{\sqrt{1 - (0.75)^2}} = 1.5119$$

Substituting these values into the first equation, we find

$$\gamma_p = m_e c^2 \frac{(\gamma_e - 1)}{m_p c^2} + 1 = \frac{0.5117(1.5119 - 1)}{937} + 1 = 1.000279$$

But $\gamma_p = \dfrac{1}{\sqrt{1 - \left(\dfrac{u_p}{c}\right)^2}}$

Therefore,

$$u_p = c\sqrt{(1 - \gamma_p^{-2})} = \boxed{0.0236c}$$

(b)   Using Equation 39.21, when $p_e = p_p$, we have

$$\gamma_p m_p u_p = \gamma_e m_e u_e$$

or

$$u_p = \left(\frac{\gamma_e}{\gamma_p}\right)\left(\frac{m_e}{m_p}\right) u_e$$

$$u_p = \left(\frac{1.5119}{1.000279}\right)\left[\frac{0.5117/c^2}{939/c^2}\right](0.75c) = 6.18 \times 10^{-4}c = \boxed{185 \text{ km/s}}$$

53. *Doppler effect for light.* If a light source moves with a speed v relative to an observer, there is a shift in the observed frequency analogous to the Doppler effect for sound waves. Show that the observed frequency f_o is related to the true frequency f through the expression

$$f_o = \sqrt{\frac{c \pm v_s}{c \mp v_s}} \, f$$

where the upper signs correspond to the source approaching the observer and the lower signs correspond to the source receding from the observer. [Hint: In the moving frame S', the period is the proper time interval and is given by $T = \frac{1}{f}$. Furthermore, the wavelength measured by the observer is $\lambda_o = (c - v_s)T_o$ where T_o is the period measured in s.]

Solution

$$f_o = \frac{1}{T_o} = \frac{c-v}{\lambda_o} = \frac{c-v}{\lambda/\gamma} = \frac{(c-v)\gamma}{c/f}$$

$$f_o = \left(\frac{c-v}{c}\right)\frac{1}{\sqrt{1-\frac{v^2}{c^2}}}f = \sqrt{\frac{c-v}{c+v}}\, f$$

If the source changes direction, $v \rightarrow -v$ and

$$f_o = \sqrt{\frac{c+v}{c-v}}\, f$$

55. A charged particle moves along a straight line in a uniform electric field E with speed v. If the motion and the electric field are both in the x direction, (a) show that the acceleration of the charge q in the x direction is given by

$$a = \frac{dv}{dt} = \frac{qE}{m}\left(1 - \frac{v^2}{c^2}\right)^{3/2}$$

(b) Discuss the significance of the dependence of the acceleration on the speed. (c) If the particle starts from rest at x = 0 at t = 0, how would you proceed to find the speed of the particle and its position after a time t has elapsed?

Solution

(a)   At any speed, the *momentum* of the particle is given by

$$p = \gamma mv = \frac{mv}{\sqrt{1 - \frac{v^2}{c^2}}}$$

Since $F = qE = \frac{dp}{dt}$,

$$qE = \frac{d}{dt}\left[mv\left(1 - \frac{v^2}{c^2}\right)^{-1/2}\right]$$

$$qE = m\left(1 - \frac{v^2}{c^2}\right)^{-1/2}\frac{dv}{dt} + \frac{1}{2}mv\left(1 - \frac{v^2}{c^2}\right)^{-3/2}\frac{2v}{c^2}\frac{dv}{dt}$$

$$\frac{qE}{m} = \frac{dv}{dt}\left[\frac{\left(1 - \frac{v^2}{c^2}\right) + \frac{v^2}{c^2}}{\left(1 - \frac{v^2}{c^2}\right)^{3/2}}\right]$$

$$a = \frac{dv}{dt} = \frac{qE}{m}\left(1 - \frac{v^2}{c^2}\right)^{3/2}$$

(b)   As $v \to c$, we see that $a \to 0$.

(c) $$\int_0^v \frac{dv}{\left(1 - \frac{v^2}{c^2}\right)^{3/2}} = \int_0^t \frac{qE}{m}\,dt; \qquad v = \frac{qEct}{\sqrt{m^2c^2 + q^2E^2t^2}}$$

Introduction to Quantum Physics

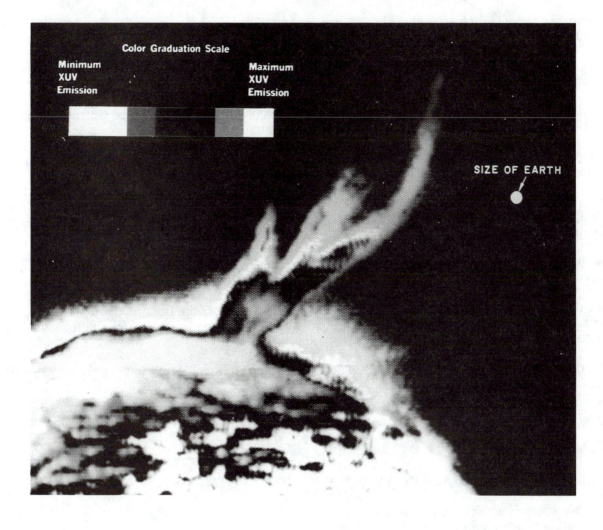

OBJECTIVES

After reading this chapter, you should be able to . . .

1. Discuss the spectral characteristics of blackbody radiation, and the limitations of the classical model predicted by the Rayleigh-Jeans law.

2. Describe the formula for blackbody radiation proposed by Planck, and the assumption made in deriving this formula.

3. Discuss the conditions under which a photoelectric effect can be observed, and those properties of photo-electric emission which cannot be explained with classical physics.

4. Describe the Einstein model for the photoelectric effect, and the predictions of the fundamental photoelectric effect equation for the maximum kinetic energy of photoelectrons.

5. Recognize that Einstein's model of the photoelectric effect involves the photon concept (E = hf), and the fact that the basic features of the photoelectric effect are consistent with this model.

6. Describe the Compton effect (the scattering of X-rays by electrons) and be able to derive the formula for the Compton shift (Eq. 41.10). Recognize that the Compton effect can only be explained using the photon concept.

7. Discuss the origin of line spectra associated with elements such as hydrogen, and the usefulness of such spectra in modern analyses.

8. State the postulates of the Bohr theory of the hydrogen atom, and use the Bohr model to derive the energy levels of hydrogen, the radii of the allowed orbits, and the allowed wavelengths corresponding to the various series in the hydrogen spectrum.

9. State the correspondence principle first postulated by Bohr, and its significance in bridging the gap between classical physics and quantum physics.

NOTES FROM SELECTED CHAPTER SECTIONS

40.1 BLACKBODY RADIATION AND PLANCK'S HYPOTHESIS

A *blackbody* is an ideal body that absorbs all radiation incident on it. Any body at some temperature T emits thermal radiation which is characterized by the properties of the body and its temperature. The spectral distribution of blackbody radiation at various temperatures is sketched in the figure. As the temperature increases, the intensity of the radiation (area under the curve) increases, while the peak of the distribution shifts to lower wavelengths.

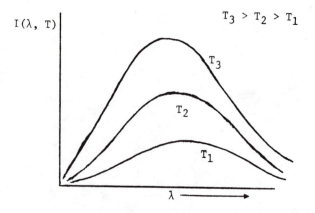

Classical theories failed to explain blackbody radiation. An empirical formula (Eq. 40.3), proposed by Max Planck, is consistent with this distribution at all wavelengths. Planck made two basic assumptions in the development of this result:

(1) The oscillators emitting the radiation could only have *discrete energies* given by $E_n = nhf$, where n is a quantum number (n = 1, 2, 3, ...), f is the oscillator frequency, and h is Planck's constant.

(2) These oscillators can emit or absorb energy in discrete units called quanta (or photons), where the energy of a light quantum obeys the relation E = hf.

Subsequent developments showed that the quantum concept was necessary in order to explain several phenomena at the atomic level, including the photoelectric effect, the Compton effect, and atomic spectra.

40.2 THE PHOTOELECTRIC EFFECT

When light is incident on certain metallic surfaces, electrons can be emitted from the surfaces. This is called the photoelectric effect, discovered by Hertz. One cannot explain many features of the photoelectric effect using classical concepts. In 1905, Einstein provided a successful explanation of the photoelectric effect by extending Planck's quantum concept to include electromagnetic fields.

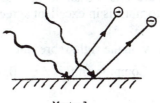

Metal

Several features of the photoelectric effect could not be explained with classical physics or with the wave theory of light. However, each of these features can be explained and understood on the basis of the photon theory of light. These observations and their explanations include:

1. No electrons are emitted if the incident light frequency falls below some cutoff frequency, f_c, which is characteristic of the material being illuminated. For example, in the case of sodium, $f_c = 5.50 \times 10^{14}$ Hz. This is inconsistent with the wave theory, which predicts that the photoelectric effect should occur at any frequency, provided the light intensity is high enough.

   The fact that the photoelectric effect is not observed below a certain cutoff frequency follows from the fact that the energy of the photon must be greater than or equal to ϕ. If the energy of the incoming photon is not equal to or greater than ϕ, the electrons will never be ejected from the surface, regardless of the intensity of the light.

2. If the light frequency exceeds the cutoff frequency, a photoelectric effect is observed and the number of photoelectrons emitted is proportional to the light intensity. However, the maximum kinetic energy of the photoelectrons is independent of light intensity, a fact that cannot be explained by the concepts of classical physics.

   The fact that K_{max} is independent of the light intensity can be understood with the following argument. If the light intensity is doubled, the number of photons is doubled, which doubles the number of photoelectrons emitted. However, their kinetic energy, which equals $hf - \phi$, depends only on the light frequency and the work function, not on the light intensity.

3. The maximum kinetic energy of the photoelectrons increases with increasing light frequency. The fact that K_{max} increases with increasing frequency is easily understood with Equation 40.9.

4. Electrons are emitted from the surface almost instantaneously (less than 10^{-9} s after the surface is illuminated), even at low light intensities. Classically, one would expect that the electrons would

require some time to absorb the incident radiation before they acquire enough kinetic energy to escape from the metal.

Finally, the fact that the electrons are emitted almost instantaneously is consistent with the particle theory of light, in which the incident energy appears in small packets and there is a one-to-one interaction between photons and electrons. This is in contrast to having the energy of the photons distributed uniformly over a large area.

40.3 THE COMPTON EFFECT

The Compton effect involves the scattering of an X-ray by an electron as shown in the figure below on the left. The scattered X-ray undergoes a *change* in wavelength $\Delta\lambda$, called the Compton shift, which cannot be explained using classical concepts. By treating the X-ray as a photon (the quantum concept), the scattering process between the photon and electron predicts a shift in photon (X-ray) wavelength given by Eq. 40.11, where θ is the angle between the incident and scattered X-ray and m is the mass of the electron. The formula is in excellent agreement with experimental results.

The figure below on the right represents Compton's data at $\theta = 90°$ for scattering of X-rays from graphite. In this case, the Compton shift is $\Delta\lambda = 0.0236$ Å and λ_o is the wavelength of the incident X-ray beam.

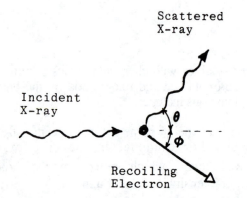

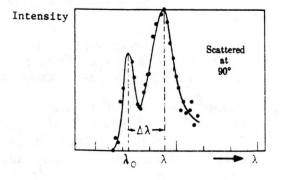

40.5 BOHR'S QUANTUM THEORY OF THE ATOM

The lines observed in the hydrogen spectrum can be arranged into series corresponding to assigned values of principal quantum numbers of the initial and final states.

For the Lyman series,
$n_f = 1$ and $n_i = 2, 3, 4, \ldots$

For the Balmer series,
$n_f = 2$ and $n_i = 3, 4, 5, \ldots$

For the Paschen series,
$n_f = 3$ and $n_i = 4, 5, 6, \ldots$

For the Brackett series
$n_f = 4$ and $n_i = 5, 6, 7, \ldots$

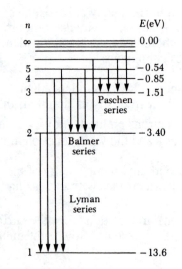

In the case of very large values of the principal quantum number, the energy differences between adjacent levels approach zero and essentially a continuous range (as opposed to a quantized set) of energy values of the emitted photon is possible. In this limit of very large quantum numbers, the classical model of the atom is reasonably accurate. This is called the correspondence principle.

The basic ideas of the Bohr theory as it applies to an atom of hydrogen are as follows:

1. The electron moves in circular orbits about the proton under the influence of the Coulomb force of attraction.

2. Only certain orbits are stable. These stable orbits are ones in which the electron does not radiate. Hence, the energy is fixed or stationary, and ordinary classical mechanics may be used to describe the electron's motion.

3. Radiation is emitted by the atom when the electron "jumps" from a more energic initial stationary state to a lower state. This "jump" cannot be visualized or treated classically. In particular, the frequency f of the photon emitted in the jump is *independent of the frequency of the electron's orbital motion*. The frequency of the light emitted is related to the change in the atom's energy.

4. The size of the allowed electron orbits is determined by an additional quantum condition imposed on the electron's orbital angular momentum. Namely, the allowed orbits are those for which the electron's orbital angular momentum about the nucleus is an integral multiple of $h/2\pi$.

In addition to providing a theoretical derivation of the line spectrum, Bohr also explained:
(a) the limited number of lines seen in the absorption spectrum of hydrogen compared to the emission spectrum,
(b) the emission of X-rays from atoms,
(c) the chemical properties of atoms in terms of the electron shell model,
(d) how atoms associate to form molecules.

EQUATIONS AND CONCEPTS

Planck developed an empirical equation which describes the intensity of blackbody radiation as a function of temperature.

$$I(\lambda, T) = \frac{2\pi hc^2}{\lambda^5 (e^{hc/\lambda kT} - 1)} \qquad (40.3)$$

In his model, Einstein assumed that light consists of a stream of particles called *photons* whose energy is given by E = hf, where h is Planck's constant.

$$E = hf \qquad (40.6)$$

In the photoelectric effect, the *maximum kinetic energy* of the ejected photoelectron is given by Eq. 40.9, where ϕ is the *work function* of the metal, which is typically a few eV. This model is in excellent agreement with experimental results, including the prediction of a *cutoff* (or threshold) wavelength above which no photoelectric effect is observed.

$$K_{max} = hf - \phi \qquad (40.9)$$

The wavelength shift in Compton scattering is a function of the angle of scattering.

$$\Delta\lambda = \frac{h}{mc}(1 - \cos\theta)$$

(40.11)

A photon of frequency f (and wavelength λ) is emitted when an electron undergoes a transition from an initial energy level to a final lower level. R is the theoretical value of the Rydberg constant.

$$\frac{1}{\lambda} = R\left(\frac{1}{n_f{}^2} - \frac{1}{n_i{}^2}\right)$$

The Bohr model of the atom assumed that the electron orbited the nucleus in a circular path under the influence of the coulomb force of attraction. Also, the angular momentum of the electron about the nucleus must be quantized in units of $\hbar$.

$$mvr = n\hbar$$
$$n = 1, 2, 3, \ldots$$

(40.21)

While in one of the allowed orbits or stationary states (determined by quantization of the orbital angular momentum), the electron does not radiate energy.

When the electron undergoes a transition from one allowed orbit to another, the frequency of the emitted photon is proportional to the difference in energies of the initial and final states.

$$hf = E_i - E_f$$

(40.20)

The electron can exist only in certain allowed orbits, the value of which can be expressed in terms of the Bohr radius, a_0.

$$r_n = \frac{n^2\hbar^2}{mke^2}$$
$$n = 1, 2, 3, \ldots$$

(40.25)

$$r_n = n^2 a_0 = n^2(0.529 \text{ Å})$$

(40.28)

The Bohr radius can be calculated in terms of fundamental constants.

$$a_0 = \frac{\hbar^2}{mke^2} = 0.529 \text{ Å}$$

(40.26)

The total energy E of the hydrogen atom (KE + PE) depends on the radius of the allowed orbit of the electron.

$$E = -\frac{ke^2}{2r}$$

(40.24)

When numerical values of the constants are used in Equation 40.24, the energy level values can be expressed in units of electron volts (eV).

$$E_n = -\frac{13.6}{n^2} \text{ eV}$$

(40.28)

EXAMPLE PROBLEM SOLUTION

Example 40.1   The cutoff wavelength for lithium is 5400 Å. What is the stopping potential for electrons emitted by lithium when the incident light has a wavelength of 3200 Å?

Solution

The work function can be calculated by using Eq. 40.10.

$$\lambda_c = \frac{hc}{\phi}$$

$$\phi = \frac{hc}{\lambda_c} = \frac{(6.63 \times 10^{-34} \text{ J·s})(3 \times 10^8 \text{ m/s})}{5.4 \times 10^{-7} \text{ m}}$$

$$\phi = 3.68 \times 10^{-19} \text{ J}$$

The maximum kinetic energy of the emitted electrons can be found by use of Eq. 40.9.

$$K_{max} = hf - \phi \quad \text{but} \quad f = \frac{c}{\lambda} \quad \text{so}$$

$$K_{max} = \frac{hc}{\lambda} - \phi$$

$$K_{max} = \frac{1.99 \times 10^{-25}}{3.2 \times 10^{-7}} - 3.68 \times 10^{-19} = 2.54 \times 10^{-19} \text{ J}$$

Express the kinetic energy in units of electron volts.

$$K_{max} = (2.54 \times 10^{-19} \text{ J}) \frac{1 \text{ eV}}{1.6 \times 10^{-19} \text{ J}} = 1.59 \text{ eV}$$

Solve for the stopping potential.

$$K_{max} = eV_o ; \quad V_o = 1.59 \text{ V}$$

ANSWERS TO SELECTED QUESTIONS

4. If the photoelectric effect is observed for one metal, can you conclude that the effect will also be observed for another metal under the same conditions? Explain.

Answer: No. The second metal may have a larger work function than the surface of the first, in which case the incident photons may not be energetic enough to eject photoelectrons.

19. An X-ray photon is scattered by an electron. What happens to the frequency of the scattered photon relative to that of the incident photon?

Answer: The X-ray photon transfers some of its energy to the electron. Thus, its frequency must be decreased.

20. Why does the existence of a cutoff frequency in the photoelectric effect favor a particle theory for light rather than a wave theory?

Answer: Wave theory predicts that the photoelectric effect should occur at any frequency, provided that the light intensity is high enough. This is in contradiction to experimental results.

22. All objects radiate energy. Why, then, are we not able to see all objects in a dark room?

Answer: All objects radiate energy, but for objects at room temperature, this energy is primarily in the infrared region of the spectrum.

24. What effect, if any, would you expect the temperature of a material to have on the ease with which electrons can be ejected from it in the photoelectric effect?

Answer: You would expect that it would be easier to eject electrons from a heated object, since the added thermal energy would increase the average kinetic energy of the electrons, making it easier for them to leave the material.

25. Some stars are observed to be reddish, and some are blue. Which stars have the higher surface temperature? Explain.

Answer: In general, the stars having the highest surface temperature produce photons with the highest energy. Thus, blue stars have a higher surface temperature than red stars.

SOLUTIONS TO SELECTED END-OF-CHAPTER PROBLEMS

12. Show that at *long* wavelengths, Planck's radiation law (Eq. 40.3) reduces to the Rayleigh-Jeans law (Eq. 40.2).

Solution

$$I(\lambda, T) = \frac{2\pi hc^2}{\lambda^5 (e^{hc/\lambda kT} - 1)}$$

As $T \to \infty$, $e^{hc/\lambda kT} \to 1 + \frac{hc}{\lambda kT}$, therefore

$$I(\lambda, T) \to \frac{2\pi hc^2}{\lambda^5 \left(1 + \frac{hc}{\lambda kT} - 1\right)} = \frac{2\pi ckT}{\lambda^4}$$

13. Show that at *short* wavelengths or *low* temperatures, Planck's radiation law (Eq. 40.3) predicts an exponential decrease in $I(\lambda, T)$ given by *Wien's radiation law*:

$$I(\lambda, T) = \frac{2\pi hc^2}{\lambda^5} e^{-hc/\lambda kT}$$

Solution

This follows from the fact that at low temperatures or short wavelengths, the exponential factor in the denominator of Planck's radiation law (Eq. 40.3) is large compared to 1, so the factor of 1 in the denominator of Eq. 40.3 can be neglected. In this approximation, one arrives at *Wien's radiation law*.

---

19. Two light sources are used in a photoelectric experiment to determine the work function for a particular metal surface. When the green light from a mercury lamp ($\lambda = 546.1$ nm) is used, a retarding potential of 1.70 V reduces the photocurrent to zero. (a) Based on this measurement, what is the work function for this metal? (b) What stopping potential would be observed when using the yellow light from a helium discharge tube ($\lambda = 587.5$ nm)?

Solution

(a) $\quad eV_o = \dfrac{hc}{\lambda} - \phi$

$$\phi = \frac{hc}{\lambda} - eV_o = \frac{1240 \text{ eV·nm}}{546.1 \text{ nm}} - 1.700 \text{ eV} = \boxed{0.571 \text{ eV}}$$

(b) $\quad eV_o = \dfrac{1240 \text{ eV·nm}}{587.5 \text{ nm}} - 0.576 \text{ eV} = 1.54 \text{ eV}$

or,

$$V_o = \boxed{1.54 \text{ V}}$$

---

31. After a 0.80-nm X-ray photon scatters from a free electron, the electron recoils with a velocity of 1.4×10^6 m/s. (a) What was the Compton shift in the photon's wavelength? (b) Through what angle was the photon scattered?

Solution

(a) The electron has a recoil kinetic energy of

$$K = \frac{1}{2} mv^2 = \frac{1}{2} (9.11 \times 10^{-31} \text{ kg})(1.4 \times 10^6 \text{ m/s})^2 = 8.93 \times 10^{-19} \text{ J} = 5.573 \text{ eV}$$

The incident photon has energy E_0, given by

$$E_0 = \frac{hc}{\lambda_o} = \frac{1240 \text{ eV·nm}}{0.80 \text{ nm}} = 1550 \text{ eV}$$

After being scattered, the photon energy is

$$E' = E_o - K = 1544.43 \text{ eV}$$

Hence, the wavelength of the scattered photon is

$$\lambda' = \frac{hc}{E'} = \frac{1240 \text{ eV·nm}}{1544.43 \text{ eV}} = 0.8029 \text{ nm}$$

and the Compton shift is

$$\Delta\lambda = \lambda' - \lambda_o = \boxed{0.0029 \text{ nm}}$$

(b)   Since $\Delta\lambda = \lambda_c(1 - \cos\theta)$,   where $\lambda_c = 0.00243$ nm,

$$1 - \cos\theta = \frac{\Delta\lambda}{\lambda_c} = \frac{0.0029 \text{ nm}}{0.00243 \text{ nm}} = 1.193$$

$$\theta = \boxed{101°}$$

---

35. A 0.0016-nm photon scatters from a free electron. For what (photon) scattering angle will the recoiling electron and scattered photon have the same kinetic energy?

Solution

The kinetic energy of the electron in $K = E_o - E'$, where E_o is the energy of the incident photon and E' is the energy of the scattered photon. Taking $K = E'$ gives $E' = E_o - E'$, or $E' = \frac{E_o}{2}$. Therefore, the wavelength of the scattered photon is

$$\lambda' = \frac{hc}{E'} = \frac{hc}{E_o/2} = 2\frac{hc}{E_o} = 2\lambda_o$$

Since $\Delta\lambda = \lambda' - \lambda_o = \lambda_c (1 - \cos\theta)$,   we find

$$\lambda' = \lambda_o + \lambda_c(1 - \cos\theta)$$

$$2\lambda_o = \lambda_o + \lambda_c(1 - \cos\theta)$$

$$1 - \cos\theta = \frac{\lambda_o}{\lambda_c} = \frac{0.0016 \text{ nm}}{0.00243 \text{ nm}} \qquad \text{or} \qquad \theta = \boxed{70.1°}$$

---

47. A photon is emitted from a hydrogen atom which undergoes a transition from the state n = 6 to the state n = 2. Calculate (a) the energy, (b) the wavelength, and (c) the frequency of the emitted photon.

Solution

(a) $\Delta E = (-13.6 \text{ eV})\left(\frac{1}{n_f^2} - \frac{1}{n_i^2}\right)$

In this case, $n_f = 2$ and $n_i = 6$, so

$$\Delta E = (-13.6 \text{ eV})\left(\frac{1}{2^2} - \frac{1}{6^2}\right) = -3.02 \text{ eV}$$

ΔE represents the *loss* in energy of the atom in this process. Hence, the emitted photon must have an energy of

$$\boxed{3.02 \text{ eV}}$$

(b) Since $E = \frac{hc}{\lambda}$ for the photon, we find

$$\lambda = \frac{hc}{E} = \frac{1240 \text{ eV·nm}}{3.02 \text{ eV}} = \boxed{411 \text{ nm}}$$

(c) From $c = \lambda f$, we obtain the frequency:

$$f = \frac{c}{\lambda} = \frac{3 \times 10^8 \text{ m/s}}{411 \times 10^{-9} \text{ m}} = \boxed{7.30 \times 10^{14} \text{ Hz}}$$

---

60. Use Bohr's model of the hydrogen atom to show that when the atom makes a transition from the state n to the state n − 1, the frequency of the emitted light is given by

$$f = \frac{2\pi^2 mk^2 e^4}{h^3}\left[\frac{2n-1}{(n-1)^2 n^2}\right]$$

Show that as n → ∞, the expression above varies as $1/n^3$ and reduces to the classical frequency one would expect the atom to emit. (Hint: To calculate the classical frequency, note that the frequency of revolution is $v/2\pi r$, where r is given by Eq. 40.25.) This is an example of the correspondence principle, which requires that the classical and quantum models agree for large values of n.

Solution

From the Bohr model, the frequency of light emitted by an atom moving from the state n_i to the state n_f is given by Eq. 40.30.

$$f = \frac{mk^2 e^4}{4\pi \hbar^3}\left(\frac{1}{n_f^2} - \frac{1}{n_i^2}\right)$$

Taking $n_i = n$ and $n_f = n - 1$, we have

$$f = \frac{mk^2e^4}{4\pi\hbar^3}\left[\frac{1}{(n-1)^2} - \frac{1}{n^2}\right] = \frac{2\pi^2mk^2e^4}{h^3}\left[\frac{2n-1}{n^2(n-1)^2}\right]$$

As $n \to \infty$, we see that

$$f \to \frac{4\pi^2mk^2e^4}{h^3}\left(\frac{1}{n^3}\right)$$

The classical frequency is $f = \frac{v}{2\pi r}$. From the Bohr model, $v^2 = \frac{ke^2}{mr}$ and $r = \frac{n^2\hbar^2}{mke^2}$, therefore

$$f = \frac{v}{2\pi r} = \frac{1}{2\pi}\sqrt{\frac{ke^2}{m}}\left(\frac{1}{r}\right)^{3/2} = \frac{1}{2\pi}\sqrt{\frac{ke^2}{m}}\left(\frac{mke^2}{n^2\hbar^2}\right)^{3/2} = \frac{4\pi^2mk^2e^4}{h^3}\left(\frac{1}{n^3}\right)$$

---

61. A muon (Problem 59) is captured by a deuteron to form a muonic atom. (a) Find the energy of the ground state and the first excited state. (b) What is the wavelength of the photon emitted when the atom makes a transition from the first excited state to the ground state?

Solution

(a) First, we find the "Bohr radius" of the atom, noting that we must use the reduced mass μ of the atom in Eq. 40.26.

$$\mu = \frac{m_1m_2}{m_1 + m_2} = \left[\frac{207(3680)}{207 + 3680}\right]m_e = 196m_e$$

Thus,

$$R_0 = \frac{\hbar^2}{\mu ke^2} = \frac{\hbar^2}{196m_eke^2} = \frac{a_0}{196}$$

where $a_0 = 0.0529$ nm (the standard Bohr radius). Substituting R_0 into Eq. 40.27 with $Z = 1$ gives

$$E_n = -\frac{ke^2}{2R_0}\left(\frac{1}{n^2}\right) = -196\frac{ke^2}{2a_0}\left(\frac{1}{n^2}\right) = -196(13.6 \text{ eV})\left(\frac{1}{n^2}\right) = (-2.67 \text{ keV})\left(\frac{1}{n^2}\right)$$

Thus,

$$E_1 = \boxed{-2.67 \text{ keV}} \quad \text{and} \quad E_2 = \boxed{-667 \text{ eV}}$$

(b) Since $\Delta E = E_2 - E_1 = 2.00$ keV, and $\Delta E = \frac{hc}{\lambda}$, we find

$$\lambda = \frac{hc}{\Delta E} = \frac{1240 \text{ eV·nm}}{2.00 \times 10^3 \text{ eV}} = \boxed{0.620 \text{ nm}}$$

67. Show that the ratio of the Compton wavelength to the de Broglie wavelength for a *relativistic* electron is given by

$$\frac{\lambda_c}{\lambda} = \left[\left(\frac{E}{mc^2}\right)^2 - 1\right]^{1/2}$$

where E is the total energy of the electron and m is its mass.

Solution

$$\lambda_c = \frac{h}{mv} \quad \text{and} \quad \lambda = \frac{h}{p}, \quad \text{therefore}$$

$$\frac{\lambda_c}{\lambda} = \frac{h/mc}{h/p} = \frac{p}{mc} \tag{1}$$

Since $E^2 = p^2c^2 + (mc^2)^2$,

$$p = \left[\frac{E^2}{c^2} - (mc)^2\right]^{1/2} \tag{2}$$

Substituting (2) into (1) gives

$$\frac{\lambda_c}{\lambda} = \frac{1}{mc}\left[\frac{E^2}{c^2} - (mc)^2\right]^{1/2} = \left[\left(\frac{E}{mc^2}\right)^2 - 1\right]^{1/2}$$

Wave Mechanics

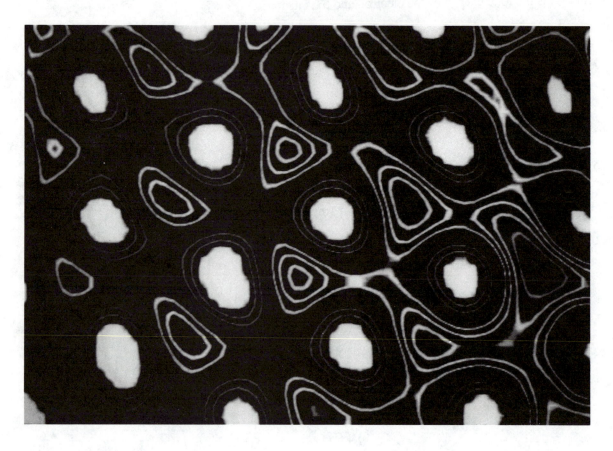

OBJECTIVES

1. Discuss the wave properties of particles, the de Broglie wavelength concept, and the dual nature of both matter and light.

2. Describe the concept of wave function for the representation of matter waves and state in equation form the normalization condition and expectation value of the coordinate.

3. Discuss the manner in which the uncertainty principle makes possible a better understanding of the dual wave-particle nature of light and matter.

4. State the time-independent form of the Schrödinger equation for a bound system of total energy, E, and discuss the required conditions on the wave function.

5. Describe the allowed wave functions and energy levels for a particle in a one-dimensional box.

6. State the Schrödinger equation for a particle in a well of finite height.

7. State the Schrödinger equation for a simple harmonic oscillator; and calculate the allowed energy levels in terms of the oscillator frequency.

NOTES FROM SELECTED CHAPTER SECTIONS

41.1 PHOTONS AND ELECTROMAGNETIC WAVES

The results of some experiments are better described on the basis of the photon model of light; other experimental outcomes are better described in terms of the wave model. The photon theory and the wave theory complement each other--light exhibits both wave and photon characteristics.

41.2 THE WAVE PROPERTIES OF PARTICLES

de Broglie postulated that a particle in motion has wave properties and a corresponding wavelength inversely proportional to the particle's momentum.

41.4 THE UNCERTAINTY PRINCIPLE

Quantum theory predicts that it is fundamentally impossible to make simultaneous measurements of a particle's position and velocity with infinite accuracy.

41.5 INTRODUCTION TO QUANTUM MECHANICS

The wave function is a complex valued quantity, the absolute square of which gives the probability of finding a particle at a given point at some instant; and the wave function contains all the information that can be known about the particle.

The *Schrödinger equation* describes the manner in which matter waves change in time and space.

The probability of finding a certain value for a quantity (position, energy) is called the *expectation value* of the quantity.

41.6 A PARTICLE IN A BOX

A particle confined to a box and represented by a well-defined de Broglie wave function is represented by a sinusoidal wave. The allowed states of the system are called *stationary states* since they represent *standing waves*.

The minimum energy which the particle can have is called the *zero-point energy*.

41.7 THE SCHRÖDINGER EQUATION

The basic problem in wave mechanics is to determine a solution to the Schrödinger equation. The solution will provide the allowed wave functions and energy levels of the system.

41.9 TUNNELING THROUGH A BARRIER

When a particle is incident onto a barrier, the height of which is greater than the energy of the particle, there is a finite probability that the particle will penetrate the barrier. In this process, called *tunneling*, part of the incident wave is transmitted and part is reflected.

EQUATIONS AND CONCEPTS

According to the de Broglie hypothesis, a material particle should have an associated wavelength λ which depends on its momentum.

$$\lambda = \frac{h}{mv} \tag{41.3}$$

Matter waves can be represented by a wave function, ψ, which, in general, depends on position and time. In the expression for the part of the wave function which depends only on position, k is the wave number.

$$\psi(x) = A \sin\left(\frac{2\pi x}{\lambda}\right) \tag{41.7}$$

The normalization condition is a statement of the requirement that the particle exists at some point (along the x-axis in the one-dimensional case) at all times.

$$\int_{-\infty}^{\infty} |\psi|^2 \, dx = 1 \tag{41.9}$$

Although it is not possible to specify the position of a particle exactly, it is possible to calculate the probability P_{ab} of finding the particle within an interval $a \leq x \leq b$.

$$P_{ab} = \int_{a}^{b} |\psi|^2 \, dx \tag{41.10}$$

The probability of measuring a certain value for the position of a particle is called the expectation value of the coordinate x.

$$<x> \equiv \int_{-\infty}^{\infty} x|\psi|^2 \, dx \tag{41.11}$$

The uncertainty principle states that if a measurement of position is made with a precision Δx and a *simultaneous* measurement of momentum is made with a precision Δp, then the product of the two uncertainties can never be smaller than a number of the order of $\hbar$.

$$\Delta x \Delta p \gtrsim \hbar$$

(41.5)

The allowed wave functions for a particle in a rigid box of width L are sinusoidal.

$$\psi(x) = A \sin\left(\frac{n\pi x}{L}\right)$$

$$n = 1, 2, 3, \ldots$$

(41.15)

The energy of a particle in a box is quantized and the least energy which the particle can have is called the zero-point energy.

$$E_n = \left(\frac{h^2}{8mL^2}\right) n^2$$

$$n = 1, 2, 3, \ldots$$

(41.17)

The time independent Schrödinger equation for a bound system (total energy E = constant) allows in principle the determination of the wave functions and energies of the allowed states if the potential energy function is known.

$$\frac{\partial^2 \psi}{\partial x^2} = -\frac{2m}{\hbar^2}(E - U)\psi$$

(41.21)

The Schrödinger equation can be written for a harmonic oscillator of total energy, E.

$$\frac{d^2 \psi}{dx^2} = -\left[\left(\frac{2mE}{\hbar^2}\right) - \left(\frac{m\omega}{\hbar}\right)^2 x^2\right]\psi$$

(41.30)

The energy levels of the harmonic oscillator are quantized.

$$E_n = \left(n + \frac{1}{2}\right)\hbar\omega$$

$$n = 0, 1, 2, 3, \ldots$$

(41.33)

EXAMPLE PROBLEM SOLUTION_____

Example 41.1   The speed of a 25-g bullet is measured with an accuracy of 0.05 percent to have a value of 300 m/s. Within what uncertainty can the position of the bullet be determined in a measurement which is simultaneous with the measurement of the speed?

Solution

Use Equation 41.5, $\Delta x \Delta p \gtrsim \hbar$, and note that $\Delta v = (0.0005)(300 \text{ m/s}) = 0.15 \text{ m/s}$, so

$$\Delta x = \frac{\hbar}{\Delta p} = \frac{\hbar}{m \Delta v} = \frac{1.056 \times 10^{-34} \text{ J} \cdot \text{s}}{(2.5 \times 10^{-2} \text{ kg})(0.15 \text{ m/s})}$$

so

$$\Delta x = 2.76 \times 10^{-32} \text{ m}$$

ANSWERS TO SELECTED QUESTIONS_____

1. Is light a wave or a particle? Support your answer by citing specific experimental evidence.

Answer: Light has a dual character. It is best described as a wave when explaining such phenomena as interference and diffraction. On the other hand, it is best described as a particle when explaining the photoelectric effect.

2. Is an electron a particle or a wave? Support your answer by citing some experimental results.

Answer: An electron also has a dual character. Its wave character is evident in describing the diffraction of electrons from a crystalline surface. Its particle nature is evident in such phenomena as atomic transitions and the photoelectric effect.

3. An electron and a proton are accelerated from rest through the same potential difference. Which particle has the longer wavelength?

Answer: Applying $\frac{1}{2}mv^2 = eV$ and $\lambda = h/p$, we find that $\lambda = h/\sqrt{2meV}$. Therefore, because the electron has the smaller mass, its wavelength is longer than that of the proton.

4. If matter has a wave nature, why is this wave-like character not observable in our daily experiences?

Answer: Macroscopic objects such as baseballs and rockets have a large mass and a *very small* de Broglie wavelength. Hence, their wave character cannot be observed in our common experiences. For example, if $m = 1$ kg and $v = 10$ m/s,

$$\lambda = \frac{h}{mv} = \frac{6.6 \times 10^{-34}}{(1)(10)} = 6.6 \times 10^{-33} \text{ m}$$

5. In what way does Bohr's model of the hydrogen atom violate the uncertainty principle?

Answer: Because the uncertainty relationship $\Delta x \Delta p_x \gtrsim \hbar$ is valid for *any* direction in space, then we can write $\Delta r \Delta p_r \gtrsim \hbar$ for the radial direction. In the Bohr model, as the electron moves in one of its circular orbits, the value of r is known exactly, so $\Delta r = 0$. Furthermore, as it moves in the circle, $p_r = 0$ and $\Delta p_r = 0$. This violates the uncertainty principle since it states that both r and p_r are known exactly.

6. Why is it impossible to simultaneously measure the position and velocity of a particle with infinite accuracy?

Answer: Such a measurement would be in violation of the uncertainty principle $\Delta x \Delta p_x \gtrsim \hbar$. That is, nature imposes a limit on the accuracy with which one can make measurements. If we wish to determine the momentum of a particle accurately, we must measure it over a long distance Δx. On the other hand, if we wish to confine the particle to a small region Δx, then we lose our ability to measure its momentum with a high accuracy.

9. The probability density at certain points for a particle in a box is zero, as seen in Figure 41.13b. Does this imply that the particle cannot move across these points? Explain.

Answer: No. The fact that $\left|\Psi\right|^2$ is zero at certain points means that the chance of finding the particle at these points is zero. It says nothing about the actual motion of the particle.

10. Discuss the relation between the zero-point energy and the uncertainty principle.

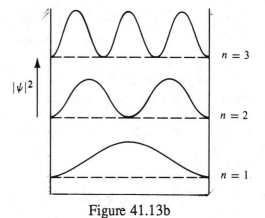

Figure 41.13b

Answer: A particle confined to a finite space cannot have zero kinetic energy. For example, if a particle is in a one-dimensional "box" of length L, then Δx is necessarily less than or equal to L. Therefore, $\Delta p \gtrsim \hbar/L$ and the *minimum* kinetic energy is about

$$E_{min} \approx \frac{\Delta p^2}{2m} = \frac{\hbar^2}{2mL^2}$$

This is the so-called zero-point energy.

11. As a particle of energy E is reflected from a potential barrier of height U, where E < U, how does the amplitude of the reflected wave change as the barrier height is reduced? How does the amplitude of the wave change?

Answer: As the barrier height is reduced, the amplitude of the reflected wave decreases while the amplitude of the transmitted wave increases.

13. In wave mechanics it is possible for the energy E of a particle to be less than the potential energy, but classically this is not possible. Explain.

Answer: If the total energy E is less than the potential energy, the kinetic energy is *negative*. Classically, the particle should turn around at the edge of the potential barrier and hence will never appear in regions where K is negative. According to quantum mechanics, a particle has a finite probability of penetrating into the classically forbidden region of negative kinetic energy.

SOLUTIONS TO SELECTED END-OF-CHAPTER PROBLEMS

9. An electron has a de Broglie wavelength equal to the diameter of the hydrogen atom. What is the kinetic energy of the electron? How does this energy compare with the ground-state energy of the hydrogen atom?

Solution

Since $\lambda = 2d = 4a_0 = 4(0.0529 \text{ nm}) = 0.2116 \text{ nm}$, and the energy of the electron is nonrelativistic, we can use $p = h/\lambda$ and

$$K = \frac{p^2}{2m} = \frac{h^2}{2m\lambda^2} = \frac{(6.626 \times 10^{-34} \text{ J} \cdot \text{s})^2}{2(9.11 \times 10^{-31} \text{ kg})(2.116 \times 10^{-10} \text{ m})^2} = 5.38 \times 10^{-18} \text{ J} = \boxed{33.6 \text{ eV}}$$

This is about 2.5 times as large as the ground state of hydrogen, which is -13.6 eV.

17. The resolving power of a microscope is proportional to the wavelength used. If one wished to use a microscope to "see" an atom, a resolution of approximately 10^{-11} m (0.1 Å) would have to be obtained. (a) If electrons are used (electron microscope), what minimum kinetic energy is required for the electrons? (b) If photons of light are used, what minimum photon energy is needed to obtain 10^{-11} m resolution?

Solution

(a) Since the de Broglie wavelength is $\lambda = h/p$,

$$p_e = \frac{h}{\lambda} = \frac{6.626 \times 10^{-34} \text{ J} \cdot \text{s}}{1 \times 10^{-11} \text{ m}} = 6.626 \times 10^{-23} \text{ kg} \cdot \text{m/s}$$

$$K_e = \frac{p_e^2}{2m} = \frac{(6.626 \times 10^{-23} \text{ kg} \cdot \text{m/s})^2}{2(9.11 \times 10^{-31} \text{ kg})} = 2.41 \times 10^{-15} \text{ J} = \boxed{15.1 \text{ keV}}$$

(b) For photons: $\quad \lambda = \frac{c}{f} = \frac{hc}{hf}$

$$E = hf = \frac{hc}{\lambda} = \frac{(6.626 \times 10^{-34} \text{ J} \cdot \text{s})(3 \times 10^8 \text{ m/s})}{1 \times 10^{-11} \text{ m}} = 1.99 \times 10^{-15} \text{ J} = \boxed{124 \text{ keV}}$$

For the photon, this would correspond to a wavelength $\lambda = 0.1$ Å which is in the x-ray range of electromagnetic spectrum.

21. A proton has a kinetic energy of 1 MeV. If its momentum is measured with an uncertainty of 5%, what is the minimum uncertainty in its position?

Solution

$$K = 1 \text{ MeV} = (1 \times 10^6 \text{ eV}) \times \left(1.6 \times 10^{-19} \frac{J}{eV}\right) = 1.6 \times 10^{-13} \text{ J}$$

$$\frac{p^2}{2m} = K = 1.6 \times 10^{-13} \text{ J}$$

$$p = \sqrt{2(1.67 \times 10^{-27} \text{ kg})(1.6 \times 10^{-13} \text{ J})} = 2.31 \times 10^{-20} \text{ kg·m/s}$$

$$\Delta p = 0.05p = 1.16 \times 10^{-21} \text{ kg·m/s}$$

Since $\Delta x \Delta p \gtrsim \hbar$,

$$\Delta x \gtrsim \frac{1.054 \times 10^{-34} \text{ J·s}}{1.16 \times 10^{-21} \text{ kg·m/s}} = \boxed{9.08 \times 10^{-14} \text{ m}}$$

Note that we have treated the problem nonrelativistically, which is justified here since the kinetic energy is only 0.11% of the rest energy.

---

27. Use the particle-in-a-box model to calculate the first three energy levels of a neutron trapped in a nucleus 2×10^{-5} nm in diameter. Are the energy level differences realistic?

Solution

$$E_n = \frac{h^2}{8mL^2} n^2$$

$$E_1 = \frac{(6.626 \times 10^{-34} \text{ J·s})^2}{8(1.67 \times 10^{-27} \text{ kg})(2 \times 10^{-14} \text{ m})^2} (1)^2 = 8.21 \times 10^{-14} \text{ J} = \boxed{0.513 \text{ MeV}}$$

$$E_2 = 4E_1 = \boxed{2.05 \text{ MeV}}$$

$$E_3 = 9E_1 = \boxed{4.62 \text{ MeV}}$$

Yes, the energy differences are of the order of 1 MeV, which is a typical energy for a γ-ray photon.

39. The wave function of a particle is given by

$$\psi(x) = A \cos (kx) + B \sin (kx)$$

where A, B, and k are constants. Show that ψ is a solution of the Schrödinger equation (Eq. 41.21), assuming the particle is free (U = 0), and find the corresponding energy E of the particle.

Solution

$$\psi(x) = A \cos kx + B \sin kx \qquad \frac{\partial \psi}{\partial x} = -kA \sin kx + kB \cos kx$$

$$\frac{\partial^2 \psi}{\partial x^2} = -k^2 A \cos kx - k^2 B \sin kx = -k^2 \psi$$

Since U = 0, $\qquad -\frac{2m}{\hbar^2}(E-U)\psi = -\frac{2mE}{\hbar^2}(A \cos kx + B \sin kx)$

The Schrödinger equation is satisfied if

$$\frac{\partial^2 \psi}{\partial x^2} = -\frac{2mE}{\hbar^2}\psi \qquad \text{or} \qquad -k^2\psi = -\frac{2mE}{\hbar^2}\psi$$

Therefore, $\qquad E = \frac{\hbar^2 k^2}{2m}$

---

57. For a particle described by a wave function $\psi(x)$, the average value, or expectation value, of a physical quantity f(x) associated with the particle is defined by

$$\langle f(x) \rangle = \int_{-\infty}^{\infty} f(x)|\psi|^2 dx$$

For a particle in a one-dimensional box extending from x = 0 to x = L, show that $\langle x^2 \rangle = \frac{L^2}{3} - \frac{L^2}{2n^2\pi^2}$.

Solution $\quad \langle x^2 \rangle = \int_{-\infty}^{\infty} x^2 |\Psi|^2 dx \qquad \Psi_n(x) = A \sin\left(\frac{n\pi x}{L}\right)$ where $A = \sqrt{\frac{2}{L}}$ is the normalization factor.

$$\langle x^2 \rangle = \left(\frac{2}{L}\right)\int_0^L x^2 \sin^2\left(\frac{n\pi x}{L}\right)dx = \frac{L^2}{3} - \frac{L^2}{2n^2\pi^2}$$

---

59. Particles incident from the left are confronted with a step potential as shown in Figure 41.25. The step has a height U_0, and the particles have energy $E > U_0$. Classically, all the particles would pass into the region of higher potential at the right. However, according to wave mechanics, one finds that a fraction of the particles are reflected at the barrier. The probability that a particle will be reflected, called the *reflection coefficient* R, is given by

$$R = \frac{(k_1 - k_2)^2}{(k_1 + k_2)^2}$$

where $k_1 = 2\pi/\lambda_1$ and $k_2 = 2\pi/\lambda_2$ are the wave numbers for the incident and transmitted particles, respectively. If $E = 2U_0$, what fraction of the incident particles are reflected? (This situation is analogous to the partial reflection and transmission of light striking an interface between two different media).

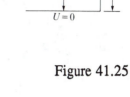

Incoming particles

Figure 41.25

Solution

$$R = \frac{(k_1 - k_2)^2}{(k_1 + k_2)^2} = \frac{(1 - k_2/k_1)^2}{(1 + k_2/k_1)^2}$$

$\dfrac{\hbar^2 k^2}{2m} = E - U$ for constant U. In this case, $\dfrac{\hbar^2 k_1^2}{2m} = E$ since U = 0 to the left of the barrier. (1)

To the right of the barrier, $\dfrac{\hbar^2 k_2^2}{2m} = E - U_0$ (2)

Dividing (2) by (1) gives

$$\frac{k_2^2}{k_1^2} = 1 - \frac{U_0}{E} = 1 - \frac{1}{2} = \frac{1}{2} \quad \text{so} \quad \frac{k_2}{k_1} = \frac{1}{\sqrt{2}}$$

and therefore

$$R = \frac{(1 - 1/\sqrt{2})^2}{(1 + 1/\sqrt{2})^2} = \left[\frac{(\sqrt{2} - 1)}{\sqrt{2} + 1)}\right]^2 = \boxed{0.0294}$$

61. An electron of momentum p is at a distance r from a stationary proton. The electron has a kinetic energy $K = p^2/2m$ and potential energy $U = -ke^2/r$. Its total energy is $E = K + U$. If the electron is bound to the proton to form a hydrogen atom, its average position is at the proton but the uncertainty in its position is approximately equal to the radius r of its orbit. The electron's average momentum will be zero, but the uncertainty in its momentum will be given by the uncertainty principle. Treat the atom as a one-dimensional system in the following: (a) Estimate the uncertainty in the electron's momentum in terms of r. (b) Estimate the electron's kinetic, potential, and total energies in terms of r. (c) The actual value of r is the one that *minimizes the total energy*, resulting in a stable atom. Find that value of r and the resulting total energy. Compare your answer with the predictions of the Bohr theory.

Solution

(a) $\Delta x \Delta p \cong \hbar$ so if $\Delta x = r$, $\Delta p \approx \dfrac{\hbar}{r}$

(b) $K = \dfrac{p^2}{2m} \approx \dfrac{(\Delta p)^2}{2m} = \dfrac{\hbar^2}{2mr^2}$

$U = -\dfrac{ke^2}{r}$

$E = \dfrac{\hbar^2}{2mr^2} - \dfrac{ke^2}{r}$

(c) To minimize E, $\dfrac{dE}{dr} = -\dfrac{\hbar^2}{mr^3} + \dfrac{ke^2}{r^2} = 0$

or

$$r = \dfrac{\hbar^2}{mke^2} = \text{Bohr radius}$$

Then

$$E = \dfrac{\hbar^2}{2m}\left(\dfrac{mke^2}{\hbar^2}\right) - ke^2\left(\dfrac{mke^2}{\hbar^2}\right) = -\dfrac{mk^2e^4}{2\hbar^2} = \boxed{-13.6\,\text{eV}}$$

This is the same value as that predicted by the Bohr theory.

CHAPTER 41

65. An electron is represented by the time-independent wave function

$$\psi(x) = \begin{cases} Ae^{-\alpha x} & \text{for } x > 0 \\ Ae^{+\alpha x} & \text{for } x < 0 \end{cases}$$

(a) Sketch the wave function as a function of x. (b) Sketch the probability that the electron is found between x and x + dx. (c) Why do you suppose this is a physically reasonable wave function? (d) Normalize the wave function. (e) Determine the probability of finding the electron somewhere in the range

$$x_1 = -\frac{1}{2\alpha} \qquad \text{to} \qquad x_2 = \frac{1}{2\alpha}$$

Solution

(a)

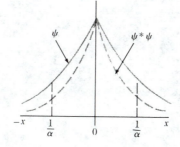

(b) $\quad P(x \to x + dx) = \Psi^*\Psi \, dx = A^2 e^{-2\alpha x} \, dx, \quad$ for $x > 0$

$$= A^2 e^{+2\alpha x} \, dx, \quad \text{for } x < 0$$

(c) $\quad \Psi$ is continuous; $\quad \Psi \to 0 \quad$ as $\quad x \to \infty; \quad$ mimics an electron bound at $x = 0 \ldots$.

(d) As Ψ is symmetric, $\quad \displaystyle\int_{-\infty}^{\infty} \Psi^*\Psi \, dx = 2 \int_{0}^{\infty} \Psi^*\Psi \, dx = 1$

or $\qquad 2A^2 \displaystyle\int_{0}^{\infty} e^{-2\alpha x} \, dx = 1 \qquad$ or $\qquad \left[\dfrac{2A^2}{(-2\alpha)}\right]\left[e^{-\infty} - e^0\right] = 1$

This gives $\quad A = \sqrt{\alpha}$

(e) $\quad P_{(-1/2\alpha) \to (1/2\alpha)} = 2 \displaystyle\int_{x=0}^{\alpha/2} (\sqrt{\alpha})^2 e^{-2\alpha x} \, dx = \left[\dfrac{2\alpha}{(-2\alpha)}\right]\left[e^{-2\alpha/2\alpha} - 1\right] = \left[1 - e^{-1}\right] = \boxed{0.632}$

Atomic Physics

OBJECTIVES

1. Understand the significance of the wave function and the associated radial probability density for the ground state of hydrogen.

2. For each of the quantum numbers, n, l (the orbital quantum number), m_l (the orbital magnetic quantum number), and m_s (the spin magnetic quantum number): (i) qualitatively describe what each implies concerning atomic structure, (ii) state the allowed values which may be assigned to each, and the number of allowed states which may exist in a particular atom corresponding to each quantum number.

3. Associate the customary shell and subshell spectroscopic notations with allowed combinations of quantum numbers n and l.

4. Calculate possible values of the orbital angular momentum, L, corresponding to a given value of the principle quantum number.

5. Describe how allowed values of the magnetic orbital quantum number, m_l, lead to a restriction on the orientation of the orbital angular momentum vector in an external magnetic field.

6. Find the allowed values for L_z (the component of the angular momentum along the direction of an external magnetic field) for given value of L.

7. Relate the intrinsic angular momentum (spin) of the electron to the observed splitting (fine structure) of atomic spectral lines.

8. State the Pauli exclusion principle and describe its relevance to the periodic table of the elements. Show how the exclusion principle leads to the known electronic ground state configuration of the light elements.

9. Describe the origin of the characteristic x-ray lines in terms of the shell structure of the atom. Calculate (approximately) the energy of an electron in the K, L, or M shell of an atom of known atomic number. Also, calculate the wavelength of an x-ray emitted as a result of transitions between these levels.

10. State the necessary conditions for laser action. Describe briefly the operation of a helium-neon gas laser in terms of energy level diagrams.

11. Describe the process of fluorescence and its application in a fluorescent light.

SKILLS

After reading the chapter in your text, review the significance and set of allowed values for each of the quantum numbers used to describe the various electronic states of electrons in an atom.

In addition to the principal quantum n (n can range from 1 to ∞), other quantum numbers are necessary to completely and accurately specify the possible energy levels in the hydrogen atom and also in more complex atoms.

All energy states with the same principal quantum number, n, form a shell. These shells are identified by the spectroscopic notation K, L, M, . . . corresponding to n = 1, 2, 3, ...

The orbital quantum number l (l can range from 0 to (n – 1), determines the allowed value of orbital angular momentum. All energy states having the same values of n and l form a subshell. The letter designations s, p, d, f, . . . correspond to values of l = 1, 2, 3, 4, ...

The magnetic orbital quantum number m_l (m_l can range from $-l$ to l), determines the possible orientations of the electron's orbital angular momentum vector in the presence of an external magnetic field.

The spin magnetic quantum number, m_s, can have only two values ($m_s = \pm\frac{1}{2}$) which in turn correspond to the two possible directions of the electron's intrinsic spin.

NOTES FROM SELECTED CHAPTER SECTIONS

It is important to understand the behavior of the hydrogen atom as an atomic system for the following reasons:

1. Much of what is learned about the hydrogen atom with its single electron can be extended to such single-electron ions as He^+ and Li^{2+}, which are hydrogen-like in their atomic structure.

2. The hydrogen atom is an ideal system for performing precise tests of theory against experiment and for improving our overall understanding of atomic structure.

3. The quantum numbers used to characterize the allowed states of hydrogen can be used to describe the allowed states of more complex atoms. This enables us to understand the periodic table of the elements, which is one of the greatest triumphs of quantum mechanics.

4. The basic ideas about atomic structure must be well understood before we attempt to deal with the complexities of molecular structures and the electronic structure of solids.

42.1 EARLY MODELS OF THE ATOM

One of the first indications that there was a need for modification of the Bohr theory became apparent when improved spectroscopic techniques were used to examine the spectral lines of hydrogen. It was found that many of the lines in the Balmer and other series were not single lines at all. Instead, each line was actually a group of lines spaced very close together. An additional difficulty arose when it was observed that, in some situations, certain single spectral lines were split into three closely spaced lines when the atoms were placed in a strong magnetic field.

42.2 THE HYDROGEN ATOM

In the three-dimensional problem of the hydrogen atom, three quantum numbers are required for each stationary state, corresponding to the three independent degrees of freedom for the electron.

The three quantum numbers which emerge from the theory are represented by the symbols n, l, and m_l. The quantum number n is called the *principal quantum number, l* is called the *orbital quantum number*, and m_l is called the *orbital magnetic quantum number*.

There are certain important relationships between these quantum numbers, as well as certain restrictions on their values. These restrictions are:

> The values of n can range from 1 to ∞.
> The values of l can range from 0 to n − 1.
> The values of m_l can range from − l to l.

For historical reasons, *all states with the same principal quantum number are said to form a shell*. These shells are identified by the letters K, L, M, . . . , which designate the states for which n = 1, 2, 3, ... Likewise, *the states having the same values of* n *and* l *are said to form a subshell*. The letters s, p, d, f, g, h, . . . are used to designate the states for which l = 0, 1, 2, 3, ...

42.3 THE SPIN MAGNETIC QUANTUM NUMBER

The *spin magnetic quantum number* m_s accounts for the two closely spaced energy states corresponding to the two possible orientations of electron spin.

42.4 THE WAVE FUNCTIONS FOR HYDROGEN

Since the potential energy of the electron in the hydrogen atom depends only on the radial distance r, the wave functions that describe the s states are *spherically symmetric.*

42.5 THE QUANTUM NUMBERS

The allowed energy depends only on the *principal quantum number* n. The allowed wave functions depend on three quantum numbers, n, l, and m_l, where l is the *orbital quantum number* and m_l is the *orbital magnetic quantum number.*

All states with the same principal quantum number n form a *shell*, identified by the letters K, L, M, . . . (corresponding to n = 1, 2, 3, . . .). All states with the same values of both n and l form a *subshell,* designated by the letters s, p, d, f, . . . (corresponding to l = 0, 1, 2, 3, . . .).

In order to completely describe a quantum state of the hydrogen atom, it is necessary to include a fourth quantum number, m_s, called the *spin magnetic quantum number.* This quantum number can have only two values, $\pm \frac{1}{2}$. In effect, this doubles the number of allowed states specified by the quantum numbers n, l, and m_l.

An atom characterized by a specific value of n can have values of orbital angular momentum L determined by the value of the orbital quantum number, l. *Space quantization,* which restricts the projection of L along the z-axis to describe values, is determined by the orbital magnetic quantum number.

42.6 ELECTRON SPIN

Fine structure refers to the splitting of spectral lines into two or more components of closely spaced lines. This is due to an intrinsic angular momentum of the electron called *electron spin.* Therefore, the total angular momentum of an electron in a particular state is the vector sum of the orbital and angular contributions.

42.7 THE EXCLUSION PRINCIPLE AND THE PERIODIC TABLE

The *exclusion principle* states that no two electrons can exist in identical quantum states. This means that no two electrons in a given atom can be characterized by the same set of quantum numbers at the same time.

Hund's rule states that when an atom has orbitals of equal energy, the order in which they are filled by electrons is such that a maximum number of electrons will have unpaired spins.

42.8 ATOMIC SPECTRA: VISIBLE AND X-RAY

An atom will emit electromagnetic radiation if an electron in an excited state makes a transition to a lower energy state. The set of wavelengths observed for a specific species by such processes is called an *emission spectrum.* Likewise, atoms with electrons in the ground state configuration can also absorb electromagnetic radiation at specific wavelengths, giving rise to an *absorption spectrum.* Such spectra can be used to identify the elements in the gases.

Since the orbital angular momentum of an atom changes when a photon is emitted or absorbed (that is, as a result of a transition) and since angular momentum must be conserved, we conclude that *the photon involved in the process must carry angular momentum.*

X-rays are emitted by atoms when an electron undergoes a transition from an outer shell into an electron vacancy in one of the inner shells. Transitions into a vacant state in the K shell give rise to the K series of spectral lines, transitions into a vacant state in the L shell create the L series of lines, and so on. The x-ray spectrum of a metal target consists of a set of sharp characteristic lines superimposed on a broad, continuous spectrum.

42.9 ATOMIC TRANSITIONS

Stimulated absorption occurs when light with photons of an energy which matches the energy separation between two atomic energy levels is absorbed by an atom.

An atom in an excited state has a certain probability of returning to its original energy state. This process is called *spontaneous emission.*

When a photon with an energy equal to the excitation energy of an excited atom is incident on the atom, it can increase the probability of de-excitation. This is called *stimulated emission* and results in a second photon of energy equal to that of the incident photon.

42.10 LASERS AND HOLOGRAPHY

The following three conditions must be satisfied in order to achieve laser action:

1. The system must be in a state of population inversion (that is, more atoms in an excited state than in the ground state).

2. The excited state of the system must be a *metastable state*, which means its lifetime must be long compared with the usually short lifetimes of excited states. When such is the case, stimulated emission will occur before spontaneous emission.

3. The emitted photons must be confined in the system long enough to allow them to stimulate further emission from other excited atoms. This is achieved by the use of reflecting mirrors at the ends of the system. One end is made totally reflecting, and the other is slightly transparent to allow the laser beam to escape.

EQUATIONS AND CONCEPTS

The potential energy of the hydrogen atom depends on the radius of the allowed orbit of the electron.

$$U(r) = -\frac{ke^2}{r} \qquad (42.1)$$

When numerical values of the constants are used, the energy level values can be expressed in units of electron volts (eV).

$$E_n = -\frac{13.6}{n^2} \text{ eV} \qquad (42.2)$$

The lowest energy state or ground state corresponds to the principle quantum number $n = 1$. The energy level approaches $E = 0$ as n approaches infinity. This is the ionization energy for the atom.

The simplest wave function for hydrogen is the one which describes the 1s state and depends only on the radial distance r.

$$\psi_{1s}(r) = \frac{1}{\sqrt{\pi a_o^3}} e^{-r/a_o}$$

(42.3)

The electron can exist only in certain allowed orbits, whose value can be expressed in terms of the Bohr radius, a_o.

$$a_o = \frac{\hbar^2}{mke^2} = 0.529 \text{ Å}$$

(42.4)

The radial probability density for the 1s state of hydrogen is defined as the probability of finding the electron in a spherical shell of radius r and thickness dr.

$$P_{1s}(r) = \left(\frac{4r^2}{a_o^3}\right) e^{-2r/a_o}$$

(42.7)

The value of the quantum number l determines the magnitude of the electron's orbital angular momentum L.

$$L = \sqrt{l(l+1)}\,\hbar$$

$(l = 0, 1, 2, \ldots, n-1)$

(42.10)

When an atom is placed in an external magnetic field, the projection (or component) of the orbital angular momentum L_z along the direction of the magnetic field is quantized.

$$L_z = m_l\,\hbar$$

$(m_l = -l, -l+1, \ldots 0, 1, 2$
$\ldots l-1, l)$

(42.11)

If an orbiting electron is placed in a weak magnetic field directed along the z-axis, the projection of the angular momentum vector of the electron L along the z-axis can have only discreet values.

$$\cos\theta = \frac{m_l}{\sqrt{l(l+1)}}$$

(42.12)

In addition to orbital angular momentum, the electron has an intrinsic angular momentum or spin angular momentum S due to spinning on its axis. This spin angular momentum is described by a single quantum number s whose value can only be $\frac{1}{2}$.

$$S = \sqrt{s(s+1)}\,\hbar$$

$$S = \frac{\sqrt{3}}{2}\,\hbar$$

(42.13)

The spin angular momentum is space quantized with respect to the direction of an external magnetic field (along the z- direction). The spin magnetic quantum number m_s can have values of $\pm\frac{1}{2}$.

$$S_z = m_s\hbar = \pm\frac{1}{2}\hbar$$

(42.14)

The spin magnetic moment μ_s of the electron is related to the spin angular momentum.

$$\mu_s = \left(-\frac{e}{m}\right)\mathbf{S}$$ (42.15)

Electron transitions between stationary states of an atom are governed by selection rules. The allowed transitions are those for which the value of the magnetic orbital quantum number changes by 1.

$$\Delta l = \pm 1$$ (42.17)

The shielding effect of the nuclear charge by inner-core electrons must be taken into account when calculating the allowed energy levels of multielectron atoms. The atomic number is replaced by an effective atomic number, Z_{eff}, which depends on the values of n and l. For K-shell electrons, $Z_{eff} \rightarrow (Z - 1)$; for M-shell electrons, $Z_{eff} \rightarrow (Z - 9)$.

$$E_n = -\frac{13.6 Z^2_{eff}}{n^2} \text{ eV}$$ (42.18)

ANSWERS TO SELECTED QUESTIONS

1. Why are three quantum numbers needed to describe the state of a one-electron atom (neglecting spin)?

Answer: Three quantum numbers are needed because there are three degrees of freedoms associated with the system.

3. Why is the direction of the orbital angular momentum of an electron opposite that of its magnetic moment?

Answer: The magnetic moment of a circulating charge q is proportional to the orbital angular momentum, according to the relation $\mu = q\mathbf{L}/2m$. Because the charge on the electron is negative (q = – e), the vectors μ and $\mathbf{L}$ point in *opposite* directions.

4. Why is an inhomogeneous magnetic field used in the Stern-Gerlach experiment?

Answer: Atoms having oppositely directed dipoles in a nonuniform field experience net forces in opposite directions which causes a displacement in opposite directions as they are passed through the magnetic field. More formally, the deflecting force on a neutral atom having a magnetic moment is proportional to the gradient of the magnetic field (that is, **B** changes along one or more directions). In a uniform **B** field, the gradient is zero in any direction, so the deflecting force is zero.

5. Could the Stern-Gerlach experiment be performed with ions rather than neutral atoms? Explain.

Answer: Practically speaking, the answer would be no. Since ions have a net charge, the magnetic force $q\mathbf{v} \times \mathbf{B}$ would deflect the beam, making it very difficult to separate ions with different orientations of magnetic moments.

6. Describe some experiments that would support the conclusion that the spin quantum number for electrons can only have the values $\pm\frac{1}{2}$.

Answer: The Stern-Gerlach experiment using hydrogen atoms (one unpaired electron) verifies that the electron spin equals $\frac{1}{2}$. Electron spin resonance experiments on atoms of ions with one unpaired electron also support this conclusion.

7. Discuss some of the consequences of the exclusion principle.

Answer: If this principle were not valid, the elements and their chemical behavior would be grossly different because every electron would end up in the lowest energy level of the atom. All matter would be nearly alike, and most materials would have a much higher density. Furthermore, the spectra of atoms and molecules would be very simple, and there would be very little color in the world.

8. Why do lithium, potassium, and sodium exhibit similar chemical properties?

Answer: The three elements have similar electronic configurations, with filled inner shells, plus an electron is an s orbital. Li $[...2s^1]$, Na $[...3s^1]$, K $[...4s^1]$.

9. From Table 42-4, we find that the ionization energies for Li, Na, K, Rb, and Cs are 5.390, 5.138, 4.339, 4.176, and 3.893 eV, respectively. Explain why these values are to be expected in terms of the atomic structures.

Answer: All elements have their valence electron in an s state. One expects the ionization energy to decrease with increasing atomic number because the inner core electrons (filled shells) provide more shielding between the nucleus and valance electron for the species with higher atomic numbers.

11. Explain why a photon must have a spin of 1.

Answer: When a photon is emitted or absorbed by an atom, the orbital angular momentum of the atom changes by one unit. Because momentum must be conserved, the photon carries with it an angular momentum equivalent to that of a particle with a spin of one unit.

12. An energy of about 21 eV is required to excite an electron in a helium atom from the 1s state to the 2s state. The same transition for the He$^+$ ion requires about twice as much energy. Explain why this is so.

Answer: In the case of the helium atom, the electron to be excited is better shielded from the positively charged nucleus by the other 1s electron, so it requires less energy to cause it to move from the 1s to the 2s state.

14. The absorption or emission spectrum of a gas consists of lines which broaden as the density of gas molecules increases. Why do you suppose this occurs?

Answer: As the density increases, the collision frequency increases, which results in a wider spread of energy levels, and a corresponding broadening in the spectral lines.

7. The wave function for an electron in a 2p-state
in hydrogen is

$$\psi_{2p} = \frac{1}{\sqrt{3}\,(2a_0)^{3/2}} \frac{r}{a_0}\, e^{-r/2a_0}$$

What is the most likely distance from the
H-nucleus to find an electron in the 2p state?
(See Figure 42.8.)

Solution

The probability density function is given by

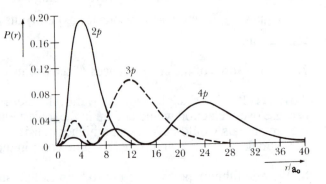

Figure 42.8

$$P = 4\pi r^2 |\psi|^2 = 4\pi r^2 \left(\frac{r^2}{24 a_0^5}\right) e^{-r/a_0}$$

The most likely position for the electron corresponds to the point at which $\frac{dP}{dr} = 0$ (where P reaches its *maximum* value).

$$\frac{dP}{dr} = \frac{4\pi}{24 a_0^5}\left[4r^3 e^{-r/a_0} + r^4\left(-\frac{1}{a_0}\right) e^{-r/a_0}\right] = 0$$

Solving for r gives

$$r = \boxed{4a_0}$$

---

8. Show that the 1s wave function for an electron in hydrogen

$$\psi(r) = \frac{1}{\sqrt{\pi a_0^3}}\, e^{-r/a_0}$$

satisfies the radially symmetric Schrödinger equation

$$-\frac{\hbar^2}{2m}\left(\frac{d^2\psi}{dr^2} + \frac{2}{r}\frac{d\psi}{dr}\right) - \frac{ke^2}{r}\,\psi = E\psi$$

Solution

$$\frac{d\psi}{dr} = \frac{1}{\sqrt{\pi a_0^3}}\frac{d}{dr}\left(e^{-r/a_0}\right) = -\frac{1}{\sqrt{\pi a_0^3}}\left(\frac{1}{a_0}\right) e^{-r/a_0}$$

Hence,

$$\frac{2}{r}\frac{d\psi}{dr} = -\frac{2}{\sqrt{\pi a_0^5}}\frac{1}{r}e^{-r/a_0} = -\frac{2}{a_0}\frac{\psi}{r} \qquad (1)$$

Likewise,

$$\frac{d^2\psi}{dr^2} = -\frac{1}{\sqrt{\pi a_0^5}}\frac{d}{dr}e^{-r/a_0} = \frac{1}{\sqrt{\pi a_0^7}}e^{-r/a_0} = \frac{1}{a_0^2}\psi \qquad (2)$$

Substituting (1) and (2) into the Schrödinger equation gives

$$-\frac{\hbar^2}{2m}\left(\frac{1}{a_0^2} - \frac{2}{a_0 r}\right)\psi - \frac{ke^2}{r}\psi = E\psi \qquad (3)$$

From Equation 42.4, we have $a_0 = \dfrac{\hbar^2}{mke^2}$ which, when substituted into (3) (where the second and third terms *cancel* each other), gives the ground state energy of hydrogen:

$$E = -\frac{ke^2}{2a_0}$$

---

29. A tungsten target is struck by electrons that have been accelerated from rest through a 40-kV potential difference. Find the shortest wavelength of the bremsstrahlung radiation emitted.

Solution

Photons produced by the bombarding electrons can have a maximum energy of 40 keV, which corresponds to radiation with the *shortest* wavelength (since $E = hc/\lambda$). Thus,

$$E_{max} = \frac{hc}{\lambda_{min}} = 40\text{ keV} = 4 \times 10^4\text{ eV}$$

$$\lambda_{min} = \frac{hc}{E_{max}} = \frac{1240\text{ eV·nm}}{4 \times 10^4\text{ eV}} = \boxed{0.0310\text{ nm}}$$

---

31. Find the wavelength of the K_α x-ray line that is emitted when electrons strike an iron target. Note that since the innermost shells are involved, Z_{eff} is approximately $Z - 1$.

Solution

The energy of a multielectron atom is approximated by Equation 42.18:

$$E_n = -\frac{13.6\,Z^2_{eff}}{n^2}\text{ eV}$$

The K_α x-ray corresponds to the transition $n = 2$ to $n = 1$, thus,

$$E_{K_\alpha} = E_2 - E_1 = 13.6 \, Z^2_{eff}\left(\frac{1}{1^2} - \frac{1}{2^2}\right) = 10.2 \, Z^2_{eff} \text{ eV}$$

Taking $Z_{eff} = Z - 1 = 26 - 1 = 25$, and noting that $\Delta E = \frac{hc}{\lambda}$ gives

$$\frac{hc}{\lambda} = 10.2 \, (25)^2 \text{ eV} = 6375 \text{ eV}$$

$$\lambda = \frac{hc}{6375 \text{ eV}} = \frac{1240 \text{ eV·nm}}{6375 \text{ eV}} = \boxed{0.195 \text{ nm}}$$

39. A dimensionless number which often appears in atomic physics is the *fine-structure constant* α, given by

$$\alpha = \frac{ke^2}{\hbar c}$$

where k is the Coulomb constant. (a) Obtain a numerical value for $1/\alpha$. (b) In scattering experiments, the "size" of the electron is the *classical electron radius*, $r_e = ke^2/mc^2$. In terms of α, what is the ratio of the Compton wavelength (Section 40.3), $\lambda_c = h/mc$, to the classical electron radius? (c) In terms of α, what is the ratio of the Bohr radius, a_0, to the Compton wavelength? (d) In terms of α, what is the ratio of the *Rydberg wavelength*, $1/R_H$, to the Bohr radius (Section 40.5)?

Solution

(a) $\dfrac{1}{\alpha} = \dfrac{\hbar c}{ke^2} = \dfrac{(1.05457 \times 10^{-34} \text{ J·s})(2.997925 \times 10^8 \text{ m/s})}{(8.9875 \times 10^9 \text{ N·m}^2/c^2)(1.60219 \times 10^{-19} \text{ C})^2} = \boxed{137.034}$

(b) $\dfrac{\lambda_c}{r_e} = \dfrac{h/mc}{ke^2/mc^2} = \dfrac{hc}{ke^2} = \boxed{\dfrac{2\pi}{\alpha}}$

(c) $\dfrac{a_0}{\lambda_c} = \dfrac{\hbar^2/mke^2}{h/mc} = \left(\dfrac{1}{2\pi}\right)\dfrac{\hbar c}{ke^2} = \boxed{\dfrac{1}{2\pi\alpha}}$

(d) $\dfrac{1}{R_H a_0} = \dfrac{4\pi c\hbar^3}{mk^2e^4}\left(\dfrac{mke^2}{\hbar^2}\right) = 4\pi\left(\dfrac{\hbar c}{ke^2}\right) = \boxed{\dfrac{4\pi}{\alpha}}$

40. Show that the average value of r for the 1s state of hydrogen has the value $3a_0/2$. (Hint: Use. Eq. 42.7.)

Solution

The average value (expectation value) of r is

$$<r> = \int_0^\infty r P_{1s}(r)\,dr$$

where

$$P_{1s}(r) = \left(\frac{4r^2}{a_0^3}\right)e^{-2r/a_0}$$

$$<r> = \frac{4}{a_0^3}\int_0^\infty r^3 e^{-2r/a_0}\,dr$$

Letting $x = \frac{2r}{a_0}$, we find

$$<r> = \frac{a_0}{4}\int_0^\infty x^3 e^{-x}\,dx$$

Integrating by parts (or using a table of integrals) gives

$$<r> = \boxed{\frac{3}{2}\,a_0}$$

---

41. Suppose that a hydrogen atom is in the 2s state. Taking $r = a_0$, calculate values for (a) $\psi_{2s}(a_0)$, (b) $|\psi_{2s}(a_0)|^2$, and (c) $P_{2s}(a_0)$. (Hint: Use Eq. 42.8.)

Solution

The wave function for the 2s state is given by Equation 42.8:

$$\psi_{2s}(r) = \frac{1}{4\sqrt{2\pi}}\left(\frac{1}{a_0}\right)^{3/2}\left(2 - \frac{r}{a_0}\right)e^{-r/2a_0}$$

(a) Taking $r = a_0 = 0.529 \times 10^{-10}$ m, we find

$$\psi_{2s}(a_0) = \frac{1}{4\sqrt{2\pi}}\left(\frac{1}{0.529 \times 10^{-10}\text{ m}}\right)^{3/2}(2-1)e^{-1/2} = \boxed{1.57 \times 10^{14}\text{ m}^{-3/2}}$$

(b) $|\psi_{2s}(a_0)|^2 = (1.57 \times 10^{14}\text{ m}^{-3/2})^2 = \boxed{2.47 \times 10^{28}\text{ m}^{-3}}$

(c)   Using Equation 42.6, and the results to (c) gives

$$P_{2s}(a_o) = 4\pi a_o^2 |\psi_{2s}(a_o)|^2 = \boxed{8.69 \times 10^8 \text{ m}^{-1}}$$

47. Consider a hydrogen atom in its ground state. (a) Treating the electron as a current loop of radius r_0, derive an expression for the magnetic field at the nucleus. (Hint: Use the Bohr theory of hydrogen and see Example 30.3.) (b) Find a numerical value for the magnetic field in this situation.

Solution

(a)   The magnetic field at the center of a circular current loop of radius r is $B = \mu_o I/2r$ (see Example 30.3). An electron moving in a circular orbit with frequency f corresponds to an *equivalent* current I given by

$$I = ef = \frac{e\omega}{2\pi}$$

From the Bohr theory, $mvr = n\hbar$, hence

$$\omega = \frac{v}{r} = \frac{n\hbar}{mr^2}$$

For the ground state, $n = 1$,   and $r = a_o$,   so

$$\omega = \frac{\hbar}{ma_o^2} \qquad \text{and} \qquad I = \frac{e\hbar}{2\pi ma_o^2}$$

Thus,

$$B = \frac{\mu_o I}{2a_o} = \frac{\mu_o e\hbar}{4\pi ma_o^3}$$

(b)   Substituting numerical values into the above expression gives

$$B = \frac{(4\pi \times 10^{-7} \text{ N/A}^2)(1.60 \times 10^{-19} \text{ C})(1.055 \times 10^{-34} \text{ J·s})}{4\pi(9.11 \times 10^{-31} \text{ kg})(0.529 \times 10^{-10} \text{ m})^3} = \boxed{12.5 \text{ T}}$$

52. According to classical physics, an accelerated charge e radiates at a rate

$$\frac{dE}{dt} = -\frac{1}{6\pi\varepsilon_o} \frac{e^2 a^2}{c^3}$$

(a) Show that an electron in a classical hydrogen atom (see Figure 42.3) will spiral into the nucleus at a rate

$$\frac{dr}{dt} = -\frac{e^4}{12\pi^2 \varepsilon_0^2 r^2 m^2 c^3}$$

(b) Find the time for the electron to reach r = 0, starting from $r_o = 2 \times 10^{-10}$ m.

Solution

(a) According to a classical model, the electron moving in a circular orbit about the proton in the hydrogen atom experiences a force ke^2/r^2, and from Newton's second law $F = ma$, its acceleration is ke^2/mr^2.

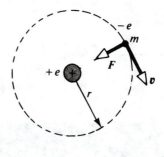

$$a = \frac{v^2}{r} = \frac{ke^2}{mr^2} = \frac{e^2}{4\pi\varepsilon_o mr^2} \tag{1}$$

We have used the fact that the Coulomb constant $k = \dfrac{1}{4\pi\varepsilon_o}$. From the Bohr model of the atom (Section 40.5), we can write the *total* energy of the atom as

$$E = -\frac{ke^2}{2r} = -\frac{e^2}{8\pi\varepsilon_o r}$$

$$\frac{dE}{dt} = \frac{e^2}{8\pi\varepsilon_o r^2} \frac{dr}{dt} = -\frac{1}{6\pi\varepsilon_o} \frac{e^2 a^2}{c^3} \tag{2}$$

Substituting (1) into (2) for a, and solving for $\dfrac{dr}{dt}$, and simplifying gives

$$\frac{dr}{dt} = -\frac{4r^2}{3c^3}\left(\frac{e^2}{4\pi\varepsilon_o mr^2}\right)^2 = -\frac{e^4}{12\pi^2\varepsilon_o^2 r^2 m^2 c^3}$$

(b) We can express $\dfrac{dr}{dt}$ in the simpler form,

$$\frac{dr}{dt} = -\frac{A}{r^2} = -\frac{3.15 \times 10^{-21}}{r^2}$$

$$-\int_{2 \times 10^{-10}\,m}^{0} r^2\,dr = 3.15 \times 10^{-21} \int_0^T dt$$

$$T = \left(3.17 \times 10^{20}\right)\frac{r^3}{3}\Bigg]_o^{2 \times 10^{-10}\,m} = \boxed{8.46 \times 10^{-10}\ s = 0.846\ ns}$$

We know that atoms 'last' much longer than 0.8 ns; thus, classical physics does not hold (fortunately) for atomic systems.

Molecules and Solids

OBJECTIVES

1. Understand the essential binding mechanisms involved in ionic, covalent, hydrogen, and Van der Waals bonding.

2. Describe in terms of appropriate quantum numbers the allowed energy levels associated with rotational and vibrational motions of molecules. Use the selection rules to determine the separation between adjacent energy levels.

3. Describe qualitatively the various bonding schemes for solids: ionic solids, convalent crystals, and metallic solids.

4. Discuss the free-electron theory of metals including the significance of the Fermi-Dirac distribution function, Fermi energy, and particle distribution by energy interval.

5. Use the band theory of solids as a basis for a qualitative discussion of the mechanisms for conduction in metals, insulators, and semiconductors.

6. Describe a p-n junction and the diffusion of electrons and holes through the junction. Discuss the fabrication and function of a junction diode and junction transistor.

NOTES FROM SELECTED CHAPTER SECTIONS

43.1 MOLECULAR BONDS

Two atoms combine to form a molecule because of a net attractive force between them when their separation is greater than their equilibrium separation in the molecule. Furthermore, the total energy of the stable bound molecule is *less* than the total energy of the separated atoms.

The potential energy for large atomic separations is negative, corresponding to a net attractive force. At the equilibrium separation, the attractive and repulsive forces just balance and the potential energy has its minimum value.

Ionic bonds are due to the Coulomb attraction between oppositely charged ions.

The energy released when an atom takes on an electron is called the *electron affinity*.

The *dissociation energy* is the energy required to separate a molecule into neutral atoms.

A *covalent bond* between two atoms results from a sharing of electrons from one or both atoms that combine to form the molecule.

In some cases, hydrogen forms a weak chemical bond called a *hydrogen bond*. Parts of negative ions are bound by the presence of the positive hydrogen ion between them.

The weak electrostatic attractions between molecules are called *Van der Waals forces*. There are three types of Van der Waals forces:

The first type, called the *dipole-dipole force*, is an interaction between two molecules each having a permanent electric dipole moment. In effect, one molecule interacts with the electric field produced by another molecule.

The second type of Van der Waals force is a *dipole-induced force* in which a polar molecule having a permanent electric dipole moment *induces* a dipole moment in a nonpolar molecule.

The third type of Van der Waals force is called the *dispersion force*. The dispersion force is an attractive force that occurs between two nonpolar molecules. In this case, the interaction results from the fact that although the average dipole moment of a nonpolar molecule is zero, the average of the square of the dipole moment is nonzero because of charge fluctuations. Consequently, two nonpolar molecules near each other tend to be correlated so as to produce an attractive force, which is the Van der Waals force.

43.2 THE ENERGY AND SPECTRA OF MOLECULES

The energy of a molecule in the gaseous phase can be divided into four categories: (1) electronic energy, due to the mutual interactions of the molecule's electrons and nuclei; (2) translational energy, due to the motion of the molecule's center of mass through space; (3) rotational energy, due to the rotation of the molecule about its center of mass; and (4) vibrational energy, due to the vibration of the molecule's constituent atoms.

The *rotational energy* of a molecule is quantized and depends on the moment of inertia of the molecule. The energy differences between adjacent rotational levels are in the *microwave range* of frequencies.

The *vibrational energy* of a molecule is also quantized. Energies associated with transitions between vibrational levels are in the infrared region.

43.3 BONDING IN SOLIDS

Ionic crystals have the following general properties:
1. They form relatively stable and hard crystals.
2. They are poor electrical conductors because there are no available free electrons.
3. They have high vaporization temperatures.
4. They are transparent to visible radiation but absorb strongly in the infrared region.
5. They are usually quite soluble in polar liquids such as water. The water molecule, which has a permanent electric dipole moment, exerts an attractive force on the charged ions, which breaks the ionic bonds and dissolves the solid.

The *covalent bond* is very strong and comparable to the ionic bond. In general, covalently bonded solids are very hard, have large bond energies, high melting points, are good insulators, and are transparent to visible light.

Metallic bonds are generally weaker than ionic or covalent bonds. The valence electrons in a metal are relatively free to move throughout the material. There are a large number of such mobile electrons in a metal, typically one or two electrons per atom. The metal structure can be viewed as a "sea" or "gas" of nearly free electrons surrounded by a lattice of positive ions. (Fig. 43.1). The binding mechanism in a metal is the attractive force between the positive ions and the electron gas.

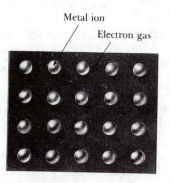

Metal ion
Electron gas

Figure 43.1

43.4 BAND THEORY OF SOLIDS

One can best understand the properties of metals, insulators, and semiconductors in terms of the *band theory of solids*. The valence band of a metal such as sodium is half-filled. Therefore, there are many electrons free to move throughout the metal and contribute to the conduction current. In an insulator at T = 0 K, the valence band is completely filled with electrons, while the conduction band is empty. The region between the valence band and the conduction band is called the *energy gap* of the material.

In the free-electron theory of metals, the valence electrons fill the quantized levels in accordance with the Pauli exclusion principle. Only those electrons having energies near the *Fermi energy* can contribute to the electrical conductivity of the metal.

43.6 CONDUCTION IN METALS, INSULATORS, AND SEMICONDUCTORS

If an electric field is applied to metal, electrons with energies near the Fermi energy require only a small amount of additional energy from the field to reach nearby empty energy states. Thus electrons are free to move with only a small applied field in a metal because there are many unoccupied states close to occupied energy states.

Although an insulator has many vacant states in the conduction band that can accept electrons, there are so few electrons actually occupying conduction band states that the overall contribution to electrical conductivity is very small, resulting in a high resistivity for insulators.

Semiconductors have a small energy gap. At T = O K, all electrons in a pure semiconductor are in the valence band and there are no electrons in the conduction band. Thus, semiconductors are *poor* conductors at low temperatures. Thermal excitation across the narrow gap is more probable at higher temperature and the conductivity of semiconductors increases rapidly with temperature.

43.7 SEMICONDUCTOR DEVICES

The band structures and electrical properties of a semiconductor can be modified by adding donor atoms with five valence electrons (such as arsenic), or by adding acceptor atoms with three valence electrons (such as indium). A semiconductor *doped* with donor impurity atoms is called an *n-type semiconductor*, while one doped with acceptor impurity atoms is called p-type.

The junction transistor consists of a semiconducting material with a very narrow n region sandwiched between two p regions. This configuration is called the *pnp transistor*. Another configuration is the npn transistor, which consists of a p region sandwiched between two n regions.

EQUATIONS AND CONCEPTS

The rotational energy of a molecule is quantized and depends on the value of the moment of inertia.

$$E_{rot} = \frac{\hbar^2}{2I} J (J + 1)$$ (43.5)

$$J = 0, 1, 2, 3, \ldots$$

For a diatomic molecule, the moment of inertia I can be written in terms of the reduced mass, μ.

$$I = \mu r^2$$ (43.3)

$$\mu = \left(\frac{m_1 m_2}{m_1 + m_2} \right)$$ (43.8)

The vibrational energy for a diatomic molecule is quantized and is characterized by the vibrational quantum number, v. The selection rule for allowed vibrational transitions is given by $\Delta v \pm 1$. The energy difference between successive vibrational levels is hf when f is the frequency of vibration.

$$E_{vib} = \left(v + \frac{1}{2}\right) \frac{h}{2\pi} \sqrt{\frac{K}{\mu}}$$

$$v = 0, 1, 2, \ldots$$

(43.10)

$$\Delta E_{vib} = hf$$

(43.11)

The ionic cohesive energy U_0 of a solid represents the energy necessary to separate the solid into a collection of positive and negative ions. In this expression, r_0 is the equilibrium ion separation and α, the Madelung constant, has a value which is characteristic of a specific crystal structure. The parameter m is a small integer.

$$U_o = -\alpha k \frac{e^2}{r_0} \left(1 - \frac{1}{m}\right)$$

(43.16)

The Fermi-Dirac distribution function f(E) gives the probability of finding an electron in a particular energy state, E.

$$f(E) = \frac{1}{e^{(E - E_F)/kT} + 1}$$

(43.17)

In the Fermi-Dirac distribution function, E_F is called the Fermi energy and is a function of the total number of electrons per unit volume, n.

$$E_F = \frac{h^2}{2m} \left(\frac{3n}{8\pi}\right)^{2/3}$$

(43.24)

The Fermi temperature T_F is defined in terms of the Fermi energy.

$$T_F = \frac{E_F}{k}$$

(43.26)

In thermal equilibrium, the number of electrons per unit volume with energy between E and E + dE is the product of the probability of finding an electron in a particular state and the density of states.

$$N(E)dE = C \frac{E^{1/2} dE}{e^{(E - E_F)/kT} + 1}$$

where

$$C = \frac{8\sqrt{2}\pi m^{3/2}}{h^3}$$

(43.21)

EXAMPLE PROBLEM SOLUTION

Example 43.1 The maximum amplitude of vibration for the HF molecule is 9.24×10^{-12} m. Calculate (a) the effective force constant for this molecule and (b) the frequency corresponding to the transition from $v = 0$ to $v = 1$.

Solution

(a)   The effective force constant can be calculated using Equation 43.10.

$$\frac{1}{2} KA^2 = \frac{h}{2\pi} \sqrt{\frac{K}{\mu}} \qquad \text{or} \qquad K = \frac{h^2}{4\pi^2 A^4 \mu}$$

In this case, $\mu = \frac{(1)(19)}{1 + 19} u (1.66 \times 10^{-27} \text{ kg/u})$

$$\mu = 1.58 \times 10^{-27} \text{ kg}$$

so

$$K = \frac{(6.625 \times 10^{-34} \text{ J·s})^2}{4\pi^2 (9.24 \times 10^{-12} \text{ m})^4 (1.58 \times 10^{-27} \text{ kg})} = 965 \text{ N/m}$$

(b)   The value of K found in part (a) can be used in Equation 43.11 to determine the fundamental frequency.

$$\frac{h}{2\pi} \sqrt{\frac{K}{\mu}} = hf \qquad \text{or} \qquad f = \frac{1}{2\pi} \sqrt{\frac{K}{\mu}}$$

so

$$f = \frac{1}{2\pi} \sqrt{\frac{965 \text{ N/m}}{1.57 \times 10^{-27} \text{ kg}}} = 1.25 \times 10^{14} \text{ Hz}$$

ANSWERS TO SELECTED QUESTIONS

1. Discuss the three major forms of excitation of a molecule (other than the translational motion) and the relative energies associated with the three excitations.

Answer: The three major forms of excitation of a molecule in order of decreasing energy are electronic, vibrational, and rotational. The frequencies associated with these excitations are of the order of 10^{15} Hz for electronic excitation, 10^{13} Hz for vibrational, and 10^{11} Hz for rotational.

2. Explain the role of the Pauli exclusion principle in describing the electrical properties of metals.

Answer: The metal is a system with a very large number of states available to the valence electrons. The electrons must fill these states in accordance with the Pauli principle, with energies ranging from $E = 0$ to $E = E_F$ (the Fermi energy). If it were not for the exclusion principle, all electrons could occupy the ground state at $T = 0$ K.

4. Table 43.5 shows that the energy gaps for semiconductors decrease with increasing temperature. What do you suppose accounts for this behavior?

Answer: The energy gap depends weakly on temperature because of thermal expansion of the lattice.

5. The resistivity of most metals increases with increasing temperature, whereas the resistivity of an intrinsic semiconductor decreases with increasing temperature. What explains this difference in behavior?

Answer: In a normal metal such as copper, which has a large number of conduction band electrons, the resistivity increases with increasing temperature due to the increase in amplitude of the lattice vibrations, and a corresponding increase in scattering of electrons from the dynamic lattice. In an intrinsic semiconductor, the resistivity decreases with increasing temperature because the average number of conduction band electrons increases with increasing temperature as some electrons are thermally excited from the top of the valence band to the conduction band.

11. When a photon is absorbed by a semiconductor, an electron-hole pair is said to be created. Give a physical explanation of this statement using the energy band model as the basis for your description.

Answer: When a photon is absorbed, an electron initially in a nearly filled valence band is given sufficient energy to make a transition to the conduction band, leaving an electron deficient center (or hole) in the valence band. This can only occur if the photon's energy is at least as large as the energy gap of the semiconductor.

12. Pentavalent atoms such as arsenic are donor atoms in a semiconductor such as silicon, while trivalent atoms such as indium are acceptors. Inspect the periodic table in Appendix C, and determine what other elements would be considered as either donors or acceptors.

Answer: The donors are phosphorus (P)...($3s^2 3p^3$) and antimony (Sb)...($5s^2 5p^3$); the acceptors are boron (B)...($2s^2 p^1$), aluminum (Al)...($3s^2 3p^1$), and gallium (Ga)...$4s^2 4p^1$).

SOLUTIONS TO SELECTED END-OF-CHAPTER PROBLEMS_____

9. The HCl molecule is excited to its first rotational energy level, corresponding to J = 1. If the distance between its nuclei is 0.1275 nm, what is the angular velocity of the molecule about its center of mass?

Solution

For the HCl molecule in the J = 1 rotational energy level, we are given $r_0 = 0.1275$ nm.

$$E_{rot} = \frac{\hbar^2}{2I} J(J + 1)$$

Taking J = 1, we have

$$E_{rot} = \frac{\hbar^2}{I} = \frac{1}{2} I\omega^2$$

or

$$\omega = \sqrt{\frac{2\hbar^2}{I^2}} = \sqrt{2}\, \frac{\hbar}{I}$$

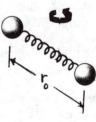

The moment of inertia of the molecule is given by Equation 43.3.

$$I = \mu r_0^2 = \left(\frac{m_1 m_2}{m_1 + m_2}\right) r_0^2 = \left[\frac{(1\ u)(35\ u)}{1\ u + 35\ u}\right] r_0^2$$

$$= (0.97222\ u)(1.66 \times 10^{-27}\ kg/u)(1.275 \times 10^{-10}\ m)^2 = 2.62 \times 10^{-47}\ kg{\cdot}m^2$$

Therefore,

$$\omega = \sqrt{2}\,\frac{\hbar}{I} = \sqrt{2}\,\frac{1.055 \times 10^{-34}\ J{\cdot}s}{2.62 \times 10^{-47}\ kg{\cdot}m^2} = \boxed{5.69 \times 10^{12}\ rad/s}$$

---

16. Consider a one-dimensional chain of alternating positive and negative ions. Show that the potential energy of an ion in this hypothetical crystal is

$$U(r) = -k\alpha\,\frac{e^2}{r}$$

where $\alpha = 2\ln 2$ (the Madelung constant), and r is the interionic spacing. Hint: Make use of the series expansion for $\ln(1+x)$.

Solution

The total potential energy is obtained by summing over all *pairs* of interactions:

Problem 16

$$U = \sum_{i \neq j} k\,\frac{q_i q_j}{r_{ij}}$$

$$= -k\frac{e^2}{r} - k\frac{e^2}{r} + k\frac{e^2}{2r} + k\frac{e^2}{2r} - k\frac{e^2}{3r} - k\frac{e^2}{3r} + k\frac{e^2}{4r} + k\frac{e^2}{4r}$$

$$U = -2k\frac{e^2}{r}\left[1 - \frac{1}{2} + \frac{1}{3} - \frac{1}{4} + \ldots\right]$$

But

$$\ln(1+x) \cong x - \frac{x^2}{2} + \frac{x^3}{3} - \frac{x^4}{4} + \ldots$$

Therefore, $x = 1$ for our series, and U becomes

$$U = -2\ln(2)\,k\frac{e^2}{r} = -\alpha\,k\frac{e^2}{r}$$

21. Calculate the energy of a conduction electron in silver at 800 K if the probability of finding the electron in that state is 0.95. Assume that the Fermi energy for silver is 5.48 eV at this temperature.

Solution

Taking $E_F = 5.48$ eV for sodium at 800 K, and given $f = 0.95$, we find

$$f = \frac{1}{e^{(E-E_F)/kT} + 1} = 0.95$$

$$e^{(E-E_F)/kT} = \frac{1}{0.95} - 1 = 0.05263$$

$$\frac{E - E_F}{kT} = \text{Ln}(0.05263) = -2.944$$

$$E - E_F = -2.944 \, kT = -2.944(1.38 \times 10^{-23} \text{ J/K})(800 \text{ K})$$

$$E - E_F = -3.25 \times 10^{-20} \text{ J} = -0.203 \text{ eV}$$

$$E = \boxed{5.28 \text{ eV}}$$

22. Show that the average kinetic energy of a conduction electron in a metal at O K is given by

$$E_{av} = \frac{3}{5} E_F$$

(Hint: In general, the average kinetic energy is given by

$$E_{av} = \frac{1}{n} \int E \, N(E) \, dE$$

where n is the density of particles, and N(E) dE is given by Equation 43.21.)

Solution

$$E_{av} = \frac{1}{n} \int_0^\infty E \, N(E) \, dE_o$$

where

$$N(E) = \frac{C E^{1/2}}{e^{(E-E_F)/kT} + 1} = C f(E) E^{1/2}$$

But at $T = 0$, $E = 0$ for $E > E_F$.

Also, $f(E) = 1$ for $E < E_F$ and $f(E) = 0$ for $E > E_F$.

So we can take $N(E) = C E^{1/2}$.

$$E_{av} = \frac{1}{n} \int_0^{E_F} C E^{3/2} \, dE = \frac{2C}{5n} E_F^{5/2}$$

But from Eq. 43.23 we have

$$\frac{C}{n} = \frac{3}{2} E_F^{-3/2}$$

so

$$E_{av} = \left(\frac{2}{5}\right)\left(\frac{3}{2}\right)\left(E_F^{-3/2}\right) E_F^{5/2} = \frac{3}{5} E_F$$

---

25. (a) Consider a system of electrons confined to a three-dimensional box. Calculate the ratio of the number of allowed energy levels at 8.5 eV to the number of allowed energy levels at 7.0 eV. (b) Copper has a Fermi energy of 7.0 eV at 300 K. Calculate the ratio of the number of occupied levels at an energy of 8.5 eV to the number of occupied levels at the Fermi energy. Compare your answer with that obtained in part (a).

Solution

(a) The density of states at the energy E is $g(E) = C E^{1/2}$. Hence, the required ratio is

$$\frac{g(8.5 \text{ eV})}{g(7.0 \text{ eV})} = \frac{C(8.5)^{1/2}}{C(7.0)^{1/2}} = \boxed{1.10}$$

(b) From Eq. 43.21, we see that the number of occupied states having energy E is

$$N(E) = \frac{C E^{1/2}}{e^{(E - E_F)/kT} + 1}$$

Hence, the required ratio is

$$\frac{N(8.5 \text{ eV})}{N(7.0 \text{ eV})} = \left(\frac{8.5}{7.0}\right)^{1/2} \left[\frac{e^{(7-7)/kT} + 1}{e^{(8.5 - 7)/kT} + 1}\right]$$

At $T = 300$ K, $kT = 0.0259$ eV, so we find

$$\frac{N(8.5 \text{ eV})}{N(7.0 \text{ eV})} + (1.10) \left(\frac{2}{e^{1.5/0.0259} + 1}\right) = \boxed{1.55 \times 10^{-25}}$$

Comparing this result with (a), we conclude that very few states with $E > E_F$ are occupied.

---

35. Show that the ionic cohesive energy of an ionically bonded solid is given by Equation 43.16. (Hint: Start with Equation 43.15, and note that $dU/dr = 0$ at $r = r_0$.)

Solution

The total potential energy is given by Equation 43.15,

$$U_{total} = -\alpha k \frac{e^2}{r} + \frac{B}{r^m}$$

The potential energy has its minimum value U_0 when $r = r_0$, where r_0 is the equilibrium spacing. At this point, the slope of the curve U versus r is *zero*. That is,

$$\frac{dU}{dr}\bigg]_{r=r_0} = 0$$

$$\frac{dU}{dr} = \frac{d}{dr}\left(-\alpha k \frac{e^2}{r} + \frac{B}{r^m}\right) = \alpha k \frac{e^2}{r^2} - \frac{mB}{r^{m+1}}$$

Taking $r = r_0$, and setting this equal to zero, we find

$$\alpha k \frac{e^2}{r_0^2} - \frac{mB}{r_0^{m+1}} = 0$$

Therefore,

$$B = \alpha \frac{ke^2}{m} r_0^{m-1}$$

Substituting this value of B into U_{total} gives

$$U_0 = -\alpha k \frac{e^2}{r_0} + \alpha \frac{ke^2}{m} r_0^{m-1}\left(\frac{1}{r^m}\right) = -\alpha \frac{ke^2}{r_0}\left(1 - \frac{1}{m}\right)$$

Superconductivity

OBJECTIVES

1. Be familiar with the historical development of the phenomenon of superconductivity.

2. Understand and be able to explain the fundamental properties and characteristics of type I superconductors in terms of critical temperature, critical magnetic field, and penetration depth.

3. Explain how materials which are type II superconductors differ from type I superconductors.

4. Describe the general features of the BCS theory of superconductivity including the concept of a Cooper pair.

5. Describe the energy gap measurements in superconductors by single particle tunneling experiments and by electromagnetic energy absorption experiments.

6. Understand Josephson tunneling as it applies to the dc Josephson effect, the ac Josephson effect, and quantum interference.

7. Be aware of the properties of high temperature superconductors which give rise to research interest in those materials.

8. Relate the empirical observations of high-temperature superconductors to the predictions of the BCS theory.

NOTES FROM SELECTED CHAPTER SECTIONS

44.2 BRIEF HISTORICAL REVIEW

The temperature at which a material becomes superconducting is called the *critical temperature*.

When superconductors in the presence of a magnetic field are cooled below their critical temperature, the *magnetic flux is expelled from the interior of the superconductor*. This behavior is known as the *Meissner effect*.

44.3 SOME PROPERTIES OF TYPE I SUPERCONDUCTORS

The dc resistance of a type I superconductor is zero below their critical temperature. Type I superconductors are characterized by low critical field values and for this reason cannot be used to construct high field magnets--superconducting magnets.

The magnetic flux in the interior of a superconductor cannot change in time. This is because the electric field in its interior must be zero.

A type I superconductor is therefore not only a perfect conductor but also a perfect diamagnet; this is an essential property of the superconducting state.

44.4 TYPE II SUPERCONDUCTORS

Type II superconductors are characterized by two critical magnetic fields. Below the lower critical field, they behave as type I superconductors. Above the upper critical field, the superconducting state is destroyed just as for type I materials. For magnetic field between the values of the upper and lower critical field values, the materials are in a *mixed state* referred to as a *vortex state*.

Most type II superconductors are compounds formed from elements of the transition and actinide series. These materials are well suited for constructing high-field *superconducting magnets* which can sustain large magnetic fields with *no power consumption*.

44.5 OTHER PROPERTIES OF SUPERCONDUCTORS

Once a current is set up in a superconducting material, *persistent currents* will continue in the superconductor for long time intervals (~ yr) *without any applied voltage* and with no measurable losses.

The *coherence length* is the distance over which the electrons in a *Cooper pair* remain together. A superconductor will be type I if the coherence length is greater than the penetration depth.

44.6 SPECIFIC HEAT

The *electronic specific heat* is the heat energy absorbed by the conduction electrons per unit increase in the temperature of the system. The temperature variation of the specific heat of a superconductor below T_c can be used to measure the energy gap of the material.

44.7 THE BCS THEORY

The central feature of the BCS theory is that two electrons in the superconductor are able to form a bound state called a *Cooper pair* if they experience an *attractive interaction*. The interaction between the two electrons occurs via the crystal lattice as it 'distorts' in the 'wake' of one of the electrons. The quantized lattice vibrations are called *phonons* and the electron-electron attractive force is called a *phonon mediated mechanism*.

The superconducting state is a state in which the electrons act collectively, rather than independently. All Cooper pairs are in the same quantum state and the system exhibits quantum effects on a *macroscopic* scale. The *condensed state of the Cooper pairs is represented by a single coherent wave function*.

44.8 ENERGY GAP MEASUREMENTS

The energy gap in superconductors can be measured by *single particle tunneling* experiments and by experiments based on *absorption of electromagnetic radiation*.

44.11 HIGH-TEMPERATURE SUPERCONDUCTIVITY

Recently, superconductivity has been found to occur in several materials at temperatures well above the boiling point of liquid nitrogen (77 K). The interest in these materials is due to the following factors:

1. The metallic oxides are relatively easy to fabricate and hence can be investigated at smaller laboratories and universities.

2. They have very high T_c values and very high upper critical magnetic fields estimated to be greater than 100 T in several materials.

3. Their properties and the mechanisms responsible for their superconducting behavior represent a great challenge to theoreticians.

4. They may be of considerable technological importance for both current applications and their potential use in nitrogen-temperature superconducting electronics and large-scale applications such as energy generation and transport, and magnetic levitation for high-speed transportation.

EQUATIONS AND CONCEPTS

The superconducting state is destroyed when the material is in the presence of a magnetic field greater than the *critical magnetic field* B_c. The temperature dependence of B_c is given by Equation 44.1.

$$B_c(T) = B_c(0)\left[1 - \left(\frac{T}{T_c}\right)^2\right] \tag{44.1}$$

The magnetic field inside a type I superconductor decreases exponentially from its external value B_0 to zero within a very thin layer. The parameter λ is called the *penetration depth*.

$$B(x) = B_o e^{-x/\lambda} \tag{44.3}$$

The penetration depth λ is a function of temperature and becomes infinite as the temperature approaches the critical temperature.

$$\lambda(T) = \lambda_o\left[1 - \left(\frac{T}{T_c}\right)^4\right]^{-1/2} \tag{44.4}$$

The superconducting state has a lower free energy per unit volume than the normal state. This is the defining equation for the critical field.

$$E_s + \left(\frac{B_c{}^2}{2\mu_o}\right) = E_n \tag{44.6}$$

The *energy gap* E_g in a superconductor represents the energy needed to break up one Cooper pair. At $T = 0\ K$, the energy gap is proportional to the critical temperature.

$$E_g = 3.53\ kT_c \tag{44.7}$$

The magnetic flux trapped in a superconductor by a persistent current is quantized in units of the magnetic flux quantum Φ_o.

$$\Phi = n\left(\frac{h}{2e}\right) \tag{44.8}$$

$$\Phi_o = \frac{h}{2e} \tag{44.9}$$

ANSWERS TO SELECTED QUESTIONS

1. Discuss the two major characteristics of a superconductor. Describe how you would measure these characteristics.

Answer: The two hallmarks of superconductivity are zero dc resistance and magnetic flux expulsion (the Meissner effect) below the critical temperature. The property of zero resistance can be established with a four-point probe resistivity measurement. The Meissner effect can be demonstrated by measuring the magnetization versus temperature and by magnetic levitation in the presence of an external magnetic field.

3. Why is it not possible to explain the property of zero resistance using a classical model of charge transport through a solid?

Answer: Classically, the electrons will always suffer collisions with other electrons, nuclei, impurities, etc., so the resistance could never be zero.

7. What are persistent currents, and how can they be set up in a superconductor?

Answer: Persistent currents are currents that flow in a superconductor *without an applied voltage*. They can exist because the dc resistance of a superconductor is zero below T_c. A persistent current can be set up in a loop by first cooling the loop in the presence of a **B** field from $T > T_c$ to $T < T_c$; when the external field is then turned off, a persistent current is established in the loop.

10. What is the isotope effect, and why does it play an important role in testing the validity of the BCS theory?

Answer: In the isotope effect, the critical temperature of many superconductors varies with different isotopes of the same element. This is strong evidence that lattice motion plays an important role in the mechanism of superconductivity, since lattice vibrational frequencies are expected to be proportional to $M^{-1/2}$.

12. What is meant when it is said that Cooper pairs act collectively? How is it possible that all Cooper pairs can occupy the same quantum state?

Answer: They act collectively in that they are all in the same quantum state (a condensation of pairs) and form a system with zero total momentum and zero spin. They are able to occupy the same state because they are bosons (zero spin particles) rather than fermions.

14. Discuss the origin of the energy gap of a superconductor, and how the energy band structure of a superconductor differs from a normal conductor.

Answer: The energy gap of a superconductor is the energy separation between the ground and excited states of the "macromolecule" associated with the collective system of Cooper pairs. In a normal conductor, Cooper pairs do not exist and hence there is no energy gap.

SOLUTIONS TO SELECTED END-OF-CHAPTER PROBLEMS

3. Determine the current generated in a superconducting ring of niobium metal 2 cm in diameter if a magnetic field of 0.02 T directed *perpendicular* to the ring is suddenly decreased to zero. The inductance of the ring is $L = 3.1 \times 10^{-8}$ H .

Solution

From Faraday's law (Eq. 32.1), we find

$$|\varepsilon| = \frac{\Delta\Phi}{\Delta t} = A\frac{\Delta B}{\Delta t} = L\frac{\Delta I}{\Delta t}$$

or

$$\Delta I = \frac{A\Delta B}{L} = \frac{\pi(0.01 \text{ m})^2(0.02 \text{ T})}{3.1\times 10^{-8} \text{ H}} = \boxed{200 \text{ A}}$$

7. *Persistent Currents.* In an experiment carried out by S. C. Collins between 1955 and 1958, a current was maintained in a superconducting lead ring for $2\frac{1}{2}$ years with no observed loss. If the inductance of the ring was 3.14×10^{-8} H, and the sensitivity of the experiment was 1 part in 10^9, determine the maximum resistance of the superconducting ring. (Hint: Treat this as a decaying current in an RL circuit, and recall that $e^{-x} \cong 1 - x$ for small x.)

Solution

If a current is set up in an RL circuit, the current decays with time according to Equation 32.11:

$$I = I_0 e^{-(R/L)t}$$

Hence,

$$\frac{I}{I_0} = e^{-(R/L)t} \cong 1 - \frac{R}{L}t$$

$$\frac{R}{L}t = 1 - \frac{I}{I_0} = \frac{I_0 - I}{I_0} = 10^{-9}$$

Since $t = 2.5 \text{ y} = 7.88 \times 10^7$ s, and $L = 3.14 \times 10^{-8}$ H,

$$R_{max} = 10^{-9}\frac{L}{t} = \frac{10^{-9}\,(3.14 \times 10^{-8}\text{ H})}{7.88 \times 10^7 \text{ s}} = \boxed{3.98 \times 10^{-25}\ \Omega}$$

15. *Cooper Pairs.* A Cooper pair of electrons in a type I superconductor has an average separation of about 10^{-4} cm. If these two electrons can interact within a volume of this diameter, how many other Cooper pairs have their centers within the volume occupied by one pair? Use the appropriate data for lead which has $n_s = 2 \times 10^{22}$ electrons/cm^3.

Solution

Each pair will occupy volume of

$$V_{pair} = \frac{4}{3}\pi R^3 = \frac{4}{3}\pi(5 \times 10^{-5}\text{ cm})^3 \cong 5.2 \times 10^{-13}\text{ cm}^3$$

Since there are 2×10^{22} electrons per cm^3 of conduction electrons, each cm^3 will contain 1×10^{22} pairs of electrons, so

$$N = \left(1 \times 10^{22}\,\frac{pairs}{cm^3}\right)(5.2 \times 10^{-13}\text{ cm}^3) = \boxed{5.2 \times 10^9\text{ pairs}}$$

That is, when the material is in the superconducting state, each Cooper pair effectively moves with 5 billion other pairs.

19. If a magnetic flux of

$$10^{-4} \, \Phi_o \left(\frac{1}{10\,000} \text{ of the flux quantum} \right)$$

can be measured with a SQUID device (Figure 44.27), what is the smallest magnetic field change ΔB that can be detected with this device, if the ring has a radius of 2 mm?

Solution   The flux quantum is given by Equation 44.9:

$$\Phi_o = \frac{h}{2e} = 2.07 \times 10^{-15} \text{ T·m}^2$$

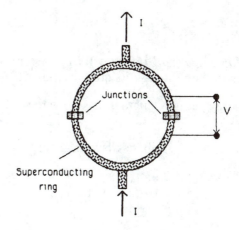

Figure 44.27

Since $\Phi = BA$, the smallest field change that can be detected is

$$\Delta B = \frac{\Delta \Phi}{A} = \frac{10^{-4} \, \Phi_o}{A} = \frac{2.07 \times 10^{-19} \text{ T·m}^2}{\pi (2 \times 10^{-3} \text{ m})^2} = \boxed{1.65 \times 10^{-14} \text{ T}}$$

---

25. *"Floating" a Wire.* Is it possible to "float" a superconducting lead wire of radius 1 mm in the magnetic field of the earth? Assume the horizontal component of the earth's magnetic field is 0.5 gauss.

Solution   If the wire can "float", the downward force of gravity must be balanced by the upward magnetic force ILB as shown. That is, ILB = Mg. Since

$$M = \rho V = \rho(\pi r^2 L)$$

we find

$$I = \frac{\rho \pi r^2 g}{B}$$

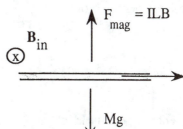

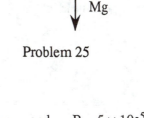

Problem 25

For lead, $\rho = 1.135 \times 10^5 \text{ kg/m}^3$, and taking $r = 10^{-3} \text{ m}$ and $B = 5 \times 10^{-5} \text{ T}$ gives

$$I = \frac{(1.135 \times 10^5 \text{ kg/m}^3)(\pi)(10^{-3} \text{ m})^2(9.8 \text{ m/s}^2)}{5 \times 10^{-5} \text{ T}} = 7000 \text{ A}$$

The magnetic field that this current would produce at the surface of the wire (from Ampere's law) is

$$B = \frac{\mu_o I}{2\pi r} = \frac{(4\pi \times 10^{-7} \text{ N/A·m}^2)(7000 \text{ A})}{2\pi (10^{-3} \text{ m})} = 2.80 \text{ T}$$

Since this *exceeds* the critical field for lead, $B_c = 0.08$ T, the lead wire *cannot* be suspended.

---

26. *Magnetic Field inside a Wire.* A type II superconducting wire of radius R carries current uniformly distributed through its cross section. If the total current carried by the wire is I, show that the magnetic energy per unit length *inside* the wire is $\dfrac{\mu_o I^2}{16\pi}$.

Solution   Ampere's law is given by

$$\oint \mathbf{B} \cdot d\mathbf{s} = \mu_o I'$$

Problem 26

where I' is the current through the area surrounded by the path of integration. Taking the path to be a circle of radius r < R, we see that I' < I, where

$$I' = \left(\frac{\pi r^2}{\pi R^2}\right) I = \frac{r^2}{R^2} I \qquad (r < R)$$

Hence,

$$\oint \mathbf{B} \cdot d\mathbf{s} = B(2\pi r) = \mu_o \frac{r^2}{R^2} I \qquad \text{or} \qquad B = \left(\frac{\mu_o I}{2\pi R^2}\right) r$$

Since the energy per unit volume is $\dfrac{B^2}{2\mu_o}$, the total energy in a length L of the wire is

$$U = \int \frac{B^2}{2\mu_o} dV = \frac{1}{2\mu_o} \int_o^R \left(\frac{\mu_o I r}{2\pi R^2}\right)^2 2\pi r L dr$$

Hence, the energy per unit length is

$$\frac{U}{L} = \frac{\mu_o I^2}{4\pi R^2} \int_o^R r^3 dr = \frac{\mu_o I^2}{16\pi}$$

29. *Entropy Difference.* The entropy difference per unit volume between the normal and superconducting states is given by

$$\frac{\Delta S}{V} = -\frac{\partial}{\partial T}\left(\frac{B^2}{2\mu_o}\right)$$

where $B^2/2\mu_o$ is the magnetic energy per unit volume required to destroy superconductivity. Determine the entropy difference between the normal and superconducting states in 1 mol of lead at 4 K if the critical magnetic field $B_c(0) = 0.08$ T and $T_c = 7.2$ K.

Solution   Since $B = B_c(0)\left[1 - \left(\frac{T}{T_c}\right)^2\right]$,   we get

$$\frac{\Delta S}{V} = -\frac{\partial}{\partial T}\left\{\frac{B_c^2(0)}{2\mu_o}\left[1 - \left(\frac{T}{T_c}\right)^2\right]^2\right\} = -\frac{B_c^2(0)}{2\mu_o}\frac{\partial}{\partial T}\left[1 - \left(\frac{T}{T_c}\right)^2\right]^2 = \frac{2B_c^2(0)}{\mu_o T_c}\left[\frac{T}{T_c} - \left(\frac{T}{T_c}\right)^3\right]$$

At $T = 0$ K, $B_c(0) = 0.08$ T   and $T_c = 7.2$ K for lead,   so

$$\frac{\Delta S}{V} = \frac{2(0.08\ \text{T})^2}{(4\pi \times 10^{-7})(7.2)}\left[\frac{4\ \text{K}}{7.2\ \text{K}} - \left(\frac{4\ \text{K}}{7.2\ \text{K}}\right)^3\right] = 543\ \text{J/m}^3\cdot\text{K}$$

Since lead has an atomic weight of 206 and a density of $\rho = 11.35$ g/cm^3, one mol occupies a volume of $\frac{206}{11.35} = 18.1$ cm$^3 = 18.1 \times 10^{-6}$ m^3. Therefore,

$$\Delta S = (543\ \text{J/m}^3\cdot\text{K})(18.1 \times 10^{-6}\ \text{m}^3/\text{mol}) = \boxed{9.83 \times 10^{-3}\ \text{J/mol}\cdot\text{K}}$$

Note that the superconducting state has *lower* entropy than the normal state.

45

Nuclear Structure

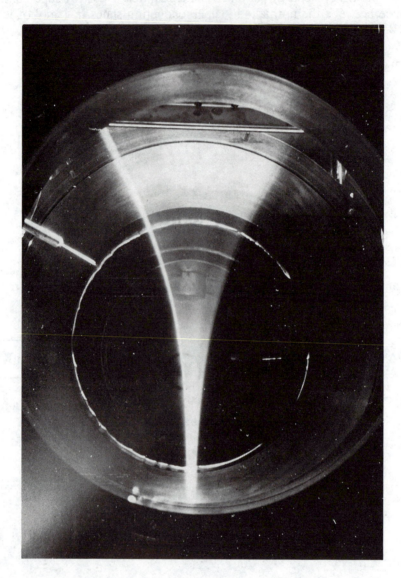

OBJECTIVES

1. Use the appropriate nomenclature in describing the static properties of nuclei.

2. Describe the experiments of Rutherford which established the nuclear character of the atom's structure.

3. Discuss nuclear stability in terms of the strong nuclear force and a plot of N vs. Z.

4. Account for nuclear binding energy in terms of the Einstein mass-energy relationship. Describe the basis for energy released by fission and fusion in terms of the shape of the curve of binding energy per nucleon vs. mass number.

5. Describe the essential features of the liquid-drop model, independent particle model, and the collective model of the nucleus.

6. Identify each of the components of radiation that are emitted by the nucleus through natural radioactive decay and describe the basic properties of each.

7. State and apply to the solution of related problems, the formula which expresses decay rate as a function of decay constant and number of radioactive nuclei.

8. Write out typical equations to illustrate the processes of transmutation by alpha and beta decay and explain why the neutrino must be considered in the analysis of beta decay.

9. Describe the process of carbon dating as a means of determining the age of ancient objects.

10. Calculate the Q value of given nuclear reactions and determine the threshold energy of endothermic reactions.

SKILLS

The rest energy of a particle is given by $E = mc^2$. It is therefore often convenient to express the unified mass unit in terms of its energy equivalent, $1\text{ u} = 1.660559 \times 10^{-27}$ kg or $1\text{ u} = 931.50\text{ MeV}/c^2$. When masses are expressed in units of u, energy values are then $E = m(931.50\text{ MeV/u})$.

Equation 45.7 can be solved for the particular time t after which the number of remaining nuclei will be some specified fraction of the original number N_0. This can be done by taking the *ln* of each side of Equation 45.7 to find $t = \left(\frac{1}{\lambda}\right) ln \left(\frac{N_0}{N}\right)$.

NOTES FROM SELECTED CHAPTER SECTIONS

45.1 SOME PROPERTIES OF NUCLEI

Important quantities in the description of nuclear properties are:
1. The *atomic number*, Z, which equals the number of protons in the nucleus.
2. The *neutron number*, N, which equals the number of neutrons in the nucleus.
3. The *mass number*, A, which equals the number of nucleons (neutrons plus protons) in the nucleus.

The nuclei of all atoms of a particular element contain the same number of protons but often contain different numbers of neutrons. Nuclei that are related in this way are called *isotopes*. The isotopes of an element have the same Z value but different N and A values.

The *atomic mass unit*, u, is defined such that the mass of the isotope ^{12}C is exactly 12 u.

Experiments have shown that most nuclei are approximately spherical and all *have nearly the same density*. The stability of nuclei is due to the *nuclear force*. This is a *short range, attractive* force which acts between all nuclear particles. Nuclei have *intrinsic angular momentum* which is quantized by the *nuclear spin quantum number* which may be integer or half integer.

The *magnetic moment of the nucleus* is measured in terms of the *nuclear magneton*. When placed in an external magnetic field, nuclear magnetic moments precess with a frequency called the *Larmor precessional frequency*.

45.2 BINDING ENERGY

The total mass of a nucleus is always less than the sum of the masses of its individual nucleons. The *binding energy* of the nucleus is mass difference multiplied by c^2.

45.3 NUCLEAR MODELS

The *liquid-drop model*, proposed by Bohr in 1936, treats the nucleons as if they were molecules in a drop of liquid. The nucleons interact strongly with each other and undergo frequent collisions as they "jiggle around" within the nucleus. This is analogous to the thermally agitated motion of molecules in a liquid. The three major effects which influence the binding energy according to the liquid-drop model are the volume effect, the surface effect, and the Coulomb effect.

The *independent-particle model,* often called the *shell model*, is based upon the assumption that *each nucleon moves in a well-defined orbit within the nucleus and in an averaged field produced by the other nucleons*. In this model, the nucleons exist in quantized energy states and there are few collisions between nucleons.

A third model of nuclear structure, known as the *collective model*, combines some features of both the liquid-drop model and the independent-particle model. The nucleus is considered to have some "extra" nucleons moving in quantized orbits, in addition to the filled core of nucleons.

45.4 RADIOACTIVITY

There are three processes by which a radioactive substance can undergo decay: alpha (α) decay, where the emitted particles are ^{4}He nuclei; beta (β) decay, in which the emitted particles are either electrons or positrons; and gamma (γ) decay, in which the emitted "rays" are high-energy photons. A positron is a particle similar to the electron in all respects except that it has a charge of $+e$ (the antimatter twin of the electron). The symbol β^- is used to designate an electron, and β^+ designates a positron.

The three types of radiation have quite different penetrating powers. Alpha particles barely penetrate a sheet of paper, beta particles can penetrate a few millimeters of aluminum, and gamma rays can penetrate several centimeters of lead.

45.5 THE DECAY PROCESS

Alpha decay can occur because, according to quantum mechanics, some nuclei have barriers that can be penetrated by the alpha particles (the tunneling process). This process is energetically more favorable for those nuclei having a large excess of neutrons. A nucleus can undergo beta decay in two ways. It can emit either an electron (β^-) and an antineutrino $(\bar{\nu})$ or a positron (β^+) and a neutrino (ν). In the electron-capture process, the nucleus of an atom absorbs one of its own electrons (usually from the K shell) and emits a neutrino.

The neutrino has the following properties:

1. It has zero electric charge.

2. It has a rest mass smaller than that of the electron, and in fact its mass may be zero (although recent experiments suggest that this may not be true).

3. It has a spin of $\frac{1}{2}$, which satisfies the law of conservation of angular momentum.

4. It interacts very weakly with matter and is therefore very difficult to detect.

In gamma decay, a nucleus in an excited state decays to its ground state and emits a gamma ray. The Q value (disintegration energy) is the energy released as a result of the decay process.

45.6 NATURAL RADIOACTIVITY

Radioactive nuclei are generally classified into two groups: (1) unstable nuclei found in nature, which give rise to what is called *natural radioactivity*, and (2) nuclei produced in the laboratory through nuclear reactions, which exhibit *artificial radioactivity*.

45.7 NUCLEAR REACTIONS

Nuclear reactions are events in which collisions change the identity or properties of nuclei. The total energy released as a result of a nuclear reaction is called the *reaction energy*, Q.

An *endothermic reaction* is one in which Q is negative and the minimum energy for which the reaction will occur is called the *threshold energy*.

EQUATIONS AND CONCEPTS

Most nuclei are approximately spherical in shape and have an average radius which is proportional to the cube root of the mass number or total number of nucleons. This means that the volume is proportional to A and that all nuclei have nearly the same density.

$$r = r_o A^{1/3} \tag{45.1}$$

The nucleus has an angular momentum and a corresponding nuclear magnetic moment associated with it. The nuclear magnetic moment is measured in terms of a unit of moment called the nuclear magneton μ_n.

$$\mu_n \equiv \frac{e\hbar}{2m_p} = 5.05 \times 10^{-27} \text{ J/T} \tag{45.3}$$

The binding energy of any nucleus can be calculated in terms of the mass of a neutral hydrogen atom, the mass of a neutron, and the atomic mass of the associated compound nucleus.

$$E_n(\text{MeV}) = \left[Zm_H + Nm_n - M\left({}^A_Z X\right) \right] \times 931 \text{ MeV/u} \tag{45.4}$$

The semiempirical binding energy formula is based on the liquid-drop model of the nucleus.

$$E_b = C_1 A - C_2 A^{2/3} - C_3 \frac{Z(Z-1)}{A^{1/3}} - C_4 \frac{(A-2Z)^2}{A} \tag{45.5}$$

The number of radioactive nuclei in a given sample which undergoes decay during a time interval Δt depends on the number of nuclei present. The number of decays depends also on the decay constant λ which is characteristic of a particular isotope.

$$\frac{dN}{dt} = -\lambda N \qquad (45.6)$$

The number of nuclei in a radioactive sample decreases exponentially with time. The plot of number of nuclei N versus elapsed time t is called a decay curve.

$$N = N_o e^{-\lambda t} \qquad (45.7)$$

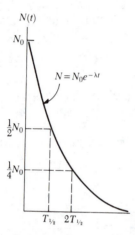

The decay rate R or activity of a sample of radioactive nuclei is defined as the number of decays per second.

$$R = \left|\frac{dN}{dt}\right| \qquad (45.8)$$

The half-life $T_{1/2}$ is the time required for half of a given number of radioactive nuclei to decay.

$$T_{1/2} = \frac{0.693}{\lambda} \qquad (45.9)$$

When a nucleus decays by **alpha emission**, the parent nucleus loses two neutrons and two protons. In order for alpha emission to occur, the mass of the parent nucleus must be greater than the combined mass of the daughter nucleus and the emitted alpha particle. The mass difference is converted into energy and appears as kinetic energy shared (unequally) by the alpha particle and the daughter nucleus.

$$^{A}_{Z}X \rightarrow ^{A-4}_{Z-2}Y + ^{4}_{2}He \qquad (45.10)$$

$$^{238}_{92}U \rightarrow ^{234}_{90}Th + ^{4}_{2}He \qquad (45.11)$$

The disintegration energy Q can be calculated in MeV when the masses are expressed in units of u.

$$Q = (M_x - M_y - M_\alpha)931.5 \text{ MeV/u} \qquad (45.14)$$

When a radioactive nucleus undergoes **beta decay**, the daughter nucleus has the same mass number as the parent nucleus but the charge number (or atomic number) increases by one. The electron that is emitted is created within the parent nucleus by a process which can be represented by a neutron transformed into a proton and an electron. The total energy released in beta decay is greater than the combined kinetic energies of the electron and the daughter nucleus. This difference in energy is associated with a third particle called a neutrino.

$$\_Z^A X \rightarrow \_{Z+1}^A Y + \beta^-$$

(45.15)

$$n \rightarrow \_1^1 p + \beta^- + \bar{\nu}$$

(45.19)

Nuclei which undergo alpha or beta decay are often left in an excited energy state. The nucleus returns to the ground state by emission of one or more photons. Gamma-ray emission results in no change in mass number or atomic number.

$$\_Z^A X^* \rightarrow \_Z^A X + \gamma$$

(45.21)

Nuclear reactions can occur when target nuclei are bombarded with energetic particles. In these reactions the structure, identity, or properties of the target nuclei are changed.

$$a + X \rightarrow Y + b$$

(45.24)

The quantity of energy required to balance the equation representing a nuclear reaction (e.g. Eq. 44.24) is called the Q value of the reaction. The Q value can be calculated in terms of the total mass of the reactants minus the total mass of the products or as the kinetic energy of the reactants. Q is positive in the case of exothermic reactions and negative for endothermic reactions.

$$Q = (M_a + M_x - M_y - M_b)c^2$$

(45.25)

TABLE 44.3   Various Decay Pathways

Alpha decay	$\_Z^A X \longrightarrow \_{Z-2}^{A-4} X + \_2^4 He$
Beta decay (β^-)	$\_Z^A X \longrightarrow \_{Z+1}^A X + \beta^- + \bar{\nu}$
Beta decay (β^+)	$\_Z^A X \longrightarrow \_{Z-1}^A X + \beta^+ + \nu$
Electron capture	$\_Z^A X + \_{-1}^0 e \longrightarrow \_{Z-1}^A X + \nu$
Gamma decay	$\_Z^A X^\circ \longrightarrow \_Z^A X + \gamma$

EXAMPLE PROBLEM SOLUTION

Example 45.1   A radioactive sample has an initial activity of 2 mCi. After a decay time of 4.5 days, the activity is found to be 1.41 mCi. Determine for this source (a) the decay constant, (b) half-life, and (c) the number of nuclei in the sample initially.

Solution

(a)   Use Equation 45.7.

$$N = N_o e^{-\lambda t}$$

$$e^{-\lambda t} = \frac{N}{N_o}$$

The ratio of the numbers of atoms equals the ratio of the activities at the corresponding times.

$$e^{-\lambda t} = \frac{1.41 \text{ mCi}}{4.5 \text{ mCi}} = 0.313$$

Take the natural logarithm of each side of the equation.

$$ln \ e^{-\lambda t} = ln \ (0.313); \qquad (-\lambda t) ln \ e = ln \ (0.313)$$

Note that $ln \ e = 1$. Use your calculator to find the value of $ln \ (0.313)$.

$$-\lambda t = -1.16$$

$$\lambda = \frac{1.16}{4.5 \text{ da}} = 0.258 \text{ da}^{-1} = 2.99 \times 10^{-6} \text{ s}^{-1}$$

(b)   We can use Equation 45.9 to find $T_{1/2}$:

$$T_{1/2} = \frac{0.693}{\lambda} = \frac{0.693}{0.258 \text{ da}^{-1}}$$

$$T_{1/2} = 2.69 \text{ da} = 2.32 \times 10^5 \text{ s}$$

(c)   Since the activity R is given by $R = \lambda N$,

$$N = \frac{R}{\lambda}$$

where $R = 2 \text{ mCi} = 2 \times 10^{-3} \text{ Ci} = (2 \times 10^{-3} \text{ Ci})(3.7 \times 10^{10} \text{ nuclei/s/Ci}) = 7.4 \times 10^7 \text{ nuclei/s}$

Substitute this value of R into the equation for N along with the value of the decay constant from part (a).

$$N = \frac{R}{\lambda} = \frac{7.4 \times 10^7 \text{ nuclei/s}}{2.99 \times 10^{-6}/\text{s}} = 2.47 \times 10^{13} \text{ nuclei}$$

ANSWERS TO SELECTED QUESTIONS

4. Why do nearly all the naturally occurring isotopes lie *above* the N = Z line in Figure 45.3?

Answer: Extra neutrons are required to overcome the increasing electrostatic repulsion of the protons.

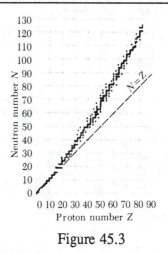

Figure 45.3

7. How many values of I_z are possible for $I = \frac{5}{2}$? How many are possible for I = 3?

Answer:

(a) $I_z = \left\{ -\frac{5}{2}, -\frac{3}{2}, -\frac{1}{2}, +\frac{1}{2}, +\frac{3}{2}, +\frac{5}{2} \right\}$: 6 values

(b) $I_z = \{ -3, -2, -1, 0, 1, 2, 3 \}$: 7 values

8. How can a neutron (which is electrically neutral) possess a magnetic moment?

Answer: For a neutron, $Q_{net} = 0$, but there exist regions of + and – charge within the neutron. (Today, we interpret these charges as belonging to the udd quarks.)

9. In nuclear magnetic resonance, if the dc magnetic field is increased, how does this change the frequency of the ac field that excites a particular transition?

Answer: $\Delta E \propto B_{DC}$. Since the photon frequency is proportional to the AC frequency, we find $f \propto B_{DC}$.

10. Would the liquid-drop or independent-particle nuclear model be more appropriate to predict the behavior of a nucleus in a fission reaction? Which would be more successful in predicting the magnetic moment of a given nucleus? Which could better explain the γ-ray spectrum of an excited nucleus?

Answer: (fission) – liquid-drop model
               (magnetic moment) – independent-particle model
               (γ-ray spectrum) – independent-particle model

13. Two samples of the same radioactive nuclide are prepared. Sample A has twice the initial activity of sample B? How does the half-life of A compare to the half-life of B? After each has passed through five half-lives, what is the ratio of their activities?

Answer: (a) They have the *same* half-lives.   (b) A will always have *twice* the activity of B.

16. Why is the electron involved in the reaction $^{14}_{6}C \rightarrow \,^{14}_{7}N + \beta^-$ written as β^-, while the electron involved in the reaction $^{7}_{4}Be + \,^{0}_{-1}e \rightarrow \,^{7}_{3}Li + \nu$ is written as $^{0}_{-1}e$?

Answer: Electrons from the atomic orbitals are referred to as $^{0}_{-1}e$, while electrons created in "beta decay" $\left[n \rightarrow p + \beta^- + \bar{\nu} \right]$ are referred to as "beta particles." (The particles are indistinguishable. They simply have different origins.)

20. Does the Q in Equation 45.25 represent the quantity (final mass − initial mass)c^2, or does it represent the quantity (initial mass − final mass)c^2?

Answer: $E_i = E_f + Q,$ so $Q = (m_i - m_f)c^2$.

SOLUTIONS TO SELECTED END-OF-CHAPTER PROBLEMS

13. (a) Use energy methods to calculate the distance of closest approach for a head-on collision between an alpha particle with an initial energy of 0.5 MeV and a gold nucleus (^{197}Au) at rest. (Assume the gold nucleus remains at rest during the collision.) (b) What minimum initial speed must the alpha particle have in order to reach a distance of 300 fm?

Solution

(a) The initial kinetic energy of the alpha particle must equal the electrostatic potential energy of the two-particle system at the distance of closest approach, r_{min}. That is,

$$K_\alpha = U = k \frac{qQ}{r_{min}}$$

or

$$r_{min} = k \frac{qQ}{r_{min}} = \frac{(9 \times 10^9 \text{ N·m}^2/\text{C}^2)(2)(79)(1.6 \times 10^{-19} \text{ C})^2}{(0.5 \text{ MeV})(1.6 \times 10^{-13} \text{ J/MeV})} = \boxed{4.55 \times 10^{-13} \text{ m}}$$

(b) Since $K_\alpha = \frac{1}{2} mv^2 = k \frac{qQ}{r_{min}}$, we find

$$v = \sqrt{\frac{2kqQ}{mr_{min}}} = \sqrt{\frac{2(9 \times 10^9 \text{ N·m}^2/\text{C}^2)(2)(79)(1.6 \times 10^{-19} \text{ C})^2}{4(1.67 \times 10^{-27} \text{ kg})(3 \times 10^{-13} \text{ m})}} = \boxed{6.03 \times 10^6 \text{ m/s}}$$

19. A pair of nuclei for which $Z_1 = N_2$ and $Z_2 = N_1$ are called *mirror isobars* (the atomic and neutron numbers are interchangeable). Binding energy measurements on these nuclei can be used to obtain evidence of the charge independence of nuclear forces (that is, proton-proton, proton-neutron, and neutron-neutron forces are approximately equal). Calculate the difference in binding energy for the two mirror nuclei $^{15}_8\text{O}$ and $^{15}_7\text{N}$.

Solution   For $^{15}_8\text{O}$ we have, using Equation 45.4,

$$E_b = [8(1.007825) \text{ u} + 7(1.008665) \text{ u} - (15.003065) \text{ u}](931.5 \text{ MeV/u}) = 111.96 \text{ MeV}$$

For $^{15}_7\text{N}$ we have

$$E_b = [7(1.007825) \text{ u} + 8(1.008665) \text{ u} - (15.000109) \text{ u}](931.5 \text{ MeV/u}) = 115.49 \text{ MeV}$$

Therefore, the difference in the two binding energies is

$$\Delta E_b = \boxed{3.53 \text{ MeV}}$$

31. A freshly prepared sample of a certain radioactive isotope has an activity of 10 mCi. After an elapsed time of 4 h, its activity is 8 mCi. (a) Find the decay constant and half-life of the isotope. (b) How many atoms of the isotope were contained in the freshly prepared sample? (c) What is the sample's activity 30 h after it is prepared?

Solution

(a) From the rate equation, $R = R_0 e^{-\lambda t}$, we can solve for λ to get

$$\lambda = \frac{1}{t} \ln\left(\frac{R_0}{R}\right) = \left(\frac{1}{4\,h}\right) \ln\left(\frac{10\,mCi}{8\,mCi}\right) = 5.58 \times 10^{-2}\,h^{-1} = \boxed{1.55 \times 10^{-5}\,s^{-1}}$$

$$T_{1/2} = \frac{\ln 2}{\lambda} = \boxed{12.4\,h}$$

(b) $R_0 = 10\,mCi = 10 \times 10^{-3} \times (3.7 \times 10^{10}\,\text{decays/s}) = 3.7 \times 10^8\,\text{decays/s}$

Since $R_0 = \lambda N_0$, we find

$$N_0 = \frac{R_0}{\lambda} = \frac{3.7 \times 10^8\,\text{decays/s}}{1.55 \times 10^{-5}\,s^{-1}} = \boxed{2.39 \times 10^{13}\,\text{atoms}}$$

(c) $R = R_0 e^{-\lambda t} = (10\,mCi)e^{-(5.58 \times 10^{-2}\,h^{-1})(30\,h)} = \boxed{1.87\,mCi}$

---

37. The radioactive isotope ^{198}Au has a half-life of 64.8 h. A sample containing this isotope has an initial activity (t = 0) of 40 μCi. Calculate the number of nuclei that will decay in the time interval between $t_1 = 10$ h and $t_2 = 12$ h.

Solution

First, let us find λ and N_0 from the given information:

$$\lambda = \frac{\ln 2}{T_{1/2}} = \frac{0.693}{64.8\,h} = 0.0107\,h^{-1} = 2.97 \times 10^{-6}\,s^{-1}$$

$$N_0 = \frac{R_0}{\lambda} = \frac{(40\,\mu Ci)(3.7 \times 10^4\,\text{decays/s/}\mu Ci)}{2.97 \times 10^{-6}\,s^{-1}} = 4.98 \times 10^{11}\,\text{nuclei}$$

Since $N = N_0 e^{-\lambda t}$, the number of nuclei which decay between times t_1 and t_2 is

$$N_1 - N_2 = N_0\left(e^{-\lambda t_1} - e^{-\lambda t_2}\right)$$

Substituting in the values for λ and N_0 gives

$$N_1 - N_2 = (4.98 \times 10^{11})\left[e^{-(0.0107\,h^{-1})(10\,h)} - e^{-(0.0107\,h^{-1})(12\,h)}\right] = \boxed{9.46 \times 10^9\,\text{nuclei}}$$

---

53. Using the appropriate reactions and Q values from Table 45.5, calculate the mass of ^{8}Be and ^{10}Be in atomic mass units to four decimal places.

Solution

^{9}Be + n $\rightarrow$ ^{10}Be + 6.810 MeV

$m(^{10}\text{Be}) = m(^9\text{Be}) + m_n - \dfrac{6.810 \text{ MeV}}{931.5 \text{ MeV/u}} = 9.01218 + 1.008665 - 0.007311 = \boxed{10.0135 \text{ u}}$

^{9}Be + 1.666 MeV $\rightarrow$ ^{8}Be + n

$m(^8\text{Be}) = m(^9\text{Be}) - m_n + \dfrac{1.666 \text{ MeV}}{931.5 \text{ MeV/u}} = 9.01218 - 1.008665 + 0.001789 = \boxed{8.0053 \text{ u}}$

57. (a) One method of producing neutrons for experimental use is based on the bombardment of light nuclei by alpha particles. In one particular arrangement, alpha particles emitted by plutonium are incident on beryllium nuclei and this results in the production of neutrons:

$$^4_2\text{He} + ^9_4\text{Be} \rightarrow ^{12}_6\text{C} + ^1_0\text{n}$$

What is the Q value for this reaction? (b) Neutrons are also often produced by small-particle accelerators. In one design, deuterons (^{2}H) that have been accelerated in a Van de Graaff generator are used to bombard other deuterium nuclei, resulting in the following reaction:

$$^2_1\text{He} + ^2_1\text{Be} \rightarrow ^3_2\text{C} + ^1_0\text{n}$$

Is this reaction exothermic or endothermic? Calculate its Q value.

Solution

(a) $Q = \left[m(^9\text{Be}) + m(^4\text{He}) - m(^{12}\text{C}) - m(^1\text{n}) \right] (931.5 \text{ MeV/u})$

$Q = \left[(9.01292 + 4.002603 - 12.0000 - 1.008665) \text{ u} \right] (931.5 \text{ MeV/u}) = \boxed{5.70 \text{ MeV}}$

(b) $Q = \left[2m(^2\text{H}) - m(^3\text{He}) - m(^1\text{n}) \right] (931.5 \text{ MeV/u})$

$Q = \left[2(2.014102) - 3.016029 - 1.008665 \text{ u} \right] (931.4 \text{ MeV/u}) = \boxed{3.27 \text{ MeV}}$

The reaction is exothermic since Q is positive.

65. (a) Find the radius of the $^{12}_{6}C$ nucleus. (b) Find the force of repulsion between a proton at the surface of a $^{12}_{6}C$ nucleus and the remaining five protons. (c) How much work (in MeV) has to be done to overcome this electrostatic repulsion to put the last proton into the nucleus? (d) Repeat (a), (b), and (c) for $^{238}_{92}U$.

Solution

(a) $R = R_o A^{1/3} = 1.2 \, A^{1/3}$ fm

When A = 12, $R = 1.2(12)^{1/3}$ fm = $\boxed{2.75 \text{ fm}}$

(b)   Since the proton interacts with Z – 1 other protons, where Z = 6, and R = 2.75 fm, we find

$$F = k\frac{(Z-1)e^2}{R^2} = \frac{(9 \times 10^9 \text{ N·m}^2/\text{C}^2)(5)(1.6 \times 10^{-19} \text{ C})^2}{(2.75 \times 10^{-15} \text{ m})^2} = \boxed{152 \text{ N}}$$

(c) $U = k\dfrac{q_1 q_2}{R} = k\dfrac{(Z-1)e^2}{R} = \dfrac{(9 \times 10^9 \text{ N·m}^2/\text{C}^2)(5)(1.6 \times 10^{-19} \text{ C})^2}{2.75 \times 10^{-15} \text{ m}} = 4.19 \times 10^{-13} \text{ J} = \boxed{2.62 \text{ MeV}}$

(d)   For $^{238}_{92}U$, we take A = 238, and Z = 92 to find R = 7.44 fm, F = 379 N, and U = 17.6 MeV

---

73. "Free neutrons" have a characteristic half-life of 12 min. What fraction of a group of free neutrons at thermal energy (0.04 eV) will decay before traveling a distance of 10 km?

Solution

The fraction that will remain is given by the ratio N/N_o, where $N/N_o = e^{-\lambda t}$ and t is the time it takes the neutron to travel a distance of d = 10 km. Since $K = \frac{1}{2} mv^2$, the time is given by

$$t = \frac{d}{v} = \frac{d}{\sqrt{\dfrac{2K}{m}}} = \frac{10 \times 10^3 \text{ m}}{\sqrt{\dfrac{2(0.04 \text{ eV})(1.6 \times 10^{-19} \text{ J/eV})}{1.675 \times 10^{-27} \text{ kg}}}} = 3.61 \text{ s}$$

$$\lambda = \frac{0.693}{T_{1/2}} = \frac{0.693}{720 \text{ s}} = 9.63 \times 10^{-4} \text{ s}^{-1}$$

Therefore

$$\lambda t = (9.63 \times 10^{-4} \text{ s}^{-1})(3.61 \text{ s}) = 3.47 \times 10^{-3},$$

so

$$\frac{N}{N_o} = e^{-\lambda t} = e^{-3.47 \times 10^{-3}} = 0.9965$$

Hence, the fraction that has decayed in this time is $1 - \dfrac{N}{N_o} = 0.0035$ or $\boxed{0.35\%}$

---

75. The decay of an unstable nucleus by alpha emission is represented by Equation 45.10. The disintegration energy Q given by Equation 45.13 must be shared by the alpha particle and the daughter nucleus in order to conserve both energy and momentum in the decay process. (a) Show that Q and K_α, the kinetic energy of the alpha particle, are related by the expression

$$Q = K_\alpha \left(1 + \frac{M_\alpha}{M}\right)$$

where M is the mass of the daughter nucleus. (b) Use the result of (a) to find the energy of the alpha particle emitted in the decay of ^{226}Ra. (See Example 45.8 for the calculation of Q.)

Solution

(a) Let us assume that the parent nucleus (mass M_p) is initially at rest, and let us denote the masses of the daughter nucleus and alpha particle by M_d and M_α, respectively. Applying the equations of conservation of momentum and energy for the alpha decay process gives

$$M_d v_d = M_\alpha v_a \qquad (1)$$

$$M_p c^2 = M_d c^2 + M_\alpha c^2 + \frac{1}{2} M_\alpha v_\alpha^2 + \frac{1}{2} M_d v_d^2 \qquad (2)$$

The disintegration energy Q is given by

$$Q = (M_p - M_d - M_\alpha)c^2 = \frac{1}{2} M_\alpha v_\alpha^2 + \frac{1}{2} M_d v_d^2 \qquad (3)$$

Eliminating v_d from Equations (1) and (3) gives

$$Q = \frac{1}{2} M_d \left(\frac{M_\alpha}{M_d} v_\alpha\right)^2 + \frac{1}{2} M_\alpha v_\alpha^2 = \frac{1}{2} \frac{M_\alpha^2}{M_d} v_\alpha^2 + \frac{1}{2} M_\alpha v_\alpha^2$$

$$= \frac{1}{2} M_\alpha v_\alpha^2 \left(1 + \frac{M_\alpha}{M_d}\right) = K_\alpha \left(1 + \frac{M_\alpha}{M_d}\right)$$

(b) $K_\alpha = \dfrac{Q}{1 + \dfrac{M_\alpha}{M_d}} = \dfrac{4.87 \text{ MeV}}{1 + \dfrac{4}{226}} = \boxed{4.79 \text{ MeV}}$

Nuclear Physics Applications

OBJECTIVES

1. Write an equation which represents a typical fission event and describe the sequence of events which occurs during the fission process.

2. Use data obtained from the binding energy curve to estimate the disintegration energy of a typical fission event.

3. Describe the basic design features and control mechanisms in a fission reactor including the functions of the moderator, control rods, and heat exchange system.

4. Identify some major safety and environmental hazards in the operation of a fission reactor.

5. Describe the basis of energy release in fusion and write out several nuclear reactions which might be used in a fusion powered reactor.

6. Describe the method of inertial and magnetic field confinement of a plasma and state Lawson's criteria for ion density and confinement time.

7. Describe briefly the basis of radiation damage in metals and in living cells.

8. Define the roentgen, rad, and rem as units of radiation exposure or dose.

9. Describe the basic principle of operation of the Geiger counter, semiconductor diode detector, scintillation detector, photographic emulsion, cloud chamber, and bubble chamber.

NOTES FROM SELECTED CHAPTER SECTIONS

46.1 INTERACTIONS INVOLVING NEUTRONS

The probability that neutrons will be captured as they move through matter *generally increases with decreasing neutron energy*. A *thermal neutron* is one which has an energy of approximately kT.

46.2 NUCLEAR FISSION

Nuclear fission occurs when a heavy nucleus, such as ^{235}U, splits, or fissions, into two smaller nuclei. In such a reaction, *the total rest mass of the products is less than the original rest mass.*

The sequence of events in the fission process is
1. The ^{235}U nucleus captures a thermal (slow-moving) neutron.
2. This capture results in the formation of $^{236}U^*$, and the excess energy of this nucleus causes it to undergo violent oscillations.
3. The $^{236}U^*$ nucleus becomes highly distorted, and the force of repulsion between protons in the two halves of the dumbbell shape tends to increase the distortion.
4. The nucleus splits into two fragments, emitting several neutrons in the process.

46.3 NUCLEAR REACTORS

A nuclear reactor is a system designed to maintain a *self-sustained chain reaction*.

The *reproduction constant* K is defined as the average number of neutrons released from each fission event that will cause another event. In a power reactor, it is necessary to maintain a value of K close to 1.

46.4 NUCLEAR FUSION

Nuclear fusion is a process in which two light nuclei combine to form a heavier nucleus. A great deal of energy is released in such a process. The major obstacle in obtaining useful energy from fusion is the large Coulomb repulsive force between the charged nuclei at close separations. Sufficient energy must be supplied to the particles to overcome this Coulomb barrier and thereby enable the nuclear attractive force to take over.

The temperature at which the power generation exceeds the loss rate is called the *critical ignition temperature*. The confinement time is the time the interacting ions are maintained at a temperature equal to or greater than the ignition temperature.

46.5 RADIATION DAMAGE IN MATTER

In biological systems, it is common to separate radiation damage into two categories, *somatic damage* and *genetic damage*. Somatic damage is radiation damage associated with all the body cells except the reproductive cells. Genetic damage affects only the reproductive cells of the person exposed to the radiation.

The roentgen (R) is defined as that amount of ionizing radiation that will produce $\frac{1}{3} \times 10^{-9}$ C of electric charge in 1 cm^3 of air under standard conditions.

One *rad* is that amount of radiation that deposits 10^{-2} J of energy into 1 kg of absorbing material.

The *RBE* (relative biological effectiveness) factor for a given type of radiation is defined as *the number of rad of x-radiation or gamma radiation that produces the same biological damage as 1 rad of the radiation being used.*

The *rem* (roentgen equivalent in man) is defined as the product of the dose in rad and the RBE factor.

EQUATIONS AND CONCEPTS

The fission of a uranium nucleus by bombardment with a low energy neutron results in the production of fission fragments and typically two or three neutrons. The energy released in the fission event appears in the form of kinetic energy of the fission fragments and the neutrons.

$$\ce{^1_0 n + ^{235}_{92}U \rightarrow X + Y + neutrons}$$ (46.2)

These fusion reactions seem to be most likely to be used as the basis of the design and operation of a fusion power reactor. The stated Q values refer to the amount of energy released per reaction.

$$^2_1\text{H} + {}^2_1\text{H} \rightarrow {}^3_2\text{He} + {}^1_0\text{n}$$
$$(Q = 3.27 \text{ MeV})$$
$$^2_1\text{H} + {}^2_1\text{H} \rightarrow {}^3_1\text{H} + {}^1_1\text{H}$$
$$(Q = 4.03 \text{ MeV})$$
$$^2_1\text{H} + {}^3_1\text{H} \rightarrow {}^4_2\text{He} + {}^1_0\text{n}$$
$$(Q = 17.59 \text{ MeV})$$

(46.5)

Lawson's criterion states the conditions under which a net power output of a fusion reactor is possible. In these expressions, n is the plasma density (number of ions per cubic cm) and τ is the plasma confinement time (the time during which the interacting ions are maintained at a temperature equal to or greater than that required for the reaction to proceed).

$$n\tau > 10^{14} \text{ s/cm}^3$$
$$(D - T \text{ reaction})$$

(46.10)

$$n\tau \geq 10^{16} \text{ s/cm}^3$$
$$(D - D \text{ reaction})$$

(46.7)

The radiation dose in rem is the product of the dose in rad and the relative biological effectiveness factor.

$$\text{Dose in rem} = \text{dose in rad} \times \text{RBE}$$

(46.8)

1 roentgen (R) is that amount of ionizating radiation that deposits 0.0876 J of energy into 1 kg of air.

1 rad is that amount of radiation that deposits 0.001 J of energy into 1 kg of any absorbing material.

RBE is a factor defined as the number of rads of X or gamma radiation that produces the same biological effect as 1 rad of the type of radiation actually being used.

ANSWERS TO SELECTED QUESTIONS

1. Explain the function of a moderator in a fission reactor.

Answer: The moderator slows down the energetic neutrons, which increases the probability that they will be captured by uranium-235 or other nuclei, and hence produce fission.

5. Discuss the advantages and disadvantages of fission reactors from the point of view of safety, pollution, and resources. Make a comparison with power generated from the burning of fossil fuels.

Answer: A fission reactor cannot operate unless it contains a large amount of fissile material, which in turn leads to a great deal of radioactive waste which must be handled, stored, etc. In fission plants, once the chain reaction is stopped, the cooling water must be circulated for a long period of time following shut-down. Hence, heat continues to be generated and the danger of radioactive contamination of the environment increases. This problem is greatly reduced in a fusion reactor. The text discusses other factors that must be taken into account. Fossil fuels present no radiation hazard; however, their limited availability and their effect on the environment are well-known problem areas.

6. In a fission reactor, nuclear reactions produce heat to drive a turbine-generator. How is this heat produced?

Answer: Heat is produced as the result of energy transfer from energetic neutrons and fusion fragments to the water molecules surrounding the reactor core, which in turn produces the steam used to drive the turbine and produce electricity. The water that circulates through a closed loop acts as both a moderator substance to slow down the neutrons and as a coolant to transfer energy to the turbines.

7. Why would a fusion reactor produce less radioactive waste than a fission reactor?

Answer: In a fission reaction, the reaction products are radioactive nuclei with long half-lives. This is not the case in fusion reactions such as deuterium-deuterium and deuterium-tritium reactions.

10. What factors make a fusion reaction difficult to achieve?

Answer: The two most difficult factors to overcome are the requirements of a high plasma density and a high plasma temperature. These two conditions must occur simultaneously.

12. Discuss the advantages and disadvantages of fusion power from the point of safety, pollution, and resources.

Answer: Fusion power plants would have to deal with some radioactivity, but the degree of radioactivity one has to contend with is small compared to fission plants. The low cost and abundance of deuterium fuel is another attractive feature of fusion power.

SOLUTIONS TO SELECTED END-OF-CHAPTER PROBLEMS

11. Inhomogeneities in the confining magnetic field make it difficult to contain plasmas. (a) Show that the magnetic field within a toroid is inversely proportional to r, where r is the distance from the central axis of the toroid. (b) Show that the fractional inhomogeneity $(B_{max} - B_{min})/B_{ave}$ in a toroidal field with rectangular windings as shown in Figure 32.32 is approximately equal to $(b - a)/r$. (c) Evaluate the fractional inhomogeneity for a toroid with $b - a = 10$ cm, first with $r = 1$ m and then with $r = 10$ m.

Solution

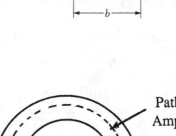

(a) $$\oint \mathbf{B} \cdot \mathbf{ds} = \mu_o NI$$

$$B \cdot 2\pi r = \mu_o NI$$

$$B = \frac{\mu_o NI}{2\pi r} \, \alpha \, \frac{1}{r}$$

(b) $$B_{max} = \frac{\mu_o NI}{2\pi a}; \qquad B_{min} = \frac{\mu_o NI}{2\pi b}$$

$$\frac{\Delta B}{B_{av}} = \frac{\frac{1}{a} - \frac{1}{b}}{\frac{1}{r_{av}}} = \left(\frac{b-a}{ab}\right) r_{av}$$

With $ab \cong r_{av}^2$, $$\frac{\Delta B}{B_{av}} \cong \frac{b-a}{r_{av}}$$

(c) $$\frac{\Delta B}{B} = \frac{0.10 \text{ m}}{1 \text{ m}} = 10\%; \qquad \frac{\Delta B}{B} = \frac{0.10 \text{ m}}{10 \text{ m}} = 1\%$$

Path for Ampere's law

Problem 11

13. To understand why containment of a plasma is necessary, consider the rate at which a plasma would be lost if it were not contained. (a) Estimate the rms speed of deuterons in a plasma at 4×10^8 K. (b) Estimate the time such a plasma would remain in a 10-cm cube if no steps were taken to contain it.

Solution

The average kinetic energy per particle $\left(\frac{1}{2} m\overline{v^2}\right)$ must equal the thermal energy $\frac{3}{2} kT$. That is,

$$\frac{1}{2} m\overline{v^2} = \frac{3}{2} kT \qquad \text{or} \qquad v_{rms} = \sqrt{\frac{3kT}{m}}$$

$$v_{rms} = \sqrt{\frac{3(1.38 \times 10^{-23} \text{ J/K})(4 \times 10^8 \text{ K})}{2(1.67 \times 10^{-27} \text{ kg})}} = \boxed{2.22 \times 10^6 \text{ m/s}}$$

$$t = \frac{x}{v} = \frac{0.1 \text{ m}}{2.22 \times 10^6 \text{ m/s}} = \boxed{4.50 \times 10^{-8} \text{ s}}$$

---

21. A "clever" technician decides to heat some water for his coffee with an X-ray machine. If the machine produces 10 rad/s, how long will it take to raise the temperature of a cup of water by 50 C°?

Solution

$$1 \text{ rad} = 10^{-2} \text{ J/kg} \qquad Q = mc\Delta T$$

$$P{\cdot}t = mc\Delta T$$

$$t = \frac{mc\Delta T}{P} = \frac{m(4186 \text{ J/kg}{\cdot}\text{C°})(50 \text{ C°})}{10(10^{-2} \text{ J/kg}{\cdot}\text{s})(m)} = 2.09 \times 10^6 \text{ s} \cong \boxed{24 \text{ days}}$$

(Note that the power P is the product of dose rate and mass.)

---

25. In a Geiger tube, the voltage between the electrodes is typically 1 kV and the current pulse discharges a 5-pF capacitor. (a) What is the energy amplification of this device for a 0.5-MeV beta ray? (b) How many electrons are avalanched by the initial electron?

Solution

(a) $\quad \dfrac{E}{E_o} = \dfrac{\frac{1}{2} CV^2}{0.5 \text{ MeV}} = \dfrac{(0.5)(5 \times 10^{-12} \text{ F})(1 \times 10^3 \text{ V})^2}{(0.5 \text{ MeV})(1.6 \times 10^{-13} \text{ J/MeV})} = \boxed{3.1 \times 10^7}$

(b) $\quad N = \dfrac{Q}{e} = \dfrac{CV}{e} = \dfrac{(5 \times 10^{-12} \text{ F})(1 \times 10^3 \text{ V})}{1.6 \times 10^{-19} \text{ C}} = \boxed{3.1 \times 10^{10} \text{ electrons}}$

---

27. The probability of a given nuclear reaction increases dramatically above the "Coulomb barrier," which is the electrostatic potential energy of the two nuclei, when their surfaces just touch. Compute the Coulomb barrier for the absorption of an alpha particle by a gold nucleus.

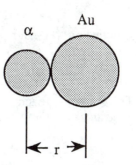

Solution We can compute the radii from $r = (1.2 \text{ fm})A^{1/3}$.

For α, $A = 4$ so $r_\alpha = (1.2 \text{ fm})(4)^{1/3} = 1.9 \text{ fm}$

For Au, $A = 197$ so $r_{Au} = (1.2 \text{ fm})(197)^{1/3} = 6.98 \text{ fm}$

Problem 27 $\quad r = r_\alpha + r_{Au}$

$$U = k\frac{q_1 q_2}{r} = \left(9 \times 10^9 \, \frac{\text{N·m}^2}{\text{C}^2}\right)\frac{(2)(79)(1.6 \times 10^{-19} \text{ C})^2}{8.88 \times 10^{-15} \text{ m}} = 4.1 \times 10^{-12} \text{ J} = \boxed{25.6 \text{ MeV}}$$

---

43. Consider the two nuclear reactions

$$\text{(I)} \quad A + B \rightarrow C + E$$

$$\text{(II)} \quad C + D \rightarrow F + G$$

(a) Show that the net Q for these two reactions ($Q_{net} = Q_I + Q_{II}$) is identical to the Q for the reaction

$$A + B + D \rightarrow E + F + G$$

(b) One chain of reactions in the proton-proton cycle in the Sun's interior is the following:

$$^1_1\text{H} + ^1_1\text{H} \rightarrow ^2_1\text{H} + ^0_1\text{e} + \nu$$

$$^1_1\text{H} + ^2_1\text{H} \rightarrow ^3_2\text{He} + \gamma$$

$$^1_1\text{H} + ^3_2\text{He} \rightarrow ^4_2\text{He} + ^0_1\text{e} + \nu$$

Based on part (a), what is Q_{net} for this sequence of three reactions?

Solution

(a) $Q_I = m_A + m_B - (m_C + m_E)$

$Q_{II} = m_C + m_D - (m_F + m_G)$

$Q_{net} = (m_A + m_B + m_D) - (m_E + m_F + m_G)$

This is identical to the Q for the reaction $A + B + D \rightarrow E + F + G$. Thus, any product (e.g., "C") that is a reactant in a subsequent reaction disappears from the energy balance.

569

(b) Eliminating $_1^2H$ and $_2^3He$ because each is used up in a subsequent reaction, the net process is

$$(4)_1^1H \rightarrow {_2^4}He + (2)\,_1^0e + 2\nu + \gamma$$

The Q for any reaction is defined as the change in the *rest* masses of the reactants and products. Since neutrinos and gamma rays both have zero rest mass,

$$Q = 4\left(_1^1H\right) - \left(_2^4He + 2\,_1^0e\right)$$

$$Q = \left[4(1.007825\ u) - 4.002603\ u\right]\left(931.5\ \frac{MeV}{u}\right) - 2(0.511\ MeV)$$

$$Q = \boxed{25.7\ MeV}$$

---

44. The carbon cycle, first proposed by Hans Bethe in 1939, is another cycle by which energy is released in stars and hydrogen is converted to helium. The carbon cycle requires higher temperatures than the proton-proton cycle. The series of reactions is

$$^{12}C + {}^1H \rightarrow {}^{13}N + \gamma$$

$$^{13}N \rightarrow {}^{13}C + \beta^+ + \nu$$

$$^{13}C + {}^1H \rightarrow {}^{14}N + \gamma$$

$$^{14}N + {}^1H \rightarrow {}^{15}O + \gamma$$

$$^{15}O \rightarrow {}^{15}N + \beta^+ + \nu$$

$$^{15}N + {}^1H \rightarrow {}^{12}C + {}^4He$$

(a) If the proton-proton cycle requires a temperature of 1.5×10^7 K, estimate the temperature required for the first step in the carbon cycle. (b) Calculate the Q value for each step in the carbon cycle and the overall energy released. (c) Do you think the energy carried off by the neutrinos is deposited in the star? Explain.

Solution

(a) Roughly, $\frac{7}{2}(15 \times 10^6)$ K or 52×10^6 K since $6 \times$ the coulombic barrier must be surmounted.

(b) $E = \Delta mc^2 = (12.00000 + 1.007825 - 13.005738\ u)(931.5\ MeV/u) = 1.943\ MeV$

$1.943 + 1.709 + 7.551 + 7.297 + 2.242 + 4.966 = \boxed{25.75\ MeV}$

The net effect is $_6^{12}C + 4p \rightarrow {_6^{12}}C + {_2^4}He$

(c) Most of the energy is lost since neutrinos have such low cross section (no charge, little mass, etc.).

Particle Physics and Cosmology

OBJECTIVES

1. Be aware of the four fundamental forces in nature and the corresponding field particles or quanta via which these forces are mediated.

2. Understand the concepts of antiparticle, pair production, and pair annihilation.

3. Understand the basis of Yukawa's prediction of the pion (pi meson) to explain the nature of the strong force. Use a Feynman diagram to describe pion exchange between a protron and a neutron.

4. Know the broad classification of particles and the characteristic properties of the several classes (relative mass value, spin, decay mode).

5. Determine whether or not a suggested decay can occur based on the conservation of bargon number and the conservation of lepton number.

6. Determine whether or not a predicted reaction/decay will occur based on the conservation of strangeness for the strong and electromagnetic interactions.

NOTES FROM SELECTED CHAPTER SECTIONS

47.2 THE FUNDAMENTAL FORCES IN NATURE

There are four fundamental forces in nature: *strong* (hadronic), *electromagnetic, weak,* and *gravitational.* The strong force is the force between nucleons that keeps the nucleus together. The weak force is responsible for beta decay. The electromagnetic and weak forces are now considered to be manifestations of a single force called the *electroweak* force.

The fundamental forces are described in terms of particle or quanta exchanges which *mediate* the forces. The electromagnetic force is mediated by photons, which are the quanta of the electromagnetic field. Likewise, the strong force is mediated by field particles called *gluons,* the weak force is mediated by particles called the W and Z *bosons,* and the gravitational force is mediated by quanta of the gravitational field called *gravitons.*

47.3 POSITRONS AND OTHER ANTIPARTICLES

An antiparticle and a particle have the same mass, but opposite charge. Furthermore, other properties may have opposite values such as lepton number and baryon number. It is possible to produce particle-antiparticle pairs in nuclear reactions if the available energy is greater than $2mc^2$, where m is the rest mass of the particle (or antiparticle).

Pair production is a process in which a gamma ray with an energy of at least 1.02 MeV interacts with a nucleus and an electron-positron pair is created.

Pair annihilation is an event in which an electron and a positron can annihilate to produce two gamma rays, each with an energy of at least 0.511 MeV.

47.4 MESONS AND THE BEGINNING OF PARTICLE PHYSICS

The interaction between two particles can be represented in a diagram called a *Feynman diagram.*

47.5   CLASSIFICATION OF PARTICLES

All particles (other than photons) can be classified into two categories: *hadrons* and *leptons*.

There are two classes of hadrons: *mesons* and *bargons* which are grouped according to their masses and spins. It is believed that hadrons are composed of units called *quarks* which are more fundamental in nature.

Leptons have no structure or size and are therefore considered to be truly elementary particles.

47.6   THE CONSERVATION LAWS

In all reactions and decays, quantities such as energy, linear momentum, angular momentum, electric charge, baryon number, and lepton number are strictly conserved. Certain particles have properties called *strangeness* and *charm*. These unusual properties are conserved only in those reactions and decays that occur via the strong force.

Whenever a nuclear reaction or decay occurs, the sum of the baryon numbers before the process must equal the sum of the baryon numbers after the process.

The sum of the electron-lepton numbers before a reaction or decay must equal the sum of the electron-lepton numbers after the reaction or decay.

47.7   STRANGE PARTICLES AND STRANGENESS

Whenever a nuclear reaction or decay occurs, the sum of the strangeness numbers before the process must equal the sum of the strangeness numbers after the process.

47.9   QUARKS - FINALLY

Recent theories in elementary particle physics have postulated that all hadrons are composed of smaller units known as *quarks*. Quarks have a fractional electric charge and a baryon number of 1/3. There are six flavors of quarks, up (u), down (d), strange (s), charmed (c), top (t), and bottom (b). All baryons contain three quarks, while all mesons contain one quark and one antiquark.

According to the theory of *quantum chromadynamics*, quarks have a property called *color*, and the strong force between quarks is referred to as the *color force*.

EQUATIONS AND CONCEPTS

Pions and muons are very unstable particles. A decay sequence is shown in Equation 47.1.

$$\pi^- \rightarrow \mu^- + \bar{\nu}$$
$$\mu^- \rightarrow e + \nu + \bar{\nu}$$

(47.1)

Hubble's law states a linear relationship between the velocity of a galaxy and its distance R from earth. The constant H is called the *Hubble parameter*.

$$v = HR$$
$$H = 17 \times 10^{-3} \text{ m/s·light year}$$

(47.4)

The *critical mass density* of the universe can be estimated based on energy considerations.

$$\rho_c = \frac{3H^2}{8\pi G}$$

(47.6)

ANSWERS TO SELECTED QUESTIONS

1. Name the four fundamental interactions and the particles that mediate each interaction.

Answer: The strong force which holds the nucleus together is mediated by *gluons*. The *electromagnetic force* which is responsible for the binding of atoms and molecules is mediated by *photons*. The *weak force* which tends to produce instability in certain nuclei is mediated by the *W and Z bosons*. The *gravitational force* which acts between any two masses is mediated by *gravitons*.

2. Discuss the quark model of hadrons, and describe the properties of quarks.

Answer: In the quark model, all hadrons are composed of smaller units called quarks. Quarks have a fractional electric charge and a baryon number of 1/3. There are six flavors of quarks; up (u), down (d), strange (s), charmed (c), top (t), and bottom (b). All baryons contain three quarks and all mesons contain one quark and one anti-quark. (see Section 47.9 for a more detailed discussion of the quark model.)

4. Describe the properties of baryons and mesons and the important differences between them.

Answer: Hadrons interact primarily through the strong force and are not elementary particles. There are two types of hadrons called baryons and mesons. Baryons have a nonzero baryon number with a spin of 1/2 or 3/2. Mesons have a baryon number of zero, and a spin of either 0 or 1. Leptons are considered to be truly elementary particles. They interact through the weak and electromagnetic forces. Some examples of leptons are the electron, muons, and neutrinos.

5. Particles known as resonances have very short lifetimes, of the order of 10^{-23} s. From this information, would you guess that they are hadrons or leptons? Explain.

Answer: Hadrons. Such particles decay into other strongly interacting particles such as p, n, and π with very short lifetimes. In fact, they decay so quickly that they cannot be detected directly. Decays which occur via the weak force have lifetimes of 10^{-13} s or longer; particles that decay via the electromagnetic force have lifetimes in the range of 10^{-16} s to 10^{-19} s.

6. The family of K mesons all decay into final states that contain no protons or neutrons. What is the baryon number of the K mesons?

Answer: The proton and neutron both have baryon numbers equal to one. Since the baryon nuber must be conserved, and the final states of the K meson decay contain no protons or neutrons, the baryon number of all K mesons must be *zero*.

7. The Ξ^o particle decays by the weak interaction according to the decay mode $\Xi^o \rightarrow \Lambda^o + \Pi^o$. Would you expect this decay to be fast or slow? Explain.

Answer: It should be slow. Decays which occur via the weak interaction are typically 10^{-10} s or longer (see Table 47.2).

10. Two protons in a nucleus interact via the strong interaction. Are they also subject to the weak interaction?

Answer: Yes, but the strong interaction predominates when the separation between the particles is small.

12. An antibaryon interacts with a meson. Can a baryon be produced in such an interaction? Explain.

Answer: No. Antibaryons have baryon number -1, baryons have baryon number +1, while mesons have baryon number 0. The interaction cannot occur since baryon number is not conserved.

14. How many quarks are there in (a) a baryon, (b) an antibaryon, (c) a meson, and (d) an antimeson? How do you account for the fact that baryons have half-integral spin while mesons have spins of 0 or 1? (Hint: Quarks have spin of 1/2.)

Answer: All baryons and antibaryons consist of three quarks. All mesons and antimesons consist of two quarks. Since quarks have spin of 1/2, it follows that all baryons (which consist of three quarks) must have half-integral spin, while all mesons (which consist of only two quarks) must have spins of 0 or 1.

15. In the theory of quantum electrodynamics, quarks come in three colors. How would you justify the statement that "all baryons and mesons are colorless"?

Answer: Each flavor of quark can have three colors, designated as red, green, and blue. Antiquarks are colored antired, antigreen, and antiblue. Baryons consist of three quarks, each having a different color. Mesons consist of a quark of one color and an antiquark with a corresponding anticolor. Thus, baryons and mesons are colorless or white.

SOLUTIONS TO SELECTED END-OF-CHAPTER PROBLEMS

3. One of the mediators of the weak interaction is the Z^o boson whose mass is 96 GeV/c². Use this information to find an approximate value for the range of the weak interaction.

Solution

The rest energy of the Z° boson is $E_0 = 96$ GeV. The maximum time a virtual Z° boson can exist is found from $\Delta E \Delta t \geq \hbar$, or

$$\Delta t \approx \frac{\hbar}{\Delta E} = \frac{1.055 \times 10^{-34} \text{ J·s}}{(96 \text{ GeV})(1.6 \times 10^{-10} \text{ J/GeV})} = 6.87 \times 10^{-27} \text{ s}$$

The maximum distance it can travel in this time is

$$d = c(\Delta t) = (3 \times 10^8 \text{ m/s})(6.87 \times 10^{-27} \text{ s}) = \boxed{2.06 \times 10^{-18} \text{ m}}$$

The distance d is an approximate value for the range of the weak interaction.

10. The following reactions or decays involves one or more neutrinos. Supply the missing neutrinos (ν_e, ν_μ, or ν_τ).

(a) $\pi^- \to \mu^- + ?$

(b) $K^+ \to \mu^+ + ?$

(c) $? + p \to n + e^+$

(d) $? + n \to p + e$

(e) $? + n \to p + \mu^-$

(f) $\mu^- \to e + ? + ?$

Solution

(a) $\pi^- \to \mu^- + \bar{\nu}_\mu$ L_μ: $0 \to 1 - 1$

(b) $K^+ \to \mu^+ + \nu_\mu$ L_μ: $0 \to -1 + 1$

(c) $\bar{\nu}_e + p \to n + e^+$ L_e: $-1 + 0 \to 0 - 1$

(d) $\nu_e + n \to p + e$ L_e: $1 + 0 \to 0 + 1$

(e) $\nu_\mu + n \to p + \mu^-$ L_μ: $1 + 0 \to 0 + 1$

(f) $\mu^- \to e + \bar{\nu}_e + \nu_\mu$ L_μ: $1 \to 0 + 0 + 1$ and

L_e: $0 \to 1 - 1 + 0$

12. Determine which of the reactions below can occur. For those that cannot occur, determine the conservation law (or laws) that each violates.

(a) $p \to \pi^+ + \pi^0$

(b) $p + p \to p + p + \pi^0$

(c) $p + p \to p + \pi^+$

(d) $\pi^+ \to \mu^+ + \nu_\mu$

(e) $n \to p + e + \bar{\nu}_e$

(f) $\pi^+ \to \mu^+ + n$

Solution

(a) $p \to \pi^+ + \pi^0$ Baryon number is violated: $1 \to 0 + 0$

(b) $p + p \to p + p + \pi^0$ This reaction can occur.

(c) $p + p \to p + \pi^+$ Baryon number is violated: $1 + 1 \to 1 + 0$

(d) $\pi^+ \to \mu^+ + \nu_\mu$ This reaction can occur.

(e) $n \to p + e + \bar{\nu}_e$ This reaction can occur.

(f) $\pi^+ \to \mu^+ + n$ Violates baryon number: $0 \to 0 + 1$ and

violates muon-lepton number: $0 \to -1 + 0$

21. A distant quasar is moving away from the earth at such high speed that the blue 434-nm hydrogen line is observed at 650 nm, in the red portion of the spectrum. (a) How fast is the quasar receding? (See the hint in Problem 20.) (b) Using Hubble's law, determine the distance from earth to this quasar?

Solution

(a) $\dfrac{\lambda'}{\lambda} = \dfrac{650 \text{ nm}}{434 \text{ nm}} = 1.498 = \sqrt{\dfrac{1 + \dfrac{v}{c}}{1 - \dfrac{v}{c}}}$

$\dfrac{1 + \dfrac{v}{c}}{1 - \dfrac{v}{c}} = 2.243$

$v = \boxed{0.383c}$ or   38.3% the speed of light

(b) Using Equation 47.4, $v = HR$, we find

$$R = \dfrac{v}{H} = \dfrac{(0.383)(3 \times 10^8 \text{ m/s})}{1.7 \times 10^{-2} \text{ m/s·light year}} = \boxed{6.7 \times 10^9 \text{ light years}}$$

23. A gamma-ray photon strikes a stationary electron. Determine the minimum gamma-ray energy to make this reaction go:

$$\gamma + e^- \rightarrow e^- + e^- + e^+$$

Solution

Using relativistic expressions for conservation of energy (1) and conservation of momentum (2), we find

$$E_\gamma + m_e c^2 = \dfrac{3 m_e c^2}{\sqrt{1 - \dfrac{u^2}{c^2}}} \tag{1}$$

$$\dfrac{E_\gamma}{c} = \dfrac{3 m_e u}{\sqrt{1 - \dfrac{u^2}{c^2}}} \tag{2}$$

Dividing (2) by (1) gives

$$\dfrac{E_\gamma}{E_\gamma + m_e c^2} = \dfrac{u}{c} = \beta \tag{3}$$

Subtracting (2) from (1) gives

$$m_ec^2 = \frac{3m_ec^2}{\sqrt{1-\beta^2}} - \frac{3m_ec^2\beta}{\sqrt{1-\beta^2}}$$

$$\frac{3-3\beta}{\sqrt{1-\beta^2}} = 1 \qquad \text{or} \qquad \beta = \frac{4}{5}$$

Therefore, from (3) we see that

$$E_\gamma = \frac{4}{5}E_\gamma + \frac{4}{5}m_ec^2 \qquad \text{or} \qquad E_\gamma = \boxed{4m_ec^2 \cong 2 \text{ MeV}}$$

---

25. The energy flux of neutrinos from the sun is estimated to be on the order of 0.4 W/m^2, at the earth's surface. Estimate the fractional mass loss of the sun over 10^9 years due to the radiation of neutrinos. (The mass of the sun is 2×10^{30} kg. The distance of the earth from the sun is 1.5×10^{11} m.)

Solution

Since the neutrino flux from the sun reaching the earth is 0.4 W/m^2, the total energy emitted per second by the sun in neutrinos is

$$(0.4 \text{ W/m}^2)(4\pi r^2) = (0.4 \text{ W/m}^2)(4\pi)(1.5 \times 10^{11} \text{ m})^2 = 1.13 \times 10^{23} \text{ W}$$

In a period of 10^9 y, the sun emits a total energy of

$$(1.13 \times 10^{23} \text{ J/s})(10^9 \text{ y})(3.156 \times 10^7 \text{ s/y}) = 3.56 \times 10^{39} \text{ J}$$

in the form of neutrinos. This energy corresponds to an annihilated mass of

$$mc^2 = 3.56 \times 10^{39} \text{ J}$$

$$m = 3.96 \times 10^{22} \text{ kg}$$

Since the sun has a mass of about 2×10^{30} kg, this corresponds to a loss of only about 1 part in 50,000,000 of the sun's mass over 10^9 y in the form of neutrinos.